CHROMATOGRAPHIC SCIENCE SERIES

A Series of Monographs

Editor: JACK CAZES
Cherry Hill, New Jersey

1. Dynamics of Chromatography, *J. Calvin Giddings*
2. Gas Chromatographic Analysis of Drugs and Pesticides, *Benjamin J. Gudzinowicz*
3. Principles of Adsorption Chromatography: The Separation of Nonionic Organic Compounds, *Lloyd R. Snyder*
4. Multicomponent Chromatography: Theory of Interference, *Friedrich Helfferich and Gerhard Klein*
5. Quantitative Analysis by Gas Chromatography, *Josef Novák*
6. High-Speed Liquid Chromatography, *Peter M. Rajcsanyi and Elisabeth Rajcsanyi*
7. Fundamentals of Integrated GC-MS (in three parts), *Benjamin J. Gudzinowicz, Michael J. Gudzinowicz, and Horace F. Martin*
8. Liquid Chromatography of Polymers and Related Materials, *Jack Cazes*
9. GLC and HPLC Determination of Therapeutic Agents (in three parts), *Part 1 edited by Kiyoshi Tsuji and Walter Morozowich, Parts 2 and 3 edited by Kiyoshi Tsuji*
10. Biological/Biomedical Applications of Liquid Chromatography, *edited by Gerald L. Hawk*
11. Chromatography in Petroleum Analysis, *edited by Klaus H. Altgelt and T. H. Gouw*
12. Biological/Biomedical Applications of Liquid Chromatography II, *edited by Gerald L. Hawk*
13. Liquid Chromatography of Polymers and Related Materials II, *edited by Jack Cazes and Xavier Delamare*
14. Introduction to Analytical Gas Chromatography: History, Principles, and Practice, *John A. Perry*
15. Applications of Glass Capillary Gas Chromatography, *edited by Walter G. Jennings*
16. Steroid Analysis by HPLC: Recent Applications, *edited by Marie P. Kautsky*
17. Thin-Layer Chromatography: Techniques and Applications, *Bernard Fried and Joseph Sherma*
18. Biological/Biomedical Applications of Liquid Chromatography III, *edited by Gerald L. Hawk*
19. Liquid Chromatography of Polymers and Related Materials III, *edited by Jack Cazes*
20. Biological/Biomedical Applications of Liquid Chromatography, *edited by Gerald L. Hawk*
21. Chromatographic Separation and Extraction with Foamed Plastics and Rubbers, *G. J. Moody and J. D. R. Thomas*

Chromatographic Analysis of Pharmaceuticals

ADDITIONAL VOLUMES IN PREPARATION

Chromatographic Analysis of Pharmaceuticals

Second Edition, Revised and Expanded

edited by

John A. Adamovics

Cytogen Corporation
Princeton, New Jersey

Marcel Dekker, Inc.　　　　New York•Basel•Hong Kong

ISBN: 0-8247-9776-0

The publisher offers discounts on this book when ordered in bulk quantities. For more information, write to Special Sales/Professional Marketing at the address below.

This book is printed on acid-free paper.

Marcel Dekker, Inc.
270 Madison Avenue, New York, New York 10016

Current printing (last digit):
10 9 8 7 6 5 4 3 2 1

PRINTED IN THE UNITED STATES OF AMERICA

Preface

The first edition of *Chromatographic Analysis of Pharmaceuticals* was published in 1990. The past years have allowed me to evaluate leads that I uncovered during the researching of the first edition, such as the first published example of the application of chromatography to pharmaceutical analysis of medicinal plants. This and other examples are found in a relatively rare book, *Uber Kapillaranalyse und ihre Anwendung in Pharmazeutichen Laboratorium* (Leipzig, 1992), by H. Platz. Capillary analysis, the chromatographic technique used, was developed by Friedlieb Runge in the mid-1850s and was later refined by Friedrich Goppelsroeder. The principle of the analysis was that substances were absorbed on filter paper directly from the solutions in which they were dissolved; they then migrated to different points on the filter paper. Capillary analysis differed from paper chromatography in that no developing solvent was used. We find that, from these humble beginnings 150 years ago, the direct descendant of this technique, paper chromatography, is still widely used in evaluating radio-pharmaceuticals.

This second edition updates and expands on coverage of the topics in the first edition. It should appeal to chemists and biochemists in pharmaceutics and biotechnology responsible for analysis of pharmaceuticals. As in the first edition, this book focuses on analysis of bulk and formulated drug products, and not on analysis of drugs in biological fluids.

The overall organization of the first edition — a series of chapters on regulatory considerations, sample treatment (manual/robotic), and chromatographic methods (TLC, GC, HPLC), followed by an applications section — has been maintained. To provide a more coherent structure to this edition, the robotics and sample treatment chapters have been consolidated, as have the chapters on gas chromatography and headspace analysis. This edition includes two new chapters, on capillary electrophoresis, and super-critical fluid chromatography. These new chapters discuss the hardware behind the technique, followed by their respective approaches to methods development along with numerous examples. All the chapters have been updated with relevant information on proteinaceous pharmaceuticals. The applications chapter has been updated to include chromatographic methods from the Chinese Pharmacopoeia and updates from *U.S. Pharmacopeia 23* and from the British and European Pharmacopoeias. Methods developed by instrument and column manufacturers are also included in an extensive table, as are up-to-date references from the chromatographic literature.

The suggestions of reviewers of the first edition have been incorporated into this edition whenever possible. This work could not have been completed in a timely manner without the cooperation of the contributors, to whom I am very grateful.

John A. Adamovics

Contents

Contributors

John A. Adamovics, Ph.D. Associate Director, Department of Analytical Chemistry and Quality Control, Cytogen Corporation, Princeton, New Jersey

James C. Eschbach Manager, Quality Control, Cytogen Corporation, Princeton, New Jersey

David L. Farb, Ph.D. Group Leader, Cytogen Corporation, Princeton, New Jersey

Nirdosh K. Jagota, Ph.D. Bristol-Myers Squibb Pharmaceutical Research Institute, New Brunswick, New Jersey

Shelley R. Rabel, Ph.D.* Department of Pharmaceutical Chemistry, University of Kansas, Lawrence, Kansas

---

Current affiliation: Department of Pharmaceutical Research & Development, DuPont Merck Pharmaceutical Company, Wilmington, Delaware

James T. Stewart, Ph.D.   Professor and Head, Department of Medicinal Chemistry, College of Pharmacy, The University of Georgia, Athens, Georgia

John F. Stobaugh, Ph.D.   Department of Pharmaceutical Chemistry, University of Kansas, Lawrence, Kansas

1

Regulatory Considerations for the Chromatographer

JOHN A. ADAMOVICS *Cytogen Corporation, Princeton, New Jersey*

I. INTRODUCTION

Analysis of pharmaceutical preparations by a chromatographic method can be traced back to at least the 1920s [1]. By 1955, descending and ascending paper chromatography had been described in the *United States Pharmacopeia* (USP) for the identification of drug products [2]. Subsequent editions introduced gas chromatographic and high-performance liquid chromatographic methods. At present, chromatographic methods have clearly become the analytical methods of choice, with over 800 cited.

The following section describes challenges presented to scientists involved in the analysis of drug candidates and final products, including the current state of validating a chromatographic method.

II. IMPURITIES

In the search for new drug candidates, scientists use molecular modeling techniques to identify potentially new structural moieties and screen natural sources or large families of synthetically related compounds, along with modifying exisiting compounds. Once a potentially new drug has been iden-

1

tified and is being scaled up from the bench to pilot plant manufacturing quantities, each batch is analyzed for identity, purity, potency, and safety. From these data, specifications are established along with a reference standard against which all future batches will be compared to ensure batch-to-batch uniformity.

A good specification is one that provides for material balance. The sum of the assay results plus the limits tests should account for 100% of the drug within the limits of accuracy and precision for the tests. Limits should be set no higher than the level which can be justified by safety data and no lower than the level achievable by the manufacturing process and analytical variation. Acceptable limits are often set for individual impurities and for the total amount of drug-related impurities. Limits should be established for by-products of the synthesis arising from side reactions, impurities in starting materials, isomerization, enantiomeric impurities, degradation products, residual solvents, and inorganic impurities. Drugs derived from biotechnological processes must also be tested for the components with which the drug has come in contact, such as the culture media proteins (albumin, transferrin, and insulin) and other additives such as testosterone. This is in addition to all the various viral and other adventitious agents whose absence must be demonstrated [3].

A 0.1% threshold for identification and isolation of impurities from all new molecular entities is under consideration by the International Conference on Harmonization as an international regulatory standard [4,5]. However, where there is evidence to suggest the presence or formation of toxic impurities, identification should be attempted. An example of this is the 1500 reports of Eosinophilia–Mylagia Syndrome and more than 30 deaths associated with one impurity present in L-tryptophan which were present at the 0.0089% level [6].

The process of qualifying an individual impurity or a given impurity profile at a specified level(s) is summarized in Table 1.1. Safety studies can be conducted on the drug containing the impurity or on the isolated impurity. Several decision trees have been proposed describing threshold levels

Table 1.1 Criteria That Can Be Used for Impurity Qualification

Impurities already present during preclinical studies and clinical trials

Structurally identical metabolites present in animal and/or human studies

Scientific literature

Evaluation for the need for safety studies of a "decision tree" for qualification and for the safety studies that should be performed [4]. For example, a 0.1% threshold would apply when the daily dose exceeds 10 mg, and a 0.5% threshold at a daily dose of less than 0.1 mg. Alternatively, when daily doses exceed 1000 mg per day, levels below 0.1% would not have to be qualified, and for daily doses less than 1000 mg, no impurities need to be qualified unless their intake exceeds 1 mg.

The USP [7] provides extensive discussion on impurities in sections 1086 (Impurities in Offical Articles), 466 (Ordinary Impurities), and 467 (Organic Volative Impurities). A total impurity level of 2.0% has been adopted as a general limit for bulk pharmaceuticals [5]. There have been no levels established for the presence of enantiomers in a drug substance/ product. This is primarily because the enantiomers may have similiar pharmacological and toxicological profiles, enantiomers may rapidly interconvert in vitro and/or in vivo, one enantiomer is shown to be pharmacologically inactive, synthesis or isloation of the perferred enantiomer is not practical, and individual enantiomers exhibit different pharmacologic profiles and the racemate produces a superior therapeutic effect relative to either enantiomer alone [8,9].

For biotechnologically derived products the acceptable levels of foreign proteins should be based on the sensitivity/selectivity of the test method, the dose to be given to a patient, the frequency of administration of the drug, the source, and the potential immunogenicity of protein contaminants [10]. Levels of specific foreign proteins range from 4 ppm to 1000 ppm.

The third category of drugs are phytotherapeutical preparations; 80% of the world population use exclusively plants for the treatment of illnesses [11]. Chromatography is relied on to guarantee preparations contain therapeutically effective doses of active drug and maintain constant batch composition. A quantitative determination of active principles is performed when possible, using pure reference standards. In many phytotherapeutic preparations, the active constituents are not known, so marker substances or typical constituents of the extract are used for the quantitative determination [11]. The Applications chapter of this book (Chapter 8) contains numerous references to the use of chromatographic methods in the control of plant extracts.

III. STABILITY

The International Conference on Harmonization (ICH) has developed guidelines for stability testing of new drug substances and products [12–14]. The guideline outlines the core stability data package required for Registration Applications.

A. Batch Selection

For both the drug substance (bulk drug) and drug product (dosage form) stability information from accelerated and long-term testing should be provided on at least three batches with a minimum of 12 months' duration at the time of submission.

The batches of drug substance must be manufactured to a minimum of pilot scale which follows the same synthetic route and method of manufacturer that is to be used on a manufacturing scale. For the drug product, two of the three batches should be at least pilot scale. The third may be smaller. As with the drug substance batches, the processes should mimic the intended drug product manufacturing procedure and quality specifications.

B. Storage Conditions

The stability storage conditions developed by the ICH are based on the four geographic regions of the world defined by climatic zones I ("temperate") and II ("subtropical"). Zones III and IV are areas with hot/dry and hot/humid climates, respectively. The stability storage conditions as listed in Table 1.2 are arrived at by running average temperatures through an Arrhenius equation and factoring in humidity and packaging.

Long-term testing for both drug substance and product will normally be every 3 months, over the first year, every 6 months over the second year, and then annually. A significant change in stability for drug substance is when the substance no longer meets specifications. For the drug product, a significant change is when there is a 5% change in potency, exceeded pH limits, dissolution failure, or physical attribute failure. If there are significant changes for all three storage temperatures, the drug substance/product should be labeled "store below 25°C." For instances where there are no significant changes label storage as 15–30°C. There should be a direct link between the label statement and the stability characteristics. The use of terms such as ambient or room temperature are unacceptable [12–14].

Table 1.2 Filing Stability Requirements at Time of Submission

- 12 months long-term data (25°C/60% RH)

- 6 months accelerated data (40°C/75% RH)

- If significant change, 6 months accelerated data (30°C/60% RH)

C. Biologics

Degradation pathways for proteins can be separated into two distinct classes; chemical and physical. Chemical instability is any process which involves modification of the protein by bond formation or cleavage. Physical instability refers to changes in the protein structure through denaturation, adsorption to surfaces, aggregation, and precipitation [15].

Stability studies to support a requested shelf life and storage condition must be run under real-time, real-temperature conditions [16,17]. The prediction of shelf life by using stability studies obtained under stress conditions and Arrhenius plots is not meaningful unless it has been demonstrated that the chemical reaction accounting for the degradation process follows first-order reaction.

IV. METHOD VALIDATION

The ultimate objective of the method validation process is to produce the best analytical results possible. To obtain such results, all of the variables of the method should be considered, including sampling procedure, sample preparation steps, type of chromatographic sorbent, mobile phase, and detection. The extent of the validation rigor depends on the purpose of the method. The primary focus of this section will be the validation of chromatographic methods.

The four most common types of analytical procedures are identification tests, including quantitative measurements for impurities, content, limit tests for the control of impurities, and quantitative measure of the active component or other selected components in the drug substance [18]. Table 1.3 describes the performance characteristics that should be evaluated for the common types of analytical procedures [18].

A. Specificity

The specificity of an analytical method is its ability to measure accurately an analyte in the presence of interferences that are known to be present in the product: synthetic precursors, excipients, enantiomers, and known (or likely) degradants that may be present. For separation techniques, this means that there is resolution of >1.5 between the analyte of interest and the interferents.

The means of satisfying the criteria of specificity differs for each type of analytical procedure: For identification, in the development phases, it would be proof of structure, whereas in quality control, it is comparison to

Table 1.3 Analytical Performance Characteristics

Type of analytical procedure; characteristics	Identification	Impurities purity test		Assay; content/potency dissolution: measurement only
		Quantiation	Limit	
Accuracy	−	+	−	+
Precision				
Repeatability	−	+	−	+
Intermed. precision	−	+[a]	−	+[a]
Reproducibility	−	+[b]	−	−[b]
Specificity	+	+	+	+[c]
Detection limit	−	+	+	−
Quantiation limit	−	+	−	−
Linearity	−	+	−	+
Range	−	+	−	+

[a]May be needed in some cases
[b]Not needed in most cases
[c]Not needed if reproducibility performed
Source: Adapted from Ref. 18.

a reference substance [17]; for a purity test, to ensure that all analytical procedures allow an accurate statement of the content of impurities of an analyte [18,19]; for assay measurements, to ensure that the signal measured comes only from the substance being analyzed [18,19].

One practical approach to·testing the specificity of an analytical method is to compare the test results of samples containing impurities versus those not containing impurities. The bias of the test is the difference in results between the two types of samples [20]. The assumption to this approach is that all the interferents are known and available to the analyst for the spiking studies.

A more universal approach to demonstrating specificity of chromatographic methods has been outlined [21]. For peak responses in high-performance liquid chromatography (HPLC), gas chromatography (GC), capillary electrophoresis (CE), or supercritical fluid chromatography (SFC) or the spots (bands) in TLC or gel electrophoresis, the primary task is to demonstrate that they represent a single component. The peak homogeneity of HPLC and GC as well CE and SFC responses can be shown by using a mass spectrometer as a specific detector. The constancy of the mass spectrum of the eluting peak with time is a demonstration of homogeneity, albeit not easily quantified [22].

Multiple ultraviolet (UV) wavelength detection has become a popular approach to evaluating chromatographic peak homogeneity. In the simplest form, the ratio between two preselected wavelengths is measured, and for a homogenous peak, the ratio remains constant. A ratio plot of pure compounds appears as a square wave, whereas an impurity distorts the square (Fig. 1.1). This technique is most useful when the spectral properties of the overlapping compounds are sufficiently different and total chromatographic overlap does not occur [23]. The ability to detect peak overlap can be enhanced by stressing (heat, light, pH, and humidity) the analyte of interest and evaluating the wavelength ratios. A degradation of 10–15% is considered adequate. The utility of this approach has been demonstrated for pipercuronium bromide [23]. Potentially, additional information about peak purity can be obtained by recording UV-vis data at the upslope, apex,

Figure 1.1 Ratio plots.

and downslope of a chromatographic response using photodiode array detection [24–27]. An example of this approach has been published for a method used in assaying an analgesic [28].

Peak purity can be assessed with a higher degree of certainty only by additional analysis using a significantly different chromatographic mode. The collected sample should also be analyzed by techniques that can be sensitive to minor structural differences such as nuclear magnetic resonance (NMR) spectroscopy [29–31].

B. Linearity

The evaluation of linearity can be best described as the characterization of the test method response curve. A plot of the test method response against analyte concentration is often expected to be linear over a specified range of concentrations. Some assays generate nonlinear curves.

The function of the standard curve is to allow the prediction of a sample concentration interpolated from the standard data. This predictive feature does not require linearity of the assay response curve, but only that it be a reasonable description of the correlation between response and concentration. Attempting a rigorous fit of a calculated curve fitting to the standard data may defeat the function because such rigorous curve fitting may emphasize the difference between the sample and the standard assay responses.

The test method response curve is characterized by comparing the goodness of fit of calculated concentrations with the actual concentrations of the standards. For a linear response, this value would be the correlation coefficient derived from a linear regression using least squares. Nonlinear response curves require curve fitting calculations with the corresponding goodness-of-fit determinations [32]. Plotting the test results graphically as a function of analyte concentration on appropriate graph paper may be an acceptable alternative to the regression line calculation.

Experimentally, linearity is determined by a series of injections of standards at six different concentrations that span 50–150% of the expected working range assay [20]. The AOAC recommends 25–200% of the nominal range of analyte [33] using standards and spiked placebo samples [34]. Response linearity for known impurities at 0.05–5.0% of the target analyte should also be evaluated [28]. A linear regression equation applied to the results should have an intercept not significantly different from zero; if it does, it should be demonstrated that there is no effect on the accuracy of the method [20].

The range of an analytical method is the interval between the upper and lower level of analyte in the sample, for which it has been demonstrated that the method has a suitable level of precision, accuracy, and linearity.

C. Limit of Measurement

There are two categories within the level of measurement, the first is the limit of detection (LOD). This is the point at which a measured value is larger than the uncertainty associated with it; for example, the amount of sample exhibiting a response three times the baseline noise [34]. The limit of detection is commonly used to substantiate that an analyte concentration is above or below a certain level, in other words, a limit test [30,35].

The second category is referred to as the limit of quantitation. This limit is the lowest concentration of analyte in a sample that can be determined with acceptable precision and accuracy; for example, the lowest amount of analyte for which duplicate injections resulted in a relative standard deviation (RSD) of $\leq 2\%$ [34]. Limit of quantitation is commonly used for impurity and degradant assays of drug substances and products [35].

The limit of measurement for an analyte is not a unique constant because of day-to-day variation in detector response. Extensive discussions of these limits have been published [36,37].

D. Precision (Random Error)

The precision of a test method expresses the closeness of agreement among a series of measurements obtained from multiple sampling of the same homogenous sample. The concept of precision is measured by standard deviations. It can be subdivided into either two or three categories. The European Community (EC) [19] divides precision into repeatability and reproducibility. Repeatability expresses precision under conditions where there is the same analyst, the same equipment, a short interval of time, and identical reagents. This is also termed intra-assay precision. Reproducibility expresses the precision when the laboratories differ, when there are reagents from different sources, different analysts, tested on different days, equipment from different manufacturers, and so on. The Food and Drug Administration (FDA) [18] uses a three-category definition of precision. The same definition is used by the EC and FDA for repeatability. The FDA differs from EC by the term "intermediate precision" (see Table 1.3) which is determined within laboratory variation: different days, different analysts, different equipment, and so forth. Reproducibility expresses the precision between laboratories (collaborative studies). Several organizations differ in their approaches to collaborative studies: the United States Pharmacopeia uses procedures validated by pubic comment and ruggedness testing rather than a collaborative study process [38], whereas the International Union of Pure and Applied Chemistry's and AOAC Offical Methods of Analysis have developed harmonized procedures for collaborative studies [39].

The reproducibility standard deviation is typically two to three times as large as that for repeatability. Precision decreases with a decrease in concentration. This dependence has been expressed as $RSD = 2^{(1-0.5 \exp \log C)}$, where RSD is expressed as a percentage and C is the concentration of the analyte [38]. For the concentration ranges typically found in pharmaceutical dosage forms ($1-10^{-3}$), the RSD under conditions of repeatability should be less than 1.0%, and less than 2.0% under conditions of reproducibility [21]. These are similiar to the 1.5% recommendation made for RSD of system repeatability after analyzing a standard solution six times [35]. For method rcpcatability, which includes sample pretreatment, six replicate assays are made with a representative sample. A RSD no greater than 2% should be obtained.

E. Accuracy

Accuracy is the closeness of agreement between what is accepted as a true value (house standard, international standard) and the value found (mean value) after several replicates. This also provides an indication of systematic error.

Two of the most common methods of determining accuracy are by comparing the proposed test procedure to a second test procedure whose accuracy is known and the recovery of drug above and below the range of use. Average recovery of the drug should be 98–102% of the theoretical value. Recoveries can be determined by either external or internal standard methods.

Quantification by external standard is the most straightforward approach because the peak response of the reference standard is compared to the peak response of the sample. The standard solution concentration should be close to that expected in the sample solution. Peak responses are measured as either peak height or area [41].

For the internal standard method, a substance is added at the earliest possible point in the analytical scheme. This compensates for sample losses during extraction, cleanup, and final chromatographic analysis. There are two variations in the use of the internal standard technique. One involves the determination of response factors which are the ratios of the analyte peak response to the internal standard peak response. The second is referred to as response ratios which are calculated by dividing the weight of the analyte by the corresponding peak response.

An internal standard must be completely resolved from all other peak responses except where mass discrimination or isotopically labeled samples are used as the internal standard. The internal standard should elute near the solute to be quantified. The detector response should be similiar in area or height to the analyte of interest. The internal standard should be similiar in terms of chemical and physical properties to the analyte being measured. Substances that are commonly used as internal standards include analogs, homologs, isomers, enantiomers, and isotopically labeled analogs of the analyte. The internal standard should not be present or be a potential degradant of the sample. Finally, the internal standard should be present in reasonably high purity.

Internal standards are often used in dissolution testing of oral dosage forms [42]. Internal standards should be avoided in stability-indicating assays due to the possible coelution with unknown degradation products.

F. Ruggedness (Robustness)

The ruggedness of an analytical method is the absence of undue adverse influence on its reliability of performance by minor changes in the laboratory environment [43]. This validation parameter is not recognized by all organizations with testing oversight, as this characteristic is implied by collaborative validation programs (see Section IV.D).

The difference in chromatographic performance between columns of the same designation (i.e., C_{18}) is the most common source of chromatographic variability. To check the column-to-column ruggedness, the specificty (selectivity) of at least three columns from three different batches supplied by one column manufacturer should be checked [44]. A similarly designated column from another manufacturer should also be evaluated. Table 1.4 lists the specifications recommended to define a liquid chromatographic column [45,46]. Testing procedures have also evolved for the evaluation of gas chromatographic capillary columns [47]. Variability is also caused by the degradation of the chromatographic column.

Besides the sorbent stability, consideration should also be given to the stability of the sample solution. The widespread use of automatic sample injectors makes it necessary to determine the length of time that a sample is stable.

V. SYSTEM SUITABILITY TESTING

After a method has been validated, an overall system suitability test should be routinely run to determine if the operating system is performing properly.

An acceptable approach is to prepare a solution containing the analyte and a suitable test compound. If the method being used is to control the

Table 1.4   HPLC Column Specifications

Column packing
   Brand name
   Chemical composition
   Particle shape
   Particle size (mean size, size distribution)
   Pore diameter (mean distribution)
   Surface area
   Maximum pressure limit
   Operating range (temperature and pH)
   Bonded phase type
   Surface coverage
   Elemental analysis
   Procedure for preparing bonded phase
   Residual hydroxy groups

Column
   Dimensions
   Type of end fitting (frit pore size)
   Selectivity
   Column efficiency
   Peak asymmetry
   Column permeability
   Reproducibility of column selectivity between columns
   Maximum operating pressure

level of impurities, the minimum resolution between the active component and the most difficult to resolve impurity should be given. The chromatographic system should demonstrate acceptable resolution of the test solution and system precision. According to the USP, a system can be considered suitable if it meets the requirements for both precision and one of the tests listed in Table 1.5. A review reflecting this approach has been published [48], as have more elaborate approaches [23].

A.   System Resolution

There are several formulas available for calculating resolution factors. The formula recommended in USP 23 for GC and HPLC is as follows:

$$R = \frac{2(t_2 - t_1)}{W_2 + W_1}$$

Table 1.5 System Suitability Tests

Resolution
Precision
Peak asymmetry factor
Column efficiency
Capacity factor

---

where t_2 and t_1 are the retention times of the two components and W_2 and W_1 are the corresponding widths at the peak base. The width is obtained by extrapolating the relatively straight sides of the peaks to the baseline.

Some computer data systems have based their resolution calculations on the peak width at half the distance from the apex to the base of the peak [49]. Peak widths have also been measured at the point of inflection.

For TLC or planar electrophoresis, resolution can be calculated by

$$R = \frac{d}{W_1 + W_2 \sqrt{2}}$$

where the distance between zone centers (d) is divided by the averages of the widths (W) of the zones [50].

Representative resolution values are tabulated in Table 1.6. Resolution values are typically greater than 1.5 and are generally expressed as a range of values.

B. Determination of System Precision

After a standard solution is injected a number of times, the relative standard deviation of the peak responses is measured as either the peak height or peak area. When using an internal standard method, the response ratio is calculated. Maximum allowable system related standard deviations made at the 99% confidence level have been tabulated [44]. For the USP monographs, unless otherwise stated, five replicate chromatograms are used if the stated limit for relative standard deviation is 2% or less. Six replicate chromatograms are used if the stated relative standard deviation is more than 2.0%. The current USP emphasis is to perform all the replicate injections prior to sample assay and during testing whenever there is a significant change in equipment, or a critical reagent, or when a malfunction is suspected.

Performing all the standard injections prior to sample assay has been controversial [51]. The main point of contention is that the analyst does

Table 1.6 Representative System Suitability Values from USP 23

	Resolution	Precision	Asymmetry factor	Theoretical plates
Cefazolin	4.0	2.0	1.5	1500
Ceftizoxime	4.0	2.0	2.0	2000
Chlorthalidone	1.5	2.0	2.0	—
Dactinomycin for injection	— —	1.0 3.0	— 2	— 1200
Dipivefrin	—	2.0	1.2	500
Ergoloid	2.5 1.35 1.0	1.5	2.5	950
Fentanyl injection	—	2.0	2.0	—
Insulin	—	1.5	2.5	—
Lidocaine	3.0	1.5	—	—
Oxycodone tablets	—	2.0	2.0	—
Oxycodone/ acetaminophen tablets	2.4	2.0	—	—
Vancomycin	3.0	—	—	1500

not have overall control of the chromatographic system from beginning to end. The recommendation is to periodically inject duplicate standard solutions which should agree to within 0.5% of their values [51]. For planar techniques such as TLC or gel electrophoresis, this is a moot point because standards can be run alongside the samples in adjacent lanes. For example, when determining the the molecular homogeneity of proteins using SDS–PAGE gel electrophoresis, the two outer lanes contain molecular-weight standards that bracket the expected masses with the reference standards of the protein of interest in the next inner lanes followed by the sample tracks in the inside lanes.

C. Asymmetry Factor (Tailing Factor)

If the peak to be quantified is asymmetric, a calculation of the asymmetry would also be useful in controlling or characterizing the chromatographic system [52]. Peak asymmetry arises from a number of factors. The increase in the peak asymmetry is responsible for a decrease in chromatographic resolution, detection limits, and precision. Measurement of peaks on solvent tails should be avoided.

The peak asymmetry factor (tailing factor) can be calculated by several different methods. By the USP,

$$T = \frac{W_{0.05}}{2f}$$

where $W_{0.05}$ is the width of the peak at 5% peak height and f is distance at 5% height from the leading edge of the peak to the distance of the peak maximum as measured at the 5% height. The system suitability test for antibiotics and antibiotic drugs recommends measurement at 10% of the peak height from the baseline [53]. Representative values from the USP are presented in Table 1.6. Values vary from 1 to 3. For a symmetrical peak, the factor is unity which increases as tailing becomes more pronounced. A variety of alternative models have been proposed to more accurately characterize peak tailing [54].

D. Column Efficiency

The resolution factor is considered to be a more discriminating measure of system suitability than column efficiency [44]. Yet, column efficiency determinations are required for the assay of antibiotics and antibiotic-containing drugs [53]. The reduced plate height (h_r) for the column is determined by first calculating the number of theoretical plates per column:

$$N = 5.545 \left(\frac{t}{W_{h/2}}\right)^2 \quad \text{or} \quad N = 16\left(\frac{t}{W}\right)^2$$

where t is the retention time of the analyte and $W_{h/2}$ is the peak width at half-height or W is the width at the base of the peak.

The height equivalent to one theoretical plate is calculated by

$$h = \frac{L}{n}$$

where L is the length of the column. Finally, the reduced plate height is determined by

$$h_r = \frac{h}{d_p}$$

where d_p is the average diameter of particles in the column.

The reduced plate has the advantage of being independent of column length and particle diameter. The resulting number can also be compared to the theoretical limiting value of 2.

The calculation of column theoretical plates by the width at half-peak height is insensitive to peak asymmetry. This is because the influence of tailing usually occurs below that measurement location. The consequence will be an overestimate of the theoretical plates for non-Gaussian peaks. Nine different calculation methods for efficiency have been compared for their sensitivity to peak asymmetry [54]. Besides being influenced by the calculation method, column efficiency is sensitive to temperature, packing type, and linear velocity of the mobile phase.

E. Column Capacity

The column capacity factor is calculated by

$$k = \frac{t_r - t_m}{t_m}$$

where the retention time of the solute is t_r and the retention time of solvent or unretained substance is t_m. The corresponding retention volume or distance can also be used, as they are directly proportional to retention time. Retention volumes are sometimes preferred, because t_r varies with flow rate. The factor is then calculated by

$$V = \frac{V_r - V_m}{V_m}$$

where V_r is the retention volume of the solute and V_m is the elution volume of an unretained substance. There is no universally accepted method for the accurate measurement of the volume of an unretained substance. Numerous methods have been proposed [54].

For TLC,

$$k' = \frac{1 - R_f}{R_f}$$

where R_f is the distance traveled by the analyte to that of the mobile phase [50].

The factors which influence the reproducibility of retention in HPLC have been studied [55]. The conclusion is that the relative method of recording retention (e.g., relative capacity factors of retention indices) is more robust for reliable interlaboratory comparisons than the use of capacity factors.

VI. PRODUCT TESTING

Product testing is one of the most important functions in pharmaceutical production and control. A significant portion of the CGMP regulations (21 CFR 211) pertains to the quality control and drug product testing.

Out-of-specification laboratory results have been given additional emphasis by the FDA, particularly after the *Barr* v. *FDA* court case [55]. An out-of-specification result falls into three catogories: laboratory error, non-process-related or operator error, and process-related or manufacturing process error. Retesting of the same sample is appropriate when the analyst error can documented. An outlier test on some chemical assays, particularily those involving extensive sample preparation and manipulation, is justifiable but is not a routine approach to rejecting results [56].

VII. CONCLUSION

There are numerous variables to consider in developing an accurate and rugged chromatorgaphic method. The extent depends on the purpose of the test: that is, stability-indicating assays are the most demanding, whereas identification tests are the least demanding.

From the six validation variables listed, specificity, accuracy of dosage form assay, and ruggedness are the most crucial. In the initial stage of developing a chromatographic method, the primary goal is to measure an analyte in the presence of interferences. The second step is to demonstrate that the analyte can be accurately measured. The ruggedness and accuracy of a method can be improved with the development of treatment steps that require minimal manual manipulation and use of column packings that do not vary from lot to lot [57].

The efforts at harmonization of the requirements among Europe, the United States, and Japan for methods validation, stability testing, and indentification of impurities are welcomed by all pharmaceutical analysts.

REFERENCES

1. H. Platz, *Uber Kapillaranalyse und ihre Anwendung in Pharmazeutischen Laboratorium*, Leipzig, 1922.
2. *The Pharmacopeia of the United States of America*, 15th revision,

United States Pharmacopeia Convention, Inc., Rockville, MD, 1955, p. 802.

3. R. W. Kozak, C. N. Dufor, and C. Scribner, Regulatory considerations when developing biological products, *Cytotechnology*, 9:203–210 (1992).

4. J. P. Boehlert, Impurities in new drug substances, PMA Fall Meeting, 1993.

5. *The Gold Sheets*, *27*(8):1 (1993).

6. Analysis of L-tryptophan for etiology of eosinophilia-myalgia syndrome, *JAMA*, *264*:2620 (1990).

7. *The Pharmacopeia of the United States of America*, 23rd revision, USP Convention, Inc., Rockville, MD, 1995.

8. B. Testa and W. F. Trager, Racemates versus enantiomers in drug development: dogmatism or pragmatism, *Chirality*, *2*:129–133 (1990).

9. PMA Ad Hoc Committee on Racemic Mixtures, Comments on enantiomerism in drug dcvclopment process, *Pharm. Tech.* *5*:46 (1990).

10. Development of compendial monographs for marcomolecular drugs and devices derived from biotechnological processes, *Pharmacopeial Form*, 4616 (1988).

11. M. Hamburger and K. Hostettmann, Analytical aspects of drugs of natural origin, *J. Pharm. Biomed. Anal.*, *7*:1337–1349 (1989).

12. M.D. VanArendonk, The new ICH stability guideline, *PMA Fall Meeting, 1993*.

13. *Fed. Reg.*, *58* (72): 21086–21091 (1993).

14. Long-term stability testing should be at 25°C/45% relative humidity, FDA Advisory Committee concurs; Testing standard should only apply to new drugs, FDA-*The Pink Sheets*, June 18, 1993, p. 9.

15. M. C. Manning, K. Patel, and R. T. Borchardt, Stability of Protein Pharmaceuticals, *Pharm. Res.*, *6*:903 (1989).

16. Ad Hoc Working Party on Biotechnology/Pharmacy Note For Guidance, Stability Testing of Biotechnolgical/Biological Products, ECC, III/3772/92-EN, June 1993.

17. *Guideline for Submitting Documentation for the Stability of Human Drugs and Biologics*, February 1987.

18. International Conference on Harmonisation; Draft Guideline on Validation of Analytical Procedures for Pharmaceuticals; Availability, *Fed. Reg.*, *59*:9750 (1994).

19. CPMP Working Party On Quality of Medicinal Products–Analytical Validation, III/844/87-EN, August 1989.

20. Current concepts for the validation of compendial assays, *Pharmacopeial Forum*, 1241 (1986).

21. M. Martin-Smith and D. R. Rudd, The importance of proper valida-

tion of the analytical methods employed in the quality control of pharmaceuticals, *Acta Pharm. Jugosl.*, *40*:7–19 (1990).

22. H.B. Woodruff, P.C. Tway, and J. Cline Love, Factor analysis of mass spectra from partially resolved chromtographic peaks using simulated data, *Anal. Chem.*, *53*:81 (1981).

23. G. Szepesi, M. Gazdag, and K. Mihalyfi, Selection of high-performance liquid chromatographic methods in pharmaceutical analysis, *J. Chromatogr.*, *464*:265–278 (1989).

24. G. T. Carter, R.E. Schiesswohl, H. Burke, and R. Young, Peak homogeneity determination for the validation of high-performance liquid chromatography assay methods, *J. Pharm. Sci.*, *71*:317 (1982).

25. A. F. Fell, H. P. Scott, R. Gill, and A. C. Moffat, Novel techniques for peak recognition and deconvolution by computer aided photodiode array detection in high-performance liquid chromatography, *J. Chromatogr.*, *282*:123 (1983).

26. J. G. D. Marr, P. Horvath, B. J. Clark, and A. F. Fell, Assessment of peak homogeneity in HPLC by computer-aided photodiode array detection, *Anal. Proc.*, *23*:254 (1986).

27. G. W. Schieffer, Limitation of assessing high-performance liquid chromatographic peak purity with photodiode array detectors, *J. Chromatogr.*, *319*:387 (1985).

28. D. A. Roston and G. M. Beck, HPLC assay validation studies for bulk samples of a new analgesic, *J. Chromatogr. Sci.*, *27*:519 (1989).

29. C. A. Johnson, Purity requirements from a pharmacopoeial point of view, *J. Pharm. Biomed. Anal.*, *4*:565 (1986).

30. S.-O. Janson, Characterization of drug purity by liquid chromatography, *J. Pharm. Biomed. Anal.*, *4*:615 (1986).

31. K. Bergstrom, Carbohydrate-purity assessment, *J. Pharm. Biomed. Anal.*, *4*:609 (1986).

32. K. Emancipator and M. H. Kroll, A quantitative measure of nonlinearity, *Clin. Chem.*, *39*:766–772 (1993).

33. Guidelines for collaborative study procedure to validate characteristics of a method, *J. Assoc. Anal. Chem.*, *72*:694–704 (1989).

34. P. A. D. Edwardson, G. Bhaskar, and J. E. Fairbrother, Method validation in pharmaceutical analysis, *J. Pharm. Biomed. Anal.*, *8*: 929–933 (1990).

35. ASTM-Task Group E1908, An evaluation of quantitative precision in high-performance liquid chromatography, *J. Chromatogr. Sci.*, *19*: 338 (1981).

36. J. E. Knoll, Estimation of the limits of detection in chromatography, *J. Chromatogr. Sci.*, *23*:422 (1985).

37. J. P. Foley and J. G. Dorsey, Clarification of the limit of detection in chromatography, *Chromatographia, 18*:503 (1984).
38. *The Referee, 13*(5):7 (1989).
39. Guidelines for Collaborative Study Procedure to Validate Characteristics of a Method of Analysis, *J. Assoc. Anal. Chem., 72*:694 (1989).
40. W. Horwitz, *The Pesticide Chemist and Modern Toxiciology*, American Chemical Society, Washington, DC, (1981), p. 411.
41. *Fed. Reg., 50*:9998 (1985).
42. R. K. Baweja, Dissolution testing of oral solid dosage forms using HPLC, *Pharm. Technol., 11*:28 (1987).
43. W. L. Paul, USP perspectives on analytical methods validation, *Pharm. Technol., 3*:129 (1991).
44. E. Debesis, J. P. Boehlert, T. E. Givand, and J. C. Sheridan, Submitting HPLC methods to the compendia and regulatory agencies, *Pharm. Technol., 9*:120 (1982).
45. D. J. Smith, The standardization of HPLC columns for drug analysis: Are C18 columns interchangeable? In *Liquid Chromatography in Pharmaceutical Development* (I. W. Wainer, ed.), Aster, Springfield, OR, 1985, p. 409.
46. R. E. Pauls and R. W. McCoy, Testing procedures for liquid chromatographic columns, *J. Chromatogr. Sci., 34*:66 (1986).
47. K. Grob, G. Grob, and K. Grob, Jr., Testing capillary gas chromatographic columns, *J. Chromtogr., 219*:13 (1980).
48. T. D. Wilson, Liquid chromatographic methods validation for pharmaceutical products, *J. Pharm. Biomed. Anal., 8*:389–400 (1990).
49. R. J. Darnowski, Quantitative chromatographic system suitability tests revisited, *Pharmacopeial Forum*, 941 (1985).
50. J. R. Conder, Peak distortion in chromatography, *HRC & CC, J. High Resolt. Chromatogr. Commun., 5*:341 (1982).
51. Part 436 — Test and methods of assay of antibiotic and antibiotic containing drugs, *Fed. Reg., 50*:999 (1985).
52. B. A. Bildingmeyer and F. V. Warren, Column efficiency measurement, *Anal. Chem., 56*:1583A (1984).
53. R. Gill, M. D. Osselton, R. M. Smith, and T. G. Hurdley, Retention reproducibilty of basic drugs in high-performance liquid chromatography on silica columns with methanol-ammonium nitrate eluent, *J. Chromatogr., 386*:54 (1987).
54. R. J. Smith, C. S. Nieass, and M. S. Wainwright, A review of methods for the determination of hold-up volume in modern liquid chromatography, *J. Liq. Chromatogr., 9*:1387 (1986).
55. R. Gill, M. D. Osselton, R. M. Smith, and T. G. Hurdley, Retention

reproducibility of basic drugs in HPLC on silica column with metha-nol-ammonium nitrate eluent, *J. Chromatogr.*, *386*:65 (1987).

56. *Guide to Inspection of Pharmaceutical Quality Control Laboratories*, FDA, Washington, DC, July 1993.

56. R. E. Madsen, *U.S.* v. *Barr Laboratories*: A Technical Perspective, *PDA J. Pharm. Sci. Technol.*, *48*:176 (1994).

57. B. S. Welinder, T. Kornfelt, and H. H. Sorensen, Stationary phases: The weak link in the LC chain? *Today's Chemist at Work*, 9:35 (1995).

2

Sample Pretreatment

JOHN A. ADAMOVICS   *Cytogen Corporation, Princeton, New Jersey*

I. INTRODUCTION

In most instances, formulated drugs cannot be chromatographically analyzed without some preliminary sample preparation. This process can generally be categorized into sampling and sample cleanup steps with the overall goal of obtaining a representative subfraction of the batch. This chapter is a discussion of manual and automated sample preparation procedures for pharmaceutical formulations.

II. SAMPLING

A. General

Samples submitted to a pharmaceutical laboratory for testing must be representative of the production lot or another bulk unit from which it was taken. This criterion helps to avoid a risk of obtaining out-of-specification results for a lot within specifications and vice versa. The Food and Drug Administration (FDA) requires that a description of a sampling plan be submitted to assure that the sample of the drug product obtained is representative of the batch [1]. The plan should include both the sampling of production batches and the selection of subsamples for analytical testing.

The plan is only applicable to batches of one particular size, so procedures for scale-up or scale-down of this sampling plan to other batch sizes must be provided. If samples are to be pooled, a justification must be given. Additional guidelines have been developed for determining whether a production lot is wellmixed or segregrated and for the estimation of the sample size and number [2].

B. Vegetable Drugs (Crude Drugs)

The United States Pharmacopeia (USP) requires that for homogenous batches of vegetable drugs, all the containers of the batch be sampled if there are 1–10 containers, 11 if there are 11–19, and for more than 19, n(# samples containers to be samples) = 10 + [N(# containers batch)/10] [3, p. 1754]. When the batch cannot be considered homogenous, it is divided into subbatches that are as homogenous as possible, then each one is sampled as a homogenous batch. Samples should be taken from the upper, middle, and lower sections of each container. If the crude material consists of component parts which are 1 cm or less in any dimension, and in the case of all powdered or ground materials, the sample is withdrawn by means of a sampling device that removes a core from the top to the bottom of the container. For materials >1 cm, sample by hand. For large bales, samples should be taken from a depth of 10 cm.

In the Chinese Pharmacopoeia [4], 5 packages are sampled if the total is <100, 5% if the total ranges from 100 to 1000, and for >1000 packages, 1% of the part in excess of 1000 are sampled. If there is sufficient sample, the quantity obtained should be 100–500 g for common drugs, 25 g for powdered, 5–10 g for precious drugs.

C. Sampling of Dosage Units

Parenterals

Generally speaking, parenteral dosage forms are homogenous or can be demonstrated to be so while validating the manufacturing process. For relatively small lots such as 3000 doses, generally two dosage units are analyzed in duplicate for each of the testing parameters and samples are set aside for reserve and stability.

Tablets and Similar Dosage Forms

The blending of a formulation containing an active ingredient with the excipients is often carried out in lot sizes which will produce thousands of tablets or similiar dosage forms. When the proportion of the active ingredient to the total mass is small, as would occur with potent drugs, it may be difficult to obtain a uniformly distributed mixture. Dosage forms of digitoxin and thinyl estradiol tablets are documented instances of heterogenous blends [5–7].

With these considerations in mind, these types of solid dosage forms can be sampled either by assaying multiple individual units or a composite sample of those individual units. Individual unit sampling should occur when the range of values in the separate units is large and/or when it is necessary to establish the variability of the units. Compositing is used when homogeneity is not a significant problem or when the unit variablilty is not important.

A number of organizations have devised procedures for tablet sampling. The Pharmacopeia of Japan [8] requires that the content of the active ingredient in each of 10 tablets be assayed. The assay result from each tablet should not deviate from the average content by more than 15%. If one tablet shows a deviation exceeding 15% but not 25%, the contents of 20 additional tablets should be assayed. From the average of these 30 determination, not more than 1 tablet should be between 15% and 25% and none should exceed 25%. The content uniformity requirements of USP ⟨905⟩ calls for assaying 10 units individually and assaying a composite specimen. The results of the two procedures are each expressed as one average dosage unit and the difference between these two numbers is evaluated. This approach is applicable to tablets, capsules, suppositories, transdermal systems, suspensions, and inhalers.

Numerous reports have noted the apparently large differences between the average composited assay value and the average assay value for the individual tablets [9]. One possible explanation for this observation is that during the ginding or blending of a composite sample segregation of the tablet components has occurred. The result of this is a nonrandomized mixture. The forces and mechanisms that come into play during particle segregation have been discussed [9] and the procedures to minimize them are discussed later in this chapter [10,11].

Other Dosage Forms

Upon standing, liquid dosage forms such as gels, lotions, and suspensions are likely to become nonhomogenous. Prior to sampling, formulations of these types must be homogenously mixed. For a suspension or syrup, withdrawing an accurate aliquot is difficult. For inhalation products, the total contents of a dosage unit are assayed. For transdermal preparations, the uniformity can be determined by punching out known surface areas of the membrane. USP ⟨905⟩ has content uniformity requirements for the above dosage forms.

III. SAMPLE PREPARATION TECHNIQUES

A. Direct Analysis

Liquid dosage forms often can be directly asayed or simply diluted with water or mobile phase prior to testing. Benzethonium chloride tincture, prilocaine hydrochloride [3, pp. 173 and 1287], and numerous biological products such as OncoScint CR/OV (a monoclonal antibody DTPA conjugate) are examples.

Volatile impurities in bulk solvents or solvents in dosage forms such as ethanol and methanol are directly analyzed by gas chromatographic methods. These methods are discussed in Chapter 4.

B. Liquid–Solid Extraction

A frequently encountered procedure is the extraction of a substance from a solid dosage form, such as in the analysis of tablets. This can be a relatively simple procedure involving the selection of a solvent or solvent combination which ideally provides good solubility of the substance of analytical interest and minimal solubility of components that interfere with the chromatographic analysis. Over the last several years there has been increased interest in extracting analytes using supercritical fluids such as carbon dioxide [12–17]. The primary limitation of this approach has been the limited solubility of most polar drugs such as antibiotics in supercritical fluids. Sulfamethoxazole and trimethoprim have been extracted with supercritical carbon dioxide from a drug formulation [18]. The ultility of supercritical chromatography is discussed in Chapter 7.

For the majority of procedures, the first step requires either the grinding or milling of the solid matrix into a fine powder followed by solvent extraction, and filtration or cenrifugation to eliminate particulates.

One problem that has been encountered in grinding tablets is the physical separation of the analyte of interest in the matrix. This phenomenon helps explain discrepancies that occur between the average of the individual tablet assay values prepared by direct dissolution and those of the corresponding tablet composites. Table 2.1 outlines various advantages and disadvantages of sample preparation procedures for overcoming segregation of tablet components [9–11].

The efficient extraction of analytes adsorbed or absorbed as in creams, ointments, and other semisolid formulations are difficult to achieve. For adsorbed analytes, displacement from the adsorption sites by a small amount of acid, base, or buffer is effective. For semisolid formulations, solvent extraction is generally performed at elevated temperatures so as to melt the solid and increase the extraction efficiency.

Table 2.1 Advantages and Disadvantages of Various Methods of Sample Preparation for Overcoming Segregational Problems Due to Grinding

Method	Summary of method	Advantages	Disadvantages
A	1. Directly dissolve tablet in suitable solvent 2. Assay aliquot of solution	Eliminates segregation	Drug must dissolve completely in solvent upon tablet disintegration.
B	1. Grind tablets to fine powdered composite 2. Dissolve powder in suitable solvent 3. Assay aliquot of solution	Eliminates segregation Drug is released independently of dissolution characteristics of tablets	Some active ingredients may remain undissolved because solubility limit of drug may be reached. False low results
C	1. Grind tablets to fine powdered composite 2. Pass powder through #60 mesh sieve 3. Assay sievings	Eliminates segregation tendencies Produces particles of uniform size	Sieving may generate electrostatic charges among particles, another cause for segregation.
D	1. Grind tablets to fine powdered composite 2. Dissolve powder in organic solvent 3. Continue grinding 4. Evaporate solvent 5. Assay residue	Eliminates free-flowing particles and segregation tendencies Facilitates dissolution of drug in solvent	Drug and other tablet ingredients may be chemically altered by the organic solvent.

Particulates from the sample matrix that are carried over during the sample preparation should be removed prior to analysis by either high-performance liquid chromatography (HPLC) or gas chromatography (GC). This is especially true for particles less than 2 μm in size. These particulates will pass through the frits on a liquid chromatographic column and settle on top of the sorbent which will eventually cause an increase in the back pressure of the chromatographic system and susequently decrease the column performance.

One efficient removal procedure is to use a 0.45-μm filter. There are basically two types of filters: depth and screen. Depth filters are randomly oriented fibers that will retain particles throughout the matrix rather than just on the surface. They have a higher load capacity than screen filters. Due to the random nature of the matrix, they have no definite upper-limit cutoff particle size retained. Their porosity is identified as a "nominal pore" size to indicate this variable.

The most common depth filter of 0.45 μm nominal porosity is glass microfiber. These filters are compatible with organic and aqueous solutions between pH ranges of 3–10.

Screen filters are polymeric membranes that have uniform distribution of pore sizes. They are relatively thin so that there is a minimal amount of liquid retention. Screen filters clog more rapidly than depth filters. Table 2.2 lists the common screen filter materials and their solvent compatabilities.

In developing a method that requires filtration, adsorption of the analyte onto the filter must be taken into account. For dilute solutions of adriamycin, >95% is adsorbed to cellulose ester membranes and about 40% to polytetrafluoroethylene membranes [19]. For more concentrated solutions, as would be encountered in bulk formulation testing, filter ad-

Table 2.2 Commonly Available Screen Materials

Membrane material	Solvent compatibility
Teflon	Organic solvents or aqueous/organic mixtures Resistant to strong acids and bases Organic and aqueous compatible pH range of 3–10
PVDF	Aqueous and organic/aqueous mixtures Resistant to strong acids and bases Low protein binding
Cellulose esters	Aqueous

sorption is not as important a concern. Nevertheless, the common practice is to discard the first several milliliters of the filtrate. For protein-based products there is significant nonspecific binding to nylon-based microporous membranes and minimal binding to hydrophilic polyvinylidene fluoride membranes [20].

C. Liquid–Liquid Extraction

In the simplest form, an aliquot of an aqueous solution is shaken with an equal volume of an immiscible organic solvent. This is an useful approach when the analyte of interest partitions itself in one layer and the interfering matrix partitions into the second layer. Because this rarely occurs, several physical and chemical factors can be changed to alter the partitioning. One approach is to add sodium chloride to the aqueous solvent to produce a saturated solution.

In aqueous solution, organic acids and bases exist in equilibrium mixtures in their neutral and ionic forms. Because the neutral and ionic forms will not have the same partition coefficient, the amount extracted depends on the acid–base equilibrium. For an efficient extraction, the analyte should be at least 95% in the extracable form. This would usually mean either as its free acid or free base. Figure 2.1 is a nomogram relating pK values to percentage of ionization at various pH values [21]. In most cases, pH adjustment of the sample to pH = pK − 2 for acidic compounds or pH = pK + 2 for basic compounds is sufficient.

Generally, a single extraction is not sufficient for drugs where the chromatographic interferences are numerous and the concentration of the analyte in the sample is low. One approach to this type of situation is to back extract the drug analyte from the organic phase into an aqueous phase of opposite pH [22]. A scheme of a back extraction for a basic drug is shown in Figure 2.2. For example, chlorpheniramine, has a pKa value of 9.1, which means that it is protonated in acidic solutions, and extract into aqueous solutions. In alkaline aqueous solutions, chlorpheniramine is extractable into an aqueous immiscible slovent. Reextraction into dilute acid would further purify the chlorpheniramine extract from coextracted neutral excipients.

Even though conventional extraction has been useful in testing of dosage forms, there are drawbacks. The primary difficulty is with the low extraction efficiencies that are common for highly ionic or amphoteric compounds. A review of 37 literature references that used conventional extraction techniques for analytes of drug products quoted recoveries of lower than 80% in 7 of the references reviewed [23].

An additional liquid extraction technique used to increase extraction

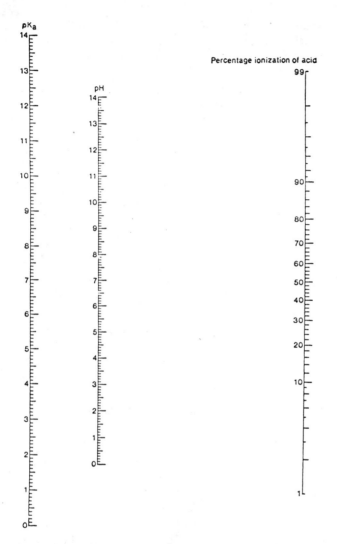

Figure 2.1 Nomogram relating pK values of acids to percentage ionization at various solutions pH values. (Reprinted from Ref. 17 with permission.)

efficiency and selectivity is ion-pair extraction. Ion-pair extraction was first used to extract strychnine from a syrup formulation [23]. This technique is based on the formation of an association complex between the ionic species and the counterion of opposite charge. Ion pairs formed between relatively large organic anions and cations often have solubility in low-polarity organic solvents. A primary requirement is that the counterion must be

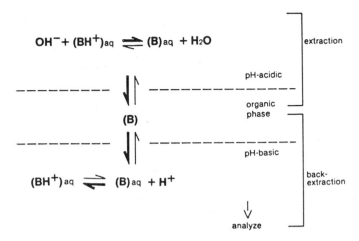

Figure 2.2 Scheme for the back-extraction procedure of a basic drug.

chosen so that the pH range of the drug and counterion overlap. Generally, there is a trade-off between extraction efficiency and selectivity [23]. The various parameters that affect ion-pair extraction have been reviewed [24].

The development of a standarized analysis strategy using ion-pair extraction from basic drugs have been reported [23]. This approach has been used to assay basic drugs in syrups, ointments, emulsions, and suppositories. Ion-pair formation with tri-*n*-octylamine extracted colorants from syrups, oral suspension, tablets, gelatin capsules, suppositories, and granules [25].

A major problem in liquid–liquid extraction for sample preparation is emulsion formation which leads to lower recoveries. Emulsions occur readily when solvents of similiar densities are mixed and when extraction solutions are highly basic.

The separatory funnel is the classical liquid–liquid apparatus used to segregate immiscible phases. The pear-shaped funnel developed by E. R. Squibb in the 1880s is still the most commonly used by chemists. Other separatory funnel designs which have higher overall efficiency have been designed but have not become popular.

D. Open-Column Chromatography

Open-column chromatographic methods are no longer widely used in quantifying drug poducts. Yet, a number of methods in the USP 23 [3] describe the use of open-column methods for sample pretreatment. Columns packed with silicates or alumnia are the most widely cited and are used to clean up amcinonide cream [3, p. 74], dexamethasone gel [3, p. 469], and lindane cream [3, p. 892]. Cumbersome and time-consuming, open-column procedures are being displaced by commerically available disposable cartridges containing a variety of sorbents and selectivities.

E.   Column Liquid–Solid Extraction

General Considerations

Disposable columns or cartridges filled with a sorbent are being used for sample cleanup and is referred to as solid-phase extraction (SPE). The packing material used in these cartridges are similiar to the material found in HPLC columns but has larger particle sizes. Analytichem International (now Varian Sample Preparation Products) introduced their Chem-Elut cartridges in the mid-1970s using diatomaceous earth as packing material. Throughout the 1980s, SPE cartridges packed with a variety of materials exhibiting a wide range of chemistries were formulated and marketed. In the early 1990s, cartridges containing rigid glass-fiber disks embedded with silica were introduced. These disks have reduced bed volumes which require substantially smaller solvent volumes (Ansys, Inc., Irvine California) [26–30].

Procedure

There are two strategies for sample cleanup using this approach. The first is to select a sample solvent that allows substances of interest to be totally retained on the extraction column sorbent while eluting substances that would interfere with the chromatographic assay. The analytes of interest are then eluted with a small volume of a solvent that will displace the analytes from the sorbent. This strategy is useful when the analyte of interest is present in a low concentration. The alternative approach is to retain the matrix interferences while eluting the desired analyte.

The first step in using SPE is to condition the sorbent with an appropriate solvent. This prewetting increases the capacity of the bonded surfaces by opening up the hydrocarbon chain of the bonded-phase sorbents [31]. For nonploar sorbents, such as C_{18}, and for the ion exchangers, one column volume of methanol followed by one column volume of distilled water is required. Excessive washing with water will reduce analyte recovery [32]. Polar sorbents such as diol, cyano, amino, and silica should be rinsed with one column volume of a nonpolar solvent such as methylene chloride. Aternate cleanup methods may have to be developed if the analyte is sensitive to lead, zinc, and copper, as silica-based sorbents are known to contain these contaminants [33].

Cartridge loadability and solvent flow rate effects must also be considered when developing cartridge-based sample preparation methods. The quantity of sorbent in the cartridge is obviously related to the loadability as the analyte's capacity factor k. The larger the k, the greater the analyte mass loading. Overloading the cartridge will cause the analyte to "break through" with an earlier retention volume. The column capacity of an analyte is also reduced by the presence of competing analytes.

Linear velocity of the solvent through the cartridge will affect the recovery and bandwidth of the analyte. For example, a flow velocity of 0.3 ml/min gave a narrow band for riboflavin and a recovery of 100%. At the excessive velocity of 27 ml/min, decreased recoveries and band broadening were observed [34].

Methods Development

As in analytical liquid chromatography (LC), analyte retention depends on sample concentration, solvent strength, and sorbent characteristics. An empirical approach to methods development initially involves screening the available sorbents. The first step is to determine which sorbents best retain the analyte. The second consideration is to evaluate the solvents needed to elute the compound and the compatibility of those sorbents to the chromatographic testing procedure. The third step is to test the blank sample matrix to evaluate the presence of possible interferents. Finally, recoveries of known quantities of analyte added to the sample matrix must be determined.

Increased solvent polarity is required to elute retained compounds from silica sorbents while decreasing solvent polarity for C_{18} sorbents. Under these conditions, most polar analytes elute last from the silica and first from the C_{18} sorbents. Methanol has been demonstrated to be superior to acetonitrile during the SPE of basic drugs such as pentacaine, propranolol, and stobadin [26], whereas a second of basic drugs indicated that there was not a significant difference [30].

Numerous examples of the ulitity and selectivity of these sorbents are given below. Table 2.3 lists nine steroids that were tested for their retentiveness on five different sorbents [31]. The steroid standards at 1 mg/ml were dissolved in methanol–water for the evaluation of a C_{18} sorbent. For all the other sorbents, the steroids were dissolved in methylene chloride. At the polarity extremes for these steroids, cholesterol (the least polar) is retained only on C_{18}, whereas hydrocortisone (most polar) is retained on all five of the sorbents tested.

Table 2.3 Retentiveness of Nine Steroids on Various Sorbents

Steroid	C_{18}	CN	Silica	Diol	NH_2
Cholesterol	+[a]	−[b]	−	−	−
Cholesterol palmitate	+	−	−	−	−
Cortisone	+	+	+	−	−
Deoxycorticosterone	+	+	+	−	−
Estradiol	−	−	+	−	−
Hydrocortisone	+	+	+	+	+
Hydrocortisone acetate	−	−	+	−	−
Prednisone	+	+	+	−	−
Progesterone	+	−	+	−	−

[a] + = retained.
[b] − = unretained.
Source: Adapted from Ref. 30.

As another example, desonide and parabens in cream and ointment formulations were cleaned up by SPE by first testing mixtures of hexane-chloroform with silica, diol, and aminopropyl sorbents [36]. The solvent combination of 20% chloroform in hexane was found to be the optimum for dissolving the ointment base and yielding high recoveries of the analytes. The silica and aminopropyl sorbents were found to give nearly identical quantitative results, whereas the diol sorbent gave lower recoveries. Table 2.4 outlines the solvent considerations needed to elute retained compounds on silica and C_{18} sorbents.

The selectivity of C_{18} sorbents has been demonstrated by the separation of a mixture of eight FD&C colorants [37]. Cartridges packed with C_{18} were washed with increasing concentrations of isopropanol in a water-isopropanol eluent. This procedure is a viable alternative to the conventional time-consuming methodology of two chromatographic columns used for the separation and identification of colorants in drugs [37]. Additional examples of published sample preparative procedures using SPE are cited in Table 2.5.

A general strategy for relatively polar analytes has been developed [38]. In this approach, the cyanopropyl-silica–bonded phase remains the

Table 2.4 Separation Guidelines Using C_{18} and Silica Solid Phases

Sorbent	Silica	C_{18}
Packing polarity	High	Low
Typical solvent polarity range	Low to medium	High to medium
Typical sample solvent	Hexane, toluene, CH_2Cl_2	H_2O, buffers
Elution solvent	Ethyl acetate, acetone, CH_3CN	H_2O/CH_3OH, H_2O/CH_3CN
Sample elution order	Least polar sample components first	Most polar sample components first
Solvent required to elute retained compounds	Increased solvent polarity	Decreased solvent polarity

Source: Adapted from Water Chromatography Division Literature.

preferred and first choice sorbent. For unretained small polar drugs, the C_{18} sorbent is the first alternative, using water as the wash solvent and either methanol (for acids) or methanol–phosphate buffer pH 3 (for bases). If enough retention is not shown on either of the above and if the drug has ionizable functions, the use of an ion-exchanging solid phase is recommended.

F. Applications—Sample Treatment

Bulk Drug

A solvent or combination of solvents must be chosen so that the analyte is soluble and compatible with chromatographic procedures. The solvents most commonly used to solubilize bulk drugs are acetone, acetonitrile, chloroform, ethanol, methanol, and water. Besides the USP, two other sources contain useful solubility data on pharmaceuticals [42,43].

Tablets and Other Solids

Solids for oral use are the most common dosage form. The preparation step generally consists of grinding or milling of the tablets. During this step, active ingredients can undergo physical separation from other tablet com-

Table 2.5 Column Liquid–Solid Extraction of Pharmaceutical Formulations

Sample	Sorbent	Procedure	Chromatographic method	Reference
Amino acids from aqueous solutions	SCX	Condition sorbent with hexane, methanol and water. Adjust sample pH to 7 and wash sorbent with water, elute with 0.1 N hydrochloric acid.	—	J.T. Baker, SPE Applications Guide
Bacitracin ointment	Diol	Heat and shake ointment with methylene chloride. Add suspension to sorbent. Dry column, elute with 0.1 N hydrocholoric acid.	Analytical column — C_8, mobile phase — phosphate buffer acetonitrile	J.T. Baker, SPE Applications Guide
Benzalkonium chloride from eye wash solutions	CN	Wash sorbent with acetonitrile and water. After adding sample, wash with 1.5 N hydrochloric acid. Air dry the sorbent, elute with methanol — 1.5 N acid (8 : 2).	Analytical column — CN, mobile phase — acetonitrile — 0.1M sodium acetate (3 : 2)	J.T. Baker, SPE Applications Guide
Carbohydrates from aqueous solutions xylose, fructose, glucose, sucrose, and lactose	CN	Condition sorbent with acetonitrile dilute sample with acetonitrile. Dilution is critical. Wash column with acetonitrile and elute with water.	—	J.T. Baker, SPE Applications Guide

Sample	Sorbent	Procedure	Analysis	Reference
Chlortetracycline ointment	Diol	Heat and shake with hexane. Add suspension to sorbent and wash with hexane. Dry sorbent, elute with 0.1N hydrochloric acid–methanol (1:1).	HPLC, C_8, 0.05M phosphate buffer–acetonitrile	J.T. Baker, SPE Applications Guide
Desonide, methyl, propyl, butyl hydroxybenzoate cream and ointment	Silica or NH_2	Sample dissolved in hexanechloroform (8:2). Add to sorbent and wash with hexane. Elute with methanol.	HPLC, C_{18}, methanol–water (3:2)	36
Estradiol valerate and testosterone enanthate from oily formulations	Silica gel	Oil dissolved with carbon tetrachloride. Add to sorbent, elute with acetonitrile.	HPLC, C_8, acetonitrile–water (7:3)	39
Flumethasone privalate ointment	Silica gel	Dissolve with hexane (1:1), heat. Add to sorbent, elute with methanol.	HPLC, C_8, tetrahydrofuran–methanol–water (3:3:4)	40
0.5% Hydrocortisone cream	Silica gel	Vortex cream with hexane–ethyl acetate (1:1). Add to extraction cartridge and wash with hexane–acetone (8:2). Dry sorbent, elute with methanol.	HPLC, C_8, methanol–water (7:3)	J.T. Baker, SPE Applications Guide
Menthol ointment formulated with tetracaine	Silica gel	Dissolve with hexane, sonicate. Add to sorbent, elute with ether.	GC	3 (p. 1505)

Table 2.5 (Continued)

Sample	Sorbent	Procedure	Chromatographic method	Reference
Methylparaben syrup and ointment	Kieselguhr	Add 0.01M hydrochloric acid to sample, add to sorbent. Elute with diethyl ether or ethanol.	GC or TLC	41
Parabens from lotions and other cream-based formulations	C_8	Agitate sample with methanol, centrifuge. Dilute supernatant with water. Condition the column with methanol than water, add sample. Elute with methanol.	HPLC, C_{18}, acetonitrile–water (45 : 55)	J.T. Baker, SPE Applications Guide
Sulfa in topical cream	C_{18}	Dissolve with tetrahydrofuran. Dilute with 0.01M ammonium phosphate to precipitate lipid components. Add sample to conditioned sorbent, elute.	HPLC, C_{18}, ammonium phosphate–methanol (4 : 1)	J.T. Baker, SPE Applications Guide
Vitamin A and vitamin E, fat-soluble vitamins	C_{18}	Add 1% acetic acid to round tablet, heat to 55°C for 2 min. Add isopropanol, add sample to sorbent that had been washed with 1% acetic acid. Wash sorbent with isopropanol–1% acetic acid (55 : 45) followed by methanol–water (8 : 2). Air dry the sorbent, elute with methylene chloride. Add BHT to enhance stability.	C_8, acetonitrile–methanol–water (47 : 47 : 6)	J.T. Baker, SPE Applications Guide & Applied Separations

Compound	Sorbent	Procedure	HPLC conditions	Reference
Vitamin B_{12} in multivitamin tablets	SAX & phenyl	Extract powder in low actinic flask with aqueous solution containing phosphate buffer–citric acid and metabisulfite. Condition sorbent with methanol, water, and extraction solvent. Fit SAX column on top of phenyl column. After applying sample, wash with extraction solvent. After removing the SAX column, phenyl column is washed with water, air dried, followed by hexane, methylene chloride, acetonitrile, and acetonitrile–methanol (95 : 5). Sample eluted with methanol–water (9 : 1).		J.T. Baker, SPE Applications Guide
Water-soluble vitamins from tablets (niacinamide, pyridoxine, riboflavin, and thiamine)	C_{18}	Extract powder with aqueous $0.01M$ sodium heptane sulfonate–acetic acid (99 : 1). Flush with nitrogen. Heat to 55°C. Wash sorbent with methanol followed by above extracting solvent. Elute with methanol.	C_8, sodium heptane sulfonate–methanol (9 : 1)	J.T. Baker, SPE Applications Guide

39

ponents. This phenomenon has led to poor reproducibility when duplicate assays from tablet composites were assayed [9–11]. Various alternative methods have been suggested; these include direct dissolution of a representative number of individual tablets in a suitable solvent, the sieving and regrinding of the ground tablets, the grinding of a composite with a suitable organic solvent and the evaporation of the solvent, and the dissolution of the composite tablet sample in a solvent. For enterically coated tablets, manual grinding with a mortar and pestle can lead to erratic results which are overcome by repeated resieving and regrinding of the particles to a uniformly sized powder. Alternatively, removing the tablet coating with an organic solvent prior to manual grinding facilitates more uniform grinding of the tablets. Direct dissolution in a suitable solvent usually produces the most accurate and precise analytical results.

As an example, ethinyl estradiol tablets are powdered and triturated with four 20-ml portions of chloroform, decanted, filtered, and analyzed by TLC [3, p. 639]. Numerous other examples can be found in the latter half of this book.

Injectables

Injectables are the next most common dosage form. A common preparation is to dilute an aliquot with mobile phase as is the case for the USP procedures for dexamethasone [3, p. 475]. Another common approach is to dilute with methanol, as is done for the assay of diazepam [3, p. 491].

Sample preparation procedures for GC are generally more involved. For example, for methadone hydrochloride, $0.5N$ sodium hydroxide is added to give the free base, followed by extraction with methylene chloride. An internal standard is added after the extract is dried with anhydrous sodium sulfate [3, p. 970]. The assay of interleukin-1α formulated with human serum albumin does not require any sample treatment prior to analysis by capillary electrophoresis [44].

Creams and Ointments

Sample preparation for complex formulations, such as creams, can frequently be as simple as dissolving the cream in the totally organic mobile phase such as the ones typically used in normal-phase chromatography. Organic solutions of fluorometholone [3, p. 677] and hydrocortisone acetate [3, p. 758] creams were assayed by HPLC and hydroquinone cream by TLC [3, p. 769]. A similiar approach has been applied to sample preparation of ointments.

A fairly common, yet labor-intensive, procedure is to heat the cream or ointment with methanol or acetonitrile until it melts, ~60°C. The melt is vigorously shaken, in some cases, with cooling in an ice–methanol bath, until it solidifies. Procedures requiring the partitioning of an ointment between a hydrocarbonlike solvent (hexanes) and polar solvent (methanol–water) have also been developed. For the GC assay of clioquinol cream, a portion of the cream is dried in a vacuum oven, and the dried sample is then derivatized [3, p. 349]. Several other examples are presented in Table 2.6.

Table 2.6 Sample Preparations Procedures for Several Representative Creams and Ointments

Drug	Procedure	Reference
Clobetasone-17-butyrate	Weigh out ointment (equivalent to 0.5 mg) in 10-ml volumetric flask. Add 6 ml methanol, place in water bath (~60°C) for 2 min, shake, add internal standard, dilute with methanol	45
Clotrimazole	Extract cream with acetonitrile/tetrahydrofuran	46
Hydrocortisone 17-butyrate	5 g cream warmed in water bath at 75°C for 15 min; 1-ml sample of the melted cream transferred into a 10-cm test tube; 5 ml methanol added, warmed (75°C) for 10 min, vortexed, centrifuged	47
Ibuprofen	Ointment was weighed into a 50-ml volumetric flask and suspended in tetrahydrofuran/0.02M phosphate buffer (pH 4), filtered	48
Methyl salicylate	Disperse ointment in 10 ml chloroform, heat to 50°C, cooled, filtered	49
Tretinoin	Cream weighed (1 mg drug), 20 ml tetrahydrofuran (stabilized), shaken, 5 ml aliquot further diluted THF aqueous phosphoric acid, filtered	3 (p. 191)

Aerosols

Aerosols used for inhalation therapy are generally packaged in containers with metered values. The standard procedure is to discharge the entire contents of the container for assay. For betamethasone dipropionate and betamethasone valerate topical aerosols, the contents are discharged into a volumetric flask and the propellants carefully boiled off. Precautions should be taken, as many of these propellants are flammable. The residue is diluted to volume with isopropanol–acetic acid (1000 : 1) and filtered [50]. Another approach is to discharge the contents into ethanol or dilute acid. An alternative is to immerse the canister in liquid nitrogen for 20 min, open the canister, evaporate the liquid contents, and dissolve the residue in dichloromethane. A unit spray sampling apparatus for pressurized metered inhalers has been described [51]. The components in an aerosol product that can be the cause of assay variance have been studied [52]. A method to quantify the volatile components of aerosol products has been developed [53].

Elixirs and Syrups

The majority of procedures simply require dilution with water or water-miscible solvents such as methanol [3, p. 515]. Several of the procedures require pH adjustment, followed by extraction with an organic solvent [3, pp. 778, 1202, 1339, 1595, 1579].

Gels

Various procedures have been used for sample treatment of gels. Gels can be dissolved in 0.001N hydrochloric acid [3, p. 564] or dispersed with acetonitrile [3, p. 179]. Gels are also partitioned between solutions of various buffers and chloroform [3, p. 466].

Lotions

Sample treatment procedures are similiar to those cited for creams and ointments [3, pp. 466, 686, 758]. Acetone and a mixture of chloroform and methanol (1 : 2) [3, p. 196] have been used to dilute lotions prior to assay. For assuring batch-to-batch uniformity a diffusion-cell system has been developed [54].

Phytotherapeutical Preparations

Medicinal plants are used as either isolated pure active constituents or complex mixtures of various constituents such as infusions, tinctures, extracts, and galenical preparations. The most common methods are partitioning among nonmiscible solvents, SPE, irreversible adsorption or precipitation of undesirable components, and acid–base extraction [55–57].

Suspensions

Suspensions are either diluted [3, pp. 739, 781, 1343] or partitioned between water and an immiscible solvent [3, pp. 631, 686, 942].

Other Formulations

Suppositories are dissolved in a separatory funnel with $0.01N$ hydrochloric acid and chloroform. After the suppository has dissolved, the chloroform layer is discarded and the aqueous layer is chromatographically assayed. Three devices have been compared as useful in vitro models for measuring drug release from a suppository [58].

An intrauterine contraceptive device is assayed for impurities by cutting off and discarding the sealed ends of the container and removing the contraceptive coil. After shaking the core with methanol and allowing the insoluble portion to settle, the extract is assayed by TLC.

The assay of transdermal preparations of scopolamine involve removing the polyester backing and extracting with chloroform at 60°C for 30 min [59]. Of the variety of different techniques evaluated for extracting triamcinolone acetonide in dermatological patches, liquid–liquid dispersion gave the best recovery and precision [60].

A generalized procedure using a method based on a reversed-phase column and three simple extraction procedures has been evaluated for 111 drugs and their various dosage forms [61].

G. Automation

When considering automating the sample preparation steps and interfacing with chromatographic systems, laboratory robotics has been the method of choice. A laboratory robotics system has a robotic arm and controller, a computer linked to a controller or connected directly to the robotic arm, and application peripherals for performing specific functions in the application process.

Over the past 10 years, several robotic systems and workstations have become available for laboratory automation development. The major difference between a robotic system and workstation is customization. Robotic systems are designed and engineered around a specific application, usually demanding a unique set of requirements. Table 2.7 lists robots that are commercially available.

The robotic workstation is designed to perform a set of common tasks

Table 2.7 Commercially Available Robotic Systems

System	Manufacturer	Type	Controller	Processer/program language	Application development/integration	Comments
ADEPT robot	Adept Technology, Inc., San Jose CA	SCARA	AdeptOne	Motorola 68000/VAL II	Third party	Limited laboratory use
CRS	CRS Plus, Inc., Burlington, Ontario, Canada	Articulated arm	CRS	NEC V30/SRS—M1A, RAPL	Third party	Moderate laboratory use
ORCA	Hewlett-Packard	Articulated arm	Direct from PC HPIB	Intel 80xx6/MDS	Third party	Part of Unified Lab
Mitsubishi RM-501 Movemaster	Mitsubishi Electronics Corp., Tokyo, Japan	Articulated arm	Direct from PC	Intel 80xx6	Third party	
Heath Robot	Heath Corp.	Articulated arm	Direct from PC	Intel 80xx6/Basic	Third party	Limited use
Maxx-5	Fisher Scientific, Pittsburgh, PA	Articulated arm	Direct from PC	Intel 80xx6	Third party	Discontinued
Zymate	Zymark Corp., Hopkinton MA	Cylindrical	Zymate controller and PC direct	Intel 80xx6/EasyLab	Complete	Largest installed base

Source: From *Analysis of Addictive and Misused Drugs*, John A. Adamovics, ed., Marcel Dekker, Inc., New York (1995).

relative to standard or prescribed procedures. Because the workstation has defined functions and a rigid design, it is manufactured in a assembly-line fashion, decreasing both size and cost. Laboratory workstations evolved around developments in robotics and microprocessor technology and a demand for bench-top automation. To meet a growing need, several workstations were developed in response to a need for SPE automation. Their architecture ranges from a robotic arm and controller integration to the dedicated accessories of an analytical instrument. Table 2.8 lists several workstations and their relative capabilities. (Tables 2.6 and 2.7 are not intended to be complete listings of all products or their capabilities.)

The Zymark and Gilson ASPEC workstations process solid-phase extraction samples sequentially without the intervention of a human analyst. These workstations are programmed to activate extraction cartridges with solvent to prepare them to receive a specimen. After the sample application, cartridges are washed to remove impurities. The analytes of interest are then eluted into collection tubes or injected onto a liquid chromatograph for sequential analysis.

The Speed Wiz is designed to operate more like a laboratory technician. Multiple cartridges are activated by pumping activation solvents simulataneously to each solid-phase cartridge. The sample is applied to the cartridge manually. After the operator changes the collection tubes, the wash and elute steps are similiar to the activation step.

The Prospekt extraction system is somewhat different from the previous solid-phase extraction workstations. It combines a high-performance liquid chromatographic autosampler with a high-pressure solid-phase cartridge unit to perform in-line solid-phase extraction sample preparation. Although the addition of a simple fraction collector would permit this system to be used in a manner similiar to the robotic workstations just described, it is primarily designed to prepare samples for direct analysis by HPLC. The Prospekt requires special cartridges that operate at pressures typical of HPLC systems (1000–4000 psi). This system requires smaller volumes of solvent for cartridge preparation and elution than those systems utilizing cartridges manufactured in syringe bodies. It also permits elution of analytes by backflushing the cartridge. This reduces solvent required for elution and minimizes the simultaneous elution of interfering compounds. Dedicated SPE workstations have been described [62].

Applications – Tablets

The various applications of the analysis of tablets for dissolution testing, content uniformity, stability-indicating assays, and routine quality control assays have all been targets of automation using a robot system.

Table 2.8 Workstations and Capabilities

Workstation	Manufacturer	Available laboratory functions	Method of control/ programming	Process method	Comments
ASPEC XL	Gilson Medical Electronics, Box 27, Middletown, WI	SPE, HPLC injection	PC or front panel/basic	Serial or batch	Standard SPE columns
ASTED	Gilson Medical Electronics, Box 27, Middletown, WI	Dialysis, SPE, concentration	PC or front panel/basic	Serial	Proprietary SPE columns only
BenchMate	Zyamark Corp., Hopkinton, MA	SPE, dilution, filtration, vortex, HPLC injection, UV visible sip	Indirect/direct PC/menu based	Serial	Built-in analytical balance
Millilab	Millipore Corp. Waters Chrom. Div., 34 Maple St., Milford, MA	SPE, dilution, filtration, HPLC injection, mixing	Indirect/direct PC/menu based	Serial or batch	Discontinued

MicroLab	Hamilton, Box 10030, Reno, NV	SPE, dilution, filtration, HPLC injection mixing	Indirect/direct PC/menu based	Serial or batch	
HP 7686 Prep-Station	Hewlett-Packard	SPE, dilution, filtration, HPLC injection, mixing	Direct PC/menu based	Serial	Dedicated accessory for HP GC/LC
Speed Wiz	Applied Separations, Box 20032 Lehigh Valley, PA	SPE	Indirect/direct PC/menu based	Batch	
Prospekt	Spark Holland	SPE	Indirect/direct PC/menu based	Serial	Marketed by Jones Chromatography

Source: From *Analysis of Addictive and Misused Drugs*, John A. Adamovics, ed., Marcel Dekker, Inc., New York (1995).

The automation of the HPLC determination of tablet content uniformity has been applied to the analysis of a variety of tablets and is used routinely for quality control purposes. The routine used is fairly typical of that described by other workers, in that the operator presents the samples to the robotic system in tubes. The system then continues the analysis as follows:

1.   Water is added, and the sample tube is placed into either a vortex mixer or sonic bath to disperse the tablet.
2.   Internal standard solution is added.
3.   The sample is further treated in the vortex mixer.
4.   The sample is transferred to the centrifuge.
5.   After centrifugation, the sample is returned to the sample rack.
6.   The sample is transferred from the centrifuge tube to an autosampler vial, being diluted where necessary.

Two early examples of this approach have been published [63,64].

A slightly more complex HPLC analysis of oral contraceptive tablets for content uniformity has been described [65]. The analysis was automated because it represented a high proportion of the laboratory workload, it was relatively routine, and it would reduce contact with steroids, therefore minimizing health risks. The last point was emphasized by placing the robotic system in a room separate from the main laboratory, with the controller outside the room. This is significantly different from the approach followed by many users of laboratory robotics, who often place the system in a main laboratory area. The robotic system performs the following steps, after the operator places the tablets into tubes in racks located on the robot table:

1.   Aqueous sodium chloride is added.
2.   Chloroform containing internal standard is added.
3.   The tablet is disrupted on a vortex mixer.
4.   A portion of the organic layer is transferred into a second tube.
5.   The second tube is transferred to the evaporator station, where the chloroform is evaporated.
6.   The samples are reconstituted with methanol–water.
7.   The samples are transferred to autosampler vials via a filter.

An approach similar to that above has been used to analyze composite samples consisting of from 5 to 32 tablets. In this case, the tablets were disintegrated by a Polytron tissue homogenizer.

However, the need for composite sample analyses should perhaps be questioned when assay automation is being considered. Composite samples are often used to obtain analytical data on batches of tablets (instead of analyzing several single tablets) because of the time (and consequently cost) savings involved. However, by automating the assay procedures and operating the equipment overnight, it may be possible (and desirable) to perform a larger number of single-tablet assays rather than analyzing fewer composite samples. This illustrates a fundamental point of assay automation using robotics—it is not necessarily the direct automation of a manual approach that will provide the best answer to a given problem. The manual approach itself, and the reasons for its existence, should perhaps be questioned prior to automation of a procedure.

An automated HPLC tablet content uniformity assay for a multipotency range of tablets (the active constituent not identified being at 0.1, 0.2, or 0.3 mg/tablet) has been described [66]. In this application, a further degree of communication was introduced between the robotic and HPLC systems, with the robot being able to alter its actions, dependent on the result of the HPLC analysis.

A generalized six-step procedure for the preparation of tablets for chromatography, incorporating most of the features of the methods discussed above, could be described as follows:

1. The operator presents tablets in a container, usually a bottle or tube, to the system which then commences the procedure sample by sample.
2. Details of the sample (e.g., analyte and potency) and the procedure to be followed are entered into the system either by the operator, or automatically from bar-coded labels on the tablet containers.
3. The container cap (if present) is removed from the container, and if a tablet weight is required, the tablet is transferred to a preweighed second tube. This second tube can then be reweighed with the tablet to obtain the sample weight, and hence a mean tablet weight can be derived through the run. If a tablet weight is not required, Step 3 could be omitted.
4. A measured amount of liquid is added to the sample, which is dispersed by vortex mixing and/or ultrasonication. An internal standard may be included at this stage if required.
5. The sample is clarified by centrifugation or filtration.
6. The clear sample solution is injected onto a chromatograph or transferred to vials for subsequent analysis, either manually or via an autosampler.

Dissolution testing for solid dosage forms generates numerous samples that can be automated using the Zymate Dissolution Testing System. This system automates the vessel cleaning, sample addition, media addition, and vessel sampling for single-point or sustained-release testing. Up to 24 lots of 6 samples may be stored on the system for the paddle method, and up to 13 lots of 6 samples for basket methods. The basket and paddle methods may also be combined. A second system, AUTO DISS®, differs from robot-assisted systems in that it can carry out all operating synchronously [67].

Analysis of Suppositories

Although sharing similarities with the tablet procedures, this differs from the typical tablet assay by using organic solvents and having two evaporation steps [68]. The unit operations carried out by the system are the following:

1. The suppository is weighed.
2. Pentane is added, and the suppository dispersed.
3. The excipient matrix is extracted with pentane.
4. The pentane extract is evaporated.
5. The residue is reconstituted in methanol.
6. The methanol solution is evaporated.
7. The residue is reconstituted with internal standard solution.
8. The sample solution is filled into capped vials.

Analysis of Parenterals

In this procedure the operator first enters the experimental parameters, including the number, size, and type of samples to be prepared, the dilution required, and the type of autosampler to be used [69]. The system then treats each sample as follows:

1. The ampule is transferred to the breaker station, where the ampule is opened.
2. A portion of the sample is transferred into a clean tube, followed by diluent, and the tube contents are mixed on a vortex mixer.
3. The programmed serial dilutions are performed.
4. The final analytical solution is transferred to the appropriate autosampler vial, which is capped.

This system has been used to prepare several parenteral products for chromatographic analysis. A similar system has been used in the laboratory for the analysis of epinephrine in multipotency combination parenteral. In this case, the test solution is presented to the system as an accurately measured sample, the system then dilutes as programmed according to the potency of the sample, mixes, centrifuges (if necessary), and injects the sample solution onto an HPLC fitted with an electrochemical detector. Exactly the same sample preparation system is also used for the dilution/ deproteination of biological fluids prior to analysis for antibiotics. This underlines another important aspect of the use of robotics in pharmaceutical analysis. Analytical procedures which traditionally have been viewed as very different and are often carried out in different laboratories, for example, the analysis of drugs in parenteral formulations and the analysis of drugs and metabolites in body fluids, if viewed in terms of the unit operations involved, become essentially identical operations. As such, a number of apparently disparate assays can be brought together and usefully automated using a single robotic system.

IV. CONCLUSIONS

Decreasing the number of manual sample treatment steps increases the overall accuracy and ruggedness of a chromatographic method. The number of sample treatment steps depends on the characteristics of the analyte, the formulation, and the chromatographic procedure. With the introduction of robotics there is a rethinking of traditional methods and procedures. For instance, the additional testing capacity that robotics/automation provides may lead to improved asssay precision and testing of several single tablets instead of analyzing a composite sample.

REFERENCES

1. *Guideline for Submitting Documentation for the Manufacture of and Controls For Drug Products*, CDER, February 1987, p.7.
2. D. Wallace and B. Kratochvil, Visman equations in the design of sampling plans for chemical analysis of segregated bulk materials, *Anal. Chem.*, *59*:226 (1987).
3. *The United States Pharmacopeia*, Twenty-Third Revision, United States Pharmacopeial Convention, Inc., Rockville, MD, 1995.
4. *Pharmacopoeia of the People's Republic of China*, English Edition, 1992, Appendix p. 25.
5. D. Barnes, A chemist's guide to regulatory drug analysis, *Assoc. Off. Anal. Chem.*, 105 (1974).
6. R. A. Nash, A validation experiment in solids blending, *Pharm. Eng.*, 5(4):24–32 (1985).
7. E. Sallam and N. Orr, Content uniformity of ethinyloestradiol tab-

lets, 10 μg: effect of variations in processing on the homogeneity after dry mixing and after tabeting, *Drug Dev. Ind. Pharm.*, *11*:607–633 (1985).

8.   *The Japanese Pharmacopoeia*, 10th edition, Yakuki Hippo, Ltd., Tokyo, 1981 (in English).

9.   G. T. Greco, I, Segregation of active constituents from tablet formulations during grinding, *Drug Dev. Ind. Pharm.*, *8*:565–578 (1982).

10. G. T. Greco, II, Segregation of active constituents from tablet formulations during grinding: significance to pharmaceutical analysis, *Drug Dev. Ind. Pharm.*, *11*:1889 (1985).

11. G. T. Greco, III. Segregation of active constituents from tablet formulations during grinding: effects on coated tablet formulations, *Drug Dev. Ind. Pharm.*, 11:1889 (1985).

12. D. R. Gere and E. M. Derrico, SFE Theory to Practice – First principles and method development, Pt. 1, *LC–GC*, *12*(5):352 (1994).

13. D. R. Gere and E. M. Derrico, SFE Theory to Practice – First Principles and Method, Part 2, *LC-GC*, *12*(6):432 (1994).

14. W. Pitkin, Fundamental considerations for SFE method development, *LC-GC*, *10*(1) (1992).

15. R. E. Majors, Supercritical fluid extraction – an introduction, *LC-GC*, *9*(2) (1991).

16. K. Jinno and M. Saito, Coupling of supercritical fluid extraction with chromatography, *Anal. Sci.*, *7*:361 (1991).

17. J. W. King and M. Hopper, Analytical supercritical fluid extraction: Current trends and future vistas, *J. AOAC Int.*, *75*:375 (1992).

18. L. J. Mulcahey and L. T. Taylor, Supercritical fluid extraction of active components in a drug formulation, *Anal. Chem.*, *64*:981 (1992).

19. A. G. Bosanquet, Stability of solutions of antineoplastic agents during preparation and storage, *Cancer Chemother. Pharmacol.*, *17*:1 (1986).

20. A. M. Pitt, The nonspecific protein binding of polymeric microporous membranes, *J. Parenteral Sci. Technol.*, *41*(2):110 (1987).

21. M. F. Hopgood, Nomogram for calculating percentage ionization of acids and bases, *J. Chromatogr.*, *47*:45 (1970).

22. B. B. Brodie, The estimation of basic organic compounds in biological material, *J. Biol. Chem.*, *168*:299 (1947).

23. G. Hoogewijs and D. L. Massart, Development of a standarized analysis strategy for basic drugs using ion-pair extraction and high-performance liquid chromatography, *J. Pharm. Biomed. Anal.*, *1*:321 (1983).

24. R. Modin and G. Schill, Selective extraction of organic compounds as ion-pairs and adducts, *Talanta*, *22*:1017 (1975).

25. M. L. Puttemans, L. Dryon, and D. L. Massart, Extraction of organic

acids by ion-pair extraction with tri-*n*-octylamine — VIII. Identification of synthetic dyes in pharmaceutical preparations, *J. Pharm. Biomed. Anal.*, *3*:503 (1985).

26. V. Marko, L. Soltes, and I. Novak, Selective solid-phase extraction of basic drugs by C₁₈-silica. Discussion of possible interactions, *J. Pharm. Biomed. Anal.*, *8*:297–301 (1990).
27. D. D. Blevins and S. K. Schultheis, Comparison of extraction disk and packed-bed cartridge technology in SPE, *LC-GC*, *12*(1) (1994).
28. J. J. Sun and J. S. Fritz, Chemically modified resins for solid-phase extraction, *J. Chromatogr.*, *590*:197 (1992).
29. R. Paschal, Extraction disks offer new technology for solid phase extraction, *Chrom Connection*, *10*:1 (1992).
30. B. Law, S. Weir, and N. A. Ward, Fundamental studies in reversed-phase liquid–solid extraction of basic; I: ionic interactions, *J. Pharm. Biomed. Anal.*, *10*:167 (1992).
31. R. D. McDowall, J. C. Pearce, and G. S. Murkitt, Liquid–solid sample preparation in drug analysis, *J. Pharm. Biomed. Anal.*, *4*:3 (1986).
32. L. S. Yago, HPLC sample preparation without the lab work, *Res. Dev.*, *9*:86 (1985).
33. R. Bagchi and P. R. Hadded, Contamination sources in the clean-up of samples for inorganic analysis, *J. Chromatogr.*, *351*:541 (1986).
34. B. A. Bildenmeyer, Guidelines for proper usage of solid-phase extraction devices, *LC*, *2*:578 (1984).
35. M. Zief, L. J. Crane, and J. Horvath, Preparation of steroid samples by solid-phase extraction, *Am. Lab.*, *15*(5):120 (1982).
36. T. T. Nguyen, R. Kriugstad, and K. E. Rasmussen, Use of extraction columns for the isolation of desonide and parabens from creams and ointments for HPLC analysis, *J. Chromatogr.*, *366*:445 (1986).
37. M. L. Young, Rapid determination of color additives using Sep-Pak C₁₈ cartridges, *J. Assoc. Off. Anal. Chem.*, *67*:1022 (1984).
38. M. Moors, B. Steenssens, I. Tielemans, and D. L. Massart, Solid-phase extraction of small drugs on apolar and ino-exchanging silica bonded phases: towards the development of a general strategy, *J. Pharm. Biomed. Anal.*, 12:463 (1994).
39. G. Carignan and B. Lodge, High-performance liquid chromatographic analysis of estradiol valerate–testosterone enanthate in oily formulations, *J. Chromatogr.*, *301*:292 (1984).
40. B. Lodge and A. Vincent, Analysis of flumethasone pivalate formulation by HPLC, *J. Chromatogr.*, *301*:477 (1984).
41. P. Majlat and E. Barthos, Quantitative gas and thin-layer chromatographic determination of methylparaben in pharmaceutical dosage forms, *J. Chromatogr.*, *294*:431 (1984).

42. A. C. Moffat (ed.), *Clarke's Isolation and Identification of Drugs*, The Pharmaceutical Press, London, 1986.
43. M. Windholz (ed.), *The Merck Index*, Merck & Co., Rahway, NJ.
44. N. A. Guzman, H. Ali, J. Moschera, K. Iqbal, and A. W. Malick, Assessment of capillary electrophoresis in pharmaceutical applications—analysis and quantification of a recombinant cytokine in an injectable dosage form, *J. Chromatogr.*, *559*:307 (1991).
45. A. G. Patel, R. B. Patel, and M. R. Patel, Liquid chromatographic determination of clobetasone-17-butyrate in ointments, *J. Assoc. Anal. Chem.*, *73*:893 (1990).
46. M. Spangler, *Supleco*, *13*(2):12.
47. S. Wanwimolruk, Rapid HPLC analysis and stability study of hydrocortisone-17-butyrate in cream preparations, *Pharm. Res.*, *8*:547 (1991).
48. V. E. Haikala, I. K. Heimonen, and H. J. Vuorela, Determination of ibuprofen in ointments by reversed-phase liquid chromatograohy, *J. Pharm. Sci.*, *80*:456 (1991).
49. S. K. Pant, P. N. Gupta, K. M. Thomas, B. K. Maitin, and C. L. Jain, Simultaneous determination of camphor, menthol, methyl salicylate, and thymol in analgesic ointments by gas-liquid chromatography, *LC–GC*, *8*:322.
50. M. Di Maso, W. C. Purdy, S. A. McClintock, and M. L. Cotton, Determination of sorbitan trioleate in metered-dose inhalers by supercritical-fluid chromatography, *J. Pharm. Biomed. Anal.*, *8*:303 (1990).
51. Uniformity of unit spray content, *Pharmacopeial Forum*, *18*:3163 (1992).
52. H. A. Havel, L. J. Beaubien, and P. D. Haaland, Analysis of the variance components in a pharmaceutical aerosol product: lodoxamide trimethamine, *J. Pharm. Sci.*, 74:978 (1985).
53. S. Cohen, Quantitative determination of volatile components in pressurized aerosols by gas chromtography, *J. Pharm. Sci.*, *57*:966 (1968).
54. V. P. Shah, J. Elkins, J. Hanus, C. Noorizadeh, and J. P. Skelly, In vitro release of hydrocortisone from topical preparations and automated procedure, *Pharm. Res.*, *8*(1) (1991).
55. P. H. List and P. C. Schmidt (eds.), *Phytopharmaceutical Technology*, CRC Press, Boca Raton, FL.
56. M. Hamgurger and K. Hostettmann, Analytical aspects of drugs of natural origin, *J. Pharm. Biomed. Anal.*, *7*:1337 (1989).
57. M. M. El-Domiaty, Improved HPLC determination of Khellin and Visnagin in Ammi visigna fruits and pharmaceutical formulations, *J. Pharm. Sci.*, *81*:475 (1992).
58. M. Yamazaki, S. Itoh, N. Sasaki, K. Tanabe, and M. Uchiyama,

Comparison of three test methods for suppositories, *Pharmacopeial Forum, 9/10*:2427 (1991).

59. E. R. Sheinin, Trends in the HPLC analysis of marketed pharmaceuticals: a regulatory view. In *Liquid Chromatography in Pharmaceutical Development* (I. W. Wainer, ed.), Aster Publishing, Springfield, OR, 1985, p.444.

60. P. A. D. Edwardson and R. S. Gardner, Problems associated with the extraction and analysis of triamcinolone acetonide in dermatological patches, *J. Chromatogr., 8*:935 (1990).

61. A. S. Sidhu, J. M. Kennedy, and S. Deeble, General method for the analysis of pharmaceutical dosage forms by HPLC, *J. Chromatogr., 391*:233 (1987).

62. R. E. Majors, Automation of solid-phase extraction, *LC–GC*, 11:336 (1993).

63. A. Greenberg and R. Young, A totally automated robotic procedure for assaying composite samples which normally require large volume dilutions. In *Advances in Laboratory Automation Robotics 1985* (J. R. Strimaitis and G. L. Hawk, eds.), Zymark Corp., Hopkinton, MA, 1985, p. 721.

64. P. Walsh, H. Abdou, R. Barnes, and B. Cohen, Laboratory robotics for tablet content uniformity testing. In *Advances in Laboratory Automation Robotics 1985* (J. R. Srimaitis and G. L. Hawk, eds.), Zymark Corp., Hopkinton, MA, 1985, p. 547.

65. J. G. Habarta, C. Hatfield, and S. J. Romano, Application of robotics for chromatographic QA sample preparation, *Am. Lab., 17*(10): 42 (1985).

66. K. J. Halloran and H. Franze, Interaction between a robotic system and liquid chromatography, HPLC control, communication and response. In *Advances in Laboratory Automation Robotics 1985* (J. R. Strimaitis and G. L. Hawk, eds.), Zymark Corp., Hopkinton, MA, 1985, p. 575.

67. E. Lamparter and Ch. Lunkenheimer, The automation of dissolution testing of solid dosage forms, *J. Pharm. Biomed. Anal., 10*:727 (1992).

68. C. Hatfield, E. Halloran, J. Habarta, S. Romeno, and W. Mason, Multi-product robotic sample preparation in the pharmaceutical quality assurance laboratory. In *Advances in Laboratory Automation Robotics 1984* (J. R. Strimaitis and G. L. Hawk, Zymark Corp., Hopkinton, MA, 1984, p. 105.

69. J. H. Johnson, R. Srinivas, and T. J. Kinzelman, Automated robotic sample preparation of pharmaceutical parenterals, *Am. Lab., 17*(9): 50 (1985).

3

Planar Chromatography

JOHN A. ADAMOVICS and JAMES C. ESCHBACH
Cytogen Corporation, Princeton, New Jersey

I. INTRODUCTION

Paper and thin-layer chromatography (TLC) along with several additional variants are generally referred to as planar chromatography. Planar chromatographic methods can be traced back to the mid-1800s with a variant of paper chromatography (capillary analysis). In Holland sometime after 1905, this technique was used to evaluate medicinal plants and was used routinely in Germany by the 1920s [1]. By the late 1930s alumina was layered on a glass plate to analyze pharmaceutical tinctures [2]. Amino acids, antibiotics, nucleotides, and radiolabeled substances were routinely being analyzed in the late 1940s.

The importance attributed to planar chromatography varies according to geographic regions. Approximately 40% of the methods in the *United States Pharmacopeia* and over 80% of the chromatographic methods described in the Japanese and Chinese pharmacopoeias are based on planar chromatography. There are a number of reasons for the popularity of the oldest chromatographic technique. One obvious strength is in qualitative drug identification where a large number of samples can be tested simultaneously. These procedures have modest demands on equipment and resources, which makes it an ideal technique for remote areas without electricity and by operators with limited training [3].

When compared with high-performance liquid chromatography (HPLC), TLC allows for greater flexibility in choice of chromatographic solvents and has versatile postchromatographic schemes which enhance the sensitivity and specificity of detection. Furthermore, several techniques are available for the optimization of the separation, such as development in two dimensions, multiple development, and sorbent impregnation. In addition, the chromatographic process can be easily followed and halted at any time. Unlike HPLC, in planar chromatography, all the components of a sample can be detected. Certainly, there are numerous disadvantages to planar chromatography which will be discussed later.

II. MATERIALS AND TECHNIQUES

A. Sorbent

For paper chromatographic procedures, the typical sorbent used is α-cellulose which has weak ion-exchange properties. For separation of lipophilic substances, these sorbent properties can be modified by esterification or silicone treatment [4]. Besides the conventional rectangular forms, unusual sorbent shapes such as triangles have been used [5]. For analysis of radiopharmaceuticals, glass microfiber impregnated with silica gel or silicic acid are used (Gelman, Ann Arbor, MI). Commercially, the technique is known as instant thin-layer chromatography (ITLC).

For TLC, silica gel is by far the most widely used sorbent, and glass is the most popular sorbent support. Adherence to the glass plate is generally accomplished by the addition of such binding agents such as calcium sulfate. The diversity of the commerically available plates are listed in Table 3.1 [6]. Most of these sorbents are carefully controlled in terms of consistent pore size, particle size, and surface area. A number of chromatographic procedures, particularily the separations that use anhydrous developing solvents, require control of the silica gel moisture content. The "ideal" is to have 11–12% water by weight [7].

Silica gel plates are modified to form a reversed-phase sorbent by being impregnated with liquid paraffin, silicone oil, or fats. Reversed-phase plates of this type are used in the identification of steroid hormones [8,9]. Table 3.2 lists various other substances that have been used for impregnation [10,11]. Sorbents that are impregnated with amminium oxalate, ammonium sulfate, magnesium acetate, sodium acetate, and silver nitrate have one time or the other been commerically available. As an example, impregnated adsorbents have been used for the resolution of sulfa drugs [19].

Chemically bonding hydrocarbon chains, using monochlorosilane and related methods, to the silica gel are more reproducible than sorbents im-

Table 3.1 Nomenclature of TLC Plates

"Sil"	A product composed of silica gel, e.g., Anasil from Analabs
G	Gypsum ($CaSO_4 \cdot \frac{1}{2} H_2O$) binder ("soft" layers)
S	Starch binder
O	Organic binder, such as polymethacrylate or polycarboxylate; layers are "hard" or abrasion resistant
H or N	No "foreign binder"; products may contain a different form of the adsorbent, e.g., colloidal or hydrated silica gel or colloidal silicic acid, to improve layer stability
HL	Hard or abrasion-resistant layer containing an inorganic hardener (Analtech)
HR	Specially washed and purified (highly refined)
P	Thicker, preparative layer or material for preparing such layers (for cellulose, see below)
P + $CaSO_4$	Preparative layer containing calcium sulfate binder
F or UV	Added fluorescent material such as Mn-activated zinc silicate
254 and 366	Used after F or UV to indicate the excitation wavelength (nm) of the added phosphor
60	Silica gel 60 (Merck) has pore size of 60 Å (10 Å $=$ 1 nm). Other pore size designations are 40, 80, 100.
D	Plates divided into a series of parallel channels
L	Layer with a preadsorbent sample dispensing area (Whatman)
K	Symbol used in all Whatman products
RP	Reversed-phase layer; RP_{18} or $RP\text{-}C_{18}$ would indicate that octadecylsilane groups are chemically bound to silica gel
4, 7, 9	These numbers after the adsorbent name usually indicate pH of a slurry

(Continued)

Table 3.1 (Continued)

E, T	Aluminum oxide layers having specific surface areas
MN 300 or 400	Machery Nagel proprietary fibrous celluloses
Avicel	American Viscose cellulose (microcrystalline)
CM	Carboxymethyl cellulose
DEAE	Diethylaminoethyl cellulose
Ecteola	Cellulose treated with ethanolamine and epichlorhydrin
PEI	Polyethyleneimine cellulose
P	Phosphorylated cellulose

Source: From Rcf. 7.

Table 3.2 Silica Plate Impregnation Materials

Impregnate	Purpose	Reference
Carbomer	Identification of neomycin sulfate	12
Tetradecane	Identification of cephradine	13
Tetradecane	Identification of cefaclor	14
2% Copper sulfate and 2% ethylenediamine	Resolution of seven barbiturates	15
0.1% Copper sulfate	Resolution of sulfa drugs	16
1% Zinc acetate	Resolution of seven antihistamines	17
EDTA	Resolution of serotonin, epinephrine, and norepinephrine	18

pregnated with hydrocarbons [20,21]. An analogy to what is commonly seen with sorbents used during HPLC analysis, TLC reversed-phase sorbents can show significant differences in chromatographic performance when compared side by side [22]. Chromatographic behavior of various pharmaceuticals indicate that there is no advantage in using smaller chained hydrocarbons such as C_2 or C_8.

Alumina, which is discussed in greater detail in Chapter 5, and cellulose are the two most often used nonsilica-based sorbents. Alumina has been used to separate compounds such as fat-soluble vitamins, alkaloids, and antibiotics. Cellulose-based ion-exchange layers have wide application for sulfonamides, nucleic acids, and steroid sulfates.

Cellulose was the first sorbent for which the resolution of racemic amino acids was demonstrated [23]. From this beginning, derivatives such as microcrystalline triacetylcellulose and β-cyclodextrin bonded to silica were developed. The most popular sorbent for the control of optical purity is a reversed-phase silica gel impregnated with a chiral selector (a proline derivative) and copper (II) ions. Separations are possible if the analytes of interest form chelate complexes with the copper ions such as D,L-Dopa and D,L-penicillamine [24]. Silica gel has also been impregnated with (−) brucine for resolving enantiomeric mixtures of amino acids [25] and a number of amino alcohol adrenergic blockers were resolved with another chiral selector [26]. A worthwhile review on enantiomer separations by TLC has been published [27].

The reader should refer to chapters within the *Handbook of Thin-Layer Chromatography* [27] for a more thorough discussion of sorbents technology.

B. Sample Application

Optimal resolution for planar methods are only obtained when the application spot size or width at the origin is as small or narrow as possible. As with any chromatographic procedure, sample and solvent overloading will decrease resolution. Studies show that in most instances automated sample application is preferred over manual application especially when applications are greater than 15 μl [28]. Inadequate manual application of a sample will cause diffusion and "double peaking." Depending on the purpose of the analysis, various sample amounts are recommended [29] and listed in Table 3.3. The design of commercially available automatic spotters has been reviewed [30].

A manual sample application of 0.5 μl is the smallest volume that can be reproducibly applied [28]. For volume applications greater than 2–10 μl, the sample should be applied stepwise with drying between each step.

Table 3.3 Recommended Sample Application Parameters

Purpose	Spot diameter (mm)	Sample concentration (%)	Sample amount (μg)
Densitomery	2 mm for 0.5 μl	0.02–0.2	0.1–1 (HPTLC plate)
	Sample volumes	–	1–10 (conventional)
Identification	3 mm for 1 μl	0.1–1	1–20
	Sample volumes	–	–
Purity testing	4 mm for 2 μl	5	100
	Sample volumes	–	–

Narrower separation zones are also obtained by "streaking" versus "spotting" application techniques [28]. Sample streaking is easily accomplished with the currently available commerical applicators.

Another approach to sample application is to use preadsorbent TLC plates. No special skills are required for sample application. These plates are precoated with two different materials. There is a lower sample application zone which is relatively inert, ideally no separating properties of its own. This zone is commonly kieselguhr (diatomaceous earth). The sample is spotted in the application zone; as the developing solvent passes through the sample spot, all the sample will move with the solvent front. At the interface between the two layers, the sample is compacted, which improves resolution [31]. Sample application on these preconcentration zones may also help to minimize degradation of labile drugs [32,33].

C. Development Techniques

Conventional and High-Performance TLC

Solvent development is usually by ascending chromatography, in which the lower edge of the sheet or plate is dipped into the developing solvent. For good chromatographic reproducibility, mobile phase chambers must be saturated with the vapors of the mobile phase [29]. A paper lining is usually inserted in the chamber and saturated with the mobile phase. Not all chromatographers consider it necessary for reproducible chromatography [29]. A review with emphasis on various chamber types and development modes has been published [34].

A mobile-phase developing distance of approximately 10–15 cm is typical, but some chromatographers prefer developing their plates 15–20 cm. For high-performance TLC (HPTLC) plates, which typically have particle sizes of 5 μm versus 20 μm for conventional plates, a developing distance of 3–6 cm is typical. Interestingly, similiar efficiencies are obtained when both types are developed 8 cm [29].

Two-Dimensional Development

For two-dimensional or bidirectional chromatography, a single sample is spotted at the corner of a sheet or plate and developed in the conventional manner [35]. After development, the sheet or plate is dried and rotated by 90° for a second solvent migration which may be the same or a different mobile-phase combination. A large variety of stationary-phase combinations have been used. Mixed sorbents or two separated sorbents using two different mobile phases have been described [36,37].

The major advantage of this technique over the one-dimensional system approach is the increased resolution and higher spot capacity. The two-dimensional technique has the potential of separating 150–300 components [35]. This is because the whole area of the plate is used, which increases the resolving power by almost the square of that obtained by one-dimensional development. This procedure can also be used for determining whether decomposition occurred during the chromatographic process. Usually one sample per plate (never more than four samples per plate) can be analyzed.

A strategy for selection of solvent systems has been developed and illustrated for 14 anesthetics [38] and 15 steroids [39]. Additional information can also be found in a dated but nevertheless useful review [36].

Overpressurized TLC

The primary distinction of overpressurized (OPTLC) from the other techniques is that the mobile phase is pumped into the sorbent. This is accomplished by covering the chromatographic plate with a plastic membrane held in place by external pressure. The pumping of the mobile phase controls the rate at which the chromatogram is developed. Consequently, the maximum resolution occurs at some optimal flow rate. OPTLC is a closed system which has also been described as corresponding to a high-performance liquid chromatographic column having a relatively thin, wide cross section [40].

Advantages of OPLC include the following: quicker development times, constant and adjustable optimized mobile-phase flow rate, reproducible R_f values, and direct correspondences of results with HPLC. The coupling of OPLC to HPTLC has also been described [41]. A variant of this approach is to use a vacuum chamber [42]. A comprehensive review of OPTLC describes general properties, methods of developing chromatograms, and the apparatus description with numerous applications [43].

Continuous Development

The chromatographic plate is placed in a specially designed developing tank with a slot on the lid for a protuding plate. Upon reaching the slot, the solvent evaporates at a continuous rate. Similar to multiple development techniques, there is a greater resolution of low-R_f solutes than with conventional development techniques. Disadvantages of the method are that there is a broadening of high-R_f zones and longer development times. The technique is described in the USP 23 for identifcation of netilmicin sulfate and for assay of antibiotics. The technique has also been used in analysis of steroids [44].

Multiple Development

This technique requires that the chromatographic sorbent be repeatedly developed in the same or different solvent. The sorbent is dried between each development. In other words, one 20-cm development has a longer development time than two 10-cm developments. This technique is most useful for resolution of components with R_f values below 0.5.

An instrumental form of multiple development is referred to as programmed multiple development. The sorbent is heat dried between each development. A instrumental variation is referred to as automatic multiple development. In this case, the mobile phase is removed from the developing chamber and the sorbent dried under vacuum. These techniques have been reviewed recently [27,45].

Circular and Anticircular Development

Proponents of these two planar chromatographic techniques claim certain advantages over the conventional rectangular procedures. For circular development (radial development), the samples are spotted around the center of the circular sorbent. The solvent is fed into the center of the plate which led to migration of the sample toward the sorbent edge. The apparent advantages of this technique are the increased resolution of low-R_f components and faster development times. As indicated by the name, anticircular development is the opposite of circular chromatography. The samples are spotted on the outside edge of a circular sorbent, with solvent flows from the edges of the plate inward to the center of the plate. This method has an advantage for separation of compounds with high-R_f values.

When comparing the above two methods to linear development, anti-circular apparently is superior in terms of sensitivity, number of samples per plate, speed of analysis, and solvent consumption. Conventional linear TLC ranked second to the anticircular techniques [46]. Also refer to the *Handbook of Thin-Layer Chromatography* [27] for additional details.

Gradient Development

Conventioal development procedures of two or more component planar mobile-phase systems form phase gradients. The most polar phase stays near the bottom of the development plate [29]. Besides these mobile-phase gradients, stationary-phase gradients based on having a sorbent of varying composition and medium gradients such as temperature have been studied [27,47].

Thin-Layer Rod/Stick Chromatography

This is a cylindrical form of planar chromatography where the support is a rodlike narrow glass or quartz tube. Stick chromatography has been used for the identification of numerous pharmaceuticals [48].

Preparative Chromatography

The traditional procedures are similiar to analytical methods with the major differences being the use of thicker sorbent layers. These procedures are generally faster and more convenient than classical column chromatography [6,49]. Preparative scale TLC usually has decreased analyte resolution compared to analytical TLC.

Significant advances have been made in instrumental preparative planar chromatography with the introduction of centrifugally accelerated layer chromatography [27]. One disadvantage of this technique is that it is limited to mobile phases with relatively low water content because that higher contents the sorbent is sloughed off. Preparative thin-layer chromatography can be a useful technique for the isolation of pharmaceutical impurities and degradants.

D. Radiopharmaceuticals

Radiochemical purity determinations consist of separating the different chemical substances containing the radionuclide. The radiochemical purity of labeled pharmaceuticals is typically determined by paper chromatography (paper impregnated with silica gel or silicic acid). The most frequently used radioisotope is technetium-99m obtained by daily elution with saline of a ^{99}Mo–^{99m}Tc generator. In the elution process, TcO$_4^-$ [Tc(VII)] and possibly reduced forms such as Tc(VI) and Tc(V), and Tc(IV) can be obtained which are known to cause problems in imaging and preparation of ^{99m}Tc kits. The presence of these impurities can be determined by using two developing solvents, acetone and saline, and paper impregnated with silica gel [50]. As another example of determination of radiochemical purity is assessing the efficacy of the radiolabeled preparations by paper impregnated with silica gel and saline as developing solvent for ^{111}In-labeled antibodies which are used in radioimmunoimaging of benign and malignant disease. ^{111}In, not bound to the antibody, migrates with the solvent front as a DTPA chelate and the antibody remains at the origin [51]. Audioradiography and other related techniques have been reviewed [27].

III.   DETECTION

After solvent development, the detection procedures can be either qualitative or quantitative. Qualitative procedures require that a drug product be identified on the chromtographic plate as having the identical R$_f$ and at about equal magnitude to a reference sample. Semiquantitative estimations can be a visual comparison of the size and intensity of the spots versus various standard concentrations. Quantitative procedures require either densitometry or the extraction of the components of interest from the sorbent followed by spectrophotometric measurement.

A.   Visualization Methods

There are various means by which pharmaceuticals can be visualized on planar sorbents. The most straightforward is the direct viewing of drugs that have color. Adriamycin, beta carotene, gentian violet, and methylene blue are the most common examples.

Many pharmaceuticals are located using short-wavelength (254 nm) ultraviolet (UV) light on sorbents impregnated with phosphor. Compounds that are naturally fluorescent can be detected under long wavelength UV (350 nm) light. In fact, over 40% of the planar methods in the USP require simple detection by UV irradiation. Certain pharmaceuticals will not absorb UV because they are in the wrong ionic form. For instance, barbiturates are not visible if the sorbent is acidic but becomes so if ammonia fumes are blown over the sorbent.

For drugs where the UV detection is inadequate, the reaction of the drug substances with various chemical reagents provide compound- or class-specific reactions. The result is either colored or fluorescent chromatographic zones which, in turn, enhance the specificity and/or sensitivity of the procedure.

Several hundred reagents are described as being useful for substance visualization [6,27,52,53]. A relatively common test is to place the sorbent into a tank of iodine crystals. The iodine vapors form weak-charge transfer complexes with unsaturated bonds of the sample. This is visibly detected as brown spots. Reaction with iodine is generally reversible but has been shown to oxidize compounds such as mercaptans and disulfides.

Another useful detection mode is to char the sample with a corrosive reagent. The most common charring reagent used in the USP is a methanolic solution of sulfuric acid which is sprayed on the chromatographic plate. After spraying, the chromatographic plate is heated above 90°C for 30 min. Black zones on a white-gray background appear when the charring is complete. Charring may also occurr by heating a sorbent impregnated with ammonium bisulfite or ammonium sulfate. For substances that are unreactive to the above procedures, 5% nitric acid is added to increase the charring reagents oxidizing capabilities. Under less rigorous conditions, heating the plates in the presence of acid vapor or ammonium bicarbonate generates fluorescent derivatives, which in most instances are many times more sensitive than those obtained by charring [52,53].

A number of spray reagents are available for visualization of nitrogenous compounds. Ninhydrin is useful for primary and secondary amines [54] and acidified iodoplatinate reacts with primary, secondary, tertiary amines, and quaternary ammonium compounds. Visualization reagents used for detection of specific drugs are listed in the last part of this book. Quantification is possible for the above detection procedures if the formed products are stable and where interferences are absent.

B. Quantification

The most commonly described USP procedure for quantification is the scrap and elution approach. Low analyte recoveries can occur but can be minimized by using polar organic solvents such as methanol, ethanol, or acetone. Generally, analytes with high-R_f values can be desorbed with high recoveries by using the mobile phase. One example of this procedure is the USP assay procedure of the steroid methyl prednisolone acetate in cream formulation. This steroid is separated from its excipients by TLC, extracted from the sorbent, derivatized, and measured spectrophotometrically.

Over the past decade, commerically available densitometers for in situ quantification have become available. This instrumentation has been reviewed along with its theoretical foundation [27]. Most scanners are capable of measuring absorbance, fluorescence, and fluorescence quenching

Table 3.4 Several Representative Examples of
Quantification of Drug Substances

Drug substance	Formulation	Reference
Amilodipine	Tablets	56
p-Aminosalicylic acid	Bulk	57
Chlordiazepoxide	Bulk/tablet	61
Chloroprocaine	Bulk/injectable	59
Diazepam	Bulk/tablet	60
Diuretics	Injectable/tablets	62
Lovastatin	Bulk	63
Rifaximine	Cream	64
Sulfonamides	Bulk/tablets	58
Vitamins	Tablets	65

along with providing spectra of the individual components. Scanners can
routinely provide reproducibility of less than 5% and often to within 1%.
Image analysis techniques using a video camera have become a cost-
effective alternative to conventional scanning densitomers. Using the best
available technolgy, TLC reproducibility is approaching that commonly
obtained with liquid chromatography [55]. Several representative examples
of quantification are listed in Table 3.4.

Specific detectors are also available for quantification of radiopharmaceuticals. These detectors use a position-sensitive proportional counter.
These detectors are sensitive to the beta and gamma nuclides listed in Table
3.5. The detector analog output can also be represented as an analog curve.
Various other detection procedures have also been used, such as flame
ionization [66], mass spectrometry [27,67], and infrared (IR) [68,69].

IV. METHODS DEVELOPMENT

Planar techniques can tolerate application of either solutions or suspensions. In addition, the solvent used to dissolve the sample need not be
compatible with the TLC mobile phase as is the case for HPLC methods.

Table 3.5 Nuclides That Can Be
Detected by Thin-Layer
Radiochromatography

Nuclide	Emitter
^{3}H	Beta
^{125}I	Gamma–beta (Auger)
^{131}I	Beta, gamma
^{14}C	Beta
^{35}S	Beta
^{32}P	Beta
^{33}P	Beta
^{99m}Tc	Gamma
^{123}I	Gamma
^{201}Tl	Gamma
^{51}Cr	Gamma
^{58}Co	Gamma
^{59}Fe	Gamma

In general, a planar method tends to have fewer sample preparative techniques than either gas chromatography (GC) or HPLC methods. The primary criteria for TLC is that the matrix should not distort or streak the analyte band or spot. One other concern should be the stability of the drug after sample application. For example, vitamin D_1 is stable on prewetted silica gel but decomposes quickly once the sorbent is dried.

To generate a stragedy for methods development, the first step is to review the pharmaceutical literature including vendor catalogs [24]. The salient features from the literature and pharmacopeia methods are listed in the last chapter. Specific methods, as those for pharmaceutical impurities [14, ⟨466⟩], colorants [70], and preservatives [71] have been complied. Numerous mobile-phase systems have been advocated for the assay of pharmaceuticals [3,52,72,73]. The *Handbook of Thin-Layer Chromatography* [27] should be referred to when methods are being sought for specific drug classes such as amino acids, antibotics, carbohydrates, steroids, and hydrophilic and lipophilic vitamins. A thin-layer chromatographic atlas for plant drug analysis is also useful [74].

For analysts who are faced with either a new drug product or unknown impurities, time-consuming choices will have to be made regarding which TLC systems to test from the hundreds proposed. Over the last several years, efforts have been directed toward choosing only a handful of

TLC systems for use in the assay of pharmaceuticals. One such attempt was based on first screening 800 pharmaceuticals with the best eight TLC systems listed in Table 3.6 [75]. Based on discriminating power of the system (i.e., the likelihood that two drugs selected at random will be separated by a TLC system), System 4 was considered the most discriminating for basic nitrogeneous drugs and System 5 for neutrals and acids [54].

A second approach to determining optimum TLC systems is based on principal components analysis. This is another statistical approach aimed at the identification of pharmaceuticals [76]. By using this approach on 360 drugs, 4 mobile phases from a set of 40 were chosen as giving the most diverse chromatographic information. Table 3.7 lists the four chosen mobile phases.

A third approach compared the five halogen-free mobile-phase systems listed in Table 3.8 using the standarized mobile phases listed in Table 3.6 [77]. The main criterion for the comparison was the number of "acceptable" spots for 22 acids and neutral drugs. The authors concluded that the mobile-phase Systems 1 and 5 in Table 3.8 gave a greater number of "acceptable" spots for acidic and neutral drugs than did the mobile phases of Table 3.5. For basic drug substances, solvents 3 and 4 in Table 3.8 were considered better than the systems in Table 3.6. In addition, halogen-

Table 3.6 Standardized TLC Systems

System	Sorbent	Mobile phase
1	Silica dipped in methanolic 0.1M KOH solution and dried.	Methanol–ammonia (100 : 1.5)
2	Same as System 1	Cyclohexane–toluene–diethylamine (75 : 15 : 10)
3	Same as System 1	Chloroform–methanol (9 : 1)
4	Same as System 1	Acetone
5	Silica	Chloroform–acetone (4 : 1)
6	Silica	Ethyl acetate–methanol–ammonium hydroxide (85 : 10 : 5)
7	Silica	Ethyl acetate
8	Silica	Chloroform–methanol (9 : 1)

Table 3.7 Standardized TLC Systems According to Principal Component Analysis

System	Sorbent	Mobile phase
1	Silica	Ethyl acetate–30% ammonia (85 : 10 : 5)
2	Silica	Cyclohexane–toluene–diethylamine (65 : 25 : 10)
3	Silica	Ethyl acetate–chloroform (1 : 1)
4	Silica dipped in methanolic 0.1M KOH solution and dried.	Acetone

containing mobile phases should be avoided due to their toxicity and disposal problems. Another important factor that should be mentioned is that the sorbents in Table 3.8 do not have to be treated with potassium hydroxide, as they do in the systems in Tables 3.6 and 3.7.

A fourth approach, which has been popular among HPLC chromatographers, is to use a simplex optimization algorithm to determine the optimal solvent [27]. This approach was used in the separation of anticancer platinum (II) complexes [78], drug screening [79], and alkaloids [80].

Table 3.8 Universal Halogen-Free TLC Mobile Phases

System	Sorbent	Mobile phase
1	Silica	Toluene–ethyl acetate–85% formic acid (50 : 45 : 5)
2	Silica	Toluene–acetone–2N acetic acid (30 : 65 : 5)
3	Silica	2-Propanol–toluene–conc. ammonium hydroxide (29 : 70 : 1)
4	Silica	2-Propanol–toluene–ethyl acetate–2N acetic acid (35 : 10 : 35 : 20)
5	Silica	Toluene–dioxane–methanol–conc. ammonium hydroxide (2 : 5 : 2 : 1)

A TLC method/approach that is low cost, maintenance free, fast and reliable, an apparatus that is made of a plastic bag, and that does not require electricity (for developing countries) has been suggested [3]. The feasibility was demonstrated by the analysis of a partial list of the essential drugs established by the World Health Organization.

In the above approaches to choosing a TLC system, silica gel has been the sorbent of choice. This is due to the greater separation potential that silica gel has over reversed-phase sorbents. For purity testing, it is advisable to use both silica gel and reversed-phase sorbents.

V. CONCLUSION

Although HPLC has superseded TLC in many application areas, conventional TLC plays a useful role where cost, rapidity, and simplicity are the overiding factors. Quantitative TLC has continued to grow in popularity. For in-depth discussions and additional references of the topics discussed in this chapter, the reader should refer to Planar Chromatography Reviews in *Analytical Chemistry* (years ending in even numbers) and *CAMAG Bibliography Services*.

REFERENCES

1. H. Platz, *Uber Kapillaranalyse und ihre Anwendung in Pharmazeutischen Laboratorium*, Director der Firma Dr. W. Schwabe, Leipzig, 1922.
2. L. S. Ettre, Evolution of liquid chromatography: a historical overview. In *High-Performance Liquid Chromatography* (C. Horvath, ed.), Academic Press, New York, 1980, p. 54.
3. A. S. Kenyon, P. E. Flinn, and T. P. Layloff, Rapid screening of pharmaceuticals by thin-layer chromatography: Analysis of essential drugs by visual methods, *J. AOAC Int.*, 78:41 (1995).
4. E. Heftmann (ed.), *A Laboratory Handbook of Chromatographic and Electrophoretic Methods, Chromatography*, Van Nostrand Reinhold, New York, 1985, p. 142.
5. K. Macek, *Pharmaceutical Applications of Thin-Layer and Paper Chromatography*, Elsevier, New York, 1972, p. 11.
6. B. Fried and J. Sherma, *Thin-Layer Chromatography*, Marcel Dekker, Inc., New York, 1986, p. 39.
7. H. R. Felton, To activate or not to activate that is the question, Analtech Technical Report No. 7905.

8. *British Pharmacopoeia*, HMSO, London, 1983, Appendix IIIA, p. A61.
9. *European Pharmacopoeia*, Maisonneuve S. A., Stainte Ruffine, France, Pt. V.3.1.2.
10. R. M. Scott, The stationary phase in TLC, *J. Liq. Chromatogr.*, *4*: 2147 (1981).
11. H. J. Issaq, Modifications of adsorbent sample and solvent in TLC, *J. Liq. Chromatogr.*, *3*:1423 (1980).
12. *European Pharmacopoeia*, Maisonneuve S. A., Stainte Ruffine, France, pt. II-5, p.197.
13. *British Pharmacopoeia*, HMSO, London, 1980, Vol. II, pp. 89 and 525.
14. *The United States Pharmacopeia*, Twenty-First Revision, United States Pharmacopeial Convention, Inc., Rockville, MD, 1985, p. 169.
15. S. P. Srivanmstava and Reena, TLC Separation of barbiturates on impregnated plates, *J. Liq. Chromatogr.*, *8*:1265 (1985).
16. S. P. Srivanmstava and Reena, TLC separation of some closely related sulfa drugs on copper sulfate impregnated plates, *Anal. Lett.*, *18*(B3):239 (1985).
17. S. P. Srivanmstava and Reena, Chromatographic separation of some antihistamines on silica gel–metal salt impregnated thin layer plate, *Anal. Lett.*, *15*(A5):451 (1982).
18. S. Ebrahimian and J. Paul, TLC separation of serotonin from epinephrine and norephinephrine, *Microchem. J.*, *26*:127 (1985).
19. P. Pandey, S. Bhattacharya, and A. Dave, Thin layer chromatographic studies of some sulfa drugs substituted pyrazoles using silics gel-G plates impregnated with various adsorbents, *J. Liq. Chromatogr.*, *15*:1665 (1992).
20. I. D. Wilson and S. Lewis, Contemporary developments in thin-layer chromatography, *J. Pharm. Biomed. Anal.*, *3*:491 (1985).
21. I. D. Wilson, S. Scalia, and E. D. Morgan, Reversed-phase thin-layer chromatography for the separation and analysis of ecdysteroids, *J. Chromatogr.*, 212:211 (1981).
22. C. A. T. Brinkman and G. de Vries, Thin-layer chromatography on chemically bonded phases: a comparison of precoated plates, *J. Chromatogr.*, *258*:43 (1983).
23. M. Kotake, T. Sakan, N. Nakamura, and S. Senoh, *JACS*, *73*:2973 (1951).
24. *Machery-Nagel Catalog*, Machery-Nagel, Duren, Germany.
25. R. Bhushan and I. Ali, TLC resolution of enantiomeric mixtures of amino acids, *Chromatographia*, *23*:141 (1987).
26. J. D. Duncan and D. W. Armstrong, Normal phase TLC separation

of enantiomers using chiral ion interaction agents, *J. Liq. Chromatogr.*, 13:1091 (1990).

27.   J. Sherma and B. Fried (ed.), *Handbook of Thin-Layer Chromatography*, Marcel Dekker, Inc., New York, 1991.

28.   J. C. Touchstone and S. S. Levin, Sample application in TLC, *J. Liq. Chromatogr.*, *3*:1953 (1980).

29.   R. A. Egli, TLC and HP TLC in the quality control of pharmaceuticals. In *Liquid Chromatography in Pharmaceutical Development* (I. W. Wainer, ed.) Aster, Springfield, OR, 1985, pp. 379 and 377.

30.   E. Beesley, Current instrumentation by TLC, *J. Liq. Chromatogr. Sci.*, *24*:525 (1985).

31.   H. R. Felton, The pre-absorbent phenomenon in TLC, Analtech Technical Report No. 8001.

32.   G. Szepesi, M. Gazdag, Zs. Pap-Sziklay, and Z. Vegh, Problems of semiquantitative TLC methods presented in the US and British Pharmacopoeia for the family testing of sulphinpyrazone, *J. Pharm. Biomed. Anal.*, *4*:123 (1986).

33.   F. Rabel and K. Palmer, Advantages of using thin-layer plates with concentration zones, *Am. Lab.*, *11*:20BB (1992).

34.   S. Nyiredy, Planar chromatography, *J. Chromatogr. Libr.*, *51A*: A109 (1992).

35.   M. Zakaria, M. F. Gonnard, and G. Guiochon, Applications of two-dimensional thin-layer chromatography, *J. Chromatogr.*, *271*:127 (1983).

36.   T. E. Beesley and E. Heilweil, Two phase, two-dimensional TLC for fingerprinting and confirmation procedures, *J. Liq. Chromatogr.*, 5: 1555 (1982).

37.   P. C. Wankat, Two-dimensional separation processes, *Separation Sci. Technol.*, *19*:801 (1984–1985).

38.   B. De Spielgeleer, W. Van den Bossche, P. De Moerloose, and D. Massart, *Chromatographia*, *23*:407 (1987).

39.   S. Habibi-Goudarzi, K. J. Ruterbories, J. E. Steinbrunner, and D. Nurok, A computer-aided survey of systems for separating steroids by two-dimensional thin-layer chromatography, *J P C*, *1*:161 (1988).

40.   R. E. Kaiser and R. I. Rieder, High pressure TL circular chromatography—a new dimension in analytical chromatography ranging from GC to LC? In *Planar Chromatography* (R. E. Kaiser, ed.), Dr. A. Huethig Verlag, New York, 1986, p. 165.

41.   H. Jork, Advances in thin layer chromatography: part 2, *Am. Lab*, 6: 24B (1993).

42.   P. Delvordre, C. Regnault, and E. Postaire, Vacuum planar chroma-

tography (VPC): a new versatile technique of forced flow planar chromatography, *J. Liq. Chromatogr.*, 15:1673 (1992).

43. Z. Witkiewics and J. Bladek, Overpressured thin-layer chromatography, *J. Chromatogr.*, *373*:111 (1986).

44. R. E. Tecklenburg, G. H. Fricke, and D. Nurok, *J. Chromatogr.*, *290*:75 (1984).

45. C. F. Poole and M. T. Belay, Progress in automated multiple development, *J. Planar Chromatogr.*, *4*:345 (1991).

46. R. E. Kaiser, HRC &CC, *J. High Resolu. Chromatogr. Chromatogr. Commun.*, *1*:164 (1978).

47. N. Grinberg, Thin-layer chromatographic separation using a temperature gradient, *J. Chromatogr.*, *333*:69 (1985).

48. K. Kawanabe, Aplication of thin-layer stick chromatographic identification test methods of drugs contained in preparations in the Pharmacopoeia of Japan, *J. Chromatogr.*, *333*:115 (1985).

49. B. Fried and J. Sherma, Preparative thin-layer chromatography. In *Planar Chromatography*, Marcel Dekker, Inc., New York, 1982, p. 161.

50. P. R. Robbins, *Chromatography of Techetium-99m Radiopharmaceuticals—A Practical Guide*, The Society of Nuclear Medicine, Inc. New York, 1985.

51. A. M. Zimmer, J. M. Kazikiewicz, S. M. Spies, and S. T. Rosen, Rapid miniaturized chromatography for ^{111}In labeled monoclonal antibodies: comparison to size exclusion HPLC, *Nucl. Med. Biol.*, *15*: 717 (1988).

52. A. C. Moffat, Thin-layer chromatography. In *Clark's Isolation and Identification of Drugs* (A. C. Moffat, ed.), The Pharmaceutical Press, London, 1986, p. 160.

53. G. Zweig and J. Sherma (eds.) *Handbook of Chromatography*, CRC Press, Boca Raton, FL, 1972, p. 103.

54. B. Basak, U. K. Bhattacharyya, and S. Laskar, Spray reagent for the detection of amino acids on thin layer chromatography plates, *Amino Acids*, *4*:193 (1993).

55. W. Naidong, S. Geelen, E. Roets, and J. Hoogmartens, Assay and purity of oxytetracycline and doxcycline by thin-layer chromatography-a comparison with liquid chromatography, *J. Pharm. Biomed. Anal.*, *8*:891 (1990).

56. T. G. Chandrashekhar, P. S. N. Rao, K. Smrita, S. K. Vyas, and C. Duff, Analysis of amlodipine besylate by HPTLC with fluorimetric detection: A sensitive method of tablets, *J P C*, *7*:458 (1994).

57. J. C. Spell and J. T. Stewart, Quantiative analysis of para-amino-

76 Adamovics and Eschbach

salicylic acid and its major impurity meta-aminophenol in para-aminosalicylic acid drug substance by HPTLC and scanning densitometry, *J P C*, 7:472 (1994).

58. H. Salomies, Quantitative HPTLC of sulfonamides in pharmaceutical preparations, *J P C*, 6:337 (1993).
59. J. C. Spell and J. T. Stewart, Quantitative analysis of chloroprocaine and its major impurity 4-amino-2-chlorobenzoic acid in drug substance and injection dosage form by HPTLC and scanning densitometry, *J P C* , 8:72 (1995).
60. D. J. White, J. T. Stewart, and I. L. Honigberg, Quantitative analysis of diazepam and related compounds in drug substances and tablet dosage form by HPTLC and scanning densitometry, *J P C*, 4:413 (1991).
61. D. J. White, J. T. Stewart, and I. L. Honigberg, Quantitative analysis of chlordiazepoxide hydrochloride and related compounds in drug substance and tablet dosage forms by HPTLC and scanning densitometry, *J P C,* 4:330 (1991).
62. L. Zivanovic, S. Agatonovic, and D. Radulovic, Determination of diuretics in pharmaceutical preparations using thin-layer chromatographic densitometry, *Pharmazie*, 44:864 (1989).
63. The isolation of lovastatin and its determination by densitometric TLC and HPLC, *J P C*, 6:404 (1993).
64. P. Corti, G. Corbini, E. Dreassi, N. Politi, and L. Montecchi, Thin layer chromatograohy in the quantitative analysis of drugs—determination of rifaximine and its oxidation products, *Analusis*, 19:257 (1991).
65. E. Postaire, M. Cisse, M. D. le Hoang, and D. Pradeau, Simultaneous determination of water-soluble vitamins by over-pressure layer chromatography and photodensitometric detection, *J. Pharm. Sci.*, 80:368 (1991).
66. N. R. Ayyangar, S. Biswas, and A. Tambe, Separation of opium alkaloids by TLC combined with flame ionization detection using the peak pyrolysis method, *J. Chromatogr.*, 547:538 (1991).
67. K. L. Busch, Mass spectrometric detection for thin-layer chromatographic separations, *Trends Anal. Chem.*, 11:314 (1992).
68. K. Wada, T. Tajima, and K. Ichimura, High sensitivity thin-layer chromatography—Fourier transform infrared spectrometry system based on zone transfer technique, *Anal. Sci.*, 7:401 (1991).
69. D. M. Mustillo and E. W. Ciurczak, The development and role of near-infrared detection in thin-layer chromatography, *Appl. Spectrosc.*, 27:125 (1992).

70. D. M. Marmion, *Handbook of U.S. Colorants for Foods, Drugs and Cosmetics*, 2nd ed., Wiley, New York, 1984.

71. G. Richard, P. Gataud, J. C. Arnaud, and P. Bore, Qualitative analysis of preservatives using high-performance thin-layer chromatography. In *Cosmetic Analysis* (P. Bore, ed.), Marcel Dekker, Inc., New York, 1985, p. 157.

72. A. C. Moffat, Optimium use of paper thin-layer and gas-liquid chromatography for the identification of basic drugs, *J. Chromatogr.*, *90*: 9 (1974).

73. J. A. Adamovics (ed.), *Analysis of Additive and Misused Drugs*, Marcel Dekker, Inc., New York, 1995, pp. 41–50 and 221–266.

74. H. Wagner, S. Bladt, and E. M. Zgainski, *Plant Drug Analysis— A Thin Layer Chromatography Atlas*, Springer-Verlag, New York, 1984.

75. A. H. Stead, R. Gill, T. Wright, J. R. Gibbs, and A. C. Moffat, Standarized thin-layer chromatographic systems for identification of drugs and poisons, *Analyst*, *107*:1106 (1982).

76. G. Musumarra, G. Scarlata, G. Cirma, G. Romano, S. Palazzo, S. Clementi, and G. Ginlietti, Qualitative organic analysis, Analysis of standarized thin-layer chromatographic data in four eluent systems, *J. Chromatogr.*, *291*:249 (1984).

77. R. A. Egli and S. Keller, Comparison of silica gel and reversed-phase TLC and LC in testing of drugs, *J. Chromatogr.*, *291*:249 (1984).

78. B. M. J. De Spiegeller, P. H. M. De Moerloose, and G. A. S. Slegers, Criterion for evaluation and optimization in thin-layer chromatography, *Anal. Chem.*, *59*:61 (1987).

79. I. Ojanpera, Toxicological drug screening by thin-layer chromatography, *Trends Anal. Chem.*, *11*:222 (1992).

80. P. M. J. Coenegracht, M. Dijkman, C. A. A. Duineveld, H. J. Metting, E. T. Elema, and Th. M. Malingre, A new quaternary mobile phase system for optimization of TLC separations of alkaloids using mixture designs and response surface modelling, *J. Liq. Chromatogr.*, *14*, 3213 (1991).

4

Gas Chromatography

JOHN A. ADAMOVICS and JAMES C. ESCHBACH
Cytogen Corporation, Princeton, New Jersey

I. INTRODUCTION

Chromatographic analyses of pharmaceutical compounds has evolved as the drug industry matured. Gas-liquid chromatography (GLC) developed from the early 1950s to the present with many concurrent innovations in chromatography columns and detection systems. Packed open tubular columns (⅛–¼ in. ID × 6′) have been replaced with capillary (0.25 mm ID) and megabore (0.5 mm ID) wall coated columns (30, 60, or 120 m) achieving greater baseline separation and selectivity. GLC has been supplanted in many areas of pharmaceutical analysis by high-performance liquid chromatography (HPLC), in particular for the assay of compounds that are thermally unstable and with molecular weights greater than 1000. GLC is the technique of choice for thermally stable, relatively volatile analytes such as residual solvents.

II. STATIONARY PHASES

The selection of a gas chromatographic stationary phase is one of the most critical aspects of this technique. There are several hundred stationary phases which have been used in packed columns [1]. Only a few common phases are usable for capillary columns which demand high thermal stability, coatability, and reproducible deposition of the stationary phases on the column wall.

The stationary phase applied to the capillary column can be nonpolar, polar, or of intermediate polarity. The most common nonpolar stationary-phase column is methylpolysiloxane (HP-1, DB-1, SE-30, CPSIL-5) and 5% phenyl–95% methylpolysiloxane (HP-5, DB-5, SE-52, CPSIL-8). Intermediate polar phases such as 50% phenyl–50% methylpolysiloxane (HP-17, DB-17, CPSIL-19) and polar phases such as polyethylene glycol (HP-20M, DB-WAX, CP-WAX, Carbowax-20M) are popular.

Phases in capillary columns are bound and cross-linked and offer the highest degree of stability and nonextractability. Cross-linking of the stationary phase immobilizes the film and reduces bleed at elevated temperatures and imparts the ability to regenerate contaminated phase by solvent rinsing. Cross-linking reactions of stationary phases is through free-radical-induced reaction of alkyl, and alkenyl substituents on the polymer or radiation can be used to initiate the reaction [2]. Capillary columns offer high stationary-phase inertness, ease of use, on-column injection capability, increased speed of analysis, and cold trapping, attributes not normally associated with packed-column chromatography.

The above phases represent the most common phases used in solving nearly all of the frequently encountered application problems. There are many other stationary phases which are produced to "tune" the phase polarity for specific applications. In addition to these phases, there are liquid crystalline, chiral, cyclodextrin, polymers such as polystyrene, divinylbenzene, molecular sieves, and alumina, which are designed for specific separation problems. The chemistry of fused silica deactivation and stationary-phase application, bonding, and cross-linking has been reviewed in detail [3,4].

A. Nonpolar and Moderately Polar Phases

Polysiloxanes are the most common phases used in the production of nonpolar to moderately polar capillary stationary phases. They offer high-temperature stability and a wide liquid range between their glass-transition and decomposition temperatures giving them wide operational temperatures (-60–300°C). Methylpolysiloxane, a nonpolar stationary phase, shows the highest operational temperatures, highest degree of inertness, and generally produces columns with the highest efficiencies. Most separations on these columns are due to differences of solute vapor pressures with some selectivity by weak dispersion interactions.

Octylmethylpolysiloxane nonpolar phases have been used successfully in GC and capillary supercritical fluid chromatography (SFC). Octylmethylpolysilane columns used in SFC are thought to offer a greater degree of inertness over that of the methylpolysiloxane, for certain compounds. The long alkyl chain reduces the chance of solute interaction with residual active sites.

The moderately polar phases, such as those containing 5–50% phenyl or biphenyl with the remainder as methylpolysiloxane, have no permanent dipole but can be temporarily polarized by a solute. This characteristic produces selectivity to polar solutes through dipole-induced dipole interaction with solutes. The thermal stability of these stationary phases is similar to that of the methylpolysiloxanes.

B. Polar Stationary Phases

The polyethylene glycols or Carbowaxes and cyanopropylpolysiloxanes are the most common of the polar stationary phases. These phases possess permanent dipoles, and acid–base interaction with solutes is common. These phases are very retentive toward solutes with polar or polarizable functionality. They have lower upper temperature stabilities and higher lower minimum operational temperatures compared to the nonpolar phases. When these phases are cross-linked, the operational upper temperature limits are about 220–270°C and the lower limits about 40–60°C.

The Carbowax column is very sensitive to oxidation when the stationary phase is exposed to traces of water or air especially at temperatures above about 160°C. A new type of cross-linking has been reported to impart resistance to oxidative degradation of the stationary phase [5–7]. Two other phases which show promise are an oligo-(ethylene oxide)-substituted polysiloxane (glyme) and an 18-crown-6-substituted polysiloxane [8]. The glyme column offers a polar phase with good operational conditions to a low of a least 20°C with the same selectivity of Carbowax. The crown polysiloxane selectivity is based on the interaction of the solute molecule with the cavity of the crown ether.

Cyanopropyl silicone, another polar stationary phase [9–13], is highly polar because pi-complex interactions take place between the nitrile groups of the stationary phase and pi-electrons of unsaturated analytes. For a long time this phase was limited to packed-column GC due to its low viscosity and poor coatability on capillary walls. New cyanoalkyl-substituted silicones have been produced with gumlike characteristics, making open tubular capillary column coating possible. The separation selectivity is dependent on the number of double bonds and location with respect to configuration geometry. These phases are stable to about 300°C.

C. Liquid-Crystal Phases

Liquid-crystalline stationary-phase selectivity is thought to occur through the geometrical interaction of the ordered crystals which are coupled to the polysiloxane backbone, with the geometrical features of the solute. Retention on these phases is highly dependent on the solute size, shape, and how well it fits into the uniform arrangements of the stationary phase. The degree of order associated with each crystalline phase varies with the temperature. Most liquid-crystal phases can be classified as semetic, nematic, or isotropic at various temperature ranges. Semetic phases are the most highly ordered, followed by the nematic state, and then the isotropic state, which has more liquidlike properties. Nematic phases favor the retention of linear molecules, whereas the isotropic phases retain planar or rodlike solutes.

A commercially available semetic phase composed of biphenylcarboxylate ester attached to the polysiloxane backbone of a fused silica column has been shown to be ideal for the separation of geometric isomers of polycyclic aromatics of various classes of compounds [14,15]. This column exhibits a wide semetic temperature range of 100–300°C and has been shown to be stable to at least 280°C. Liquid-crystal stationary phases have also been employed in SFC [16].

D. Cyclodextrin Phases

There has been a great deal of interest in the use of cyclodextrins as stationary phases [13]. Cyclodextrins are cyclic oligosaccharides composed of varying numbers of glucopyranose units. Cyclodextrins are available in the alpha, beta, and gamma forms. The glucopyra-nose units are linked together so all secondary hydroxyl groups are situated on one of the two edges of a formed ring, with primary hydroxyls on the outer edges. The cavity formed is lined with glycosidic oxygen bridges which are rich in electron density.

Gamma cyclodextrin is the largest of the three compounds, composed of eight glucose units, with a molecular weight of 1297, a cavity diameter of 10 Å, and a cavity volume of 510 Å^3. Alpha is the smallest, composed of six glucose units, with a molecular weight of 972, a cavity diameter of 5–6 Å and a cavity volume of 176 Å^3. Beta cyclodextrin, which is intermediate in size, is composed of seven glucose units and has a 1135 molecular weight, a cavity diameter of 6–8 Å, and a cavity volume of 346 Å^3. It is these cavities that result in the ability of cyclodextrin stationary phases to separate enantiomers. In fact, cyclodextrins can be a host molecule for compounds having the size of one or two benzene rings or a side chain of comparable size.

Octakis (2,3,6-tri-*O*-methyl-gamma-cyclodextrin) was used to separate enantiomers of methyl esters of deltametrinic acid and permetrinic acid; the positional isomers of nitrotoluene were also separated on the same column [17,18]. Various alkyl- and dialkyl-benzenes have been separated on beta- and gamma-cyclodextrin [19]. A complete review of the use of cyclodextrins in chromatography has been published by Hinze [20]. Cyclodextrins have been analyzed by packed-column gas chromatography as their dimethylsilyl ethers [21].

E. Absorption

Capillary columns coated with aluminum oxide and molecular sieves can be described as absorption phases [22]. These columns are also known as porous-layer open tubular (PLOT) columns. Aluminum oxide PLOT columns deactivated with potassium chloride have been very effective at the separation of C_{1-10} hydrocarbons.

These columns offer excellent loadability and easy separability of light hydrocarbon isomers. Some nonlinear absorption is experienced with these columns, and as a result, some peaks will exhibit tailing effects. Water absorption by the alumina can reduce the selectivity of the column. Conditioning the column at 200°C is generally sufficient to remove the surface-absorbed water.

Polar compounds such as alcohols, aldehydes, and ketones are strongly absorbed to the active sites of the alumina. In most cases, they will not be easily desorbed from the alumina even after a conditioning at 200°C. Only after long conditioning, column back-flushing, and cutting, will the selectivity of the column be returned. Isomers of perflouroalkanes and flourobenzenes differing by a 1°C boiling point have been separated on an aluminum oxide PLOT column [23].

Molecular sieves have been applied to fused silica columns, 10 m × 0.5 mm ID. Molecular sieve 13 × has great retentivity for hydrocarbons and is ideal for the separation of naphthene/paraffin mixtures. Molsieve 5A has been applied to 0.32-mm-ID columns to give a porous layer of about 30 μm for the analysis of permanent gases.

A good application is the determination of headspace oxygen in lyophilized and rubber-stoppered pharmaceutical bottles, vials, and pouches. The efficiency of nitrogen purging during bulk drug packaging can be determined by simply injecting a portion of the headspace of the sealed vial against air as the standard. A small dab of silicone rubber cement is applied to multilayer aluminum–polyproplene pouches and is allowed to cure overnight. A syringe fitted with a 22-gauge needle pierces the pouch through the silicone dab, and the inside pouch atmosphere can be sampled.

F. Porous Polymers

The porous polymer stationary phases which for many years have been
available in packed gas chromatography columns has only recently become
available as a coated capillary [24]. These cross-linked porous polymer
columns are produced by copolymerizing styrene and divinylbenzene. The
pore size and surface are varied by altering the amount of divinylbenzene
added to the polymer. These PLOT capillary columns exhibit the same
separative characteristics as Poropak Q packed columns.

Columns lengths of 10–25 m with a 0.32-mm ID and 10-μm film
thickness are available commercially. These phases are ideal for the separation òf analytes in aqueous solutions or trace analysis of residual water,
because the hydrophobic nature of the polymer allows water to be eluted as
a sharp peak. The upper operational temperature of 250°C makes these
phases a good choice for the separation of polar light hydrocarbons and
alcohols. At subambient temperatures oxygenated gases such as CO and
CO_2 are separated without tailing.

Porous polymers containing various metal chelates bound to nitrogen
functionalities have been used to separate oxygen from argon, nitrogen,
and carbon monoxide [25]. The porous polymer is synthesized with pyridyl
functional groups, which serve as an axial base for the metal chelate in
coordinate bond formation between the metal chelate and the polymer. It
also serves to activate the metal complex for oxygen coordination.

III. HARDWARE

A. Capillary Inlets

There are several types of capillary inlets in use today which can be divided
into two categories: direct on-column injection and split/splitless injection
techniques.

Direct injection simply transfers the entire sample to the analytical
column. Direct injection techniques include direct flash vaporization (hot
direct injection) and cold on-column. Direct flash vaporization has become
popular used with wide-bore capillary columns (>530 μm ID) having phase
ratios less than 80 and high sample volume capacity and can easily accom-
modate injections of up to 10 μl of sample.

Specially designed direct flash injection port liners, called Uniliners,
made by Restek Corporation (Port Matilda, PA) can be used for both flash
vaporization and wide-bore on-column injections. In the direct flash mode,
an injection syringe with up to a 22-gauge needle can be used to seal the
vaporization chamber while a 530-μm ID column butts to the bottom of the

liner. This design eliminates exposure of the sample to the polyamide outer coating and reduces sample exposure to the metal surface of the syringe. With the use of a syringe of the proper needle gauge, the tightly controlled radial restriction of the liner seals the expansion chamber and reduces flashback spillover into the injector cavity, allowing for full automation using a standard syringe.

The liner may also be inverted, reversing the location of its radial restrictions. If the narrower restriction is placed at the top and a wide-bore column into the bottom restriction, the column will enter into the expansion chamber and butt to the narrower end. In this mode, a 26-gauge needle can be used to perform wide-bore on-column injections. The advantage of this mode of sample introduction is that exposure to active sites of the glass liner is eliminated. The disadvantage is that small samples must be used relative to the direct flash. The effect of depositing particles directly on the column is more severe for the on-column procedure, along with wider solvent peaks.

Cold on-column injections allow the inlet and/or inlet section of the column to be cooled or maintained at lower temperatures than the oven. Cold on-column injection suffers from the possibility of sample zone band broading due to excessive solvent flooding. The advantage of cold on-column injection is that it reduces sample discrimination or sample loss, due to syringe needle heating during injection.

For flash vaporization injection, the effect of injection speed, choice of solvent, and injector temperature greatly affect the solvent peak width. Injection of a low-boiling solvent at high injector temperatures can cause the entire expansion volume of the liner to be exceeded by the expanded solvent. It must be remembered that 1 μl of liquid solvent injected at 250°C at 10 psig head pressure is converted to some 200–1000 μl of gas. Thus, as the injection volume increases, the injection rate should be adjusted so as not to exceed the expansion volume available in the port liner. Generally, nonpolar solvents perform best in the direct flash mode. The injection rate should follow the following formula [26]:

$$\text{Injection rate} \geq \frac{(\text{Expansion volume sample} + \text{Solvent}) - \text{Liner volume}}{\text{Column flow rate}}$$

On-column injection into columns of 320 μm or less is more difficult and not easily automated. This usually requires the use of a fused silica needle of sufficiently small outer diameter or a needle capable of entry into the analytical column. The column is usually placed into a specially designed injection port fitted with "duck bill" or isolation-valve-type septa. A second approach which has shown great success is the use of a precolumn of a wide-bore capillary (530 μm ID) which is connected to the analytical column (320 μm ID or less). This technique is usually called retention gap. Sample injection volumes of up to 100 μl can be made.

Compared with on-column injection into 320-μm-ID columns, the advantage of the retention gap approach is that large sample volumes can be injected. The effects of particulates deposited on the precolumn is not severe. The approach can be automated and the length of the flooded zone is more controllable. The length of the flooded zone in the column inlet and the evaluation of different retention gaps have been studied [27]. A comprehensive review of the retention gap technique has been published [28].

Splitless injection involves keeping the injector split vent closed during the time the sample is deposited on the column, after which the vent is reopened and the inlet purged with carrier gas. In splitless injection, the inlet temperature is elevated with respect to the column temperature. The sample is focused at the head of the column with the aid of the "solvent effect." The solvent effect is the vaporization of sample and solvent matrix in the injection port, followed by trapping of the analyte in the condensing solvent at the head of the column. This trapping of the analyte serves to refocus the sample bandwidth and is only achieved after proper selection of the solvent, column and injector temperatures. Splitless injection techniques have been reviewed in References 29 and 30.

Splitless, direct, and cold-on-column techniques all utilize the solvent effect to maximize sample loading and minimize sample bandwidths. There is a wealth of information on how to best utilize the solvent effect to minimize the starting sample bandwidths in the splitless mode of injection. Several articles review the proper use of the solvent effect [31–36]. Splitless injection is ideal for dilute clean samples; it, however, is not suited for heat-sensitive samples. Classical split injection is discussed in a comprehensive review recently published [37]. The solvent effect in split injection has been discussed in two articles [38,39].

In programmed temperature vaporization (PTV) injection, the sample is introduced into an injector kept below the boiling point of the sample solvent. After withdrawal of the syringe needle, the vaporization chamber packed with glass wool is heated rapidly after the sample solvent has evaporated (splitless mode). In the split PTV injection mode, the sample is vaporized after the needle is removed while keeping the split vent open. This flash heating vaporizes the sample, driving it into the column. The advantage to PTV injection is that it subjects samples to less thermal stress compared to vaporization techniques of injection. Also, the splitless PTV mode is less prone to be affected by a sample matrix than classical splitless techniques. For a complete review of the PTV injection, see Refs. 40–42. The performance of PTV injection has been compared with hot-splitless and on-column [43] injection, and with classical splitless injection [44].

Several published reports review the basic types of inlets available for capillary gas chromatography [45–50]. Reference 41 can be a useful guide to the proper selection of an inlet system and the parameters for its optimum performance (Table 4.1).

B. Capillary Columns

Capillary columns are usually composed of fused silica with polyamide outer coating to give flexibility and reduce breakage of the capillary during handling. Capillary columns can be categorized into three classes. The megabore or wide-bore columns are greater than 0.32 mm ID, whereas normal bore or high-resolution columns are 0.32–0.22 mm ID, and microbore or high-speed columns are 0.2–0.1 mm ID.

The high-speed columns are generally used where the highest efficiencies and speeds are required, such as gas chromatography–mass spectrometry (GC–MS) applications where reduced run times can increase source and electron multiplier lifetimes. A second advantages of fused silica capillaries in GC–MS applications have been lower temperatures and consequent minor change in vacuum during temperature programming, resulting in better sensitivity. Other advantages of fused silica capillaries in GC–MS application have been discussed by Settiage and Jaeger [51].

High-speed columns yield 5000–10,000 plates per meter, whereas most high-resolution columns give 3000–5000 plates per meter. Most applications of capillary columns are with column lengths of 10–50 m and are somewhat dependent on the complexity of the sample and the number of components of interest. A general rule is that samples with 20–50 components are best handled on columns of 20–30 m, whereas samples of 50 or more components will require 30–50 m of column.

Capillary supercritical fluid chromatography (SFC) columns are 0.1–0.025 mm ID and 3–20 m in length. Good reviews of the technique of SFC have been recently published [52–55]. It was reported that the optimum inner diameter for capillary SFC based on plate height, linear velocity, analysis time, and column length was around 0.050 mm.

One type of column is the wall-coated open tubular column (WCOT) in which the stationary phase is applied and bound directly to the walls of the column. Porous-layer open tubular columns (PLOT) are columns in which the stationary phase is deposited on fine particles of absorbent, absorbed on the walls of the column, increasing the available surface area of the column wall. Support-coated open tubular (SCOT) columns are those

Table 4.1 Selection of the Injection System and Optimum Performance
Parameters

Split injection	Application for • relatively concentrated solutions: 1% to 20 ppm (FID*) per component • analysis of undiluted samples • headspace analysis • rapid, fully isothermal analysis Sample handling; high flexibility regarding sample concentration, solvent, and column temperature; optimal reproducibility of absolute retention times; demand for analyses requiring high accuracy; high risk of systematic errors.
Splitless injection	Application for • dilute samples; 50–0.5 ppm (FID) per component • dirty samples, especially if accuracy of the results is not very important Produces relatively accurate results for volatile solutes; problems with quantitation of high-boiling solutes (matrix effects!); requires reconcentration of the initial bands by cold trapping or solvent effects, which often forces cooling of the column for the injection. This is time-consuming and causes problems with absolute retention times.
PTV injection	Produces better quantitative results than classical vaporizing injection but is not yet sufficiently explored to be classified as a routine method with known working rules.
On-column injection	Diluted samples: 300–0.01 ppm (FID) per component; optimum method for producing highly accurate results; not suitable for very dirty samples (samples containing more than 0.1% of involatile by-products); requires cooling below solvent boiling point.

Table 4.1 (Continued)

Sample volume	If evaporation inside the syringe needle cannot be avoided, use 5- or 10-μl syringes, resulting in a minimum sample volume injected corresponding to the needle volume. Injection of 0.5–1-μl volumes improves elution from the syringe needle. Splitless injection: maximum sample volume around 2 μl (1 μl plus needle volume).
	Split injection: small sample volumes tend to reduce problems.
Length of syringe needle	Long needles, releasing the sample near the column entrance, for spitless injection and for split injection with low split flow rates.
	Short needles, providing a long way for the sample to evaporate, for split injection with high split flow rates. Short syringe needles help to avoid sample evaporation inside the syringe needle.
	Packing material strongly promotes decomposition of labile solute material and tends to retain (adsorb) high-boiling components. Split-injection: promoted evaporation may improve or worsen quantitative results (to be tested).
	Splitless injection: a light plug of glass wool reduces matrix effects for solutes having fairly high boiling point.
Carrier gas flow rate	Splitless injection: high carrier gas flow rates improve the efficiency of the sample transfer; below 2.5 ml/min sample transfer becomes unsatisfactory (below 1.5 ml/min if solvent recondensation accelerates the sample transfer).
	Split injection: use high carrier gas flow rates to obtain maximum sensitivity; low carrier gas flow rates for very high split ratios. High carrier gas flow rates strongly favor use of hydrogen as carrier gas as well as columns

(Continued)

Table 4.1   (Continued)

	with up to 0.35 mm ID (wider bore columns provide strongly reduced separation efficiencies of only weakly increased flow rates).
Column temperature during injection	Split injection: relatively unimportant; column temperatures below the solvent boiling point promote the recondensation effect, causing more sample material to enter the column than expected from the present split ratio.
	Splitless injection: reconcentration of the bands broadened in time requires either lowering of the column temperature at least 60–90°C below the elution temperature of the solutes of interest (cold trapping) or keeping of the column at least 20–25°C below the solvent boiling point to create solvent effects.
Injector temperature	Minimum injector temperature if sample evaporation inside the syringe needle is to be avoided.
	If sample evaporation inside the syringe needle is unavoidable, the maximum injection temperature which can be tolerated without degrading solute material.
	Splitless injection: high injector temperature improves sample transfer and reduces matrix effects.
	Split injection: high injector temperature promotes evaporation, which may or may not be advantageous.
Width of injector insert	Sample vapors must not overfill the injector insert.
	Splitless injection: inserts with 3.5–4 mm ID.
	Split injection: wide-bore inserts (3–5 mm ID) reduce deviations between the preset and the true split ratio and improve evaporation (prolonged evaporation time).
	Narrow inserts (~2 mm ID) reduce dilution of sample vapors with carrier gas; of interest for analyses requiring maximum sensitivity.

Table 4.1   (Continued)

Packed inserts	Packing the vaporizing chamber, e.g. with (silanized) glass wool, improves sample evaporation and hinders involatile by-products from entering the column.

*FID = flame-ionization detection
Source: From Reference 41, with permission.

in which the stationary phase is deposited on a solid support coating the column wall.

A fourth type of column is the whisker walled (WW) in which the wall of the column has been etched by chemical means, leaving behind whiskers on the surface of the column. These projections of the fused silica significantly increase the available surface area of the column. Wall-coated, porous-layer, and support-coated capillary columns have all been available as whisker walled and have been given the acronyms WWCOT, WWPLOT, and WWSCOT, respectively.

The film-thickness stationary phase of these columns is usually 0.1–10 μm and can be broken into three film-thickness ranges. Thin-film columns are usually 0.1–0.2 μm offering the greatest stationary-phase stability. They have smaller sample capacity compared to the thicker films but are the film thickness of choice for high-temperature work. Thick-film columns are generally 0.6–10 μm and offer high sample capacity, better retentivity to volatile compounds, and a high degree of inertness, but have a larger amount of bleed at the higher temperatures compared to the thinner-film columns.

The medium-film thickness is about 0.3–0.6 μm and generally offers the best compromise of sample capacity, retentivity, and phase stability. The phase ratio determines the capacity of the column and influences its retentivity of solutes. The phase ratio (β) can be defined as the ratio of the inner column radius to that of the product of twice the stationary-phase film thickness or $\beta = r/2d_f$. We can now also use phase ratios to group film thicknesses and now say that thick-film columns have phase ratios of less than about 80. (In capillary SFC the typical stationary-phase film thicknesses are 0.1–0.3 μm.) The effective phase ratio can change in capillary SFC, depending on the characteristics of the stationary phase and the operating density [57]. The change in phase ratio can be attributable to a swelling of the stationary phase under certain SFC conditions.

C. Detection

There are different types of detectors available for gas chromatography designed for specific analytical uses. The thermal conductivity detector is a universal detector which will respond to everything including water. Selective detectors such as electron capture and flame ionization respond to certain functional or elemental characteristics of the analyte. Specific detectors respond in such a way as to give specific qualitative information concerning the analyte's structure. Specific detectors include Fourier transform infrared, flame photometric, and mass selective detection. The following subsections highlight most of the detectors available and outline useful and interesting applications.

Electron Capture

Electron-capture detectors show great sensitivity to halogenated compounds. In electron-capture detectors, the carrier gas is ionized by beta particles from a radioactive source (usually tritium or nickel-63), to produce a plasma of positive ions, radicals, and thermal electrons. Thermal electrons are formed as the result of the collision of high-energy electrons and the carrier gas.

Electron-absorbing compounds react with the thermal electrons to produce negative ions of higher mass. When a potential difference is applied to the detector collector, thermal electrons are collected to produce the standing current of the detector. Thus, the reduction in standing current due to the combination of thermal electrons and electron-capturing compounds provides the analytical signal. The other possible reaction that can take place in the detector is the interaction of an excited carrier molecule with a sample to produce an electron. This reaction increases the standing current and results in negative peaks.

To reduce the likelihood of these types of reactions, the detector gas of choice is 5% methane in argon. The methane serves to increase the energy-reducing collisions and prevent the high-energy collisions that form electrons. Thus, the low-energy reactions are favored and detector noise is minimized. It has also been shown that the response characteristics of the detector can be altered dramatically by the addition of oxygen or nitrous oxide to the carrier gas [58]. These dopants react to negative ions, which act as a catalyst to electron capture and thus enhance response with certain molecules.

Poole [59] has outlined some molecular features governing the response of electron-capture detectors to organic compounds, which can be used as a guide to judge response and selection of the proper derivative:

1. Alcohols, amines, phenols, aliphatic saturated aldehydes, thioethers, ethers, fatty acid esters, hydrocarbons, aromatics, vinyl-type fluororinated, and those with one chlorine atom all give a low response.
2. A high response is given by halocarbon compounds, nitroaromatics, and conjugated compounds containing two groups which in their own right are not strongly electron attracting but become so when connected by specific bridges.
3. Compounds with a halogen atom attached to a vinyl carbon have a lower response than the corresponding saturated compounds.
4. Greater sensitivity is obtained if the halogen atom is attached to an allyl carbon atom than the corresponding saturated compound.
5. Response for the halogen decrease in the following order I > Br > Cl > F and increase synergistically with multiple substitution on the same carbon.

Several reviews of the electron-capture detector have been published [59–61]. The fundamental properties of derivatization techniques to enhance electron-capture detection have been published [62,63]. There have been many reported pharmaceutical applications of the electron capture detector; a few selected interesting applications are listed in Table 4.2.

Thermionic

The most popular thermionic detector (TID) is the nitrogen–phosphorus detector (NPD). The NPD is specific for compounds containing nitrogen or phosphorus. The detector uses a thermionic emission source in the form of a bead or cylinder composed of a ceramic material impregnated with an alkyl-metal. The sample impinges on the electrically heated and now molten potassium and rubidium metal salts of the active element. Samples which contain N or P are ionized and the resulting current measured. In this mode, the detector is usually operated at 600–800°C with hydrogen flows about 10 times less than those used for flame-ionization detection (FID).

There is no sustaining flame in this operational mode, as there is in flame-ionization detection, and most hydrocarbons give little response because they are not ionized. The NPD has the highest sensitivity to N and P compounds, with limits of detection of about 1–10 pg. The detector can also be operated with hydrogen and air ratios to provide a self-sustaining flame. This mode is called the flame thermal-ionization detector (FTID).

In the FTID mode, the detector is specific for nitrogen- and halogen-containing compounds, with limits of detection to about 1 ng. By operating the detector with air only as the detector gas, the detector response to halogens are increased compared to FTID and weak response to nitrogen

Table 4.2 Pharmaceutical Applications
Using Electron-Capture Detection

Analyte	Reference
Amino acids	64
Amines, β-aminoalcohols	65
Prostaglandin	66
Antiarrhythmics	67
Arylalkylamines	68
ACE inhibitors	69, 70
Prostaglandins	71
Thromboxane antagonist	72
Anilines	73
Opiates	74
Propylnorapomorphine	75
Phenols	76
Chlorinated phenols	77
Carboxylic acid, phenols	78
Phenols	79
Methylene chloride	80
Tyosyl peptide	81
Iodine	82
Reduced sulfur	83
Beta-blockers	84
Propane-, butane-diols	85

compounds can be obtained. The most sensitive response to nitrogen compounds is obtained when the detector is operated with nitrogen as a detector
gas. Typical limits of detection of detection of nitrogen compounds can be
achieved in the range 0.1–1 pg.

Organolead compounds may be detected by turning off the heating to
the thermionic source and running in the FTID mode. In this mode, the
combustion of organolead compounds lead to long-lived negative-ion products which are detected at the TID collector.

When using a TID, the gas flow in the detector greatly affects the
response curve for many compounds. When performing trace analysis, it is
worth taking the time to generate detector gas flow versus response curves
to obtain optimal sensitivity. Chlorinated solvents and silanizing reagents
can deplete the alkali source and should be avoided. Glassware should be
rinsed free of any traces of phosphate detergents. Phosphoric-acid-treated

columns, glass wool, and stationary phases with high nitrogen content should not be used as they can generate a large background signal. Three recent reviews of thermionic detection have been published [83–85]. Several interesting applications are listed in Table 4.3.

Photoionization

When a compound absorbs the energy of a photon of light it becomes ionized and gives up an electron. This is the basis for the photoionization detector (PID). The capillary column effluent passes into a chamber containing an ultraviolet (UV) lamp and a pair of electrodes. As the UV lamp ionizes the compound, the ionization current is measured.

The PID allows for the detection of aromatics, ketones, aidehydes, esters, amines, organosulfur compounds, and inorganics such as ammonia, hydrogen sulfide, HI, HCl, chlorine, iodine, and phosphine. The detector will respond to all compounds with ionization potentials within the range of the UV light source, or any compound with ionization potentials of less than 12 eV will respond.

The advantage to the detector is that some common solvents such as methanol, chloroform, methylene chloride, carbon tetrachloride, and acetonitrile give little or no response if a lamp with an ionization energy of 10.2 eV is used. The most common lamps available are 9.5, 10.0, 10.2, 10.9, and 11.7 eV. To enhance the selectivity of the detector, a lamp is chosen which is just capable of ionizing the analyte of interest.

Table 4.3 Pharmaceutical Applications Using Thermionic Detection

Analyte	Reference
Methylpyrazole	86
Antiarrhythmics	87
N–P compounds	88
Aminobenzoic acid	89
Reducing disaccharides	90
Antihistamines	91
Symphathomimetic amines, psychomotor stimulants, CNS stimulants, narcotic analgesics	92
Benzodiazepines	93
Barbiturates	94

A major advantage to this technique is that inorganics can be detected to low levels (1–2 pg) using a nondestructive detector. This means that the PID can be connected in series with other detectors and is ideal for odor analysis. The sensitivity of the detector is directly related to the efficiency of ionization of the compound. The PID is about 5–10 times more sensitive to aliphatic hydrocarbons, 50–100 times more sensitive to ketones than FID, and 30 times more sensitive to sulfur compounds than flame photometric detection. Several reviews on the PID and its sensitivity have been published [94–97].

Flame Photometric

The flame photometric detector (FPD) uses the principle that when compounds containing sulfur or phosphorus are burned in a hydrogen–oxygen flame, excited species are formed, which decay and yield a specific chemiluminescent emission. The detector is composed of a dual-stacked flame jet and a photomultiplier tube. By selecting either a 393- or 526-nm bandpass interference filter between the flame and photomultiplier, sulfur or phosphorus detection is selected. The dual-flame arrangement enhances detector response because the first flame is where most of the combustion of the column effluent takes place, whereas the second flame is where the emission takes place. This minimizes emission quenching that can occur when solvents and sulfur or phosphorus species are in the flame simultaneously. The second type of quenching is observed at high concentrations of the heteroatom species in the flame. At high concentrations, the energy absorption due to collisional effects, chemical reactions between species, or reabsorption can reduce photon emissions.

Gas flows as well as hydrogen–air or hydrogen–oxygen flow ratios are critical to maximum response. Sensitivity on the sulfur mode decreases with increases in detector temperature, whereas in the phosphorus mode it increases with increased detector temperature.

The response of the detector in the phosphorus mode is linear with respect to concentration. In the sulfur mode, the square root of the response is proportional to concentration. The selectivity for sulfur or phosphorus to hydrocarbons is about 10^4–10^5 to 1, thus the presence of most solvents is not a problem. The reactions that occur in the flame are being studied. The species most commonly responsible for emissions in the sulfur mode is S_2, whereas in the phosphorus mode it is HPO. The typical sensitivity of the FPD is about 10–20 pg of a sulfur-containing compound and about 0.4–0.9 pg of a phosphorus-containing compound.

Detection difficulties in the sulfur mode are quite frequently attributed to problems with the detector. The analyst must always keep in mind that there is a possibility for absorption and oxidation with sulfur species [98]. A recent review on the sulfur detection mode of the FPD has been published [99]. The separation of trace amounts of seven volatile reactive sulfur gases has been achieved [100]. Carbon disulfide has been determined to 1 pmol/liter in water [101].

Electrolytic Conductivity

The electrolytic conductivity (ELCD) detector is specific for the detection of sulfur, nitrogen, and halogens. The detector is composed of a furnace capable of temperatures of at least 1000°C; effluent from the GC column enters the furnace and is pyrolyzed in a hydrogen- or oxygen-rich atmosphere. The decomposition takes place (reduction or oxidation) and several reactor species are produced. The effluent is passed through a scrubber tube to remove the unwanted species. The scrubbed effluent is brought into contact with a deionized alcohol–water mixture stream (conductivity liquid). The gas–liquid contact time is sufficient that the species enter the conductivity solution, which is pumped at 4–5 ml/min through a conductivity cell. The presence of these species in the conductivity liquid changes its conductivity and results in the analytical signal.

When the detector is operated in the (X = halogen) reductive mode with hydrogen as a reaction gas, H_2S, HX, NH_3, and CH_4 are the major reaction products of the decomposition of sulfur-, halogen-, and nitrogen-containing compounds. If a nickel furnace tube is used and a scrubber containing $Sr(OH)_2$ or $AgNO_3$ is used, HX will be removed. In addition, H_2S gives little or no response; thus, the only response is from the nitrogen-containing compounds. If the scrubber is removed and the nickel furnace tube is replaced with a quartz tube, no NH_3 or CH_4 is produced; consequently, the only response will be from halogen-containing compounds.

In the oxidative mode using air as the furnace reaction gas, sulfur-, halogen-, and nitrogen-containing compounds produce SO_2 and SO_3, HX, CO_2, H_2O, and N_2 products. Carbon dioxide gives little response because the gas–liquid contact time is short and it is poorly soluble in the alcoholic conductivity solution. Water and N_2 also give CaO scrubber, and as before, HX can be removed with a $AgNO_3$ or $Sr(OH)_2$ scrubber. The oxidative mode is the usual mode for selective detection of sulfur-containing compounds.

The electrolytic conductivity detector is a good alternative to the FPD for selective sulfur detection. The ELCD has a larger linear dynamic range and a linear response to concentration profile. The ELCD in most cases appears, under ideal conditions, to yield slightly lower detection limits for sulfur (about 1–2 pg S/sec), but with much less interference from hydrocarbons compared to the FPD. The performance of the ELCD compared to FPD and the performance in the sulfur mode in the presence of hydrocarbons have been published [102,103]. A useful article for troubleshooting operator problems has been published [104]. The use of the ELCD for nitrogen-selective detection has been reviewed [105]. The ELCD has been used for the determination of barbiturates without sample cleanup [106]. A report dealing with the detection of benzodiazepines, tricyclic antidepressants, phenothiazines, and volatile chlorinated hydrocarbons in serum, plasma, and water has been published [107].

Chemiluminescence

The principle of chemiluminescence detection is a chemical reaction forming a species in the electronically excited state that emits a photon of measurable light on returning to their ground state.

The oldest chemiluminescent detector was the thermal energy analyzer (TEA), which was specific for N-nitroso compounds. N-nitroso compounds such as nitrosamines are catalytically pyrolyzed and produce nitric oxide which reacts with ozone to produce nitrogen dioxide in the excited state, which decays to the ground state with the emission of a photon. A photomultiplier in the reaction chamber measures the emission. Nitrosodimethylamines have been detected to about 30–40 pg [108].

More recently, chemiluminescence detectors based on redox reactions have made possible the detection of many classes of compounds not detected by flame ionization. In the redox chemiluminescence detector (RCD), the effluent from the column is mixed with nitrogen dioxide and passed across a catalyst containing elemental gold at 200–400°C. Responsive compounds reduce the nitrogen dioxide to nitric oxide. The nitric oxide is reacted with ozone to give the chemiluminescent emission. The RCD yields a response from compounds capable of undergoing dehydrogenation or oxidation and produces sensitive emissions from alcohols, aldehydes, ketones, acids, amines, olifins, aromatic compounds, sulfides, and thiols. The RCD gives little or no response to water, dichloromethane, pentane, octane, carbon dioxide, oxygen, nitrogen, and most chlorinated hydrocarbons.

The usefulness of the detector is for those compounds giving low response to the FID, such as ammonia, hydrogen sulfide, carbon disulfide, hydrogen peroxide, carbon monoxide, formaldehyde, and formic acid, which all give apparently good response to the RCD. By changing the catalyst from gold to palladium, saturated hydrocarbons can be detected. The specificity of the detector decreases as the catalyst temperature increases and as gold is substituted for palladium.

The sulfur chemiluminescence detector (SCD) is based on the reaction of compounds containing a sulfur–carbon bond and fluorine. In the SCD, an electrical discharge tube converts sulfur hexafluoride into flouride, which enters a vacuum chamber containing a photomultiplier tube; the GC column enters the chamber via a heated transfer line. The vacuum pump keeps the chamber at low pressure. In this chamber, fluorine and sulfur-containing compounds react to form HF in the excited state, which decays to the ground state through the emission of a photon of light. Most sulfides, thiols, disulfides, and heterocyclic sulfur compounds can be detected in the mid to low picogram range. This detector gives little or no response to saturated hydrocarbons, methylene chloride, acetonitrile, methanol, and carbon tetrachloride. Weak responses are seen for compounds with C–H bonds such as alkenes and organics with amine groups. The advantage of the SCD over the FPD is that there is a linear response with respect to concentration and that there is no quenching due to solvent. Limits of detection for ethyl sulfide are about 5 pg. Reviews on the use of chemiluminescence detectors have been published [109–111].

Helium Ionization

The helium ionization detector (HID) is a sensitive universal detector. In the detector, Ti_3H_2 or Sc_3H_3 is used as an ionization source of helium. Helium is ionized to the metastable state and possesses an ionization potential of 19.8 eV. As metastable helium has a higher ionization potential than most species except for neon, it will be able to transfer its excitation energy to all other atoms. As other species enter the ionization field the metastable helium will transfer its excitation energy to other species of lower ionization potential, and an increase in ionization will be measured over the standing current.

The detector requires a helium source of at least 99.9999% pure, because the purity of the detector gas will affect the detector response, its background current and the polarity of the response for certain compounds. With very high-purity helium, the detector will respond negatively to hydrogen, argon, nitrogen, oxygen, and carbon tetrafluoride. The magnitude of the negative response will decrease as the purity of the helium decreases, until the minimum in the background current is reached. At the minimum in the background current, all gases, except neon, will give a positive response accompanied with a decrease in the overall sensitivity of the detector.

The HID is about 30–50 times more sensitive than the FID, with typical detection limits of low parts per billion of most gases. The HID has been used to detect nitrogen oxides, sulfur gases, alcohols, aldehydes, ketones, and hydrocarbons. The analysis of impurities in bulk gases and liquids is an ideal application for this detector.

Formaldehyde, which is difficult to detect at trace levels without derivatization, was determined in air to about 200 ppb with the HID. Reviews of the performance characteristics and applications of the HID have been published [113,114].

Mass Selective

The mass spectrometer when used as a detector for GC is the only universal detector capable of providing structural data for unknown identification. By using a mass spectrometer to monitor a single ion or few characteristic ions of an analyte, the limits of detection are improved. The term *mass selective detection* can refer to a mass spectrophotometer performing selected ion monitoring (SIM) as opposed to operation in the normal scanning mode. Typical limits of detection for most compounds are less than 10^{-12} g of analyte.

Comprehensive reviews of the use of the mass spectrometer as a detector in GC have been published [115–118]. The vast majority of references have been for the detection of pharmaceuticals and their metabolites in biological matrices.

Fourier Transform Infrared

The use of an on-line Fourier transform infrared (FTIR) detector with GC has allowed for the identification of unknowns and the distinction between structurally similar compounds. Many compounds with structural similarities cannot be identified by electron impact mass spectrometry because the fragmentation patterns are (or are nearly) identical. An example is the identification of positional isomers of substituted chlorobenzenes, whose mass spectra are identical. In these cases, chemical ionization can be used to highlight structural differences. The infrared detector (IRD) gives quite different spectra for positional isomers, and when compared to library spectra of authentic compounds, it gives unequivocal identification.

The FTIR is also useful in the identification of unknown solvents when performing trace analysis for residual solvents in bulks. The FTIR also must be looked upon as a complement to data collected by GC–MS. Reviews on the performance and application of the FTIR to various problems have been published [119,120]. Reviews on the use of the FTIR in combination with mass spectrometry have been published [121–123].

Atomic Spectroscopy

Atomic spectroscopy as a means of detection in gas chromatography is becoming popular because it offers the possible selective detection of a variety of metals, organometallic compounds, and selected elements. The basic approaches to GC–atomic spectroscopy detection include plasma emission, atomic absorption, and fluorescence.

Microwave-induced plasma (MIP), direct-current plasma (DCP), and inductively coupled plasma (ICP) have also been successfully utilized. The abundance of emission lines offer the possibility of multielement detection. The high source temperature results in strong emissions and therefore low levels of detection. Atomic absorption (AA) and atomic fluorescence (AF) offer potentially greater selectivity because specific line sources are utilized. On the other hand, the resonance time in the flame is short, and the limit of detectability in atomic absorption is not as good as emission techniques. The linearity of the detector is narrower with atomic absorption than emission and fluorescence techniques.

The microwave-induced plasma (MID) operating with helium at atmospheric pressure is quickly becoming a valuable means of element-selective detection of carbon, halogens, hydrogen, oxygen, nitrogen, and many organometallics. A TM_010 cavity is popular and is used with helium flows of about 60 ml/min. Nitrogen (about 1 ml/min) has been used as a scavenger gas to reduce carbon deposits on the plasma containment tube. Other cavities have been used to detect other elements, but the TM_010 cavity has been the cavity of choice for capillary applications [124]. Sensitivity is influenced by the choice of carrier gas and microwave power, but limits of detection have been determined for fluorine to be about 5 pg/sec [125]. When a rapid scanning instrument was used for bromine and chlorine, limits of detection were reported to be about 200–300 pg/sec and 50–150 pg/sec for bromine and chlorine, respectively. Ten elements have been simultaneously determined by GC–GC–microwave plasma emission spectroscopy [126].

The ICP is composed of a torch containing the plasma of gases. A radiofrequency (RF) is transferred by induction to the plasma through a coil wrapped around the torch. When the coil is energized with 0.5-5 kW of RF-power, a magnetic field is induced in the torch, heating the plasma gases to 5000°K. The torch is composed of several tubes, each carrying different gas flow velocities. Usually, the outer stream is high flow and serves to dissipate heat given off through the touch wall and also helps sustain the plasma. The center gas stream carries the sample through the RF coil region of rapid heating and ionization. As the sample ions and atomic species pass through the plasma, the atomic species return to their ground state with the emission of a characteristic radiation. Because of the high cost of operation of the ICP, applications of GC–ICP are not as frequent as the other techniques. The ICP is much more tolerant of organic solvents compared to the other techniques because of the high plasma temperatures.

In CG–direct-current plasma (DCP), a direct-current arc is struck between two electrodes as an inert gas sweeps between the electrodes carrying the sample. Carrier gases such as helium, argon, and nitrogen have been used.

Gas chromatography–atomic absorption (AA) has gained popularity because the interfacing is quite simple. In its crudest form, the effluent from the GC column is directly connected to the nebulization chamber of the AA. Here, the effluent is allowed to be swept into the flame by the oxidant and flame gases. There have been several recent reviews of the technique [127,128].

Atomic fluorescence spectroscopy (AFS) has also been used as a means of detection in gas chromatography. Alkylmercury compounds have been determined in air by cold-vapor GC–AFS with limits of detection of about 0.3–2.0 pg [129].

A comprehensive review of directly coupled gas chromatography–atomic spectroscopy applications has been published [128]. This review list over 100 references classified according to the detection technique and is highly recommended. Another excellent review outlines the advances in interfacing and plasma detection [130]. A review of the gas chromatographic detection of selected trace elements (mercury, lead, tin, selenium, and arsenic) has been published. This article reviews the many different detection methods available including atomic emission techniques [131].

D. Liquid Chromatography–Gas Chromatography

The number of articles dealing with the on-line coupling of the two most widely used separative techniques, liquid and gas chromatography, are few. Many analyses of complex mixtures or trace analyses in complex matrices utilize a liquid chromatograph for sample cleanup or for analyte concentration prior to gas chromatographic analysis.

Successful transfer of large volumes of liquid chromatography (LC) effluent to GC requires that the solvent must be evaporated some place in the inlet system. The two most common approaches to the evaporation are the retention gap and concurrent solvent evaporation. In concurrent solvent evaporation, the column oven is kept above the boiling point of the LC solvent. Using a valve–loop interface, LC effluent up to several milliliters is driven by the carrier gas into a precolumn. In this case, the eluent evaporates from the front of the liquid plug. At the head of the liquid plug, the high-boiling components are deposited, whereas the volatiles are evaporated along with the solvent.

A second approach takes full advantage of the retention gap by the addition of a small amount of cosolvent. The cosolvent is a higher-boiling solvent compared to the bulk eluent and serves to trap the volatiles while the bulk solvent evaporates. Thus, the sample is focused and the chromatography starts with sharp bands of analyte. The effects of the cosolvent and concurrent solvent evaporation have been reviewed [132], along with the minimum temperature need for concurrent solvent evaporation [133].

The application of the loop-type interface for LC–GC for multifraction introduction has been introduced [134]. The use of microbore LC columns have been used as a means to reduce the injection volumes of solvent [135,136].

Two approaches to the venting of the solvent prior to the detector have been presented in detail [137]. Packed GC columns coupled to capillary columns have been used for the total transfer of effluent from the LC [138]. The current status of LC–GC has been reviewed [139]. The use and performance of the ELCD, NPD, and FPD GC detectors in liquid chromatography has also been reviewed [140]. Even though the majority of applications are not directly related to the analysis of pharmaceuticals, they may nevertheless be useful [141–146].

E. Headspace Analysis

Headspace sampling is useful for those samples where

Direct injection would reduce column life because the matrix is corrosive or contains components which would remain on the column
Extensive sample preparation would be required before injection to remove the major components which would interfere with the analysis
Degradation of a component of the matrix in the injection port or on the column would generate degradants which would interfere with the analysis of the components of interest

Additional advantages are realized with headspace sampling. Sample preparation time is minimized because the sample in many cases is simply placed in a vial which is then sealed and capped. The compound(s) of interest may be released from the matrix by heat or chemical reaction, and aliquots of the headspace gas are collected for assay. Columns last longer because a gaseous sample is much cleaner than a liquid sample. The solvent peak is much smaller for a vapor sample than for a solution sample.

Static Sampling

A sample is placed in a glass vial that is closed with a septum and thermostated until an equilibrium is established between the sample and the vapor phase. A known aliquot of the gas is then transferred by a gas-tight syringe to a gas chromatograph and analyzed. The volume of the sample is determined primarily from practical considerations and ease of handling. The concentration of the compound of interest in the gas phase is related to the concentration in the sample by the partition coefficient. The partition coefficient is included in a calibration factor obtained on a standard. The analysis can easily be automated where a series of samples is to be analyzed, resulting in improved precision.

Improvement in sensitivity can be obtained by increasing the temperature of the sample or by the salting-out effect, which is particularly useful for compounds such as phenols and fatty acids which form strong hydrogen bounds in aqueous solutions. With some compounds, the use of a more sensitive detector such as an electron-capture detector or an element-specific detector will enhance sensitivity.

Volatiles in solid samples will yield good chromatograms when analyzed by headspace chromatography. However, for purposes of calibration, it is difficult to mix a certain amount of a volatile compound into a sample homogeneously. In addition, an excessive period may be required to equilibrate between the solid and the gas phase, for example, monomers arising from polymers. One solution to this problem is to dissolve the sample in a suitable solvent. The solvent should have a longer retention time than the volatile compounds of interest. Back-flushing techniques can be used to rapidly remove the solvent from the column. Suitable solvents are water, benzyl alcohol, dimethylformamide, and high-boiling hydrocarbons. However, if the highest sensitivity is required, solvents should be avoided, as dilution reduces the detection limit. An alternative to the use of solvent is to heat the polymer above the glass-transition point. Another alternative is the use of dynamic sampling, which will be described later.

Multiple Stage

Quantitative analysis is best performed on liquid samples or on solutions prepared from solid samples. This approach is not possible when a suitable solvent cannot be found. Multiple static extractions can be conducted in these situations.

Dynamic Sampling

Samples are purged with an inert gas, and volatiles are cold-trapped or absorbed on a packing such as charcoal or Tenax. The trap or packing is then rapidly heated to transfer the volatiles to the chromatographic column.

This technique is useful when a solid sample cannot be dissolved or heated above a transition temperature. However, this exhaustive extraction can be time-consuming.

IV. APPLICATIONS

A. Separation of Enantiomers

It has long been recognized that biological activity of certain chiral compounds varies and is related to their stereochemistry. Biological activity of certain enantiomers can vary dramatically and not only be biologically active but toxic as well. It is for these reasons that the separation of enantiomers is so important (see additional discussion in Chapter). There are two approaches to the separation of enantiomers by GC.

The first is the use of chiral derivatizing reagents followed by separation of the resulting diastereoisomers on a nonchiral column. In this approach, the chiral reagent must be both chemically and optically pure. The material must be carefully characterized in terms of enantiomeric purity and must not exhibit racemization during storage. In Table 4 for a racemic mixture containing the (R) and (S) enantiomers, if the chiral reagent containing (R′) and trace of (S′) is reacted with the racemic mixture, several products will be formed.

Table 4.4 Quantitation of Enantiomeric Separations Using Chiral Reagents on a Nonchiral Column

Racemic mixture:	R	R
Chiral reagent:	R′	S′
Products formed:	RS′ + RR′	SS′ + SR′
Peaks separated:	—	—
Peak 1 components:	RS′ + SR′	—
Peak 2 components:	RR′ + SS′	—

To quantitate the percentage of S in the racemic mixture:

$$\%S = \frac{P(A1 + A2) - A1}{(A1 + A2)(2P - 1)} \times 100$$

where

P = purity of the chiral reagent, expressed as (percent/100)
A1 = area of the first peak
A2 = area of the second peak

The (R) compound will react with the reagent to form (RS′) and (RR′), whereas the (S′) portion of the racemic mixture will react with the reagent to form (SS′) and (SR′). Because the separation is carried out on a nonchiral column, only two peaks will be apparent; that is, (RS′) and (SR′) will coelute, and (RR′) and (SS) will coelute. The percentage of the S component in the mixture is then determined by the formula shown in Table 4.4.

From the formula shown in Table 4.4, the optical purity of the chiral derivatizing reagent is very important. If the reagent is 99% pure, the minimum detectable trace enantiomer is approximately 0.3%. The choice of chiral reagent is very important because it must impart a sufficient difference in functionality to the enantiomers to resolve the diasteroisomer products formed. Second, the reaction must be both quantitative and produce stable derivatives resistant to racemization. A good practice to confirm the identity of the peaks after reaction with chiral reagents is to react a single sample of high optical purity with both (R) and (S) chiral reagents. Suppose there is a sample which is predominantly (S) and react it with a (S) chiral reagent. The major peak should be the coelution of (SS) and (RR). If this same sample is then reacted with (R) reagent, the major peak is composed of (SR) and (RR). Thus, the major peaks should reverse in elution order and confirm the correct peak.

There are several types of chiral derivatizing reagents commonly used depending on the functional group involved. For amines, the formation of an amide from reaction with an acyl halide [147,148], chloroformate reaction to form a carbamate [149], and reaction with isocyanate to form the corresponding urea are common reactions [150]. Carboxyl groups can be effectively esterified with chiral alcohols [151–153]. Isocynates have been used as reagents for enantiomer separation of amino acids, *N*-methylamino acids, and 3-hydroxy acids [154]. In addition to the above-mentioned reactions, many others have been used in the formation of derivatives for use on a variety of packed and capillary columns. For a more comprehensive list, refer to References 155–159.

The second general approach and in most instances the preferred method of enantiomeric separation is the use of nonchiral derivatization reagents followed by separation on a chiral stationary phase. This direct method allows the analyst a greater selection of derivatizing reagents, consequently making method development easier. The derivatizing reagents do not have to be as stringently characterized and monitored for enantomeric purity changes. More importantly, the reaction need not be quantitative. The disadvantage of this approach to enantomeric separation is that most chiral stationary phases have low upper temperature limits (200–240°C max). Therefore, one must choose a derivative that will not only allow for the introduction of the functionality for separation of entantiomers but also produce a derivative with volatility within the operational range of the column. Quantitation in these cases is much easier because the enantiomers are directly resolved.

There have been many reported chiral stationary phases for use in both packed and capillary gas chromatography. Most of these phases are of the carbonyl-bis-L-valine isopropyl ester, diamide, and peptide phase types. The most common phase is Chirasil-Val from Alltech Applied Science Laboratories (State College, PA). This phase is ideal for the separation of a variety of enantiomers including amino acids, sugars, amines, and peptides. The phase is composed of L-valine-tert-butylamide linked through a caroxamide group to a polysiloxane backbone every seven dimethylsiloxane units apart.

B. Excipients, Preservatives, and Pharmaceuticals

The last half of this book includes an extensive listing of gas chromatographic methods used to analyze pharmaceuticals and excipients in a wide variety of formulations. Additional applications are listed in Table 4.5.

C. Headspace Analysis

Ethylene Oxide – Single Stage

Romano et al. [186] developed a headspace method for the analysis of residual ethylene oxide in sterilized materials. A weighed portion of sample was heated at 100°C for 15 min. Duplicate headspace samples were removed with a gas-tight syringe (no differences were found between hot and cold sampling) and injected into a column packed with Porapak R. A flame-ionization detector was used and the results of the two injections were averaged. The vial was purged, recapped, and reheated under the above conditions. Duplicate samples were again withdrawn and analyzed. The sum of the two averages represented the ethylene oxide content of the sample. An external standard was used for calibration (Figure 4.1).

Samples of materials which were sterilized by ethylene oxide were halved; one portion was analyzed by the headspace method and the other by an extraction method using dimethylformamide. Good agreement was obtained between the methods with the exception of cotton which was neither swelled nor dissolved by dimethylformamide. The headspace method for cotton gave considerably higher values than did the extraction method (i.e., 494 ppm versus 325 ppm) even when the extraction was carried out over a 3-day period. The authors speculated that when undissolved

Table 4.5 Gas Chromatographic Methods
Used in the Analysis of Pharmaceuticals and
Excipients

Analyte	Reference
Alkaloids	159
Antiarrhythmics	164
Antibiotics	160, 161
Antidepressants	162, 163
Antiepileptics	165
Arsenic	166
Carbohydrates (glycoproteins)	167
Creams	168
Cytotoxic Drugs	169
EDTA	170
Fatty acids	171
General	172–174
Germicidal Phenols	175
Iodide	176
Lithium	177
Parabens	178
Psychotropics	179
Residual Solvents	180
Steroids	181, 182
Stilbesterols	183
Surfactants	184
Vitamins	185

polar solids are being extracted, ethylene oxide partitions between the solid and liquid phases and an equilibrium are established. The precision of the method was determined by checking paired polyester halves against each other. An average deviation of 3.2 ppm between paired halves was found at levels of 60–84 ppm.

Gramiccioni et al. [187] reported the determination of residual ethylene oxide in sterilized polypropylene syringes and in materials such as plasticized PVC, polyurethane, and para rubber. The sterilized object was cut into small pieces, weighed, and placed into a flask containing *N,N*-dimethylacetamide (DMA). The flask was capped and shaken to make the sample homogeneous. After 24 hr it was shaken again and a sample was

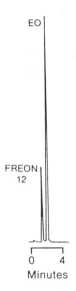

Figure 4.1 Analysis of a polyester sample for ethylene oxide (EO) showing Freon 12 used as a diluent in the sterilization process. (From Reference 186.)

transferred to a vial which was subsequently sealed. The vial was thermostated at 65°C for 1 hr to reach equilibrium. Headspace analysis was conducted using a 24% FFAP on 100–110 mesh Anakrom column at 50°C. The detection limit of 0.1 μg EO/ml DMA corresponded to 2 μg/g of sterilized object. Recovery over the range of the procedure of 0.1–0.2 μg EO/ml DMA was 98% (Figure 4.2).

Bellenger et al. [188] analyzed ethylene oxide in nonreusable plastic medical devices using methanol as an internal standard. A 200-mg sample, cut into small pieces, was treated with dimethylformamide and mixed with methanol internal standard in a vial. The sample was heated at 100°C for 10 min and the headspace gases chromatographed on a Chromosorb 102 column at 140°C. The analysis range was linear up to 100 ppm. For levels less than 10 ppm, the results agreed satisfactorily with those obtained by a colorimetric method. For levels greater than 10 ppm, the headspace technique yielded values greater than those of the colorimetric method. The authors explained that this variance was due to saturation of the scrubbers utilized in the trap used in the colorimetric method. The solubility of the test material also affected the result by salting out the methanol. It was necessary, therefore, to limit the amount of sample to 200 mg.

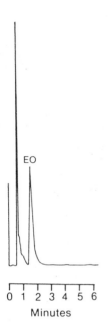

EO

0 1 2 3 4 5 6
Minutes

Figure 4.2   Chromatogram of an ethylene oxide (EO) sterilized sample. (From Reference 187.)

In another investigation, ethylene oxide in polyvinylchloride was determined by dissolving 65 mg of sample in 1 ml of dimethylacetamide [189]. Headspace analysis was conducted on a glass column packed with Porapak T under isothermal conditions. The solvent was removed by back-flushing. An external standard was used for calibration. A vinylchloride monomer was also detected in this analysis (Figure 4.3).

A statistical evaluation of methods using headspace gas chromatography for the determination of ethylene oxide in plastic surgical items was performed by Kaye and Nevell [190]. Two methods were evaluated: an external standard method using ethylene oxide in air, and an internal standard method using a dilute aqueous solution with acetone as the internal standard. Carbowax 20M (10%) on a Chromosorb column at 120°C was used for the external standard method. For the internal standard method, a Chromosorb 101 (80–100 mesh) column was used at 125°C. Sealed vials, empty or containing preweighed plastic samples, were evacuated, and portions of calibrating solution or internal standard solution were introduced. Each vial was placed in a heated block at 120°C for 10 min, and a sample of headspace gas was drawn for analysis. For the external standard method,

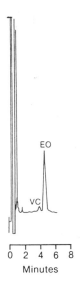

0   2   4   6   8
Minutes

Figure 4.3   Analysis of ethylene oxide (EO) in a PVC sample. (From Reference 189.)

each vial containing a weighed plastic sample was placed in an air-circulating oven at 120°C for 15 min. A sample of headspace was taken immediately after removing the vial from the oven. Studies were conducted on high-level (80 μg/g) and low-level (12 μg/g) materials. The two methods gave similar results. Determinations using either method were reliable to within 3% for residual levels of 80 μg/g or 7% for residual levels of 12 μg/g.

Residuals

Boyer and Probecker [191] determined organic solvents in several pharmaceutical forms using a Perkin-Elmer HS-6 headspace sampler. Typically, the samples were heated at 90°C for 10 min to establish equilibrium. Headspace samples were injected onto a Chromosorb 102 column. Ten injections of a mixed ethanol–acetone standard using methanol as the internal standard gave better precision than manual injections as measured by the relative standard deviation; 1.63% and 2.48% for ethanol and acetone, respectively, using the sampler as compared to 4.77% and 3.93% by manual injection, respectively. Methods were reported for acetone and ethanol in dry forms such as tablets and microgranules, ethanol of crystallization in raw materials, and ethanol in syrups. Denaturants such as *n*-butanol and isopropanol in ethyl alcohol were determined using ethyl acetate as the internal standard.

Litchman and Upton [192] reported the determination of triethylamine in streptomycin sulfate and in methacycline hydrochloride to levels as low as 0.05%. A weighed sample was treated with 1M sodium hydroxide solution at 60°C for 1 hr. A headspace sample was manually withdrawn and analyzed on a polystyrene column at 160°C using a flame-ionization detector. The levels of triethylamine found ranged from 0.15% to 0.36% for streptomycin sulfate and from 0.06% to 0.13% for methacycline hydrochloride. Recoveries were better than 94%. The precision of the determination, based on five replicate weighings of sample, was 2% for streptomycin sulfate and 5% for methacycline hydrochloride.

Bicchi and Bertolino [193] analyzed a variety of pharmaceuticals for residual solvents. Samples were equilibrated directly or dissolved in a suitable solvent with a boiling point higher than that of the residual solvent to be determined. Equilibration conditions were 90 or 100°C for 20 min. A Perkin-Elmer HS-6 headspace sampler was used. The chromatographic phase chosen was a 6′ × ⅛ in. column packed with Carbopack coated with 0.1% SP 1000. Residual ethanol in phenobarbital sodium was determined by a direct desorption method. An internal standard, *t*-butanol, was used. Typically, 0.44% of ethanol was detected (compared to a detection limit of 0.02 ppm). The standard deviation of six determinations was 0.026. Pharmaceutical preparations which were analyzed by the solution method included lidocaine hydrochloride, calcium pantothenate, methyl nicotinate, sodium ascorbate, nicotinamide, and phenylbutazone. Acetone, ethanol, and isopropanol were determined with typical concentrations ranging from 14 ppm for ethanol to 0.27% for acetone. Detection limits were as low as 0.03 ppm (methanol in methyl nicotinate).

Ethanol

Kojima [194] reported the determination of ethanol in various samples of tinctures. The sample was dissolved in *n*-propanol (as the internal standard) at a concentration of 1.0–5.0% (v/v). A portion was equilibrated at 50°C, and 2 ml of the headspace gas was manually injected onto a column of either 5% polyethylene glycol 20 M on Chanelite CS (60–80) mesh and was assayed using a flame-ionization detector. Interfering peaks were not detected in the five tinctures studied. Ethanol contents ranged from 65% to 90% and results were in good agreement with those obtained by conventional methods. Kojima [195] subsequently expanded the method to the determination of ethanol in a variety of liquid and solid drug forms where the content ranged from 2% to 73%.

1-Menthol, d,l-Camphor, and Methyl Salicylate

Nakajima and Yasuda [196] have successfully applied headspace gas chromatography to the analysis of 1-menthol, *d,l*-camphor, and methyl salicylate. Sample portions with ethyl salicylate as internal standard were added in 1-ml measures to 50 ml of 30% ethanol in a 100-ml vial, which was subsequently sealed. After shaking for 30 min, the vial was equilibrated in a constant-temperature water bath at room temperature for 30 min. Headspace gas (1 ml) was withdrawn with a gas-tight syringe and injected onto a 1.5 m × 3 mm Gaschrom Q (80–100 mesh) column coated with 2% DCQF-1 and 1.5% OV-17. Standard solutions were analyzed in a similar manner.

When 1 ml of gas was injected seven times from a single vial of mixed standard and internal standard solution, the coefficients of variation of the peak heights were 3.18%, 2.96%, and 0.85% for 1-menthol, camophon, and methyl salicylate, respectively.

When 1 ml of gas was injected from each of seven vials, the coefficients of variation of the peak heights were 4.64%, 2.24%, and 0.71% for 1-menthol, camphor, and methyl salicylate, respectively.

Recoveries were better than 97% in a variety of preparations. The authors found that the method could not be applied to samples containing castor oil.

Choline

Sauceman et al. [197] reported the determination of choline, (β-hydroxethyl)-trimethylammonium hydroxide, in liquid and powder formula products. Ethyl ether was added to the sample as an internal standard. The sample was digested under alkaline conditions for 24 hr at 120°C. Under these conditions, choline undergoes the Hofmann elimination reaction to form trimethylamine. The equilibrated headspace was sampled and analyzed on a 28% Pennwalt 223 + 4% KOH on Gas Chrom R column. A typical chromatogram is shown in Figure 3. Ten replicate injections of headspace gas from a single standard gave a relative standard deviation of 3.21%. No interfering peaks were produced when samples were digested in the absence of potassium hydroxide. Other quaternary N compounds did not interfere with the analysis. Typical RDSs on the analysis of 10 replicate samples of Enfamil and Pro Sobee were 9.6% and 7.7%, respectively. Sampling was performed by both a manual method and an automated method using a headspace sampler at 40°C. Better method precision was obtained with the automated method (CV of 1.8%) than with the manual method (CV of 3.2%). A throughput of up to 400 samples a week was possible with the automated method.

Camphor

Ettre et al. [198] reported the determination of camphor in an ointment using the method of standard additions. Camphor, 5 mg, was added to a 1-g sample of a rub-in ointment. Volatiles were chromatographed on a

Carbowax 20M column (Figure 4.4). The concentration of camphor in the sample was 1.1%.

Dimethylnitrosamine

The determination of trace levels of dimethylnitrosamine (DMNA) in pharmaceuticals containing aminophenazone has been reported [199]. A tablet was pulverized and suspended in a headspace vial in a solution of $2M$ H_2SO_4 (to remove volatile amines) to which had been added solid potassium sulfate (for a salting-out effect). The vial was heated at 120°C for 1 hr. Headspace gases were injected onto a 5% Carbowax 10M on Chromosorb G, AW–DMCS column and detected using a nitrogen phosphorous detector (Figure 4.5). Calibration was carried out by the method of addition. The detection limit using this method was 20–40 ppb. A typical level of DMNA found was 75 ppb.

Oxygen

Lowering of oxygen levels is one way to increase the shelf life of pharmaceutical products. Lyman et al. [200] developed a method for the determination of oxygen in both aqueous and nonaqueous products. The method was applied to liquids and to solids with a melting point of 75°C or less. A known amount of sample (2–3 g) in a 20-ml vial was first purged in an ice-water bath. The sample was then heated at 75°C with stirring and degassed

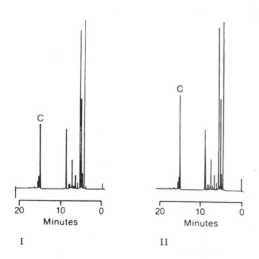

Figure 4.4 Determination of camphor (C) in a rub-in ointment. I — sample; II — sample plus 5 mg camphor. (From Reference 198.)

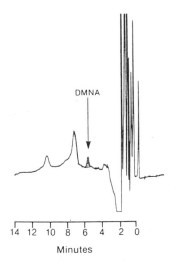

Figure 4.5 Analysis of 75 ppb dimethylnitrosamine (DMNA) in a tablet containing aminophenazone. (From Reference 199.)

for 6 min (glass-coated stirring bars were most effective; Teflon contributed oxygen to the system). The headspace was then purged onto a molecular sieve column and analyzed using a thermal conductivity detector. The purge time was carefully controlled at 50 sec to get the maximum amount of oxygen onto the column with the least amount of tailing and band broadening.

The volume of the sample vial was chosen as 20 ml to ensure a good purge efficiency and to handle samples of 2–3 g. A calibration curve was obtained by analyzing sample vials which had been spiked with known amounts of oxygen gas. The accuracy of the method was determined by analyzing air-saturated water and comparing results with literature data on the amount of oxygen in water at known temperatures and pressure. The amount of oxygen found averaged about 90% of the theoretical value. The quantitation limit was 1 ppm for a 3-g sample. Precision of the method depended on the type of sample. Air-saturated water produced a coefficient of variation of about 4%. The method was developed to solve a problem with a cream product. Each of the ingredients was analyzed for oxygen. Totaling the contribution of the four ingredients gave a theoretical oxygen concentration of about 12 ppm. The final product assayed between 25 and 30 ppm. The source of the higher levels was believed to be due to air bubbles trapped in the cream. Each step of the manufacturing process was monitored using this method to isolate trouble spots. The final product level ultimately was reduced to 5 ppm, which resulted in an increase of shelf life to more than 2 years.

Fyhr et al. [201] reviewed several commercially available oxygen analyzers intended for the analysis of oxygen in the headspace of vials. However, preliminary validation revealed insufficient reproducibility and linearity. The authors developed headspace analysis systems. Sample volumes down to about 2.5 ml could be used without significant errors. Sample recovery was in the range 100–102%. It was necessary to measure the headspace pressure and volume in order to be able to present the assay in partial oxygen pressure or in millimoles of oxygen. Up to 40 vials per hour could be analyzed using this technique.

Ethylene Oxide – Multiple Stage

Multiple static extractions can be conducted until the sample is exhaustively extracted. The result is the sum of the individual extractions. This, however, can be a time-consuming process.

Kolb and Pospisil [202] have shown that quantitative results can be obtained after several extractions because the extraction follows an exponential relationship. This approach has been termed discontinuous gas extraction. These workers determined the amount of ethylene oxide in a sample of sterilized gloves. Volatiles were chromatographed on a Chromosorb 102, 60–80 mesh column using a flame-ionization detector. A typical chromatogram is shown in Figure 4.6. The calculated amount of ethylene oxide (four extractions) was 5.4 ppm.

Kolb [203] describes a stepwise gas-extraction procedure called multiple headspace extraction (MHE). Using this method, Kolb found that the determination can be performed with only two extractions. The volume of the sample was compensated for by adding a similar volume of an inert material such as glass beads. Ethylene oxide in surgical silk sutures was determined by this procedure. The extrapolated total area (four steps) was nearly identical to the total area value obtained using the two-step MHE process, 184 versus 183, respectively.

Residual Solvents – Multiple Stage

Methylene chloride in a tablet was analyzed by Kolb [203] using the multiple headspace extraction method (three steps). The sample was analyzed as a dry powdered material using a glass capillary column, Marlophen 87, isothermally at 35°C. A concentration of 35 ppm was found, which was in reasonable agreement with that obtained (40 ppm) when the sample was dissolved in water and analyzed by normal headspace analysis using the method of standard addition for quantitation. The extrapolated total area

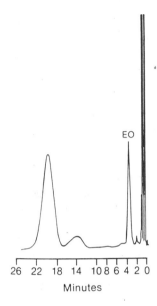

EO

26 22 18 14 10 8 6 4 2 0

Minutes

Figure 4.6 Headspace analysis of ethylene oxide (EO) from sterilized gloves. (From Reference 203.)

(four steps) was similar to the total area value obtained using the two-step MHE process.

Residuals—Dynamic Sampling

Wampler et al. [204] used dynamic headspace analysis to determine the presence of three types of volatile materials in pharmaceuticals: naturally occurring volatiles in raw materials, processing agents, and decomposition products due either to the chemical instability of the compound or to bacterial action. Thermal desorption was accomplished using a Chemical Data Systems Model 320 sample concentrator with Tenax traps. A capillary gas chromatograph equipped with a 50 m × 0.25 mm fused silica capillary column (SE-54) and a flame-ionization detector was utilized. Aspirin samples which are either past their prime or which have been improperly stored degrade to give acetic acid which produces a vinegary smell. One crushed aspirin tablet was subjected to analysis at desorption temperatures ranging from room temperature to 70°C (Figure 4.7). The authors subjected a powdered pharmaceutical to analysis for residual solvents. The most intense peak in the chromatogram was toluene which was used as a solvent in the manufacturing process. The amount of toluene was quantitated using

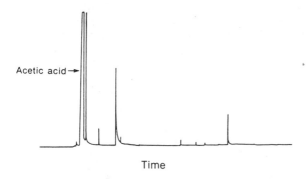

Acetic acid →

Time

Figure 4.7   Volatiles in a buffered aspirin tablet. (From Reference 204.)

benzene as an internal standard. One microliter of benzene (1% in methanol) was added to the powdered sample before analysis. In the example
given, a toluene concentration of 0.0086% was found. Spiked samples
showed the recovery of toluene to be 95%. The method gave a relative
standard deviation of 1.5%.

Letavernier et al. [205] analyzed residual solvents from processing
operations in film-coated tablets. They also determined residual solvents
which arise from migration from packaging materials into pharmaceutical
products. A weighed sample (35 mg to 1 g) was heated and volatiles swept
with nitrogen gas onto a Tenax trap refrigerated with liquid nitrogen. After
a specified time, the Tenax trap was rapidly heated (maximum of 300°C)
to desorb volatiles which were swept onto a Porapak Q column. The authors were able to fingerprint solvents from several types of coated tablets.
Cyclohexanone could be detected at a level of 0.2 mg/g of sample.

Characterization of Bacteria

Zechman et al. [206] characterized pathogenic bacteria by analysis of headspace volatiles. Cultures of microorganisms were heated with magnetic stirring for 20 min of 37°C. Volatiles were swept onto a Tenax trap. The
organics were desorbed at 200–220°C onto a CPWAX-57CB chemically
bonded high-capacity fused silica capillary column with a 1.3-μm film
thickness. A flame-ionization detector was used. Anisole was added to each
culture before analysis to serve as an internal standard for calculating relative retention times and to monitor transfer efficiency. Chromatograms
were found to be reproducible in retention times and relative appearance of
the profiles. Volatile bacterial metabolites consisted of three to six major
constituents. The prominent constituents produced by several strains of
bacteria were as follows:

Staphylococcus aureus: isobutanol, isopentanol, and acetone
Pseudomonas aeruginosa: isobutanol, butanol, and isopentanol
Pseudamonas mirabilis and *Klebsiella pneumoniae*: isobutanol, isopentyl
   acetate, and 9 isopentanol

Heated Thiamine Solutions

Reineccius and Liardon [207] studied volatiles evolved from heated thiamine solutions. Samples of 2% thiamine hydrochloride in various $0.2M$ buffers were heated under various conditions. A temperature of 40°C and a sampling time of 45 min were found to minimize artifact formation and yet produce sufficient volatiles for analysis. Nitrogen was used as the purge gas at a flow rate of 50 ml/min. Several materials were evaluated as absorbents, with graphite found to be the optimum. A microwave desorption system was used to rapidly desorb the trapped volatiles onto a fused silica capillary column. Twenty-five compounds were identified in the headspace of the heated thiamine solutions.

Organic Volatile Impurities

The United States Pharmacopeia (USP) test ⟨467⟩ describes three different approaches to measuring organic volatile impurities in pharmaceuticals. Method I uses a wide-bore coated open tubular column (G-27, 5% phenyl–95% methylpolysiloxane) with a silica guard column deactivated with phenylmethyl siloxane and a flame-ionization detector. The samples are dissolved in water and about 1 μl is injected. Limits are set for benzene, chloroform, 1,4-dioxane, methylene chloride, and trichloroethylene. Methods V and VI are nearly identical to method I except for varying the chromatographic conditions. For the measurement of methylene chloride in coated tablets, the headspace techniques described above are recommended.

V. CONCLUSION

The use of capillary columns is becoming increasingly common particularly for the resolution of very complex mixtures. Gas chromatography has found its niche in the monitoring of certain impurities, measuring and characterizing excipients, preservatives, and active drugs. In assays where sensitivity is required, gas chromatographic methods are still unsurpassed.

   Chapter 8 is a comprehensive listing of published GC methods that have been used in the assay and identity of drug products, and several excellent reviews [208–214] should also be consulted for more detailed discussions.

REFERENCES

1. L. Blomberg, Stationary phases for capillary gas chromatography, *Trends Anal. Chem.*, *6*(2):41–45 (1987).
2. J. Hubbal, P. MiDauro, et al., Developments in crosslinking of stationary phases for capillary gas chromatography by cobalt-60 gamma radiation, *J. Chromatogr. Sci.*, *22*:185–191 (1984).
3. B. Jones, K. Markides, et al., Contemporary capillary column technology for chromatography, *Chromatogr. Forum*, 38–44 (May–June 1986).
4. B. Tarbet, J. Bradshaw, et al., The chemistry of capillary column technology, LC-GC, 6:3, 232–248 (1988).
5. P. Silvis, J. Walsh, et al., Application of bonded Carbowax capillary GC columns, *Am. Lab.*, 41–47 (February 1987).
6. L. Sojak, I. Ostrovsky, et al., Separation and identification of C_{14}–C_{17} alkylbenzenes from dehydrogenation of n-alkanes by capillary gas chromatography using liquid crystals as the stationary phase, *Ropa. Uhlie*, *25*(3):149–157 (1983).
7. J. Kramer, R. Fouchard, et al., Difference in chromatographic properties of fused silica capillary columns, coated, crosslinked, bonded or crosslinked and bonded with polyethylene glycols (Carbowas 20 M) using complex fatty acid methyl ester mixtures, *J. Chromatogr. Sci.*, *23*:54–56 (1985).
8. C. Pouse, A. Finlinson, et al., Comparison of oligo (ethylene oxide)-substituted polysiloxane with poly-ethylene glycol) as stationary phases for capillary gas chromatography, *Anal. Chem.*, *60*:901–905 (1988).
9. B. Richter, J. Kuei, et al., Polysiloxane stationary phases containing tolyl and cyanopropyl groups: oxidation during cross-linking, *J. Chromatogr.*, *279*:21 (1983).
10. B. Richter, J. Kuel, et al., Nonextracable cyanopropyl poly-siloxane stationary phases for capillary chromatography, *Chromatographia*, *17*:570 (1983).
11. B. Jones, J. Kuei, et al., Characterization and evaluation of cyanopropyl polysiloxane stationary phases for gas chromatography, *J. Chromatogr.*, *298*:389 (1984).
12. F. David, P. Sandra, et al., OH-terminated cyanopropyl silicones, *HRC&CC J. High Res. Chromatogr. Chromatogr. Commun.*, *11*: 256–263 (1988).

13. J. Szejtli, *Cyclodexdins and Their Inclusion Complexes*, Akademiai Kiado, Budapest, Hungary.
14. K. Markides, M. Nishioka, et al., Smectic biphenylcarboxylates ester liquid crystalline polysiloxane stationary phase for capillary gas chromatography, *Anal. Chem.*, *57*:1296–1299 (1985).
15. K. Markides, H. Chang, et al., Smectic biphenylcarboxylates ester liquid crystalline polysiloxane stationary phases for capillary gas chromatography, *HRC&CC J. High Res. Chromatogr. Chromatog. Commun.*, *8*:516–520 (1985).
16. S. Rokushika, K. Naikwadi, et al., Liquid crystal stationary phases for gas chromatography and supercritical fluid chromatography, *HRC&CC J. High Res. Chromatogr. Chromatog. Commun.*, *8*:480–484 (1985).
17. G. Alexander, Z. Juvancz, et al., Cyclodextrins and their derivatives as stationary phases in GC capillary columns, *HRC&CC J. High Res. Chromatogr. Chromatogr. Commun.*, *11*:110–113 (1988).
18. Z. Juvancz, G. Alexander, et al., Permethylated β-cyclodextrin as stationary phase in capillary gas chromatography, *HRC&CC J. High Res. Chromatogr. Chromatogr. Commun.*, *10*:105–107 (1987).
19. E. Smolkova-Keulemansova, E. Newmannova, et al., Study of the stereospecific properties of cyclodextrins as gas-solid chromatographic stationary phases, *J. Chromatogr.*, *365*:279–288 (1986).
20. W. Hinze, *Separ. Purif. Meth.*, *10*(2):159–237 (1981).
21. J. Beadle, Analysis of cyclodextrin mixtures by gas chromatography of their dimethylsilyl ethers, *J. Chromatogr.*, *42*:201–206 (1969).
22. J. Zeeuw, L. Henrich, et al., Absorption chromatography on PLOT (porus layer open tubular) columns: a new look at the future of capillary GC, *J. Chromatogr. Sci.*, *25*:71–82 (1987).
23. L. Ghaoui, E. Dessai, et al., Analysis of perfluoroalkanes and fluorobenzenes using an aluminum oxide porous-layer open tubular column, *Chromatographia*, *20*(2):75–78 (1985).
24. J. de Zeeuw, R. C. de Nijs, et al., PoraPLOT Q: A porous layer open tubular column coated with styrene-divinylbenzene copolymer, *HRC&CC J. High Res. Chromatogr. Chromatogr. Commun.*, *11*: 162–167 (1988).
25. J. Gillis, R. Sievers, et al., Selective retention of oxygen using chromatographic columns containing metal chelate polymers, *Anal. Chem.*, *57*:1572–1577 (1985).
26. P. Silvis, Restek Corporation, personnel communication.
27. K. Grob, H. Neukom, et al., Length of the flooded zone in the column inlet and evaluation of different retention gaps for capillary

gas chromatography, *HRC&CC J. High Res. Chromatogr. Chromatogr. Commun.*, *7*:319–326 (1984).

28. K. Grob Jr., G. Karrer, et al., On-column injection of large sample volumes using the retention gap technique in capillary gas chromatography, *J. Chromatogr.*, *334*:129–155, (1985).

29. F. Yang, A. Brown, et al., Splitless sampling for capillary gas chromatography, *J. Chromatogr.*, *158*:91–109 (1978).

30. K. Grob, *Classical Split and Splitless Injection in Capillary Gas Chromatography, Part B*, Heidelberg, Basel, pp. 97–212.

31. R. Snell, J. Danielson, et al., Parameters affecting the quantitation performance of cold on-column and splitless injection systems used in capillary gas chromatography, *J. Chromatogr. Sci.*, *25*:225–230 (1987).

32. K. Grob and K. Grob, Jr., Isothermal analysis on capillary columns without stream splitting, *J. Chromatogr.*, *94*:53–64 (1974).

33. K. Grob, Jr., Solvent effects in capillary gas chromatography, *J. Chromatogr.*, *279*:225–232 (1983).

34. K. Grob Jr., Solvent trapping in capillary gas chromatography, *J. Chromatogr.*, 253:17–22 (1982).

35. W. Cretney, F. McLaughlin, et al., Implications of the solvent effect in quantitative capillary gas chromatography of minor constituents in mixtures, *HRC&CC J. High Res. Chromatogr. Chromatogr. Commun.*, *10*:428–434 (1987).

36. K. Grob and K. Grob, Jr., Splitless injection and the solvent effect, *HRC&CC J. High Res. Chromatogr. Chromatogr. Commun.*, *1*:57–64 (1978).

37. K. Grob, Classical split and splitless injection. In *Capillary Gas Chromatography, Part A*, Heidelberg, Basel, pp. 8–89.

38. R. Miller and W. Jennings, Normal and reverse solvent effects in split injection, *HRC&CC J. High Res. Chromatogr. Chromatogr. Commun.*, *2*:72–73 (1979).

39. L. Ghaoui, Solvent effect in split injection in capillary gas chromatography, *HRC&CC J. High Res. Chromatogr. Chromatogr. Commun.*, *11*:410–413 (1988).

40. F. Poy, L. Cobelli, Sample introduction. In *Gas Chromatography*, Heidelberg, Basel, pp. 77–97.

41. K. Grob, Classical split and splitless injection. In *Capillary Gas Chromatography*, *Part C*, Heidelberg, Basel, pp. 219–278.

42. J. Hinshaw, Programmed temperature injection in open tubular gas chromatography, *Instrum. Res.*, 14–24 (1985).

43. H. Stan and H. Muller, Evaluation of automated and manual hot splitless, cold-splitless (PTV) and on-column injection technique us-

ing capillary gas chromatography for the analysis of organophospho-
rus pesticides, *HRC&CC J. High Res. Chromatogr. Chromatogr.
Commun.*, *11*:140–143 (1988).

44. K. Grob, T. Laubli, et al., Splitless injection-development and state
of the art, *HRC&CC J. High Res. Chromatogr. Chromatogr. Commun.*, *11*:462–470 (1988).

45. G. Schomburg, H. Bahiau, et al., Sampling techniques in capillary
gas chromatography, *J. Chromatogr.*, 142:87–102 (1977).

46. D. McMahon, A collaborative study to evaluate quantitation utilizing different injection modes for capillary GC, *J. Chromatogr. Sci.*,
23:137–143 (1985).

47. J. Purcell, Quantitative capillary gas chromatographic analysis,
Chromatographia, *15*:546–558 (1982).

48. J. Hinshaw, Capillary inlet systems for gas chromatographic trace
analysis, *J. Chromatogr. Sci.*, *26*:142–145 (1988).

49. B. Middleditch, A. Ziatkis, et al., Trace analysis of volatile polar
organics: problems and prospects, *J. Chromatogr. Sci.*, *26*:150–152
(1988).

50. Hinshaw, Modern inlets for capillary gas chromatography, *J. Chromatogr. Sci.*, *25*:49–55 (1987).

51. J. Settlage and H. Jaeger, Advantages of fused silica capillary gas
chromatography for GC/MS applications, *J. Chromatogr. Sci.*, *22*:
192–197 (1984).

52. C. White and R. Houck, Supercritical fluid chromatography and
some of its applications: a review, *HRC&CC J. High Res. Chromatogr., Chromatogr. Commun.*, *9*:4–17 (1986).

53. D. Later, B. Richter, et al., Capillary supercritical fluid chromatography: instrumentation and applications, *Am. Lab.*, (August 1986).

54. J. Fjeldsted and M. Lee, Capillary supercritical fluid chromatography, *Anal. Chem.*, *56*(4): 619A–628A (1984).

55. D. Gere, Supercritical fluid chromatography, *Science*, *222*:253–259
(1983).

56. S. Fields, R. Kong, et al., Effect of column diameter on efficiency in
capillary supercritical fluid chromatography, *HRC&CC J. High Res.
Chromotogr. Chromatogr. Commun.*, *7*:312–318 (1984).

57. S. Springston, P. David, et al., Stationary phase phenomena in capillary supercritical fluid chromatography, *Anal. Chem.*, *58*:997–1002
(1986).

58. P. Grimsrud and C. Valkenburg, New schemes for the electron capture sensitization of aromatic hydrocarbons, *J. Chromatogr.*, *302*:
243–256 (1984).

59. C. F. Poole, The electron capture detector in capillary column gas

chromatography, *HRC&CC J. High Res. Chromatogr. Chromatogr. Commun.*, *5*:454–471 (1982).

60. E. Pellizzari, Electron capture detection in gas chromatography, *J. Chromatogr.*, *98*:323–361 (1974).
61. H. Bente, Electron capture detectors, *Am. Lab.* (August 1978).
62. A. Ziatkis and C. Poole, Derivatization techniques for the electron capture detector, *Anal. Chem.*, *52*:1002–1016 (1980).
63. C. Poole and S. Poole, Derivatization as an approach to trace analysis by gas chromatography with electron capture detection, *J. Chromatogr. Sci.*, *25*: (1987).
64. J. Yeung, G. Baker, et al., Simple automated gas chromatographic analysis of amino acids and its application to brain tissue and urine, *J. Chromatogr.*, *378*:293–304 (1986).
65. D. Chambers, Extractive derivatization of primary amines and beta-aminoalcohols with aromatic aldehydes to form Schiff base or oxazolidine products for analysis by electron capture gas chromatography, *Diss. Abstr. Int. b.* *44*(7):2148, (1984).
66. J. Wickramasinghe, W. Morozowich, et al., Detection of prostaglandin F_2 as pentafluouobenzyl ester by electron capture GLC, *J. Pharm. Sci.*, *62*:1428–1431 (1973).
67. A. Sedman and J. Gal, Simultaneous determination of the enantiomers of tocainide in blood plasma using gas liquid chromatography with electron capture detection, *J. Chromatogr.*, *306*:155–164 (1984).
68. I. Martin, G. Baker, et al., Gas chromatography with electron capture detection for measurement of bioactive amines in biological samples, *Methodol. Surv. Biochem. Anal.*, *14*:331–336 (1984).
69. M. Bathala, S. Weinstein, et al., Quantitative determination of captopril in blood and captopril and its disulfide metabolites in plasma by gas chromatography, *J. Pharm. Sci.*, *73*:3 (1984).
70. M. Berens and S. Salmon, Quantitative analysis of prostaglandins in cell culture medium by high resolution gas chromatography with electron capture detection, *J. Chromatogr.*, *307*:251–260 (1984).
71. M. Jemal, E. Ivaskiv, et al., Determination of a thromboxane Z_2 recpetor antagonist in human plasma by capillary gas chromatography with electron capture detection, *J. Chromatogr.*, *381*:424–430 (1986).
72. M. Jemal and A. Cohen, Determination of S-benzoyl captopril in human urine by capillary gas chromatography with electron capture detection, *J. Chromatogr.*, *342*:186–192 (1985).
73. I. DeLeon, J. Brown, et al., Trace analysis of 2,6-disubstituted ani-

lines in blood by capillary gas chromatography, *J. Anal. Toxicol.*, *7*: 185–197 (1985).

74. P. Edlund, Determination of opiates in biological samples by gas capillary chromatography with election capture detection, *J. Chromatogr.*, *206*:109–116 (1981).

75. H. Maksoud, S. Kuttab, et al., Analysis of N-propylinorapomorphines in plasma and tissue by capillary gas chromatography-electron capture detection, *J. Chromatogr.*, *274*:149–159 (1983).

76. C. Chriswell, R. Chang, et al., Chromatographic determination of phenols in water, *Anal. Chem.*, *47*:1325–1329 (1975).

77. T. Edgerton, R. Moseman, et al., Determination of trace amounts of chlorinated phenols in human urine by gas chromatography, *Anal. Chem.*, *52*:1774–1777 (1980).

78. E. Fogelelqvist, B. Joefasson, et al., Determination of carboxylic acids and phenols in water by extractive alkylation using pentafluorobenzylation, gas capillary GC and electron capture detection, *HRC&CC J. High Res. Chromatogr. Chromatogr. Commun.*, *3*: 568–574 (1980).

79. M. Lehtonen, Gas chromatographic determination of phenols as 2,4-dinitrophenyl ethers using glass capillary columns and an electron capture detector, *J. Chromatogr.*, *202*:413–421 (1980).

80. A. Oomens and L. Noten, Picomole amounts of methyl chloride by reaction gas chromatography, *HRC&CC J. High Res. Chromatogr. Chromatogr. Commun.*, *7*:280 (1984).

81. A. Sentissi, M. Joppich, et al., Pentafluorobenzenesulfonyl chloride: a new electophoric derivatizing reagent with application to tyosyl peptide determination by gas chromatography with electron capture detection, *Anal. Chem.*, *56*:2512–2517 (1984).

82. D. Doedens, Iodide determination in blood by capillary gas chromatography, *J. Anal. Toxicol.*, *9*:109–111 (1985).

83. P. Patterson, A comparison of different methods of ionizing GC effluents, *J. Chromatogr. Sci.*, *24*:466–472 (1986).

84. P. Patterson, Recent advances in thermionic ionization detection for gas chromatography, *J. Chromatogr. Sci.*, *24*:4152 (1986).

85. R. Hall, The nitrogen detector in gas chromatography, *CRC Crit. Rev. Anal. Chem.*, *7*:324–381 (1978).

86. R. Achari and M. Mayersohn, Analysis of 4-methylpyrazole in plasma and urine by gas chromatography with nitrogen-selective detection, *J. Pharm. Sci.*, *73*:690–692 (1984).

87. J. Scavone, G. Meneilly, et al., Gas chromatographic analysis of underivatized tocainide, *J. Chromatogr.*, *419*:339–344 (1987).

88. G. Verga, High resolution gas chromatographic determination of nitrogen and phosphorus compounds in complex organic mixtures by tunable selective thermionic detector, *J. Chromatogr.*, *279*:657–665 (1983).

89. J. Libeer, S. Scharpe, et al., Simultaneous determination of p-aminobenzoic acid and p-aminohippuric acid in serum and urine by capillary gas chromatography with use of a nitrogen-phosphorus detector, *Clin. Chim. Acta*, *115*:119–123 (1981).

90. H. Chaves and A. Riscado, Capillary gas chromatography of reducing disaccharides with nitrogen-selective detection and selective ion monitoring of permethylated deoxy(methylmethoxyamino) alditol glycosides, *J. Chromatogr.*, *367*:135–143 (1986).

91. L. Rushing, A. Gosnell, et al., Separation and detection of doxylamine and its chromatography utilizing nitrogen/phosphorus detection, *HRC&CC J. High Res. Chromatogr. Chromatogr. Commun.*, *9*(8):435–440 (1986).

92. A. Krylov and O. Davydovz, Screening analysis of several pharmacological preparations in biological fluids, *Lab. Del.*, *12*:732–734 (1982).

93. H. Karnes and D. Farthing, Improved method for the determination of oxazepam in plasma using capillary gas chromatography and nitrogen-phosphorus detection, *LC–GC*, *5*:978–979 (1988).

94. P. Verner, Photoionization detection and its application in gas chromatography, *Chromatogr. Rev.*, 249–264 (1984).

95. A. Freedman, Photoionization detector response, *J. Chromatogr.*, *236*:11–15 (1982).

96. M. Langhorst, Photoionization detector sensitivity of organic compounds, *J. Chromatogr. Sci.*, *19* (1981).

97. J. Driscoll, Review of photoionization detection in gas chromatography: the first decade, *J. Chromatogr. Sci.*, *23*:488–492 (1985).

98. G. Liu, Oxidation of thiols within gas chromatographic columns, *J. Chromatogr.*, *441*:372–375 (1988).

99. S. Farwell and C. Barinaga, Sulfur selective detection with FPD: current enigmas, practical usage and future directions, *J. Chromatogr. Sci.*, *24*:483–494 (1986).

100. C. Barinaga and S. Farwell, Noncryogenic FSOT column separation of sulfur containing gases, *HRC&CC J. High Res. Chromatogr. Chromatogr. Commun.*, *10*:538–543 (1987).

101. K. Kim and M. Andreae, Determination of carbon disulfide in natural waters by absorbent preconcentration and gas chromatography with flame photometric detection, *Anal. Chem.*, *59*:2670–2673 (1987).

102. S. Gluck, Performance of the model 700A Hall electrolytic conductivity detector as a sulfur selective detector., *J. Chromatogr. Sci.*, *20*: 103-108 (1982).
103. B. Ehrlich, R. Hall, et al., Sulfur detection in hydrocarbon matrices. A comparison of the flame photometric detector and the 700A Hall electrolytic conductivity detector, *J. Chromatogr. Sci.*, *19*:245-249 (1981).
104. V. Inouye, H. Kanai, et al., Guide for troubleshooting operator problems in the Hall 700 and 700A electrolytic conductivity detectors, *J. Chromatogr. Sci.*, *22*:262-263 (1984).
105. R. Hall, The nitrogen detector in gas chromatography, part III, electrochemical detectors, *CRC Crit. Rev. Anal. Chem.*, *7*:345-363 (1978).
106. R. Hall and C. Risk, Rapid and selective determination of barbiturates by gas chromatography using the electrolytic conductivity detector, *J. Chromatogr. Sci.*, *13*:519-524 (1975).
107. B. Pape, Analytical toxicology: applications of the element-selective electrolytic conductivity detector for gas chromatography, *Clin. Chem.*, *22*:739-748 (1976).
108. B. Budevska, N. Rizov, et al., Photolytic chemiluminescence detector for gas chromatographic analysis of N-nitroso compounds, *J. Chromatogr.*, *351*:501-505 (1986).
109. S. Nyarady, R. Barkley, et al., Redox chemiluminescence detector: application to gas chromatography, *Anal. Chem.*, *57*:2074-2079 (1985).
110. R. Hutte, R. Sievers, et al., Gas chromatography detectors based on chemiluminescence, *J. Chromatogr. Sci.*, *24*:499-505 (1986).
111. S. Montzka and R. Sievers, Redox chemiluminescence detectors for chromatography, *Chromatography*, *2*:22-26 (1987).
112. P. Andrawes, Analysis of formaldehyde in pure air by gas chromatography and helium ionization detection, *J. Chromatogr. Sci.*, *22*: 506-508 (1984).
113. F. Andrawes and R. Ramsey, The helium ionization detector, *J. Chromatogr. Sci.*, *24*:513-518 (1986).
114. F. Andrawes and S. Greenhouse, Applications of the helium ionization detector in trace analysis, *J. Chromatogr. Sci.*, *26*:153-159 (1988).
115. R. Milberg and J. Cook, The mass spectrometer as a detector for gas-liquid chromatography. In *GLC and HPLC Determination of Therapeutic Agents*, Marcel Dekker, New York, pp. 235-258.
116. M. Grayson, The mass spectrometer as a detector for gas chromatography, *J. Chromatogr. Sci.*, *24*:529-542 (1986).

117. P. Quinn, B. Kuhnert, et al., Measurement of meperidine and nor-meperidine in human breast milk by selected ion monitoring, *Biomed. Mass Spectrom.*, *13*:133–135 (1986).
118. K. Biemann, The mass spectrometer as a detector in chromatography, *J. Chromatogr. Libr.*, *32*:43–54 (1985).
119. K. Krishnan, Advances in capillary gas chromatography–Fourier transform interferometry, *Fourier Transform Infrared Spectrosc.*, *4*: 97–145 (1985).
120. S. Smith and G. Adams, Chromatographic performance and capillary gas chromatography–Fourier transform infrared spectroscopy, *J. Chromatogr.*, *279*:623–630 (1983).
121. J. Cooper, I. Bowater, et al., Gas chromatography/Fourier transform infrared/mass spectrometry using a mass selective detector, *Anal. Chem.*, *58*(13):2791–2796 (1986).
122. E. Olson and J. Diehl, Serially interfaced gas chromatography/Fourier transform infrared spectrometer/ion trap mass spectrometer system, *Anal. Chem.*, *59*:443–448 (1987).
123. J. Demirgian, Gas chromatography–Fourier transform infrared spectroscopy–mass spectrometry, *Trend Anal. Chem.*, *6*:58–64 (1987).
124. J. Brill, B. Narayanan, et al., Selective determination of organofluorine compounds by capillary column gas chromatography with an atmospheric pressure helium microwave-induced plasma detector, *HRC&CC J. High Res. Chromatogr. Chromatogr. Commun.*, *11*: 368–374 (1988).
125. M. Zerezghi, K. Mulligan, et al., Application of a rapid scanning plasma emission detector and gas chromatography for multi-element quantification of halogenated hydrocarbons, *J. Chromatogr. Sci.*, *22*:348–352 (1984).
126. W. Yu, Q. Qu, et al., Development and application of a proto-type (GC)^2–MES hyphenated apparatus, *Proc. Int. Symp. Capillary Chromatogr.*, *4*:445–464 (1981).
127. J. Wu, Developments of GC-AA and HPLC-AA for metal speciation studies, *Diss. Abstr. Int. B.*, *46*:3032 (1986).
128. L. Ebdon, S. Hill, et al., Directly coupled atomic spectroscopy—a review, *Analyst (London)*, *111*:1113–1138 (1986).
129. N. Bloom and W. Fitzgerald, Determination of volatile mercury species at the picogram level by low temperature gas chromatography with cold-vapor atomic fluorescence detection, *Anal. Chim. Acta*, *208*:151–161 (1988).
130. P. Uden, Element-selective chromatographic detection by atomic emission spectroscopy, *Chromatogr. Forum*, 17–26 (Nov.–Dec. 1986).

131. C. Cappon, GLC speciation of selected trace elements, *LC–GC*, 5: 400–418 (1987).
132. K. Grob and E. Muller, Co-solvent effects for preventing broading or loss of early peaks when using concurrent solvent evaporation in capillary GC, *HRC&CC J. High Res. Chromatogr. Chromatogr. Commun.*, 11:388–394 (1988).
133. K. Grob and T. Laubli, Minimum column temperature required for concurrent solvent evaporation in coupled HPLC-GC, *HRC&CC J. High Res. Chromatogr. Chromatogr. Commun.*, 10:435–440 (1987).
134. V. Hakkinen, M. Virolaninen, et al., New on-line HPLC-GC coupling system using a 10 port valve interface, *HRC&CC J. High Res. Chromatogr. Chromatogr. Commun.*, 11:214–216 (1988).
135. D. Duquet, C. Dewaele, et al., Coupling micro-LC and capillary GC as a powerful tool for the analysis of complex mixtures, *HRC&CC J. High Res. Chromatogr. Chromatogr. Commun.*, 11:252–256 (1988).
136. F. Maris, E. Noroozian, et al., Determination of polychlorinated biphenyls in sediment by on-line narrow-bore column liquid chromatography/capillary gas chromatography, *J. High Res. Chromatogr. Chromatogr. Commun.*, 11:197–202 (1988).
137. B. Pacciarelli, B. Muller, et al., GC column effluent splitter for problematic solvents introduced in large volumes: determination of di-(2-ethylhexyl)phthalate in triglyceride matrices as an application, *HRC&CC J. High Res. Chromatogr. Chromatogr. Commun.*, 11: 135–139 (1988).
138. A. Heim, The analysis of trace components using the total transfer technique in coupled column systems, Anal. Volatiles: Methods Appl., Proc. Int. Workshop 1983, pp. 171–182.
139. T. Raglione, N. Sagliano, et al., Multidimensional LC–LC and LC–GC separations, *LC–GC*, 4:328–338 (1986).
140. U. Brinkman, Selective GC detectors for use in liquid chromatography, *LC–GC*, 5:6 (1987).
141. E. Fogelqvist, M. Drysell, et al., On-line liquid-liquid extraction in a segmented flow directly coupled to on-column injection into a gas chromatograph, *Anal. Chem.*, 58:1516–1520 (1986).
142. T. Noy, E. Weiss, et al., On-line combination of liquid chromatography and capillary gas chromatography. Preconcentration and analysis of organic compounds in aqueous samples, *HRC&CC J. High Res. Chromatogr. Chromatogr. Commun.*, 11:181–186 (1988).
143. K. Grob, E. Muller, et al., Coupled HPLC–GC for determining PCBs in fish, *HRC&CC J. High Res. Chromatogr. Chromatogr. Commun.*, 10:416–417 (1987).

144. F. Munari and K. Grob, Automated on-line HPLC-HRGC: Instrumental aspects and application for the determination of heroin metabolites in urine, *HRC&CC J. High Res. Chromatogr. Chromatogr. Commun.*, *11*:172–176 (1988).

145. K. Grob, D. Frohlich, et al., Coupling of high-performance liquid chromatography with capillary gas chromatography, *J. Chromatogr.*, *295*:55–61 (1984).

146. T. Raglione and R. Hartwick, Liquid chromatography–gas chromatography using microbore high-performance liquid chromatography with bundled capillary stream splitter, *Anal. Chem.*, *58*:2680–2683 (1986).

147. S. Matin, M. Rowland, et al., Synthesis of N-pentafluorobenzyl-S-(-)prolyl-1-imidazolide, a new electron capture sensitive reagent for determination of enantiomeric composition, *J. Pharm. Sci.*, *62*:821–823 (1973).

148. W. Pirkle and J. Hauske, Broad spectrum methods for the resolution of optical isomers. A discussion of the reasons underlying the chromatographic separability of some diastereomeric carbamates, *J. Organ. Chem.*, *42*:1839–1844 (1977).

149. M. Wilson and T. Walle, Silica gel high performance liquid chromatography for the simultaneous determination of propranolol and 4-hydroxypropranolol enantiomers after chiral derivatization, *J. Chromatogr.*, *310*:424–430 (1984).

150. A. Sedman and J. Gal, Resolution of the enantomers of propranolol and other beta-adrenergic antagonists by high performance liquid chromatography, *J. Chromatogr.*, *278*:199–203 (1983).

151. J. P. Kamerling, M. Duran, et al., Determination of the absolute configuration of some biologically important urinary 2-hydroxydicarboxylic acids by capillary gas–liquid chromatography, *J. Chromatogr.*, *222*:276–283 (1981).

152. W. A. Konig and I. Benecke, Gas chromatographic separation of chiral 2-hydroxy acids and 2-alkyl-substituted carboxylic acids, *J. Chromatogr.*, *195*:292–296 (1980).

153. M. Jemal and A. Cohen, Determination of enantiomeric purity of Z-oxylysine by capillary gas chromatography, *J. Chromatogr.*, *394*:388–394 (1987).

154. W. Konig, I. Benecke, et al., Isocyanates as reagents for enantiomer separation; application to amino acids, N-methylamino acids and 3-hydroxy acids, *J. Chromatogr.*, *279*:555–562 (1983).

155. W. Konig, Sterochemical aspects of pharmaceuticals. In *Drug Stereochemistry* (I. Wainer, ed.), Marcel Dekker, New York, pp. 113–145.

156. D. Knapp, Derivatives for chromatographic separation of optical

isomers. In *Handbook of Analytical Derivatization Reactions*, John Wiley & Sons, New York, pp. 405–436.

157. W. Konig, The practice of enantiomer separation by capillary gas chromatography. In *Chromatographic Method* (W. Bertsch, ed.), Alfred Huthig Verlag, New York.

158. R. Souter, Stereoisomer separations by gas chromatography. In *Chromatographic Separations of Stereoisomers*, CRC Press, Boca Raton, FL, pp. 11–85.

159. M. Zief and L. Crane (eds.), *Chromatographic Chiral Separations*, Marcel Dekker, New York, 1987.

160. T. Alexander, Gas chromatographic analysis [of antibiotics], *Drugs. Pharm. Sci., Mod. Anal. Antibot., 27*:1–18 (1986).

161. A. Aszalos, *Modern Analysis of Antibiotics*, Marcel Dekker, New York, 1986.

162. T. Norman and K. Maguire, Analysis of tricyclic antidepressant drugs in plasma and serum by chromatographic techniques, *J. Chromatogr., 340*:173–197 (1985).

163. R. Braithwaite, Tricyclic antidepressants: analytical techniques, *Ther. Drug Monit., 3*:239–254 (1981).

164. C. Kumana, Therapeutic drug monitoring-antiarrhythmics, *Ther. Drug Monit., 3*:370–390 (1981).

165. J. Burke and J. Thenot, Determination of antiepileptic drugs, *J. Chromatogr., 340*:199–241 (1985).

166. K. Dix, C. Cappon, et al., Arsenic speciation by capillary gas–liquid chromatography, *J. Chromatogr. Sci., 25*:164–169 (1987).

167. M. Chaplin, A rapid and sensitive method for the analysis of carbohydrate compounds in glycoproteins using gas–liquid chromatography, *Anal. Biochem., 123*(2):336–341 (1985).

168. F. Van de Vaart, A. Indemans, et al., The application of chromatography to the analysis of pharmaceutical creams, *Chromatographia, 16*:247–250 (1982).

169. W. Aherne, Cytotoxic drugs: analytical techniques, *Ther. Drug. Monit., 3*:482–491 (1981).

170. M. Ribick, M. Jemal, et al., Determination of ehtylenediametetraacetic acid in aqueous rinses of detergent-washed rubber stoppers of pharmaceutical vials using solid phase extraction and capillary gas chromatography, *J. Pharm. Biomed. Anal., 5*:687–694 (1987).

171. J. Craske and C. Bannon, Gas liquid chromatography analysis of the fatty acid composition of fats and oils: a total system for high accuracy, *JAOCS, 64*:1413–1417 (1987).

172. E. Reid, *Assay of Drugs and Other Trace Components in Biological Fluids*, Elsevier/North-Holland, New York.

173. K. Tsuji, *GLC and HPLC Determination of Therapeutic Agents*,
      Marcel Dekker, New York, 1978.
174. D. Jack, *Drug Analysis by Gas Chromatography*, Academic Press,
      New York.
175. R. Cline, L. Yert, et al., Determination of germicidal phenols in
      blood by capillary gas chromatography, *J. Chromatogr.*, *307*:420–
      425 (1984).
176. D. Doedens, Iodide determination in blood by gas chromatography,
      J. Anal. Toxicol., *9*:109–111 (1985).
177. D. Fry, Lithium-analytical techniques, *Ther. Drug. Monit.*, *3*:217–
      223 (1981).
178. F. Croo, J. DeSchutter, et al., Gas chromatographic determination
      of parabens in various pharmaceutical dosage forms, *Chromato-
      graphia*, *18*:260–264 (1984).
179. R. Coutts and G. Baker, *Gas Chromatography, Handbook Neuro-
      chemistry*, 2nd ed. (A. Lajtha, ed.), Pienum, New York.
180. F. Matsui, G. Lovering, et al., Gas chromatographic method for
      solvent residues in drug raw materials, *J. Pharm. Sci.*, *73*:1664–1666
      (1984).
181. S. Gorog, *Analysis of Steroid Hormone Drugs*, Akademial Kiado,
      Budapest.
182. S. Gorog, Steroid analysis in pharmaceutical industry, *Trends Anal.
      Chem.*, *3*:157–161 (1984).
183. H. Duerbeck and I. Bueker, Recent improvements in the determina-
      tion of stilbesterols and synthetic androgens, *Anal. Chem. Symp.
      Ser.*, *Adv. Steroid Anal. 23*:399–411 (1985).
184. D. Campeau, I. Gruda, et al., Analysis of amphoteric surfactants of
      the alkyaminopropylglycine type by gas chromatography, *J. Chro-
      matogr.*, *405*:305–310 (1987).
185. J. Augustin, *Methods of Vitamin Assay*, John Wiley and Sons, New
      York.
186. S. Romano, J. Renner, and P. Leitner, Gas chromatographic deter-
      mination of residual ethylene oxide by head space analysis, *Anal.
      Chem.*, *45*:2327–2330.
187. L. Gramiccioni, M. Milana, and S. DiMarzio, A head space gas
      chromatographic method for the determination of traces of ethylene
      oxide in sterilized medical devices, *Microchem. J.*, *32*:89–93 (1985).
188. P. Bellenger, F. Pradier, M. Sinegre, and D. Pradeau, Determina-
      tion of residual ethylene oxide in non-reusable plastic medical devices
      by the head-space technique, *Sci. Technol. Pharm.*, *12*:37–39 (1983).
189. Perkin Elmer Corporation, Applications of Gas Chromatographic
      Head Space Analysis, Technical Note 16/1978.

190. M. Kaye and T. Nevell, Statistical evaluation of methods using headspace gas chromatography for the determination of ethylene oxide, *Analyst*, *110*:1067–1071 (1985).

191. J. Boyer and M. Probecker, A simple method of quantitative analysis for the determination of solvents in pharmaceutical forms, *Labo-Pharma-Probl. Tech.*, *344*:525 (1984).

192. M. Litchman and R. Upton, Headspace GLC determination of triethylamine in pharmaceuticals, *J. Pharm. Sci.*, *62*:1140–1142 (1973).

193. C. Bicchi and A. Bertolino, Determination of residual solvents in drugs by headspace gas chromatography, *N Farmacu.*, *37*:88–97 (1982).

194. M. Kojima, A simple quantitative determination of ethanol in drugs by gas chromatography using the headspace analytical technique. I. Tinctures, *Yakugaku Zasshi*, *96*:1365–1369 (1976).

195. M. Kojima, A simple quantitative determination of ethanol in drugs by gas chromatography using the headspace analytical technique. II. Liquid drugs except tinctures, *Yakugaku Zasshi*, 97:1142–1146 (1977).

196. K. Nakajima and T. Yasuda, Simultaneous determination of 1-menthol, dl-camphor and methyl salicylate in pharmaceutical preparations for external application using headspace gas chromatography, *Chiba-Ken Eisei Kenkyusho Kenkyu Haboku*, *8*:14–18 (1984).

197. J. Sauceman, C. Winstead, and T. Jones, Quantitative gas chromatographic headspace determination of choline in adult and infant formula products, *J. Assoc. Off. Anal. Chem.*, *67*:982–985 (1984).

198. L. Ettre, B. Kolb, and S. Hurt, Techniques of headspace gas chromatography, *Am. Lab.*, 76–83 (October 1983).

199. Perkin Elmer Corporation, Application of Gas Chromatographic Headspace Analysis, Technical Note 15/1977.

200. G. Lyman, R. Johnson, and B. Kho, Gas–solid chromatographic determination of oxygen in various pharmaceutical forms, *J. Assoc. Off. Anal. Chem.*, *64*:177–180 (1981).

201. P. Fyhr, D. Behr, L. Rydmag, and A. Brodin, Headspace sampling and analysis of oxygen, *J. Parent. Sci. Technol.*, *41*:26–30 (1987).

202. B. Kolb and P. Pospisil, A gas chromatographic assay for quantitative analysis of volatiles in solid materials by discontinuous gas extraction, *Chromatographia*, *10*:705–711 (1977).

203. B. Kolb, Multiple headspace extraction — a procedure for eliminating the influence of the sample matrix in quantitative headspace gas chromatography, *Chromatographia*, *15*:587–594 (1982).

204. T. Wampler, W. Bowe, and E. Levy, Dynamic headspace analyses of residual volatiles in pharmaceuticals, *J. Chromatogr. Sci.*, *23*:64–67 (1985).

205. J. Letavernier, M. Aubert, G. Ripoche, and F. Pellerin, Research of solvent residues of plastic material by headspace gas chromatography, *Ann. Pharm. Fr.*, *43*:117–122 (1985).

206. J. Zechman, S. Aldinger, and J. LaBows, Jr., Characterization of pathogenic bacteria by automated headspace concentration—gas chromatography, *J. Chromatogr.*, *377*:49–57 (1986).

207. G. Reineccius and R. Liardon, The use of charcoal traps and microwave desorption for the analysis of headspace volatiles above heated thiamine solutions, Top. Flavour Res. Proc. Int. Conf., 1985, pp. 125–136.

208. *The United States Pharmacopeia*, Twenty-First Revision, United States Pharmacopeial Convention, Inc., Rockville, MD, 1984.

209. H. Leach and J. D. Ramsey, Gas chromatography. In *Clark's Isolation and Identification of Drugs* (A. C. Moffat, ed.), The Pharmaceutical Press, London, 1986, p. 178.

210. D. B. Jack, *Drug Analysis by Gas Chromatography*, Academic Press, New York, 1984.

211. C. F. Poole and S. A. Schuette, *Contemporary Practice of Chromatography*, Elsevier, New York, 1984, p. 145.

212. R. E. Clement, F. I. Onuska, F. J. Yang, G. A. Eiceman, and H. H. Hill, Gas chromatography, *Anal. Chem.*, *58*:321R (1986).

213. T. Daldrup, F. Susanto, and P. Michalke, Combination of TLC, GLC and HPLC for a rapid detection of drugs and related compounds, *Fresenius Z. Anal. Chem.*, *308*:413 (1981).

214. R. W. Souter, *Chromatographic Separations of Stereoisomers*, CRC Press, Boca Raton, FL, 1985, p. 11.

5

High-Performance Liquid Chromatography

JOHN A. ADAMOVICS and DAVID L. FARB   *Cytogen Corporation, Princeton, New Jersey*

I. INTRODUCTION

Historically, high-performance liquid chromatography (HPLC) can be traced back to the amino acid analyzers of the early 1960s. By 1975, liquid chromatographic instrumentation was described in the *United States Pharmacopeia*. Since that time, HPLC has become the most popular chromatographic technique in the pharmaceutical laboratory.

This chapter is intended to be a practical overview of the liquid chromatography sorbents, instrumentation, and the various method development approaches used in pharmaceutical laboratories for both relatively small molecules and biomolecules.

II. SORBENTS

A. Silica Gel

More than 90% of the column packings are based on silica gel and its bonded phases [1]. The primary reasons for its widespread use is because of high surface area and porosity, easy preparation, adjustable polarity, and good mechanical strength. Silica gel optimized for chromatography should have the characteristics listed in Table 5.1 [2].

Table 5.1 Characteristics of Optimized Silica Gel

	Typical values	Ideal value
Specific surface area (m^2/g)	150–400	200
Mean pore diameter (nm)	6–10 or 30	< 10
Specific pore volume (ml/g)	~ 0.2–1	0.7
Trace metal content (ppm)	< 1000	< 1000
Surface pH neutral	Acid to basic	
Mean particle size (µm)	3.5 or 10	–
Apparent density (g/ml)	0.4–0.6	0.45

Source: Adapted from Reference 1.

A silica gel with a higher surface area gives a higher capacity ratio, as demonstrated by the determination of famotidine [3] in the presence of its potential degradants and preservatives. The capacity factor increased by more than a factor of 3 when the silica surface area increased from 200 to 350 m^2/g.

The resolution of relatively low-molecular-weight molecules decreases sharply as a function of both pore diameter and pore volume. For proteins, the resolution increases as the pore diameter and pore volume of the silica gel is increased [4,5].

Iron and other metal oxide impurities are commonly found in relatively large amounts (0.3% by weight) in commercial-grade silica gel [6]. These metal impurities are believed to be the major contributing factor to the poor chromatographic performance of band broadening and tailing of basic analytes [6–10]. Fortunately, most of these metal impurities which are believed to cause higher acidities of certain silanols and consequently increased interactions with bases can be removed from silica gel with acid washes [7,11]. The important consideration is that the silica should be as pure as possible.

The importance of small, < 10 µm, particle size has been known for some time [13]. The smaller the liquid chromatographic packing size, the higher the chromatographic efficiency. The optimum particle size with regard to analysis time, plate number, and pressure drop is 2–4 µm [6]. For sorbents used in preparative procedures, particle sizes greater than 10 µm are generally used primarily because of their greater loading capacity. The 15-µm particle seems to represent a reasonable compromise between resolution and loading capacity for preparative procedures.

Apparent density is related to the specific pore volume and can be measured relatively easily. Silica gels with low apparent density do not have the structural strength to withstand high pressures generated during packing; consequently, they do not pack efficiently [2].

B. Reversed Phase

The use of silica-based reversed-phase sorbents still predominates in the pharmaceutical laboratory . The organic bonding reactions used to generate these phases have been reviewed [1]. The historic disadvantage of these packings has been chemical stability where eluents with pH values above 7 were not recommended due to dissolution of the silica, and below pH 2, due to cleavage of siloxane linkages. Column degradation also occurs with high salt concentrations and in the presence of some ion-pairing reagents.

Numerous alternatives have been used to minimize sorbent degradation. One of the first approaches was to react (end cap) the unreacted silanol groups with a smaller silane such as trimethylchlorosilane. Manufacturers have reported enhanced stability of these end-capped packings up to pH 10. End-capping may also decrease the peak tailing of some polar analytes. Also, coating the silica with a polymethyloctadecylsiloxane generated by gamma rays or free radical generators have reportedly given packings greater stability in various solvents and at high pH [14]. Increased stability has also been generated by using zirconia-cladded [15] and alumina-doped [16] silicas. In general, the working pH range increases when the bonding density and/or purity of the silica is used [1].

The most popular alternative to silica-based sorbents is fully polymeric reversed-phase poly(styrene-divinylbenzene) (PRP) which is stable over a pH range of 1–13 and behaves much like a high-carbon-content C_{18} column [17–19]. These gels are also a basis for a number of other sorbents, the most familiar being the ion-exchange supports [20,21]. The first modern liquid chromatography utilized sulfonated poly(styrene-divinylbenzene) polymeric ion-exchange packing. These polymers have also been derivatized on C_{18} to give a sorbent which is chemically stable over a broad pH range and has characteristics similar to C_{18} [22].

C. Other Silica-Based Sorbents

A vast range of materials have been bonded to silica such as phenyl, cyano, nitro, amino, and diol functionalities. The variety of silica-based sorbents are listed in the *United States Pharmacopeia* (USP) [23] where it is noted that of the 33 column packings, 19 are silica based. The synthesis and characteristics of over 80 nonconventional bonded silicas have been reviewed [24]. In spite of the number of chromatographic packings, 50% of the chromatographers develop methods using a C_{18} sorbent [1]. Sorbents that are used to resolve enantiomers and biomolecules will be discussed later in this chapter.

D. Nonsilica-Based Sorbents

Alumina

During the last 20 years, the chromatographic use of aluminum oxide has steadily decreased even though alumina partially or totally overcomes the difficulty arising from the relatively low pH stability of silica [25,26]. Alumina oxide is stable over a broader pH range (2–12) than silica gel but generally has a lower specific surface area, typically 70 m^2/g [27,28].

Like silica, alumina can also act like an ion exchanger. Alumina has amphoteric properties with both cation and anion-exchange properties ranging over a broad pH range. At solvent pH values below that of the pK of the alumina surface, the alumina surface has a positive net charge. At a higher pH, the surface charge is negative (Figure 5.1). The zero-point charge (no net charge) of alumina occurs at a pH of 9.2 but can be shifted to 6.5 in acetate buffer and 3.5 in citrate buffer [29]. Similar to what has

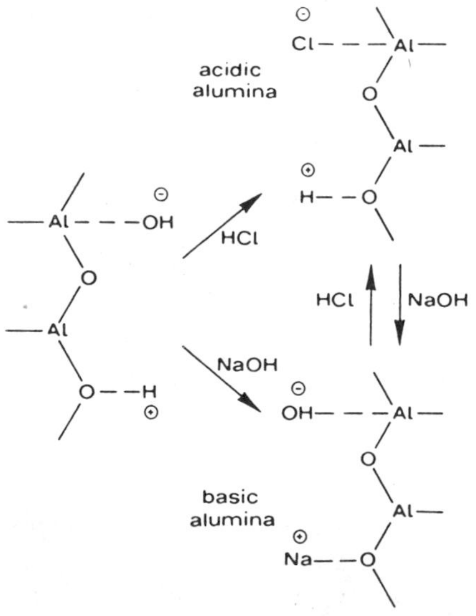

Figure 5.1 Alumina is an anion exchanger in acidic solvents and a cation exchanger in neutral or basic solvents. (Reproduced with permission from Reference 28.)

occurred with silica, alumina-based reversed phase has been prepared by the free-radical polymerization of cross-link polymers to alumina particles or by chemically bonding alkyl groups to alumina [30]. Cyano bonded to alumina has been shown to be useful in resolving various pencillins, cephalosporins, macrolide antibiotics, and tricyclic antidepressants [26].

Carbon

Similiar to alumina, graphitized carbon has a wide range of pH stability. The advantages of a carbon-based sorbent is that it has a highly stereoselective surface for resolving isomeric compounds and is inherently reproducible [31–33]. The primary disadvantage is that the carbon particles are not easily modified. Steroids [34,35], adamantanes [36], amino acids [37], and analgesics [33] have been chromatographed on various carbon packings. Chiral compounds have been resolved when cyclodextrins have been added to the mobile phase [38].

A partial listing of the variety of resins used for biomolecules is listed in Table 5.5.

E. Column Hardware

Due to its ability to withstand high pressure, its relative low cost, and inertness, stainless steel has become the standard material of columns and other chromatographic components. However, under certain circumstances, stainless steel has been shown to interact with the sample and the mobile phase [39]. The best known example is chloride salt corrosion of stainless steel. Data indicate that nearly all common eluents dissolve iron from stainless steel [39]. It appears that proteins also adsorb to stainless steel [39]. The adsorption process is fast, whereas desorption is slow, a result which leads to variable protein recoveries. A number of manufacturers are offering alternatives to stainless components with Teflon™-lined columns and Teflon frits. Titanium is being explored as an alternative to stainless steel. A cheaper and simpler procedure is to oxidize the surface of the stainless steel with $6N$ nitric acid. This procedure should be repeated about every 6 months.

Columns made of glass and Kel-F with operating pressures of up to 4000 psi have been manufactured. Radially compressed plastic cartridges containing sorbent are also a viable alternative [40].

Various advantages can be realized by varying the column dimensions from the conventional 4.6 mm ID × 25 cm. One approach has been to use shorter columns of 3–10 cm packed with 3-μm or 5-μm particles. When operating at high flow rates (2–5 ml/min), this approach has become known as fast HPLC [41].

Increased productivity can be the main advantage of the shorter columns. For instance, content uniformity testing was performed on 42 samples with the conventional chromatographic run time of 20 min, and for the fast chromatographic system, a run time of only 30 sec [42]. Speed can be used to the pharmaceutical analyst's advantage if the analysis time is limited by the chromatographic run time, but this is not generally the case. For instance, most autosamplers require up to 1 min to inject the sample, and sample preparation, especially for complex matrices, involve several time-consuming steps. The extent to which fast chromatography is used will also be limited by the analyst's capability to review and keep up with the data output [43].

A second alternative to the conventional chromatographic column is to use a narrower column bore, typically 1–3 mm ID. For these narrow-bore columns, the primary advantage is an increase in mass sensitivity [44] along with reducing the volume of the eluent. In contrast to the 1–2-mm-ID columns the 3.0-mm-ID columns can be used with conventional HPLC equipment. The most likely role for narrow-bore columns will be in pharmaceutical analysis as an interface to detectors such as a mass spectrometer.

III. INSTRUMENTATION

Liquid chromatography (LC) instrumentation is continuously being refined and improved. Advances in electronics are incorporated into pumps, system injectors, data handling, and detectors. The reader is directed to the annual reviews from PITTCON published in *LC–GC Magazine* or *American Laboratory* or the Column Liquid Chromatography: Equipment and Instrument review in *Analytical Chemistry* for the fundamental developments in instrumentation.

A. Detection

The detection systems employed by HPLC are based on instrumentation designed to respond a particular physical or chemical property of the sample component being eluted. By far, the most common types are spectrophotometric detectors.

Over the past decade, a considerable amount of research and development has been expended in producing HPLC–mass spectrometers (MS). The application of LC–MS to pharmaceutical analysis is discussed later in this chapter.

Spectrophotometric Detectors

The simplest spectrophotometric detector is the fixed-wavelength variety which most commonly utilizes a low-pressure mercury source with a high-energy output at several discrete lines. As will be seen in Chapter 8 of this book, this detector is by far the most commonly used detector. The primary reason for its popularity is that this detector is both durable and inexpensive.

Most of these fixed-wavelength detectors can be filtered with a series of filters and or phosphors to detect at wavelengths varying from 280 to 546 nm. Wavelengths in the range 214–229 nm can be monitored by the use of zinc and cadmium source lamps, respectively [46].

The use of multichannel UV detectors has increased dramatically over the last 10 years. These utilize either a deuterium or xenon lamp source. The individual wavelengths are isolated by a monochromator. This offers the advantage of being able to tune the exact wavelength or lambda maximum absorbance for a compound of interest. Differences in UV absorptivity can be exploited. For example, by using detection at 206 nm for the antibiotic aztreonam, both aztreonam and arginine, its weakly UV-absorbing counterion, can be detected [47]. At 254 nm, only aztreonam is detected. If interfering analytes have different UV maximum, monitoring the wavelength maximum for the analyte of interest will enhance the selectivity. A variant of this approach is to use absorbance ratioing as a means of identifying related substances. The ratio response provides a screen for determining which related substances have significant changes in spectral characteristics. This has been used on the impurities of vancomycin and daptomycin [48]. Absorbance ratioing has also been used for peak purity assessment of sulfasalazine [49].

A more popular approach for detemining peak purity has been the use of diode-array detectors which were first introduced in 1982 [50]. Peak homogeneity of similiar benzodiazepines [51] and theophylline have been determined by this technique [52]. Rapid-scan detectors are also useful in confirming the presence of known components such as colorants, which are commonly used in drug products [53,54] and identification of related substances [55,56].

There are relatively few examples where wavelengths below 200 nm have been used to monitor chromatographic eluants. One example is the antibiotic cortalcerone which is resolved from numerous monosaccharides using an amino-phase column with 70% aqueous acetonitrile and UV detection at 195 nm. Several techniques have been developed for detection of low-UV-absorbing or non-UV-absorbing analytes. One approach is to form an ion pair between any ionic analytes of interest and an UV-absorbing species dissolved in the mobile phase [58–60]. A second method uses a UV-absorbing mobile phase which results in negative peaks when the non-absorbing analyte elutes. This is referred to as indirect photometric chromatography (IPC). IPC has been used to analyze a lactated Ringer's irrigating solution and buffered solutions of cephalosporin for chloride, phosphate, bisulfite, and sulfate [61] and other inorganic anions [62]. A similar approach was used for the detection of hydrogen peroxide in the cephalothin matrix [63] and ethanol, glycerol, isopropanol, propylene glycol, and N,N-dimethylacetamide in various formulations [64].

Inorganic anions, such as chloride and sulfate, can be detected at UV wavelengths as low as 195 nm [65]. Detection at wavelengths below 195 nm requires transparent or nonabsorbing mobile phases with dissolved oxygen removed. The detection of these anions is important in numerous pharmaceutical limits tests.

Fluorescence

Fluorescence is a luminescence phenomenon that occurs when a compound absorbs radiation, UV or visible, and then emits it at a longer wavelength. There are relatively few drugs that have such strong native fluorescence [66]. For these compounds, fluorescence detection can usually achieve increased specificity and sensitivity over that obtained with UV detection.

The choice of the mobile phase is very important, as fluorescence is sensitive to fluorescence quenchers. Highly polar solvents, buffers, and halide ions quench fluorescence. The pH of the mobile phase is also important to fluorescence efficiency; for example, quinine and quinidine only display fluorescence in strongly acidic conditions, whereas oxybarbiturates are only fluorescent in a strongly alkaline solution [67,68]. Due to the stability of the chromatographic sorbents, the use of very acidic or basic mobile phase may not be possible. One alternative is to alter the effluent pH postcolumn. Postcolumn addition of sulfuric acid has been used for the assay of ethynodiol diacetate and mestranol in tablets [69]. Another example is the determination of tetracycline antibiotics in capsules and syrup where EDTA and calcium chloride were added to enhance fluorescence [70].

Fluorescence detection can be further extended by precolumn derivatization of amino acids using phenylisothiocyanate (PITC) as shown in Figure 5.2. This derivatization procedure along with others using reagents such as orthophthalaldehyde (OPA) can be easily adapted to automation using an autoinjector [71]. Postcolumn derivatization using OPA has been used to analyze neomycin drug products [72], especially amino bisphosphonate [73]. A variety of precolumn and postcolumn fluorogenic derivatization procedures have been used for aminoglycoside antibiotics [74]. UV irradiation of tamoxifen postcolumn yields highly fluorescent phenanthrene derivatives that can be easily detected.

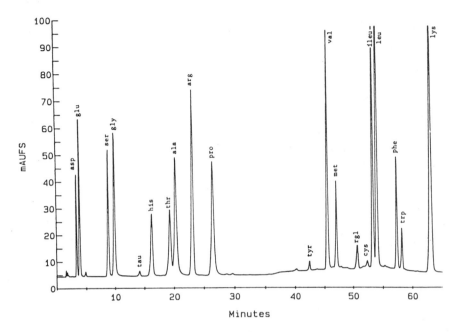

Figure 5.2 Chromatographic analysis of amino acid IV solution that had been derivatized with PITC. Samples assayed on a Pico-Tag Free Amino Acid analysis column at 47°C using gradient elution. (Courtesy of Bob Pfeifer and Mary Dwyer, Waters, Division of Millipore.)

The advantages and disadvantages of precolumn and postcolumn procedures have been compared. In general, precolumn procedures have a number of significant advantages over postcolumn procedures [66].

Electrochemical

Several detectors employ the measurement of an electrochemical property to monitor a liquid chromatographic effluent. The two most commonly used are the conductivity and amperometric detectors.

Conductivity detectors monitor differences in the equivalent conductances of the sample ion and the competing ions in the mobile phase. There are two distinct strategies for maximizing the conductivity response signal.

The first approach uses a suppressor device which is located between the analytical column and the detector cell. This device chemically removes the mobile-phase buffer counterions, thus reducing the background conductivity. This type of detector increases postcolumn dead volume and puts limitations on the types of mobile phases that can be used. In addition, the cost of the suppressor and the frequency of its replacement can be significant.

A simpler and technologically superior approach is the measurement of the direct electrical conductance. The background conductivity of the mobile phase is electronically subtracted, not requiring a suppressor device. One example of direct conductivity detection is the simultaneous determination of potassium nitrate and sodium monofluorophosphate in dentrifices [76]. Alendronate, a bisphonate, can be directly detected in intravenous solutions and tablets using an anion-exchange column and conductivity detection [77]. Another example, from one of the author's (JA) laboratory is shown in Figure 5.3. Direct conductivity detection makes it possible to selectively detect choline in the presence of an equal molar amount of an antibiotic which is not detected.

Applicability of the conductivity detector can be extended by chemical derivatization or by the use of postcolumn photochemical reactions [78]. The use of a photochemical reaction detector, also known as a photoconductivity detector, can also be very selective. Only certain organic compounds such as trinitroglycerin, chloramphenicol, and hydrochlorothiazide will undergo photolytic decomposition to produce ionic species.

The second type of electrochemical detector is amperometric, which can be either by direct current or pulsed. The direct-current mode is used

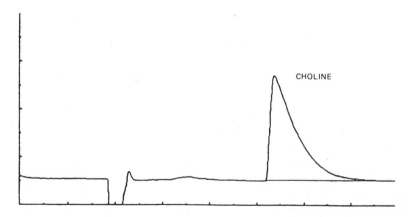

Figure 5.3 Chromatogram of choline in the presence of equal molar quantity of antibiotic. A silica gel packed column with the mobile phase of aqueous 1% phosphoric acid with direct conductivity detection. The antibiotic is not detected by the conductivity detector.

for compounds which can be electrolytically oxidized or reduced at a working electrode. The oxidative mode is generally restricted to compounds containing an easily oxidized moiety, such as a phenol. This mode has been limited to relatively low applied potentials due to oxidation of the aqueous mobile phase. Even with these restrictions, a number of drugs, including doxorubicin, haloperidol, morphine, naloxone, paracetamol, phenothiazines, salicylic acid, tricyclic antidepressants [79], melphalan and chlorambucil in tablets [80], and aztreonam [81], can be detected.

The selectivity of amperometric detection has been useful in simplifying the sample pretreatment steps in the determination of a number of drug products [82–86]. A method requiring no sample preparation using an amperometric detector and UV detector in series was developed for lidocaine hydrochloride injectable solutions [87]. The drug epinephrine is quantified with the amperometric detector, whereas lidocaine and methyl paraben are detected by ultraviolet light. Disodium EDTA had to be added to the mobile phase to eliminate a peak response from iron leached from the stainless steel.

Iodide in vitamin tablets can be found by amperometric detection [88]. Nonaqueous eluents of methanol-containing ammonium perchlorate, which are relatively oxidant resistant, have been used in conjunction with a silica column to detect a wide range of drugs [89]. The use of higher potentials not possible in totally aqueous mobile phases allows for the detection of secondary and tertiary aliphatic amines; 462 drugs have been detected in this manner. For compounds that are not electroactive, a procedure using a postcolumn photolysis can generate electroactive species [90] for penicillins [91], proteins [92], and barbiturates [93].

Operation of amperometric detectors in the reductive mode is more difficult because dissolved oxygen in the sample and mobile phase will interfere. Nevertheless, the quinone of doxorubicin and the nitro group of chloramphenicol have been reduced [94].

Pulsed amperometric detection (PAD) was developed because the Pt or Au electrodes under constant potential is rapidly fouled by electroactive, aliphatic, organic compounds. The electrode is cleaned and reactivated through use of a triple-step potential waveform. The triple-step process starts with the application of the detection potential followed by a large positive oxidative potential where the elecrode surface is cleaned and then reduced. Aminoglycosides have been detected using PAD [95–97], but, by far, the greatest application of this detector has been for the detection of monosaccharides and their oligomers in glycoproteins. This specific application has generated much interest since it was demonstrated that changes in glycosylation due to changes in manufacturing of a protein may impact on the potency of the drug product. Several examples are erthropoietin [98], human serum transferrin, human immunoglobulin, and human α_1-antitrysin [99], interleukin-3 [100], and OKT3* [101].

Refractive Index

The refractive index detector was one of the first on-line HPLC detectors used. Detection is based on changes in the refractive index of the effluent when an analyte is present versus the solvent alone. This detector is commonly used when an analyte does not have a suitable UV chromophore. An example would be the detection of carbohydrates in drug preparations or acetylcholine in an ophthalmic solution [102,103]. Figure 5.4 is a chromatogram of propylene glycol, propylene carbonate, and resorcinol in anhydrous ointment detected by a refractive index detector [104].

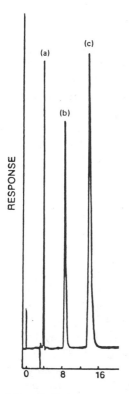

Figure 5.4 Chromatogram of commercial anhydrous ointment where the peak responses are propylene glycol (a), propylene carbonate (b), and resorcinol (c). (Courtesy of *Journal of Pharmaceutical Sciences*.)

Radioisotopes

Two basic approaches have been used for monitoring of radioisotopes in liquid chromatographic effluents. Radioisotopes in liquid chromatographic eluents can be either fraction collected or detected by an on-line detector [105,106]. Not surprisingly, fraction collection followed by liquid scintillation counting is more expensive and time-consuming than the use of an on-line flow-through detector. The one advantage of off-line detection is a greater counting precision.

On-line detection can be classified as either homogeneous or heterogeneous. In the homogeneous system, the effluent is mixed with a liquid scintillation cocktail before passing through a flow cell that is positioned in a scintillation counter. In the heterogeneous system, the effluent passes through a flow cell packed with a solid scintillator.

Radioisotope detectors commonly have an energy range of 0–1500 keV, which allows the user to adjust for low-, medium-, and high-energy radioisotopes. These detectors can be used for detection of beta-emitting radioisotopes and low-energy gamma emitters such as the important radio-pharmaceuticals ^{32}P, ^{99m}Tc, ^{111}In, and ^{125}I [107–109]. Radioistope detectors in series with UV detectors are used to evaluate the stability of antibody-^{111}In which are used for in vivo tumor imaging and for EDTMP-^{153}Sm chelates which are used for the relief of bone pain caused by tumor metasis.

Atomic Absorption and Emission

Most of the existing methods for the assay of metal-containing drugs are generally based on nonspecific nonchromatographic methods [110]. Chromatographic interfaces that are coupled to an atomic absorption detector for the specific detection of mercury-containing drugs have been specifically designed [111]. As another example, Ridaura, a gold-containing antirheumatic drug, was detected by interfacing a liquid chromatograph with an emission detector [112].

Polarimeter

Polarimetry coupled to a liquid chromatograph has been used to determine the optical purity of drugs [113–121]. The advantages of this approach for determining optical purity is that the enantiomers do not need to be resolved and only eluates which give rise to optical rotation or circular dichroism are detected. This advantage makes on-line polarimetry inherently more accurate than conventional polarimetry because minor interfering components can be chromatographically resolved. The shape of the chromatographic peak when a UV detector is added in series to the polarimetry is a useful indicator of homogeneity; that is, a pure enantiomer will exhibit a trace that is of the same shape on both detectors. The disadvantages are that optically pure standards are required for quantification and that the polarimetry is 100 times less sensitive than UV detection. On-line polarimetry has been used in studying ethambutol [120], ephedrine, camphor [119], nicotine [117], and lorazepam [116]. Chromatographic methods for resolving enantiomers are discussed in Section IV.

Light Scattering

The principle of operation and basic theory of this detection mode has been reviewed [44]. This detection mode has been applied to the analysis of steroids in bulk drugs and in formulations [122]. The main attribute of this detection mode is that it can directly measure mass concentrations of all components including unknown impurities. A variant of this technique, low-angle laser light-scattering photometer, is used as a molecular-weight detector for size-exclusion chromatography [123,124]. This can be particularly useful for determining the physical stability of protein products which tend to noncovalently polymerize when heated or shaken.

Evaporative Light Scattering

The eluate is nebulized and the aerosol passes through a heated tube where the mobile phase is vaporized. The remaining microparticles are then passed through a beam of light. The incident light is scattered by the particles [125]. This detector detects any sample less volatile than the mobile phase, such as lipids, carbohydrates, surfactants, and polymers. Unlike RI and low-wavelength UV, gradient conditions can be used.

Fourier Transform Infrared

There continues to be major problems with coupling HPLC to FTIR (Fourier transform infrared) due to the interference caused by water. The interface is the critical component in the system [126]. The two basic types of interfaces are continuous and capture. A continuous interface has been developed that uses a liquid–liquid extraction. In this approach, the analytes are extracted from the mobile phase by mixing postcolumn with a stream of IR (infrared) transparent, water-immiscible solvent. In the capure technique, the eluent is deposited on a continuously moving, IR transparent, inert substrate from which the eluent can be easily removed by evaporation. These techniques have been applied to identification of racemic precursors of diltizam, AZT derivatives, and steroids [127].

Mass Spectrometry

Mass spectrometry is the most universal of detectors because it detects most organic compounds and is highly selective when selective ionization techniques are employed. Off-line LC–MS where the analyte is collected, concentrated, and analyzed by mass spectrometry is a relatively common practice in the pharmaceutical industry. When the compound in the collected fraction is unstable, on-line LC–MS techniques are preferred. There are eight LC–MS interfaces that have been reviewed [128] and their performance characteristics tabulated (Table 5.2).

Liquid chromatography thermospray – mass spectrometry has been utilized to determine the identity and purity of taxol, solvent-induced degradation of cloxacillin [129], benzodiazepines, and ethanolamine-type antihistamines [130]. Electrospray LC–MS has been used for peptide mapping of the human growth hormone [131].

B. Preparative Scale

There are two distinct instances where isolating quantities of purified material occur in the development of a potential drug product. The first instance occurs during the purification of the bulk drug which generally involves processing kilograms of sample. The second instance occurs with the isolation of impurities and degradants from the bulk and formulated product which may only require micrograms of an analyte for structure elucidation.

Isolation of material from an HPLC separation is greatly simplified when chromatographic conditions can be easily translated from the analytical scale to the preparative scale. If it is necessary to maintain equivalent analysis time, the linear velocity of the mobile phase must be kept the same.

The correlation between analytical and preparative procedures is found by calculating the square of the ratio of analytical to preparative column inner diameters and multiplying that value by the flow rate of the analytical system:

$$\text{Flow rate preparative} = \text{Flow rate analytical} \left(\frac{\text{Diameter preparative column}}{\text{Diameter analytical column}} \right)^2$$

The same factor is used for the scale-up of the sample load. The translation of an analytical method to a preparative system is obviously dependent on the sorbent characteristics of the preparative column. Several reviews on preparative methods, column-packing techniques, theory, and equipment design have been published [132–135,145]. Specific examples of the application of preparative procedures are listed in Table 5.3.

Table 5.2 The Performance of LC–MS Interfaces

Interface	MBI	DLI	TSP	API–HPN	API–ESP	API–HIGH flow ISP	PB	CF–FAB
HPLC column	Conventional (1 : 10) microbore	Conventional (1 : 20) microbore	Conventional	Conventional	Microbore	Conventional	Conventional (1 : 10) microbore	Conventional (1 : 100) microbore
Eluate flow	1.5–2 ml/min (nonpolar mobile phase) 0.1–0.3 ml/min (polar mobile phase)	1–50 µl/min	0.1–2 ml/min	2 ml/min	5–10 µl/min	0.1–2 ml/min	0.1–0.5 ml/min	1–5 µl/min
Involatile compounds	No (with exception of FAB)	Yes	Yes	Yes	Yes	Yes	No	No
Thermally unstable compounds	No	Yes	Yes	Yes	Yes	Yes	No	Yes
Ionization modes	EI, CI (volatile) FAB (Involatile)	CI (Conventional) EI (Microbore)	CI	CI	CI	CI	CI, EI	CI, EI

Information obtained	Structural/molecular	Molecular	Molecular	Molecular	Molecular	Molecular/structural	Structural/molecular	Structural/molecular CI reagent buffer or gas buffer solvent buffer solvent solvent buffer/solvent gas gas
Advantages	CI, EI, FAB	Involatile and thermally unstable	Conventional HPLC column Involatile and thermally unstable	Instrument stability	Instrument stability High MW compounds	Conventional HPLC column Structural information	CI, EI	CI, EI
Disadvantages	Volatility range	Stability problems Lack of sensitivity	Stability problems	CI only	CI only	CI only	Thermally stable and volatile	Low flow rate

Table 5.3 Several Representative Preparative Liquid Chromatographic Procedures

Compound	Comment	Reference
AZT phosphate diglyceride	Purified for structure confirmation, efficacy studies	136
Aztreonam	Isolation of impurities and degradants	137
Bacitracin	Assayed for microbial activity	138
Benzodiazepinone	Enantiomer separation	139
Cephacetrile	Purification of bulk material	140
Cefonicid	Purification of bulk material	141
Steroids	Comparison of preparative equipment	142
Numerous drugs	Review of preparative procedures in pharmaceutical industry	143
Vitamin D_2	Resolved from previtamin D, recycle	144

Biomolecules

Chromatography isolation procedures for biotechnology products use both low pressure and high pressure. The approach depends largely on the stability of the molecular structures to the combined shear forces of high flow rates and hydrophobic surfaces which tend to unfold proteins. Due to Food and Drug Administration (FDA) requirements to address the removal of biological contaminants and the need to add a variety of biologic impurities into most biotechnology fermentation cultures, multistep column isolation procedures are routinely used to ensure complete removal of residual materials such as viruses, nucleic acids, endotoxin, surfactants, stationary-phase leechates, and proteinaceous cofactors which are used for cell expression or during purification [146]. Table 5.4 lists the parameters used in selecting resins.

For larger protein products that maintain their native conformation through a complex set of secondary and tertiary structural interactions, the

Table 5.4   Resin Selection Criteria

Economic	Chemical	Physical
Supply	Recovery of mass	Recovery of activity and
Longevity	Clearance of DNA,	structural integrity
Sanitation steps	virsus, endotoxin,	
Specificity	proteins, residual	
Throughput	leechates	
Capacity		

recovery of an active, unaltered structure often is the greatest challenge facing any chromatographic preparation. To eliminate unrelated proteins, DNA, and viruses, for example, product-specific affinity chromatography preparations are preferable. Resins constructed of high-affinity ligands show very high specificity, capacity, and throughput; however, product recovery is usually dependent on the relatively harsh conditions (to proteins) for elution. One manufacture of recombinant human antihemophilic factor has overcome this limitation in using monoclonal antibodies for the affinity ligand by initially screening their murine hybridoma cells for a clone-producing low-affinity antibody (personal communication). As a result, their purification step sacrifices throughput and capacity but is effective in clearing unwanted contaminants while preserving the product's molecular integrity.

Affinity separations include both group-specific and product-specific classes of interactions. Protein A has been extensively employed as a group-specific ligand to purify monoclonal antibodies, but its specificity is insufficient for some materials. For example, if a monoclonal is harvested in media containing 5% fetal bovine serum, the polyclonal bovine antibodies are not differentiated by a linear gradient on a protein A column but completely separated using step elution on a preparative cation exchanger column (ABx) [147].

High-affinity resins are perhaps most useful in removing problematic contaminants from the product stream. An example is the use of a column of polymyxin B (theoretical capacity of 6 mg/cc) to remove 90–98% of the residual endotoxin [148] from protein preparations.

A list of commercially available, group-specific affinity resins is shown in Table 5.5 [149]. The same suppliers also provide a variety of hydrophobic, reversed-phase, size-exclusion, and ion-exchange resins.

Table 5.5 Chromatographic Resins Used for the Purification of
Biomolecules

Supplier	Matrix	Ligand
Amicon	Matrex cellufine Agarose	Formyl, amino, carboxyl, sulfate, dyes
Bioprobe	Polymer Agarose	Protein A, AL (protein G mimic), hydrazide, FMP
Bioprocessing	Controlled pore glass	Protein A, gelatin, lysine
Bio-Rad	Macroporous hydrophilic	Protein A
Chromatochem	Polymer-coated silica	Protein A/G
Cuno	Cellulose/acylic polymer composite	Protein A/G, benzamidine
DuPont	Fluoropolymer	Dyes, antibodies
Genex	Agarose	Protein G
3M	Polymeric	Protein A, antibodies, lectins
PerSeptive Biosystems	Poros polymeric aggregates	Protein A, trypsin
Pharmacia LKB	Cross-linked agarose	Blue dye
Repligen	Agarose	Protein A
Sepragen	Cellulose	Protein A, hydroxylapatite
Sterogene	Hardened agarose Cellulose	Preactivated
TosoHass	Tresyl toyopearl	Heparin, cibacron blue, iminodiacetate

Hydrophobic, reverse phase. Although many reversed-phase (RP) applications have been developed for the analysis of biotechnology products, a few preparative methods are employed in either the exchange of counterions or purification of some peptide hormones. Peptide products consisting of noncovalently linked subunits, however, are more susceptible to dissociation due to the strong apolar environments encountered with typical reverse-phase HPLC. Nevertheless, the activities of three such hormones were preserved by use of a short-chain aliphatic (C_4), a derivatized solid phase, and a neutral mobile phase [150]. Hydrophobic interaction methods are being proposed as a replacement to RP HPLC for the chromatography of large biomolecules. As this technique is largely dependent on the aqueous solubility of each protein or peptide, it shows promise in discriminating subtle differences in hydrophobic tendency.

Size exclusion. Peptide hormones of similar molecular weight (MW) can differ significantly in retention over size-exclusion chromatography unless the mobile phase was carefully selected to overcome all ionic and hydrophobic interactions with the resin support – silanol groups for silica and aromatic groups for polydivinylbenzene. For recombinant G-CSF, IFN-alpha, and interleukin-2, this multimodal interaction of the size-exclusion resin was used to advantage in their selective isolation [151]. When size-exclusion HPLC is employed for the analytical characterization of many biomolecules, special detection techniques using either low-angle light scatter or mass spectroscopy can be used to establish the correct molecular weight of eluting biomolecules.

Ion exchange, metal chelation. Polypeptides that differ by a single change in their amino acids have been differentiated by ion-exchange HPLC. Using a 1-in.-diameter sulfopropyl cation-exchange column and a low-ionic-strength gradient from acid to neutral pH buffers, recombinant human interleukin 1-alpha (pI 5.3) and 1-beta (pI 6.7) were separated from their desamido (pI 5.1) and N-met (pI 6.5) forms [152]. For many proteins, a combination of ion-exchange types and adsorption versus flow-through techniques are used to maximize the removal of trace proteins and contaminants that may otherwise be carried with the product. A typical flow chart uses both size-exclusion and ion-exchange methods along with virus-inactivation steps for the isolation of human immunoglobin G from plasma, this is illustrated in Figure 5.5 [153].

An overview of the use of liquid chromatography in biotechnology presents additional examples [154].

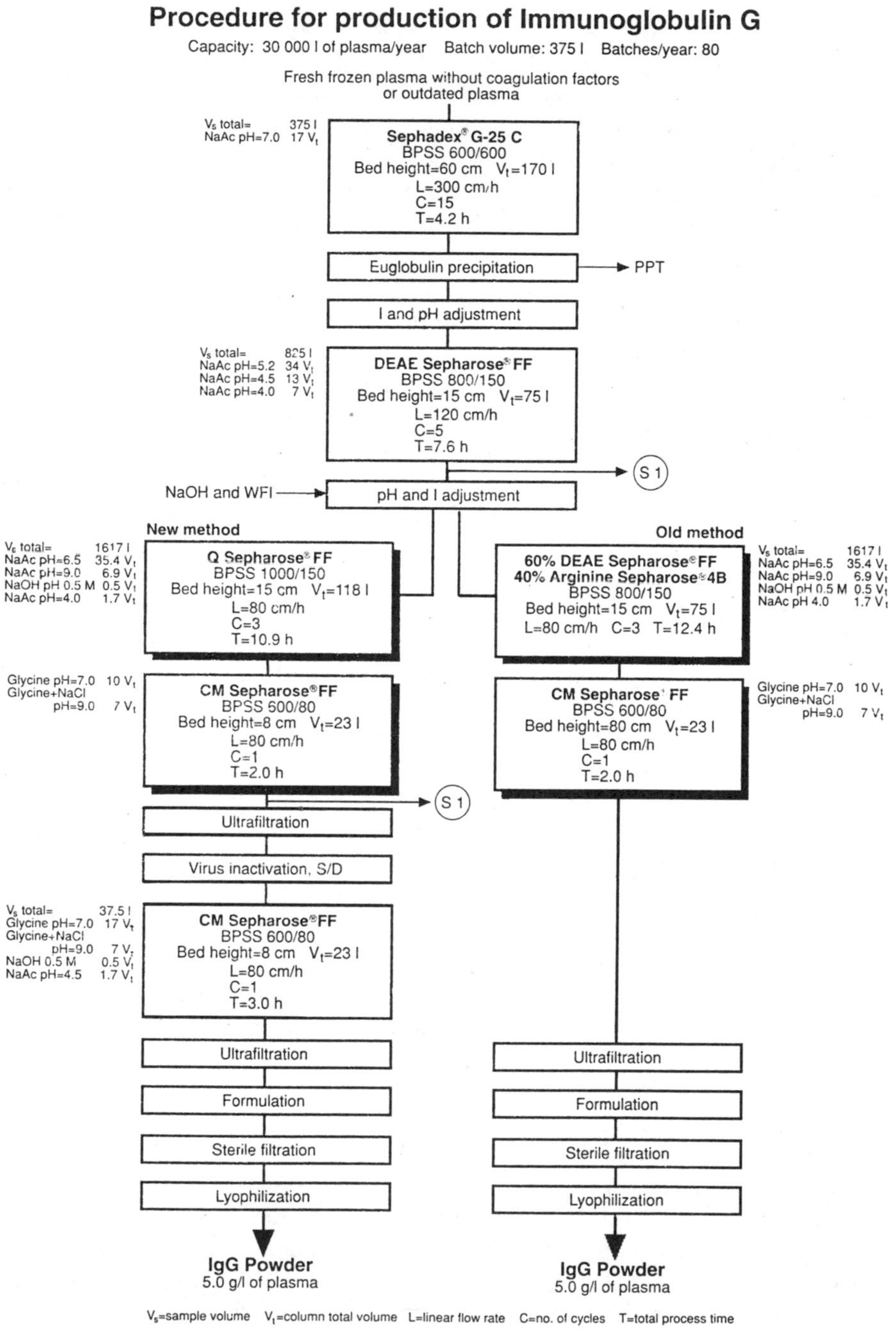

Procedure for production of Immunoglobulin G

Capacity: 30 000 l of plasma/year   Batch volume: 375 l   Batches/year: 80

Figure 5.5   Typical flow chart for production of IgG (Pharmacia).

IV. METHOD DEVELOPMENT

The primary purpose of this section is to compare and contrast the numerous approaches that have been used to develop a HPLC method. The characteristics of a validated method are described in Chapter 1.

One of the more notable attempts at simplifying and systematizing the development of liquid chromatographic methods has been to use computer-based expert systems; four of these systems have been reviewed [155]. An expert system ideally allows a computer to simulate the train of thought of an expert. The logical structure of an expert system can be described as a decision tree, where, at each node, there are a number of possible connections with the next node. A chromatographic problem in this context is a search for a path along this decision tree from some initial state to a final validated method [156,157]. To keep expert systems within feasible dimensions of available computer technology, the elimination of redundant possibilities is necessary. One approach is to limit the number of sorbents. Sorbent redundancy and the superiority of one sorbent over another for resolution of pharmaceuticals has been investigated for TLC (thin-layer chromatography) [158], GC [159], and HPLC [160]. A review of this book's application section (Chapter 8) clearly indicates that there is no single correct approach to chromatographic methods development. Consequently, the configuration of an expert system would most likely reflect the philosophy of one "expert" and not the widely varying methods that are found in the published literature. An approach to resolving this problem has been attempted by integrating several expert sytems [161]. The various expert systems available have been used for analysis of analgesics [162], steroids [163], basic drugs [164], diclofenac and its bromide impurity [165], nifedipine [166], sulphonamides [167], and zalospirone [168]. Despite the rapid advances made, computer-assisted method development has had relatively little impact on everyday work. The use of chemometrics undoubtedly has a future as an aid to method development. The following subsections review approaches to developing methods for enantiomers, biomolecules, drugs containing basic functionality, and neutral drugs.

A. Enantiomers

The majority of the new drugs approved and the most often prescribed drugs in the United States have at least one asymmetric center [169]. Approximately half of these chiral drugs are of a racemic composition. Many of the optically pure drug substances are of natural origin, while the majority of the synthetically derived products are in racemic form [172]. Enantiomers can differ in their pharmacological activity or one may be inactive or be toxic. Some examples are listed in Table 5.6 and reviewed in literature [170,171]. Clearly, the presence of an unwanted enantiomer could lead to unwanted side effects or to a less potent formulation. The three general approaches to resolving enantiomers are: derivatization of enatiomers to form diastereomers which are followed by separation on an achiral column; forming diastereometric complexes with a chiral selector which is added to the mobile phase; and separation of the enantiomers on a chiral sorbent (also refer to discussion on p. 105).

The advantages of using the indirect mode over the direct mode is that there is an improved peak symmetry and resolution over the direct mode [172]. Disadvantages are the need for a derivatizable functional group, knowledge of chiral purity, and stability of the derivatizing agent and quantitative derivatization conditions [172]. The major disadvantages of the direct mode of analysis is poor selectivity between achiral impurities and the enantiomers with limited ability to control analyte elution. Direct chiral separations are generally more susceptible to resolution changes than achiral columns, with certain chiral sorbents showing lot to lot variability [172] and limited column lifetimes, which make it difficult for them to withstand long-term stability studies.

Diastereomeric Derivatives

Enantiomers are derivatized with an optically pure chiral derivatization reagent to form a pair of diastereomers. The ability to resolve the diastereomeric derivatives on an achiral sorbent is enhanced when the chiral centers of the enantiomers and the derivatives are in close proximity [181]. Two different separation mechanisms have been proposed. One postulates that the diastereomers are separated by differences in molecular structure and polarity [182]. The other possible mechanism is based on differences in the diastereomer energies of adsorption [183]. Table 5.7 lists the chiral reagents that have been used for separation of enantiomers as diastereomers.

One of the most commonly used class of derivatization agents for diasteromer formation are isothiocyanates and isocyanates. Enantiomers of β-blockers, amphetamine, epinephrine, methamphetamine, and mexiletine have been resolved after derivatization with these agents. Isothiocyanates produce thiourea derivatives upon reaction with primary and secondary amines. Thiourea derivatives also provide a strong UV absorbance for the detection of enantiomers lacking a strong UV chromophore. Isocyanates produce ureas when reacted with amines. The physical properties of these ureas are similar to thiourea derivatives. Isocyanates will also react with alcohols to yield carbamates.

Optically pure chiral o-phthaladehyde (OPA)-related reagents would

Table 5.6 Pharmacological Activities of Enantiomeric Forms

Drug	Comment	Reference
Amphetamine	(S)-form is a CNS stimulant, whereas (R)-form has little activity	173
Epinephrine	One of the enantiomers is 10 times as active as a vasoconstrictor	174
Ibuprofen	S-isomer active, R-isomer inactive	175
Ketoprofen	(S)-(+) enantiomer is analgesic/anti-inflammatory, R-(−) isomer active against bone loss in periodontal disease	176
Labetalol	R,R-stereoisomer has beta-blocking activity, SR diastereomer has alpha-antagonist activity, two other isomers are inactive	177
Methamphetamine	(S)-form is a controlled substance, whereas (R) form is used as a nasal decongestant	178
Nebivolol	(+)-β-blocker and (−) is a vasodilator	179
Penicillamine	L-isomer is toxic	174
Penicillin V	L-isomer has little if any antibiotic activity	—
Propoxyphene	One isomer is an antitussive, whereas the other is inactive	174
Propranolol	Only (S)-form has β-adrenergic blocking activity	178
Synephrine	One isomer has 60 times more pressor activity	173
Warfarin	(S)-form is 5 times more potent as by a blood anticoagulant	180

Table 5.7 Chiral Derivatization Reagents for Separation of Enantiomers as Diastereomers

Compound	Chiral Reagent	Reference
Acebutolol	R-(−), S-(+)-1-(Napthyl)ethyl isocyante	185
	R-1-Phenylethyl isocyanate	186
Alprenolol	L-*N*-tert-butoxycarbonylleucine	187
Amphetamines	R-(−)-Benoxaprofen	188
3-Aminoquinuclidine	S-(−)-1-Phenylethyl isocyanate	189
	R-1-(1-Napthyl)ethyl isocyanate	
	RR-(+)-*O,O*-Dibenzoyltartaric acid	
	SS-(−)-*O,O*-Dibenzoyltartaric acid	
Atenolol	S-α-Phenylethyl isocyanate	190
	(−)-Menthyl chloroformate	191
	RR-*O,O*,-Di-*p*-toluoyltartaric acid anhydride	192
	2,3,4,6-Tetra-*O*-acetyl-β-D-glucopyranosyl isocyanate	
Baclofen	L-*N*-Acetylcysteine + *O*-pthaldialdehyde	193
	S-(+)-Naproxen	194
Betaxolol	R-(−)-1-(Napthyl)ethyl isocyanate	195
Bupranolol	2,3,4-Tri-*O*-acetyl-α-D-arabinopyranosyl isothiocyanate	196
Carprofen	L-Leucinamide	196
Chloroamphetamine	2,3,4,-Tri-*O*-acetyl-α-D-arabinopyranosyl isothiocyanate	197
Diacetolol	R-1-Phenylethyl isocyanate	186

Table 5.7 (Continued)

Compound	Chiral Reagent	Reference
Diltiazem	S-(−)-N-1-(2-Naphthylsulphonyl)-2-pyrrolidine carbonyl chloride	198
	(+)-2-(2-Naphthyl)propionyl chloride	
Encainide	(−)-Menthyl chloroformate	199
Ephedrine	L-1 - [(4-Nitrophenyl)sulphonyl] chloride	200
	2,3,4,6-Tetra-*O*-acetyl-β-D-glucopyranosyl isothiocyanate	201
Epinephrine	2,3,4,6-Tetra-*O*-acetyl-β-D-glucopyranosyl isothiocyanate	242
	2,3,4,-Tri-*O*-acetyl-α-D-arabinopyranosyl isothiocyanate	202
Etodolac	S-1-Phenylethylamine	203
Fenoprofen	R-1-Phenylethylamine	204
	L-Leucinamide	205
Flavodilol	(−)-Menthyl chloroformate	206
Flecainide	(−)-Menthyl chloroformate	207
	L-1-[(4-Nitrophenyl)-sulphonyl]prolyl chloride	208
	R-1-(2-Napthyl)ethyl isothiocyanate	209
	S-1-(1-Napthyl)ethyl isothiocyanate	
Flunoxaprofen	S-1-Phenylethylamine	210
Flurbiprofen	L-Leucinamide	211
	R-1-Phenylethyl isocyanate	212
Gossypol	(+)-Dehydrobietylamine	213
	R-(−)-2-Amino-1-propanol	214

(Continued)

Table 5.7 (Continued)

Compound	Chiral Reagent	Reference
Ibuprofen	L-1-[(4-Nitrophenyl)-sulphonyl]prolyl chloride	215
	R-1-(Napthyl)ethyl isocyanate	191
	R-(+)-1-Ferrocenylethylamine	243
	S-(+)-Ferrocenylpropylamine	243
Ketoprofen	L-1-[(4-Nitrophenyl)-sulphonyl]prolyl chloride	204
	L-Leucinamide	218
Lombricine	L-Butoxycarbonylcysteine + O-phthaldialdehyde	216
Loxaprofen	S-1-(4-Dimethylaminoaphthyl)-ethylamine	217
Metaproterinol	(−)-Menthyl chloroformate	218
Methylphenidate	d-10-Camphorsulphonic acid	244
	(−)-Heptafluoro-butyrylthioprolyl chloride	244
Metroprolol	(−)-Menthyl chloroformate	219
	S-Ethyl 3-(chloroformoxy) butyrate	220
	S-(+)-Benoxaprofen	221
	L-N-tert-Butoxycarbonylleucine	187
	S-tert-Butyl 3-(chloroformoxyl)-butyrate	220
	2,3,4-Tri-O-acetyl-α-D-arabinopyranosyl isothiocyanate	188
Mexiletine	R-1-(2-Napthyl)ethyl isothiocyanate	209
	S-1-(1-Napthyl)ethyl isothiocyanate	209
	2,3,4,6-Tetra-O-acetyl-β-D-glucopyranosyl isothiocyanate	
Naproxen	L-Phenylalanine-β-napthylamide	223
	L-Alanine-β-napthylamide	223
	R-(+)-1-Ferrocenylethylamine	243

Table 5.7 (Continued)

Compound	Chiral Reagent	Reference
	S-(+)-Ferrocenylpropylamine	243
	S-(−)-N-1-(2-Napthyl-sulphonyl)-2-pyrrolidine carbonyl	222
	S-1-(4-Dimethylaminonaphthyl)-ethylamine	196
Norephedrine	2,3,4,6-Tetra-*O*-acetyl-β-D-glucopyranosyl isothiocyanate	202
	2,3,4-Tri-acetyl-α-D-arabinopyranosyl isothiocyanate	200
Oxprenolol	S-(+)-Benoxaprofen	221
Pindolol	2,3,4,6-Tetra-*O*-acetyl-β-D-glucopyranosyl isothiocyanate	196
Piprofen	R-1-Phenylethylamine	224
Practolol	2,3,4-Tri-*O*-acetyl-α-D-arabinopyranosyl isothiocyanate	196
Prenalterol sulphate	2,3,4,6-Tetra-*O*-acetyl-β-D-glucopyranosyl isothiocyanate	225
Prenylamine	R-(−)-1-(Napthyl)ethyl isocyanate	226
Pronethalol	2,3,4,6-Tetra-*O*-acetyl-β-D-glucopyranosyl isothiocyanate	196
Propafenone	R-1-(2-Napthyl)ethyl isothiocyanate	209
	S-1-(1-Napthyl)ethyl isothiocyanate	209
Propranolol	2,3,4,6-Tetra-*O*-acetyl-β-D-glucopyranosyl isothiocyanate	225
	R-(+)-1-Phenylethyl isocyanate	227

(Continued)

Table 5.7 (Continued)

Compound	Chiral Reagent	Reference
	L-*N*-Trifluoroacetylprolyl chloride	228
	L-*N*-tert-Butoxycarbonylleucine	230
	L-*N*-tert-Butoxycarbonylalanine	232
	R-1-Phenylethyl isocyanate	231
	(−)-Menthyl chloroformate	206
	S(−)-Flunoxaprofen isocyanate	232
	RR-*O*,*O*-Diacetyltartaric acid anhydride	233
Proxyphylline	(−)-Camphanoyl chloride	234
Pseudoephedrine	2,3,4,6-tetra-*O*-acetyl-β-D-glucopyranosyl isothiocyanate	201
	L-1-[(4-Nitrophenyl)-sulphonyl]prolyl chloride	200
Salsolinol	S-1-(1-Napthyl)ethyl isothiocyanate	235
	L-*N*-Trifluoroacetylptolyl chloride	236
Solatol	(−)-Menthyl chloroformate	219
	2,3,4,6-Tetra-*O*-acetyl-β-D-glucopyranosyl isothiocyanate	188
Tiaprofenic acid	L-Leucinamide	205
Thyroxine	L-*N*-Acetylcysteine	237
Tocainide	R-(−)-O-Methylmandelic acid	238
	S-1-(1-Napthyl)ethyl isothiocyanate	209
	R-1-(2-Napthyl)ethyl isothiocyanate	209
	S-(+)-1-(Napthyl)ethyl isocyanate	239
Toliprolol	(−)-Menthyl chloroformate	219
Warfarin	L-Carbobenzylproline	240

be expected to be useful fluorescent derivatization agents for pharmaceuticals containing amines. The primary application of this derivatization agent has been used to resolve α-amino acids and lombrincine. Homochiral amines such as S or R-1- phenylethylamine and dehydroabiethylamine have been used to derivatize enantiomeric organic acids such as fenoprofen, ibuprofen, ketoprofen, loxoprofen, and naproxen.

Although these chiral derivatization reagents have demonstrated their utility for indirect chiral separation, the arguments against them include length of time involved for derivatization, possibilty of racemization, and presence of optically active contaminants, variability of the formation rate of the diasteromers [184].

Mobile-Phase Additives

A number of enantiomers have been resolved by forming diastereomeric complexes with a chiral selector which had been added to the mobile phase. These complexes can be resolved on conventional achiral sorbents. Metal chelates, ion-pairing agents, and proteins have been used as chiral selectors.

Copper chelates of amino acid enantiomers such as proline or phenylalanine have been used to resolve enantiomers of amino acids and structurally related compounds [241,245]. Other metals such as zinc and cadmium have also been used. Metal chelates have been used to resolve α-amino-α-hydroxy carboxy acids and α-methyl-α-amino acid enantiomers [246]. One example of pharmaceutical interest is the resolution of D-penicillamine from the L-antipod [247] and resolution of L,D-thyroxine [248].

The elution order of the resolved enantiomers can be controlled by the ligand. Generally, the D-ligand selector gives an elution order reversed from the L-ligand selector elution order, but this is not always the case [249].

Diastereomeric complexes can also be formed by ion-pairing of an enantiomer with a chiral counterion. In order to form this diastereomeric complex, it has been postulated that at least three interaction points between the ion pair are required [250]. Nearly all of these form weak complexes in aqueous mobile phases. Consequently, the chromatographic methods that have been developed have been either silica or diol columns with low-polarity mobile phases. Enantiomeric amines, such as the beta-blockers, have been optically resolved when (+)-10-camphorsulfonic acid was used as the chiral counterion [251]. Enantiomers of norephedrine, ephedrine, pseudoephedrine, and phenyramidol have all been resolved from their respective enantiomers with *n*-dibutyltartrate [252]. Enantiomers of naproxen, a chiral carboxylic acid, are resolved from each other by either using quinidine or quinine in the mobile phase [253]. In these studies, silica

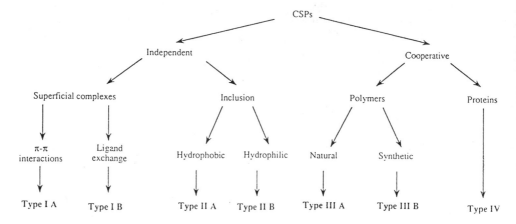

Figure 5.6 Chiral stationary phases: classification according to chiral recognition mechanisms and chemical structures. (From Reference 256.)

sorbents gave greater resolution than either diol or cyano sorbents [254]. The protein albumin has been used as a chiral complexing agent for the separation of carboxylic acid enantiomers and local anesthetics. The stereoselectivity was found to be dependent on albumin concentration and pH [255].

An unexpected but possibly related phenomenon is the separation of enantiomers of nicotine in a totally achiral system [256]. The mechanism is unclear but may involve the formation of in situ diastereometric dimers, where a dimer formed from the same two enantiomers could possibly resolve from a racemic dimer.

Chiral Stationary Phases

Approximately 70 chiral stationary phases (CSPs) have been marketed since 1981 [256]. A classification scheme has been proposed for the numerous commercially available CSPs which takes into account chiral recognition mechanism and chemical structure (Figure 5.6).

The majority of the type IA CSPs are based on amino acid derivatives. The separation mechanisms are based on hydrogen bonding, charge-transfer, dipole stacking, and stearic interactions. The majority of these phases are covalently bonded with a few being ionically attached. The ionic phases are restricted to mobile phases containing less than 20% propanol in hexane [257–259]. Enantiomeric purity of amphetamine tablets [260], decongestant dextromethamphetamine [261], etodalac [262], and various β-lactams [263] have been examined using these chiral sorbents. Derivatization is usually required to achieve resolution. These sorbents have good stability and are compatible with all conventional mobile phases. The analytes are limited to low to medium polarity [256].

Type IB sorbents are chiral ligand exchangers. Several columns are commercially available with either proline, hydroxyproline, or valine and Cu(II) bonded to silica [256]. The binding is via a 3-glycidoxpropyl spacer; Cu(II) needs to be added to the mobile phase to minimize the loss of copper from the sorbent. Silica modified by L-(+)-tartaric acid has also been synthesized. These columns generally have poor efficiency and analytes are limited to bidentate solutes [256].

Type II sorbents are based on an inclusion mechanism. Chiral recognition by optically active polymers is based solely on the helicity of that polymer. Optically active polymers can be prepared by the asymmetric polymerization of triphenylmethyl methacrylate using a chiral anionic initiator [264]. Helical polymers are unique from the previously discussed chromatographic approaches because polar functional groups are not required for resolution [265]. These commercially available sorbents have been used to resolve enantiomers of α-tocopherol [266]. The distinction between this group (IIb) and the sorbents containing cavities is vague (IIa).

The chiral recognition of these types of CSPs is based on the partial insertion of an enantiomer into a chiral cavity. Completely enveloped molecules cannot be separated. Of the various sorbents in this category, cyclodextrin-bonded phases have been the most extensively studied [267]. Cyclodextrins are cyclic, nonreducing oligosaccharides that contain 6–12 glucose units, all in the chair conformation. Cyclodextrin is linked to silica via various coupling techniques [267]. Derivatized celluloses have also been used as CSPs [272–275]. Substituted polyacrylamides are a third class of CSPs which resolve enantiomers by inclusion into asymmetric cavities [271]. Dynamically coating permethylated β-cyclodextrins and related derivatives on a bare silica surface has also been shown to be enantioselective [272–275]. Table 8 lists the various chiral drugs that have been resolved using these phases.

Natural polymers like cellulose and amylose comprise the Type IIIA CSPs, but the mechanical stability of these packings is not sufficiently adequate to be used as a chromatographic sorbent. More satisfactory sorbents have been obtained by chemically modifying them as ester or carbamate derivatives and then coating them onto large-pore silica (300 Å) [276]. These CSPs are marketed under the trade names ChiralCel (cellulose) and ChiralPak (amylose). These packings have a wide scope of applications, good stability, and use on a preparative scale.

Type IV CSPs are proteins immobilized primarily on silica. The solute-CSP complexes are mainly due to ionic and hydrophobic interactions

Table 5.8 Representative Enantiomeric Drugs Resolved
on Type II CSPs

Benzothiadiazine diuretics	Mephobarbital
Chlorpheniramine	Oxazaphosphorines
Chlorthalidone	Phenothiazines
Hexobarbital	Propranolol
Ketoprofen	Terfenadine
Mephenytoin	Thalidomide

and been used for a wide array of pharmaceuticals. Four are commercially available: bovine serum albumin (BSA), α_1-acid glycoprotein, ovomucoid (OVM), and human serum albumin (HSA) [253,277–287].

Resolution of the enantiomeric compounds, aromatic amino acids, amino acid derivatives, aromatic sulfoxides, coumarin derivatives, benzoin, and benzoin derivatives have been accomplished on the albumin columns. The α-acid glycoprotein protein column has been used to resolve 50 enantiomeric drugs [288]. The mobile-phase requirements of these sorbents have been reviewed [288]. Although the selectivity of these sorbents are often outstanding, the solute capacity is only 1 nmol per injection.

Enantiomer Separation Strategy

Figure 5.7 presents a guide to choosing the CSP best fitted to the racemate structure. Optimization strategies for various CSPs, including the effect of organic modifier and eluent pH on enantioselectivity, have been reported [289–292].

B. Biomolecules

For the characterization and routine analysis of biopharmaceutical products, HPLC is being chosen to replace other more time-consuming determinations of identity, purity, and potency. The limitations previously noted for preparative purifications of biomolecules are not as critical in analytical applications. Instead, the standard analytical criteria of precision, linearity, accuracy, limit of detection, ruggedness, and specificity are the major concerns in HPLC analyses of biomolecules. Selectivity of a particular bioanalytical separation is ideally dependent on three primary physiochemical macromolecular parameters (size and shape, charge density, and surface hydrophobicity); however, due to competing chemical phenomena, each method must also be evaluated with regard to protein self-aggregation,

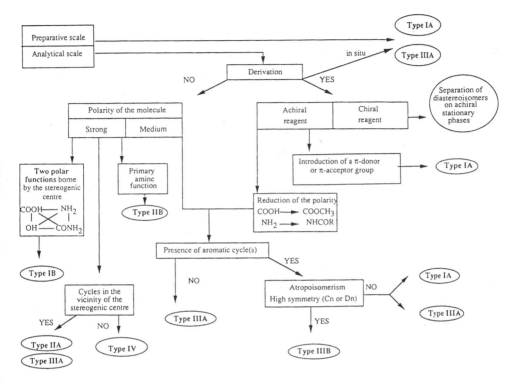

Figure 5.7 Enantiomer separation strategy: choice of CSP according to solute polarity. (From Reference 256.)

deamidation, oxidation, alkylation of amines, and hydrolytic cleavage. There are four basic approaches for separation: reversed-phase, size-exclusion, ion-exchange, and affinity chromatography. Each will be discussed in general terms and also with specific examples.

Due to the tendency of larger proteins and peptides to unfold on hydrophobic surfaces and then to elute as multiple undifferentiated conformations, reversed-phase techniques are limited to smaller materials (usually <10 kD or 80 residues) with limited secondary structures. However, it is still the method of choice for the characterization of specific peptide fragments from larger proteins (i.e., peptide mapping). Appropriate stability formulations for recombinant interferon gamma and a recombinant plasminogen activator were developed by analyzing their trypsin fragmentation profiles on reversed-phase (C_{18}) HPLC [293]. Stability changes due to deamidation and oxidation were identified by an altered retention of the affected peptides and the appearance of new mass ions in their mass spectra. A useful high-sensitivity peptide mapping technique was used for recombinant human erythropoietin to confirm the identity with limited amounts of natural product that was isolated in minute quantities from normal urine. Both materials were iodinated using I-125 NaI and chloramine T, proteolytically fragmented with trypsin, and then separated by reversed-phase (C_{18}) HPLC [294]. Confirmation of structural identity was established with superimposable chromatograms, because single amino acid substitutions, oxidation, deamidation, and alkylation modifications are all readily differentiated by reversed-phase (C_{18} or C_8) HPLC using TFA and gradient elution with increasing acetonitrile.

Reversed-phase analyses of small intact polypeptide products are especially useful in determining minute changes in related impurities due to degradation. The stability of human insulin in different infusion admixtures was assayed by reversed-phase HPLC using repeated injections over 6-hr periods to establish its incompatibility with sodium bisulfite and stabilization by glucose [295]. The content of desamido degradants and stability of bovine, porcine, and human insulins in pharmaceutical delivery systems were measured at a sensitivity limit of 0.05 μg per injection using reversed-phase (C_{18}) HPLC or reversed-phase (CN) HPLC with an octanesulphonate ion-pair agent [296].

Proteolytic secretion variants of recombinant human growth hormone were isolated by anion-exchange HPLC; but compared analytically to the reference standard using reversed-phase (C_4) HPLC with an ammonium bicarbonate/acetonitrile mobile phase rather than the more commonly used TFA/acetonitrile mobile phase [297].

In contrast to reversed-phase applications, size-exclusion HPLC applications are primarily suited for molecules that range from 15 kD to nearly 1 million MW. With the use of a physiological buffered mobile phase, size-exclusion HPLC is dominant in determining the composition of aggregated forms of biotechnology products. The significance of this parameter is paramount, as both in vitro activity and in vivo circulation of proteins are often dependent on a specific monomer or aggregated species. The stability of recombinant-derived analogs of alpha interferon, gamma interferon, and interleukin-2 was evaluated for degradation into inactive aggregate forms using silica-based size-exclusion HPLC at a low pH [298]. The results obtained from HPLC using the nonphysiological mobile phase do not necessarily reflect the potency or aggregate composition at physiological pH. Unfortunately, low yield and altered retention times were observed for the recombinant products at neutral pH. Without independent equivalency data, this HPLC method would be very difficult to validate for use in either purity or potency analyses.

The analytical results obtained from size-exclusion HPLC, which was validated for analysis of human growth hormone potency, correlated with and were more precise than the traditional bioassay results obtained with hypophysectomized rats [299]. For both human insulin and human growth hormone, size-exclusion HPLC was employed with a physiological mobile phase similar to the formulation vehicle to analyze for noncovalent dimers. Because fractions containing the dimeric molecules were inactive, the quantity of monomer determined by HPLC was used indirectly to establish potency and dosage. As mentioned previously, reversed-phase (C_{18}) HPLC was employed to determine trace amounts of monomeric impurities (desamido and sulfoxide derivatives); reversed-phase (C_8) HPLC of the peptide fragments obtained through proteolysis with *S. aureus* or trypsin was compared to that of the reference material to determine identity [300].

Size-exclusion HPLC is particularily useful in either direct pharmaco-dynamic studies of the radiolabeled product or indirect studies that employ a labeled monoclonal antibody. In order to observe shifts in apparent MW due to noncovalent binding interactions, the mobile phase for these analyses should be a physiological buffer and the ligand size cannot be less than half that of the labeled protein. In cases where complexation may interfer with in vivo targeting, size-exclusion HPLC can be used prior to clinical administration of the potential or existing biotechnology product to establish the most effective regimen or dose.

Ion-exchange HPLC accommodates biomolecules of all sizes, shapes, and charge as long as the pH and salt conditions promote good solubilization without self-aggregation or chemical derivatization. The pH of the mobile-phase buffer is selected in an anion-exchange column to provide a net negative protein charge, and in a cation exchanger, a net positive protein charge. Differences in protein or peptide charge densities are usually resolved by applying a gradient of increasing salt rather than pH. Furthermore, separation can be enhanced by selecting a buffer whose pH is close to the protein pI and applying a shallow gradient at a lower ionic strength that still promotes ionic interactions. With extremely low ionic strengths, however, mixed-mode interactions can be expected with proteins, including cationic, anionic, and hydrophobic forces. Some knowledge of the pK of groups involved is very helpful in selecting optimum conditions for ion-exchange HPLC. For example, deamidated molecules with a pI near the pK of carboxyl groups would be difficult to resolve, as the effective charge difference is less in buffers with a pH near the carboxyl group pK. Finally, the degree that a single charge difference is resolvable depends on the total number of charge groups on the protein surface. This latter limitation is significant for larger proteins where isoform heterogeneity becomes extremely difficult to resolve for proteins over 50 kD MW. Nevertheless, chromatofocusing techniques, which essentially involve a very gradual

change in pH stabilized by high concentrations of many different amphoteric buffer molecules, have successfully used ion-exchange HPLC to resolved the isoforms of monoclonal antibodies. The major utility of ion-
exchange techniques for biotechnology products appears in establishing
their amino acid and carbohydrate compositions and sequence. For amino
acid analyses, ion-exchanger resins have a long well-established history of
use. For carbohydrate and oligosaccharide analyses, anion-exchange separations are a more recent and complementary application, because peptide
residues containing carbohydrate cannot be sequenced directly.

Two classes of carbohydrates are resolvable using anion-exchange
HPLC: negatively charged species containing single or multiple sialic acid
residues and neutral species. Because the neutral monosaccharrides and
oligosaccharides are all partially ionized at high pH, they show slightly
different anion-exchanger retentions due to their different pKa values.
Chromatographic detection of carbohydrates may be accomplished by at-
taching a chromophore to the reducing end or by using a pulsed amperometric detector.

Changes in bioengineered expression systems can alter the number
and composition of carbohydrate residues attached to recombinant or
monoclonal antibody products [301]. The use of high pH anion-exchange
HPLC and pulsed amperometric detection (Dionex Corp) has enabled the
analytical profiling or "mapping" of the characteristic biantennary carbohydrate structures following their enzymatic release from both murine and
humanized monoclonal antibodies [302].

For a recombinant human granulocyte colony-stimulating factor multiple HPLC analyses have been used to determine glycosylation. Cation-
exchange (sulfopropyl) HPLC of the intact molecule was performed at a
pH of 5.4 to determine the content of di- and mono-sialylation versus
asialo- or aglyco-species [303]. Peptide fragments of each species were then
generated using *S. aureus* protease and CNBr digestions and separated on
reversed-phase (C$_4$) HPLC to identify the specific site of carbohydrate
attachment.

Affinity chromatography techniques have shown less utility in analytical testing than in preparative separations for a variety of reasons, including
cost and the difficulty of validating consistent operation as the column
changes over time. Protein A affinity has been commonly used to quantitate the total antibody content of either ascites or cell culture fluids. To
provide guidance in the development of a purification process, specific
immunoaffinity resins are either available or can be readily prepared to
quantitate the levels of unrelated protein contaminants. To rapidly determine what the active species in a mixture is, a monoclonal antibody that

interferes with or inhibits the bioassay is often used. Inactive variants of human growth hormone were confirmed using a tandem chromatography technique which utilized an immunoaffinity column and a size-exclusion column [304]. In this case, polyclonal antibodies were used to differentiate the more readily bound monomer from the noncovalent dimer form. Although most large protein products are heterogeneous mixtures which are not easily differentiated chromatographically, inhibitory monoclonal antibody columns may be used in a potency assay to selectively differentiate those modifications that affect the active site of an enzyme or hormone. For radiolabeled monoclonal antibody products, affinity chromatography has been routinely applied in both open (low-pressure) columns and in TLC format using immobilized antigen to differentiate active from inactive substances.

The FDA requirements for "well-characterized" biopharmaceutical products have been discussed at length [305]. A central concern is that tests for identity, purity, impurities, and potency (mass of the active substance) should be sensitive, quantitative, and validated. Although HPLC measurements of mass eliminate much of the variability inherit in bioassays and are preferable for determination of dosage, whenever possible the product's potency should also be determined by a cell-based bioassay. In addition, variations in the primary structure, including posttranslational and process-related modifications, need to be detectible and evaluated. The tests used to characterize a biotechnology product and process or to establish release and stability specifications are uniquely determined by the particular cellular or formulation instability noted for each molecule. For example, monoclonal antibodies from (unstable) murine hybridoma cell lines can switch large sections of DNA (or peptide domains), but the resulting protein differences are readily detected either immunochemically or by ion-exchange (ABx) HPLC. Minor heterogeneity in the carbohydrate composition, in deamidation and in the C-terminal lysine residue are consistent and reproducible characteristics of monoclonals, which are either inconsequential or detectible through potency (antigen-binding) studies. On the other hand, production of recombinant products often involve incorrectly folded or aggregated structures, as well as single amino acid substitutions, more random attachment of carbohydrate (or no carbohydrate in the case of prokaryote expression), oxidations of methionine, and N-terminal or lysine alkylations. Each of these chemical modifications are often detectible using tryptic fragmentation, RP HPLC to separate the resulting peptides, and mass spectroscopy to detect differences in their molecular composition. The significance of HPLC techniques for many biopharmaceutical products is summarized in Table 5.9.

Table 5.9 Characterization Techniques for Biomolecules

Analytical parameter	Monoclonal antibody	Recombinant
Characterization	Carbohydrate map N-terminal sequence	Tryptic map (LC/MS) Peptide sequences
Identity	ABx-exchange (IEF–PAGE)	Ion exchange
Purity and impurities	Size exclusion (SDS-PAGE) (ELISA)	Size exclusion Reversed phase (SDS-PAGE)
Potency	Antigen binding (UV absorbance)	(Cell transduction)

C. Drugs Containing a Basic Functionality

Reversed Phase

As discussed in the beginning of this chapter, reversed-phase packings are the most widely used for analysis of basic drugs, but they generally give asymetric peaks. For quantitative analysis, an asymmetry factor of less than 1.5 is preferred.

Numerous studies have been performed on the relationship of peak symmetry and various commerically available reversed-phase packings which has led to several generalizations [306]. For compounds having a $pK_a < 6$, little or no asymmetry problems (factor < 1.5 and plate numbers 4000–6000) were observed. Asymmetric peaks were obtained for compounds with $pK_a > 6$. Structural parameters such as flexibility of the protonated N atom also seems to be a contributing factor. When tailing occurs, the addition of silanol-blocking agents, such as triethylamine, is the most effective approach to minimizing its occurrence. In addition, the use of electrostatically shielded phases improved peak shapes. Finally, polymeric-based stationary phases generally give an acceptable symmetry but a low plate number.

Silica

Over the last 10 years there has been an increasing number of publications that have demonstrated the ulility of using aqueous eluents with nonbonded silica for the analysis of not only basic analytes but also neutral and acidic pharmaceuticals.This approach was first demonstrated in 1975 for the screening and quantification of basic drugs in biological fluids [307]. Table 5.10 is a listing of the published drug product assays using this approach.

The second example in Table 10 demonstrates the advantages of this silica approach [308]. Using a mobile phase of methanol–water (75 : 25) buffered with ammonium phosphate at pH = 7.8, various syrups and tablets were analyzed for antihistamines, antitussives, and decongestants. A comparison between reversed-phase and silica methods of similar cough syrups clearly demonstrates that peak responses obtained by the aqueous silica method are more symmetric than the reversed-phase methods (compare Figure 5.8 with Figure 5.9). In addition, the sample preparation procedures in the silica method are relatively simple, requiring dilution for syrup formulations and dissolution for tablets.

Changes in mobile-phase components such as pH, ionic strength, and water content have been systematically studied [3,310,316,317]. These studies indicate that retention of basic analytes is mediated primarily by the cation-exchange properties of the silica [2]. Interestingly, it has been suggested from retention data of various pharmaceuticals that the retention mechanisms of silica with aqueous eluents and reversed-phase systems are similar [317,318]. Due to the ion-exchange properties of silica, mobile-phase pH adjustments are useful in changing the retention of ionic compounds.

In order to unambiguously ascertain the influence of surface silanols, the quaternary ammonium compound emepronium was studied [320]. The

Table 5.10 HPLC Assay for Silica of Drugs Containing Basic Functionalities

		References
Antihistamines, antitussives, and decongestants	Syrup and tablets	308, 309
Catecholamines	Various preparations	310
Famotidine	Oral dosages	3
Hydroxyzine hydrochloride	Syrups	311
Imidazoline derivatives	Capsules, ointments, and nasal drops	312
Prazosin hydrochloride	Capsules	313
Pseudoephedrine hydrochloride	Syrup and tablets	314
Succinylchloine	Injection	315
Triprolidine hydorchloride	Syrup	315

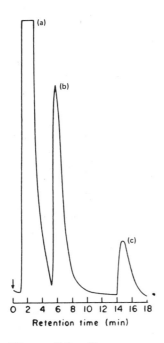

Retention time (min)

Figure 5.8 Chromatogram of cold syrup using C_{18} column with detection at 254 nm (a), diphenhydramine (b), and chlorpheniramine (c). (From Reference 319.)

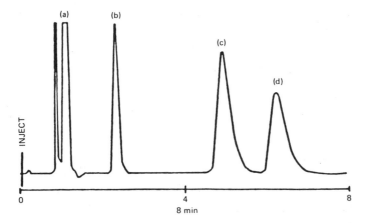

8 min

Figure 5.9 Resolution of acetaminophen (a), phenylpropanolamine (b), chlorpheniramine (c), and dextromethorphan (d) by silica gel using methanol, water, and phosphate buffer at pH-7.8. (Courtesy of *Journal of Pharmaceutical Sciences.*)

retention of emepronium increased with the increase of the mobile phase pH (0–11). This retention behavior was attributed to the various types of silanols present on silica with the strongly acidic sites being ionized even at low pHs and the weakly acidic sites ionized at neutral or basic pHs [320].

For other drugs that have tertiary, secondary, and primary nitrogens, the pKa values of the individual compounds are important in predicting retention when varying the pH of the mobile phase [320]. For example, the retention of phenylpropanolamine with a pKa of 9.0 rapidly decreases as it becomes the free base at a pH greater than 8. For lidocaine with a pKa of 7.9, the formation of its free base and decrease of retention occur at a pH greater than 7. In compounds such as benzocaine, which has a pKa at 2.8, little or no change occurs with pH variation in the mobile phase [3].

The pH of the mobile phase can also affect the peak shape in aqueous silica methods. Differences in solvation, which depends on the degree of protonation, has been cited as a possible explanation [320].

The nature of the ionic components of the mobile phase will affect analyte retention, as would be expected by an ion-exchange mechanism. In other words, retention of ionic analytes can be increased by ions of the opposite charge and decreased by ions of the same charge [317].

Addition of surfactants, such as cetyltrimethylammonium bromide (CTMA), causes silica to mimic separations obtained with reversed-phase sorbents. The impurities of propranolol and pharmaceutical preparations of catecholamines have been chromatographically studied using this surfactant [310,316]. The one apparent advantage of using these surfactants is that the brand-to-brand variations in selectivity commonly seen for bonded phases is avoided [310]. For basic analytes, the addition of either methanol or acetonitrile changes the sorbent selectivity. Commonly, a retention minimum occurs at about 50% organic solvent content with increases in retention at either increased or decreased organic content [312].

Additional interactions between silica and neutral and acidic analytes have been observed [3,317,321]. One example is the resolution of methylparaben from propylparaben, benzoic acid, and famotidine [3]. Retention mechanism studies appear to show that interactions with the siloxane bridges are important [322].

Antibiotics containing acidic functionalities have also been successfully chromatographed with aqueous silica systems [317,321,322].

Alumina

Bases that have been assayed by aqueous alumina methods are listed in Table 5.11 [27–29,47].

The amphoteric character of alumina leads to a more complex reten-

Table 5.11 Drugs Analyzed by
Alumina Using Aqueous Mobile Phases

Acetylcodeine	Narcotine
Atropine	Papaverine
Codeine	Procaine
Morphine	Strychnine
Naphazoline	Thebaine

tion mechanisms than that for silica. When the net charge on an alumina surface is zero, it is referred to as the zero point of charge (ZPC). By lowering the pH, the net charge on the surface becomes positive (anion exchanger), and at a pH higher than that of the ZPC, the alumina surface becomes negative (cation exchanger). ZPC is not a constant value but depends on the nature of the buffer being used in the mobile phase. For example, citrate buffer gives the alumina a ZPC at pH = 3.5 [28]. When the pH of the mobile phase incorporating citrate buffer exceeds 3.5, alumina becomes a cation exchanger. The ZPC of other buffers is listed in Table 5.12.

In developing a mobile-phase system for basic drugs, for example, brucine and dihydromorphine, the following consideration must be kept in mind. These two bases are positively charged below a pH of 6.5 and would be best chromatographed with a cation exchanger. According to the above discussions, using citrate buffer at a pH above 3.5, alumina becomes a cation exchanger. The other buffers listed in Table 5.11 could also be used, but their influence on chromatographic behavior of these base compounds is minimized because they would be anion exchangers below pH = 6.5.

Table 5.12 Zero-Point Charge of
Alumina in Various Buffers

Buffer	ZPC
Citrate	3.5
Acetate	6.5
Phosphate	6.5
Borate	8.3
Carbonate	9.2

Source: Adapted from Reference 28.

Increasing the ionic strength of the mobile phase generally decreases the retention of most analytes on ion exchange. This seems to be primarily due to increased competition of ions for the ion-exchange sites [29].

The addition of organic solvents such as methanol or acetonitrile improves selectivity of the sorbent. For the basic analytes studied, retention increased with decreasing acetonitrile content. For a decreasing methanol content, certain analytes increased in retention and others decreased [29].

D. Neutral Drugs—Steroids

Both nonbonded silica and bonded-phase sorbents have been utilized in the analysis of steroids. For the silica column, the mobile-phase composition is water-saturated butyl chloride : butyl chloride : tetrahydrofuran : methanol : glacial acetic acid [323]. This procedure has been adopted as method for 17 drugs formulated as tablets, suspensions, creams, lotions, ointments, and injectables. The apparent advantage of this approach over other possible approaches such as reversed phase is the following: a minimal number of assay variables for a relatively large number of drug products, consequently an increase in laboratories efficiency. For example, fluorometholone cream needs only to be dissolved in acetonitrile and extracted with hexane prior to assay by the above silica procedure [324]. This is in contrast to a reversed-phase method for a related glucocorticoid, betamethasone dipropionate, which requires the addition of methanol, heating, shaking, reheating, shaking, freezing, and centrifugation [325]. Consistent with the above examples, one proposal suggests that the above silica method be used only for oil-based formulations and that aqueous-based formulations should be analyzed by alkyl-bonded methods using aqueous mobile phases [326].

A third proposal for standardization of steroid analysis has been published. Once again, the procedure requires the use of silica but with aqueous mobile phases containing cetyltrimethylammonium bromide as a mobile-phase additive [327].

The authors compared this aqueous silica method with alkyl-bonded procedures by testing 12 different corticosteroids on 8 different silica sorbents and on 6 different reversed-phase sorbents. The variations in selectivity of the 12 corticosteroids among the silica columns were found to be substantially less than those based on the reversed-phase sorbents. Based solely on this comparison, silica-based separations using aqueous mobile phases would be preferred over reversed-phase sorbents for assay of steroids [328]. The arguments on which sorbent is best for the assay of steroids will more than likely continue.

E. Multidimension Column Techniques

Multidimensional chromatography has been called column switching, cou-
pled-column chromatography, recycle chromatography, and mode sequencing, among other terms.

There are two basic approaches: off-line and on-line. The off-line
method, as discussed in the chapters on sample pretreatment, are most
often used because they involve either manually or automatically collecting
a fraction from a sample cleanup sorbent. The appropriate fraction is transferred and then assayed by a second chromatographic method. The manual
steps are time-consuming and potentially introduce significant error to the
precision and accuracy of the method. The on-line method, when fully
automated, would have the chromatography system perform sample pretreatment by column switching between two or more columns.

For the on-line procedures there are numerous combinations of sorbents and mobile phases. The primary objective is to partially separate the
component(s) of interest on one column, followed by diversion of those
fractions of interest onto a second column. The sorbents used for each
column can be different, but the mobile phases must be miscible. To
achieve column switching, high-pressure, low-internal-volume, valves are
used. Numerous valve configurations are used and these are discussed
below.

The analysis of the *o*-hydroxybenzoate preservatives in pharmaceutical syrups and parenterals appears to be the first example of on-line cou-
pled-column analysis. These formulations were injected directly onto a
short column of Amberlite XAD-2 resin using a mobile phase with which
these preservatives are strongly retained. After washing the short column
with bisulfite reagent, which elutes interfering aldehydes and acids, the
short column is switched in-stream to a longer analytical column using a
mobile phase that elutes the preservatives. The procedure was shown to be
free of interferences from common drugs, dyes, flavoring agents, and other
excipients [329].

Calcium pantohenate (CP), the calcium salt of vitamin B_3, is a component of a variety of multivitamin formulations [330]. Column switching
using a C_{18} guard column with a C_8 analytical column has been utilized to
remove interfering material. The guard column retains the formulation
excipients which are back-flushed off the guard column (Figure 5.10).

In another method, creams and ointments of a developmental corticosteroid are simply dissolved in tetrahydrofuran–isopropanol (30 : 60), clarified by centrifugation, and injected directly onto the HPLC column [331].
The automated switching valves direct the analytes plus an internal standard through the guard column to the reversed-phase analytical column,

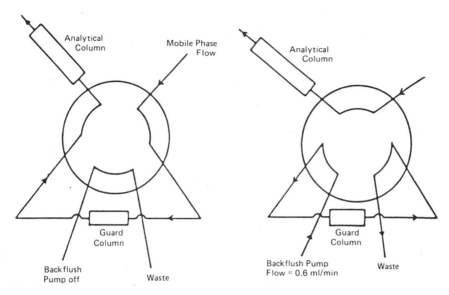

Figure 5.10 Typical valve configuration for multidimensional chromatography where the interfering matrix is retained on the guard column, whereas the analytes of interest are resolved on the analytical column.

while retained excipient materials are back-flushed to waste. Column switching reduces required sample preparation times by a factor of ~3.

A nearly identical procedure is used in the analysis of hydrocortisone and sulconazole nitrate in a topical cream [332].

Another useful approach to analyzing creams is to initially dissolve the formulation in tetrahydrofuran. The first column, which is packed with a gel permeation sorbent, resolves the analyte of interest from the cream excipients. The second column, which is packed with a C_{18} sorbent, further fractionates the analytes [333,334].

Nine structurally diverse substances of an analgesic tablet can be resolved by two columns containing C_8 material. The separation is better and faster than that obtained with a gradient elution method [335].

A three-column ion-exchange and reversed-phase system has been developed for the assay of enprostil in soft elastic gelatin capsules [336]. A two-column approach has also been used to resolve D,L-amino acids in complex matrices [337]. Erythromycin A and its known impurities were resolved using two C_{18} columns [338].

The major disadvantage of column switching is that some excipients may become irreversibly bound to the sorbents, consequently requiring relatively frequent changes in the sorbents. Under these circumstances, the off-line sample pretreatment cleanup procedures described in Chapter 2 should be used.

F. Antibiotics

A review of the chromatographic behavior of 90 penicillin and cephalsporins and their correlation to hydrophobicity has been published [339]. The current state of chromatographic methods submitted to the USP for complex antibiotics has been reviewed [340]. A comparison has been made between poly(styrene-divinylbenzene) stationary phases and silica-based reversed-phase sorbents for analysis of erthromycin and minocycline, and it was concluded that the nonsilica-based packings are more stable and reproducible [341]. The comparative retentions of ampicillin, amoxicillin, and pencillin G was determined on C_{18}, cyanopropyl–silica and poly(styrene-divinylbenzene), and nonbonded silica [342].

G. Size Exclusion

This technique for separating molecules according to size has been extensively used for the determination of polymeric molecular-weight distributions. The one big advantage this technique has over other chromatographic modes is that under ideal conditions, the analyst has merely to dissolve the sample in the mobile phase and inject it. The only decision is the choosing of the optimum pore size, which can be selected by knowing the molecular-weight operating range of the packed column and of the sample. The molecular weight of heparin solution has been determined in this manner [343]. Molecules (< 2000 Da) can be resolved if they have at least a 10% difference in size. This approach is not widely used; however, there are several examples in pharmaceutical analysis which demonstrate its utility. Aspirin in tablets has been easily resolved from salicyclic acid on a size-exclusion column using chloroform as the mobile phase [344]. In a similiar manner, tolnaftate was resolved from BHT using methylene chloride as the solvent [345].

Besides using a simple mobile phase requiring little or no method development, the sample preparation steps can be simplified. This has been shown in the analysis of benzocaine ointment which is dissolved in tetrahydrofuran, filtered, and injected (Figure 5.11). Using an identical approach, a solution of the corticosteroid halcinonide is resolved from the excipi-

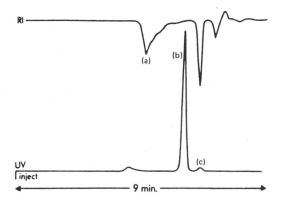

Figure 5.11 Analysis of benzocaine ointment using a size-exclusion column with UV and RI detection: Tween 85 (a), benzocaine (b), and benzyl alcohol (c). (Courtesy of Millipore Corporation, Waters Chromatography Division.)

ent antioxidant BHT. This approach has also been demonstrated for the assay of other steroidal creams [345].

H. Generalized Approaches

At this point, if the chromatographic problem is still unresolved, more general approaches should be investigated. A number of strategies have been developed that ideally would provide "first-run problem solving." The criterion is to elute all components by broad eluent polarity changes following by the detection of all the components. One approach is to use a gradient elution; optimized mobile-phase conditions for isocratic separations have been determined in this manner [346].

A second approach is to use TLC as a scouting technique. A good correlation from TLC to HPLC is generally obtained when mobile phases contain solvents equal to or less polar than ethyl acetate [347]. For solvents more polar than ethyl acetate, the correlation is not as useful because the more polar solvent absorbs preferentially to the TLC sorbent [347]. Under these circumstances, compensation for this change in TLC can be made by using a LC solvent mixture less polar than the TLC solvent mixture. Correlations of mobility between TLC and HPLC separations for cephalosporin antibiotics and steroidal hormones have been reported [348–350]. If all else fails a three part series on the basics of separations should be consulted [351–353].

V. CONCLUSION

The last decade has seen a tremendous improvement in HPLC instrumentation. However, the most significant changes have occurred in the column sorbent technology, detection, and automation. HPLC has become the primary tool in the pharmaceutical industry in characterizing the purity and impurities of pharmaceutical products.

In the next several years, advances in column technology particularly for the resolution of enantiomers and proteins will continue. This author also expects that sorbent manufacturers will be paying greater attention to minimizing the all too common column-to-column variations.

REFERENCES

1. A. Berthod, Silica: backbone material of liquid chromatorgaphic column packings, *J.Chromatogr.*, *549*:1–28 (1991).
2. M. Verzele, C. Deqaele, and D. Duquet, Quality criteria and structure of silica gel column packing material, *J. Chromatogr.*, *329*:351 (1985).
3. S. E. Buffar and D. J. Mazzo, Reversed-phase determination of famotidine, potential degradates and preservatives in pharmaceutical formulations by high-performance liquid chromatography using silica as a stationary phase, *J. Chromatogr.*, *353*:243 (1986).
4. B. W. Sands, Y. S. Kim, and J. L. Bass, Characterization of bonded-phase silica gels with different pore diameters, *J. Chromatogr.*, *360*: 353 (1986).
5. F. A. Buytenhuys and F. P. B. Van Der Maeden, Gel permeation chromatography or unmodified silica using aqueous solvents, *J. Chromatogr.*, *149*:489 (1978).
6. J. Nawrocki, D. Moir and W. Szczepaniak, Trace metal impurities in silica as a cause of strongly interacting silanols, *Chromatographia*, *28*: 143 (1989).
7. M. DePotter, M. Verzele, and J. Sijsels, Trace elements in HPLC silica gel, *HRC & CC J. High Res. Chromatogr. Chromatogr. Commun.*, *2*:151 (1979).
8. M. Verzele, Trace metals in silica gel-based HPLC packing materials, *LC Mag.*, *1*:217 (1983).
9. H. Muller, *Chromatogr. GIT 3*(Suppl.):43 (1983).
10. M. Verzele and C. Dewaele, Stationary phase characterizations in HPLC. A test for trace metal activity in octadecyl bonded silica gel, *J. Chromatogr.*, *217*:399 (1981).

11. E. Stahl, *Thin-Layer Chromatography. A Laboratory Handbook*, Springer-Verlag, Berlin, 1969, p. 1041.
12. H. Engehardt and H. Muller, Chromatographic characterization of silica surfaces *J. Chromatogr.*, *218*:395 (1981).
13. J. A. Knox, *High Performance Liquid Chromatography*, Edinburgh University Press, Edinburgh, 1978.
14. G. Schomberg, J. Koehlen, H. Figge, A. Deege, and U. Bren-Vogel Sang, Immobilization of stationary liquids on silica particles by γ-radiation, *Chromatographic*, *18*:265 (1984).
15. D. A. Hanggi and N. R. Marks, Introduction to chromatography on polybutadiene-coated zirconia, *LC-GC*, *11*:128 (1993).
16. J. Falkenhagen and P.G. Dietrich, New aspects concerning the stability of silica-based HPLC separation phases, *Amer. Lab.*, 48 (September 1994).
17. F. Nevejans and M. Verzele, On the structure and the chromatographic behaviour of polystyrene phases, *J. Chromatogr.*, *406*: 325–342 (1987).
18. J. R. Benson and D. J. Woo, Polymeric columns for liquid chromatography, *J. Chromatogr. Sci.*, *22*:386 (1984).
19. D. P. Lee, Reversed-phase HPLC pH 1 to 13, *J. Chromatogr. Sci.*, *20*:203 (1982).
20. R. E. Major, New approaches to reversed-phase chromatography columns, *LC–GC Mag.*, *4*:872 (1986).
21. D. P. Lee, A new anion exchange phase for ion chromatography, *J. Chromatogr. Sci.*, *22*:327 (1984).
22. D. Mac Blane, N. Kitagawa, and J. R. Benson, A C-18 derivatized polymer for reversed-phase chromatography, *Am. Lab.*, *2*:134 (1987).
23. *The United States Pharmacopeia*, Twenty-Third Revision, United States Pharmacopeia Convention, Inc., Rockville, MD, 1995, p. 2024.
24. A. Foucault and R. Rosset, Greffages non conventionnels des gels de silice: synthese, proprietes et applications des silices "exotiques" en chromatographie, *Analusis*, *17*:485–507 (1989).
25. T. Cserhati, Retention characteristics of an aluminium oxide HPLC, *Chromatographia*, *29*:593–596 (1990).
26. K. B. Holland, J. M. Washington, D. C. Moe, and C. M. Conroy, Alumina-based stationary phases for HPLC applications, *Am. Lab.*, 51 (February 1992).
27. B. C. Laurent, H. A. H. Billiet, and L. de Galan, On the use of alumina in HPLC with aqueous mobile phases at extreme pH, *Chromatographia*, *17*:253 (1983).

28. H. Billiet, C. Laurent, and L. de Galan, A reappreciation of alumina in HPLC, *TRAC*, *4*:100 (1985).

29. H. Lingeman, H. A. Van Munster, J. H. Beyner, W. J. M. Underberg, and A. Hulshoff, High-performance liquid chromatographic analysis of basic compounds on non-modified silica gel and aluminum oxide with aqueous solvent mixtures, *J. Chromatogr.*, *352*:261 (1986).

30. J. J. Pesek and H. D. Lin, *Chromatographia*, *28*:565 (1989).

31. B. Kaur, The use of porous graphitic carbon in high performance liquid chromatography, *LC–GC*, *8*:468 (1990).

32. J. A. Knox, B. Haur, and G. R. Millward, Structure and performance of porous graphitic carbon in liquid chromatography, *J. Chromatogr.*, *3*:353 (1986).

33. K. K. Unger, Porous carbon packings for liquid chromatography, *Anal. Chem.*, *55*:361A (1983).

34. H. Colin, N. Ward, and G. Cuiochon, Comparison of some packings for reversed-phase high-performance liquid chromatography, I. Some considerations, *J. Chromatogr.*, *149*:169 (1978).

35. H. Colin and G. Guiochon, Comparison of some packings for reversed-phase high performance liquid chromatography, II. Some considerations, *J. Chromatogr.*, *158*:182 (1978).

36. D. Prusova, H. Colin, and G. Guiochon, Liquid chromatography of adamantanes on carbon adsorbents, *J. Chromatogr.*, *234*:1 (1982).

37. E. Smolkova, J. Zima, F. P. Dousek, J. Jansta, and S. Pizak, A PTFE-based carbon adsorbent in HPLC, *J. Chromatogr.*, *191*:61 (1980).

38. B. J. Clark, Resolution of chiral compounds by HPLC using mobile phase additives and a porous graphitic carbon stationary phase, *J. Pharm. Biomed. Anal.*, *7*:1883–1888 (1989).

39. Metal Components, A potential source of interference in HPLC analysis, *All-Chrom Newslett.*, *25*:1 (1986); *LC–GC*, *7*:742 (1989).

40. J. S. Landy, J. L. Ward, and J. G. Dorsey, A critical evaluation of some stainless steel and radially compressed reversed-phase HPLC columns, *J. Chromatogr. Sci.*, *21*:45 (1983).

41. M. Konishi, Y. Mori, and T. Amano, High-performance packed glass-lined stainless steel capillary column for microcolumn liquid chromatography, *Anal. Chem.*, *57*:2235–2239 (1985).

42. J. C. Gfeller, R. Haas, J. M. Troendle, and F. Erni, Practical aspects of speed in high-performance liquid chromatography for the analysis of pharmaceutical preparation, *J. Chromatogr.*, *294*:247 (1984).

43. S. J. Van der Wal, Practical aspects of high-speed liquid chromatography, *LC Mag.*, *3*:488 (1985).

44. R. Gill and B. Law, Appraisal of narrow bore (1 mm I.D.) high-performance liquid chromatography with view to the requirements of routine drug analysis, *J. Chromatogr., 354*:185 (1986).
45. R. E. Majors, Trends in HPLC column usage, *LC Mag., 2*:660 (1984).
46. J. W. Dolan and V. Berry, Optical detectors, part II: Fixed-wavelength UV detectors, *LC Mag., 2*:365 (1984).
47. J. S. Adamovics, Return to unmodified silica and alumina chromatography, *LC Mag., 2*:393 (1984).
48. E. L. Inman, M. D. Lantz, and M. M. Strohl, Absorbance ratioing as a screen for related substances of pharmaceuticals, *J. Chromatogr. Sci., 28*: 578 (1990).
49. J. B. Castledine, A. F. Fell, R. Modin, and B. Sellberg, Absorbance-ratio matrix correlation: A novel approach to the use of absorbance ratios for the assessment of peak purity in liquid chromatography, 3rd Internat. Symp. of Pharm. Biomed. Anal., 1991.
50. L. Huber and S. A. George (eds.), *Diode Array Detection in HPLC*, Marcel Dekker, Inc., New York, 1993.
51. H. K. Chan and G. P. Carr, Evaluation of a photodiode array detector for the verification of peak homogeneity in high-performance liquid chromatography, *J. Pharm. Biomed. Anal., 8*:271–277 (1990).
52. D. Lincoln, A. F. Fell, N. H. Anderson, and D. England, Assessment of chromatographic peak purity of drugs by multivariate analysis of diode-array and mass spectrometric data, *J. Pharm. Biomed. Anal., 10*:837–844 (1992).
53. R. B. Taylor, D. G. Durham, A. S. Ashivji, and R. Reid, Development of a stability-indicating assay for nafimidone by high-performance liquid chromatography, *J. Chromatogr., 353*:51 (1986).
54. J. W. M. Wegener, H. J. M. Grunbaun, R. J. Fordham, and W. Karcher, A combined UV–VIS spectrophotometric method for the identification of cosmetic dyes, *J. Liquid Chromatogr., 7*:809 (1984).
55. G. Cavina, L. Valvo, B. Gallinella, R. Porra, and A. L. Savella, The identification of related substances in 9α-fluoroprednisolone-21 acetate by means of high-performance liquid chromatography with diode array detector and mass spectrometry, *J. Pharm. Biomed. Anal., 19*:437 (1992).
56. J. Wang and D. E. Moore, A study of the photodegradation of benzydamine in pharmaceutical formulations using HPLC with diode array detection, *J. Pharm. Biomed. Anal., 10*:535 (1992).

57. J. Gabriel, O. Vacek, E. Kubatova and J. Volc, High-performance liquid chromatographic determination of the antibiotic cortalcerone, *J. Chromatogr.*, *542*: 200–203 (1991).
58. K. Slais and M. Krejci, Vacant peaks in liquid chromatography, *J. Chromatogr.*, *91*:161 (1974).
59. H. Small and T. E. Miller, Indirect photometric chromatography, *J. Chromatogr.*, *91*:161 (1974).
60. J. Crommen and P. Herne, Indirect UV detection of hydrophilic ionized compounds in reversed-phase liquid chromatography by use of a UV-absorbing ion of the same charge, *J. Pharm. Biomed. Anal.*, *2*:241 (1984).
61. D. R. Jenke and N. Raghavan, Application of indirect photometric chromatography of pharmaceutical analysis, *J. Chromatogr. Sci.*, *23*:75 (1985).
62. J. H. Zou, S. Motomizu, and H. Fukutomi, Reversed-phase ion-interaction chromatorgaphy of inorganic anions with tetraalkylammonium ions and divalent organic anions using indirect photometric detection, *Analyst*, *116*:1399 (1991).
63. D. R. Jenke, Indirect determination of hydrogen peroxide by ion chromatography, *J. Chromatogr. Sci.*, *24*:352 (1986).
64. M. Massaccesi, Indirect UV detection in HPLC determination of UV-transparent non-electrolytes in pharmaceutical dosage forms, *Il Farmaco*, *47*(Suppl. 5):737–767 (1992).
65. R. G. Gerritse and J. A. Adenly, Rapid determination in water of chloride, sulphate, sulphite, selenite, selenate and arsenate among other inorganic and organic solutes by ion chromatography with UV detection below 195 nm, *J. Chromatogr.*, *347*:419 (1985).
66. R. Weinberger, Drug determination in biological fluids by liquid chromatography–fluorescence. In *Therapeutic Drug Monitoring and Toxicology by Liquid Chromatography* (S. H. Y. Wong, ed.), Marcel Dekker, Inc., New York, 1985, p. 151.
67. R. J. Flanagan and I. Jane, High-performance liquid chromatographic analysis of basic drugs on silica columns using non-aqueous ionic eluents, *J. Chromatogr.*, *323*:173 (1985).
68. L. S. King, Analysis of drugs encountered in fatal poinsoning using high-performance liquid chromatography and fluorescence detection, *J. Chromatogr.*, *208*:113 (1981).
69. *The United States Pharmacopeia*, Twenty-First Revision, United States Pharmacopeia Convention, Inc., Rockville, MD, 1985, p. 418.
70. K. Iwaki, N. Okumura, and M. Yamazaki, Determination of tetracycline antibiotics by reversed-phase high-performance liquid chroma-

tography with fluorescence detection, *J. Chromatogr.*, *623*:153–158 (1992).

71. B. Karol, J. C. Hodgen, P. Y. Howard, D. M. Ball, C. Cloete, and L. De Jager, An automated device for in situ pre-column derivatization and injection of amino acids for HPLC analysis, *J. Chromatogr. Sci.*, *21*:503 (1983).

72. J. A. Apffel, J. Van Der Lowz, K. R. Lammers, W. Th. Kok, U. A. Th. Brinkman, R. W. Frei, and C. Brugess, Analysis of neomycins A, B, and C by high-performance liquid chromatography with post-column reaction detection, *J. Pharm. Biomed. Anal.*, *3*:259 (1985).

73. E. Kwong, A. M. Y. Chiu, S. A. McClintock, and M. L. Cotton, HPLC analysis of an amino bisphosphonate in pharmaceutical formulations using postcolumn derivatization and fluorescence detection, *J. Chromatogr. Sci.*, *28*:563 (1990).

74. P. Gambardella, R. Punziano, M. Gionti, C. Guadalupi, G. Mancini, and A. Mangia, Quantitative determination and separation of analogues of aminoglucoside antibiotics by high-performance liquid chromatography, *J. Chromatogr.*, *348*:229 (1985).

75. J. Salamoun, M. Macka, M. Nechvatal, M. Matousek, and L. Knesel, Identification of products formed during UV irradiation of tamoxifen and their use for fluorescence detection in high-performance liquid chromatography, *J. Chromatography.*, *514*:1790–1807 (1990).

76. S. S. Chen, H. Lulla, F. J. Sena, and V. Reynoso, Simultaneous determination of potassium nitrate and sodium monofluorophosphate in dentifrices using single column ion chromatography, *J. Chromatogr. Sci.*, *23*:355 (1985).

77. E. W. Tsai, D. P. Ip, and M. A. Brooks, determination of alendronate in pharmaceutical dosage formulations by ion chromatography with conductivity detection, *J. Chromatogr.*, *596*:217–224 (1992).

78. W. A. McKinley, Application of the photo-conductivity detector to the liquid chromatographic analysis of pharmaceuticals in biological fluids, *J. Anal. Toxicol.* *5*:209 (1981).

79. C. Lavrich and P. T. Kissinger, Liquid chromatography electro-chemistry: potential utility for therapeutic drug monitoring. In *Therapeutic Drug Monitoring and Toxicology by Liquid Chromatography* (S. A. Y. Wong, ed.), Marcel Dekker, Inc., New York, 1985, p. 191.

80. F. Malecki and J. C. Crawhall, Determination of melphalan and chlorambucil in tablet dosage form using high-performance liquid chromatography and amperometric detection, *Anal. Lett.*, *23*:1685–1693 (1990).

81. X. Huang, W. Th. Kok, and H. Fabre, Determination of aztreonam by liquid chromatography with UV and amperometric detection, *J. Liquid Chromatogr.*, *14*:2721–2733 (1991).

82. T. A. Getek, A. C. Haneke, and G. B. Salzer, Determination of gentamicin sulfate C-1a, C-2 and C-1 components by ion-pair LC with electrochemical detection, *J. Assoc. Off. Anal. Chem.*, *66*:172 (1983).

83. R. V. Smith and D. W. Humphrey, Determination of apomorphine in tablets using high-performance liquid chromatography with electrochemical detection, *Anal. Lett.*, *14*:601 (1981).

84. K. Kamata, T. Hagiwara, M. Takahashi, S. Uehara, K. Nakayama, and K. Akiyama, Determination of biotin in multivitamin pharmaceutical preparations by HPLC with electrochemical detection, *J. Chromatogr.*, *356*:326 (1986).

85. J. T. Stewart and P. D. Arp, LCEC determination of propantheline bromide and xanthanoic acid in tablet dosage form, *LC–GC Mag.*, *4*:918 (1986).

86. J. P. Hart, M. D. Norman, and C. J. Lacey, Voltammetric behavior of vitamins D_2 and D_3 at a glassy carbon electrode and their determination in pharmaceutical products by using liquid chromatography with amperometric detection, *Analyst*, *117*:1441 (1992).

87. S. M. Waraszkiewicz, E. A. Milano, and R. L. DiRubio, A stability indicating LCEC method for epinephrine and its application to the analysis of lidocaine hydrochloride injectable solutions, *Curr. Sep.*, *4*(2):23 (1982).

88. W. J. Hurst and R. A. Martin, Using a silver electrode for the electrochemical detection of iodide, *Curr. Sep.*, *7*:9 (1985).

89. I. Jane, A. McKinnon, and R. J. Flanagan, High-performance liquid chromatographic analysis of basic drugs on silica columns using nonacqueous ionic eluents, *J. Chromatogr.*, *323*:191 (1985).

90. I. R. Krull, C. M. Selavka, K. Bratin, and I. Lurie, Trace analysis of drugs and biologically active materials using LC-photolysis-electrochemistry, *Curr. Sep.*, *70*:11 (1985).

91. C. M. Selavka, I. S. Krull, and K. Bratin, Analysis for penicillins and cefoperazone by HPLC-photolysis-electrochemical detection (HPLC-hv-EC), *J. Pharm. Biomed. Anal.*, *4*:83 (1986).

92. L. Dou, A. Holmberg, and I. S. Krull, Electrochemical detection of proteins in high-performance liquid chromatography using on-line, postcolumn photolysis, *Anal. Biochem.*, *197*:377–383 (1991).

93. C. M. selvaka and I. S. Krull, Trace determination of barbiturates with LC–photolysis electrochemical detection. In *Analytical Meth-*

ods in Forensic Chemistry (M. Ho, ed.), Ellis Horwood, London, 1990.

94. K. Bratin, P. T. Kissinger, and C. S. Brunlett, Reductive mode thin-layer amperometric detector for liquid chromatography, *J. Liquid Chromatogr.*, *4*:1777 (1981).
95. J. A. Polta, D. C. Johnson, and K. E. Merkel, Liquid chromatographic separation of aminoglycosides with pulsed amperometric detection, *J. Chromatogr.*, *324*:407–414 (1985).
96. D. A. Roston and R. R. Rhinebarger, Evaluation of HPLC with pulsed-amperometric detection for analysis of an aminosugar drug substance, *J. Liquid Chromatogr.*, *14*:539–556 (1991).
97. L. G. McLauglin and J. D. Henion, Determination of aminoglycoside antibiotics by reversed-phase ion-pair high-performance liquid chromatography coupled with pulsed amperometry and ion spray mass spectrometry, *J. Chromatogr.*, *591*:195–206 (1992).
98. Carbohydrate analysis of recombinant-derived erythropoietin, *LC-GC*, *11*:216 (1993).
99. R. Kumarasamy, Oligosaccharide mapping of therapeutic glycoproteins by high-pH anion-exchange high-performance liquid chromatography *J. Chromatogr.*, *512*:149–155 (1990).
100. M. Svoboda, M. Przyblski, J. Schreurs, A. Miyajima, K. Hogeland, and M. Deinzer, Mass spectrometric determination of glycosylation sites and oligosaccharide composition of insect-expressed mouse interleukin-3, *J. Chromatogr.*, *562*:403–419 (1991).
101. H. Krotkiewski, G. Gronberg, B. Krotkiewska, B. Nilsson, and S. Svensson, The carbohydrate structures of a mouse monoclonal IgG antibody OKT3, *J. Biol. Chem.*, *265*:20195–29291 (1990).
102. B. B. Wheals and P. C. White, In situ modification of silica with amines and its use in separating sugars by HPLC, *J. Chromatogr.*, *176*:421 (1979).
103. *The United States Pharmacopeia*, Twenty-First Revision, United States Pharmacopeia Convention, Inc., Rockville, MD, 1985, p. 18.
104. H. Cheng and R. R. Grade, Determination of propylene carbonate in pharmaceutical formulations using liquid chromatography, *J. Pharm. Sci.*, *74*:695 (1985).
105. R. F. Roberts and M. J. Fields, Monitoring radioactive compounds in high-performance liquid chromatographic eluates: fraction collection versus on-line detection, *J. Chromatogr.*, *342*:25 (1985).
106. M. J. Kessler, Quantitation of radiolabeled compounds eluting from the HPLC system, *J. Chromatogr. Sci.*, *20*:523 (1982).
107. T. E. Boothe, A. M. Emran, R. D. Finn, P. J. Kothari, and M. M.

Vora, Chromatography of radiolabeled anions using reversed-phase liquid chromatographic columns, *J. Chromatogr.*, *333*:269 (1985).

108. F. Hasaini, Quality assurance of HPLC procedures with radiotracers, *LC Mag.*, *1*:114 (1983).

109. R. E. Needham and M. F. Delaney, Liquid chromatographic separation of rhenium analogues of technetium radiopharmaceuticals, *LC Mag.*, *2*:760 (1984).

110. P.C. White, Recent developments in detection techniques for high-performance liquid chromatography, *Analyst*, *109*:677 (1984).

111. W. Holak, Analysis of mercury-containing drugs by HPLC with atomic absorption detection, *J. Liquid Chromatogr.*, *8*:73 (1985).

112. P. C. Uden, Specific element chromatographic detection by plasma spectral emission as applied to organometallics and gold containing drugs, *Methodol. Surv. Biochem. Anal.*, *14*:123 (1984).

113. J. L. D. Cesare and L. S. Ettre, New ways to increase the specificity of detection in liquid chromatography, *J. Chromatogr.*, *251*:1 (1982).

114. E. S. Yeung, New detection schemes in liquid chromatography, *J. Pharm. Biomed. Anal.*, *2*:235 (1984).

115. B. H. Reitsma and E. S. Yeung, High-performance liquid chromatographic determination of enantiomeric ratios of amino acids without chiral separation, *J. Chromatogr.*, *362*:353 (1986).

116. P. Salvadori, C. Bertucci, and C. Rosini, Circular dichroism detection in HPLC, *Chirality*, *3*:376–385 (1991).

117. T. A. Perfetti and J. K. Swadesh, On-line determination of the optical purity of nicotine, *J. Chromatogr.*, *543*:129–135 (1991).

118. C. Bertucci, E. Domenici, and P. Salvadori, Circular dichroism detection in HPLC: Evaluation of the anistropy factor, *J. Pharm. Biomed. Anal.*, *8*:843–846 (1990).

119. J. K. Swadesh, Applications of an on-line detector of optical activity, *Am. Lab.*, 72 (February 1990).

120. B. Blessington, A. Beiraghi, T. W. Lo, A. Drake, and G. Jonas, Chiral HPLC–CD studies of the antituberculosis drug (+)-ethambutol, *Chirality*, *4*:227 (1992).

121. A. Mannschreck, Chiroptical detection during liquid chromatography, *Chirality*, *4*:163 (1992).

122. P. A. Asmus and J. B. Landis, Analysis of steroids in bulk pharmaceuticals by liquid chromatography with light scattering detection, *J. Chromatogr.*, *316*:461–472 (1984).

123. H. H. Stuting, I. S. Krull, and R. Mhatre, High performance liquid chromatography of biopolymers, *LC–GC*, *7*:402 (1989).

124. P.J. Wyatt, C. Jackson, and G. K. Wyatt, Absolute GPC determina-

tion of molecular weights and sizes from light scattering, *Am. Lab.*, (May/June 1988).

125. M. Lafosse, C. Elfakir, L. Morin-Allory, and M. Dreux, The advanyages of evaporative light scattering detection in pharmaceutical analysis by HPLC and supercritical fluid chromatography, *J. High Res. Chromatogr.*, *15*:312 (1992).

126. J. E. DiNunzio, Pharmaceutical applications of HPLC interfaced with Fourier transform infrared spectroscopy, *J. Chromatogr.*, *626*: 97 (1992).

127. LC-Transform™, Lab Connections In

128. J. F. Garcia and D. Barcelo, An overview of LC–MS interfacing with selected applications, *HRC& CC, J. High. Res. Chromatogr. Chromatogr. Commun.*, *16*:633 (1993).

129. K. L. Tyczkowska, R. D. Voyksner, and A. L. Aronson, Solvent degradation of cloxacillin in vitro — tentative identification of degradation products using thermospray LC–MS, *J. Chromatogr.*, *594*: 195 (1992).

130. W. A. Korfmacher, T. A. Getek, E. B. Hansen, Jr., and J. Bloom, Characterization of diphenhydramine, doxylamine, and carbinoxamine using HPLC-thermospray MS, *LC–GC*, *8*:538 (1990).

131. Automated peptide mapping and sequencing by LC/ESI/MS techniques, *Genetic Eng. News*, *1*:17 (May 1994).

132. M. Verzele and C. Dewaele, Preparative liquid chromatography, *LC Mag.*, *3*:22 (1985).

133. G. Guiochon and H. Colin, Theoretical concepts and optimization in preparative scale liquid chromatography, *Chromatogr. Forum.*, 21, (Sept.–Oct. 1986).

134. Preparative and process liquid chromatography, *Anal. Chem.*, *57*: 998A (1985).

135. S. Golshan-Shirazi and G. Guiochon, Optimization in preparative liquid chromatorgaphy, *Am. Lab.*, 26 (June 1990).

136. J. V. Amari, P. R. Brown, P. E. Pivarnik, R. K. Sehgal, and J. G. Turcotte, Isolation of experimental anti-AIDS glycerophospholipids by micro-preparative reversed-phase HPLC, *J. Chromatogr.*, *590*: 153 (1992).

137. J. Adamovics and S. Unger, Preparative liquid chromatography of pharmaceuticals using silica gel with aqueous eluents, *J. Liquid Chromatogr.*, *9*:141 (1986).

138. R. G. Bell, Preparative HPLC separation and isolation of bacitracin components and their relationship to microbiological activity, *J. Chromatogr.*, *590*:163 (1992).

139. A. Katti, P. Erlandsson, and R. Dappenm Application of prepara-

tive liquid chromatography to the isolation of enantiomers of a ben-
zodiazepinone derivative, *J. Chromatogr.*, *590*:127 (1992).

140. S. A. Matlin, and L. Chan, Preparative HPLC. Part 2. Purification
      of beta-lactam derivatives using laboratory-assembled equipment,
      HRC&CC J. High Res. Chromatogr. Chromatogr. Commun., *8*:23,
      (1985).
141. R. Cantwell, R. Calderone, and M. Sienko, Process scale-up of a
      β-lactam antibiotic purification by HPLC, *J. Chromatogr.*, *316*:133
      (1984).
142. S. A. Matlin and L. Chan, Preparative HPLC. Part I: A comparison
      of three types of equipment for the purification of steroids,
      HRC&CC J. High Res. Chromatogr. Chromatogr. Commun., *7*:570
      (1984).
143. R. Sitrin, P. De Phillips, J. Dingerdissen, K. Erbard, and J. Filan,
      Preparative liquid chromatography, *LC Mag.*, *4*:530 (1986).
144. W. S. Letter, Preparative isolation of vitamin D_2 from previtamin D_2
      by recycle HPLC, *J. Chromatogr.*, *590*:169 (1992).
145. Issues of Industrial Chromatography News
146. M. S. Verall and M. J. Hudson (eds.), *Separations for Biotechnol-
      ogy*, Ellis Horwood Ltd., London, 1987, p. 502.
147. B. J. Moellering and C. P. Prior, Separation of clinical grade chim-
      eric antibodies from serum-derived immunoglogulins, *Biopharma-
      cology* (1990).
148. K. W. Talmadge and C. J. Siebert, Efficient endotoxin removal with
      a new sanitizable affinity column: Affi-Prep Polymyxin, *J. Chroma-
      togr.*, 175–185 (1989).
149. P. Knight, Bioseparations: media and modes, *Bio/Technolology 6*:
      220–201 (1990).
150. J. Hiyama and A. G. C. Renwick, Separation of human glycoprotein
      hormones and their subunits by reversed-phase liquid chromtogra-
      phy, *J. Chromatogr.*, *529*:33–41 (1990).
151. E. Watson and W. C. Kenney, Hydrophobic retention of proteins
      using high-performance size-exclusion chromatography, *Biotechnol.
      Appl. Biochem.*, *10*:551–554 (1988).
152. S. P. Monkarsh, E. A. Russoman, and S. K. Roy, Separation of
      interleukins by a preparative chromatofocusing-like method, *J.
      Chromatogr.*, *631*:227–280 (1993).
153. Faster processing of IgG with a Q Sepharose Fast Flow, *Downstream
      20*:61–17 (1995).
154. J. F. Kennedy, Z. S. Rivera, and C. A. White, The use of HPLC in
      biotechnology, *J. Biotechnol.*, *9*:83 (1989).
155. A. Drouen, J. W. Dolan, L. R. Synder, A. Poile, and P. J. Schoen-

makers, Software for chromatographic method development, *LC–GC*, *9*:714 (1991).

156.   L. R. Synder and D. C. Lommen, The use of a computer to select optimized conditions for HPLC separation, *Pharm. Biomed. Anal.*, *9*:611 (1991).

157.   Y. Zhang, H. Zou, and P. Lu, Advances in expert systems for HPLC, *J. Chromatogr.*, *515*:13–26 (1990).

158.   A. H. Stead, Standardised thin-layer chromatographic system for the identification of drugs and poisons, *Analyst*, *107*:1106 (1982).

159.   S. R. Lowry, G. L. Ritter, H. B. Woodruff, and T. L. Isenhour, Selecting liquid phases for multiple column gas chromatography from their eigenvector projections, *J. Chromatogr. Sci.*, *14*:126 (1976).

160.   D. Helton, M. Ready, and D. Sacks, Advantages of a common column and common mobile phase system for steroid analysis by normal phase high-performance liquid chromatography by W. F. Beyer — comments received, *Pharmacopeial Forum*, *2794* (1983).

161.   P. Conti, T. Hamoir, M. De Smet, N. Vanden Driessche, F. Maris, H. Hindriks, P. J. Schoenmakers, and D. L. Massart, Integrating expert systems for HPLC method development, *Chemometrics Intell. Lab. Syst.*, *11*:27 (1991).

162.   G. Szepesi and K. Valko, Predicitin of initial HPLC conditions for selectivity optimizations in pharmaceutical analysis by an expert system approach, *J. Chromatogr.*, *550*:87–100 (1991).

163.   N. G. Mellish, Computer-assisted HPLC method development in a pharmaceutical laboratory, *LC–GC*, *9*:845 (1990).

164.   F. Maris, R. Hindriks, J. Vink, A. Peeters, N. Vanden Driessche, and L. Massart, Validation of an expert system for the selection of initial HPLC conditions for analysis of basic drugs, *J. Chromatogr.*, *506*:211 (1990).

165.   P. F. Vanbel, J. A. Gilliard, and B. Tiliquin, Chemometric optimization in drug analysis by HPLC: a critical evaluation of the quality criteria used in the analysis of drug purity, *Chromatographia*, *36*:120 (1993).

166.   J. C. Berridge, P. Jones, and A. S. Roberts-McIntosh, Chemometrics in pharmaceutical analysis, *J. Pharm. Biomed. Anal.*, *9*:507 (1991).

167.   J. Wieling, J. Schepers, J. Hempenius, C. K. Mensink, and J. H. G. Jonkman, Optimization of chromatographic selectivity of twelve sulphonamides in reversed-phase HPLC using mixture designs and multi-criteria decision making, *J. Chromatogr.*, *545*:101 (1991).

168.   L. Wrisley, Use of computer simulations in the development of gra-

dient and isocratic HPLC methods for analysis of drug compounds and synthetic intermediates, *J. Chromatogr.*, *628*: 191 (1993).

169. T. C. Daniel and E. C. Jorgensen, *Textbook of Organic Medicinal and Pharmaceutical Chemistry*, 7th ed. (C. O. Wilsar, O. G. Gisvald, and R. F. Doerge, eds.), J. B. Lippencott, Philadelphia, 1977.

170. Y. W. F. Lam, Stereoselectivity: an issue of significant importance in clinical pharmacology, *Pharmacotherapy*, *8*:147 (1988).

171. G. T. Tucker and M. S. Lennard, Enantimer specific pharmacokinetics, *Pharmacol. Ther.*, *45*:309 91990).

172. T. J. Wozniak, R. J. Bopp, and E. C. Jensen, Chiral drugs: an industrial prespective, *J. Pharm. Biomed. Anal.*, *9*:363 (1991).

173. R. W. Souter, The determination of isomeric purity. In *Modern Methods of Pharmaceuticals Analysis* (R. E. Schirmer, ed.), CRC Press, Boca Raton, FL, 1982, p. 173.

174. K. Gunther, J. Martens, and M. Schickendanz, Resolution of optical isomers by thin-layer chromatography, enantiomeric purity of D-penicillamine, *Arch. Pharm.* (*Weinheim*), *319*:461 (1986).

175. *C & E News*, Sept. 28, 1992, p. 47.

176. *C & E News*, Sept. 27, 1993, p. 38.

177. R. T. Brittain, G. M. Drew, and G. P. Ley, The alpha- and beta-adrenoceptor blocking potencies of labetatol and its inidividual stereoisomers in dogs, *Br. J. Pharmacol.*, *77*:105 (1982).

178. J. A. Thompson, J. L. Holtzman, M. Isuru, and C. L. Lerman, Procedure for the chiral derivatization and chromatographic resolution of R(+) and S(−) propranolol, *J. Chromatogr.*, *238*:470 (1982).

179. *CN & E News*, Sept. 28, 1992, p. 56.

180. C. Banfield and M. Rowland, Stereospecific high-performance liquid chromatographic analysis of warfarin in plasma, *J. Pharm. Sci.*, *238*:470 (1982).

181. G. Helmchen, G. Nill, D. Flockerzi, W. Schuhle, and M. Youssef, Extreme liquid chromatographic separation effects in the case of diastereometric amides containing polar substituents, *Angew. Chem. Int. Ed. Engl.*, *18*:62 (1979).

182. W. Pirkle and J. Hauske, Broad spectrum methods for the resolution of optical isomers: A discussion of the reasons underlying the chromatographic separability of some diastereomeric carbamates, *J. Organ. Chem.*, *42*:1839 (1977).

183. W. Dieterle and J. W. Faigle, Multiple inverse isotope dilution assay for the stereospecific determination of R(+) and S(−) oxprenolol in biological fluids, *J. Chromatogr.*, *259*:311–318 (1983).

184. N. R. Srinivas and L. N. Igwemezie, Chiral separation by HPLC.

I. Review on indirect separation of enantiomers as diastereomeric derivatives using UV, fluorescence and electrochemical detection, *Biomed. Chromatogr.*, *6*:163 (1992).

185. P. M. Miller, R. T. Foster, F. T. Pasutto, and F. Jamali, *J. Chromatogr.*, *526*:129 (1990).

186. A. A. Gulaid, G. W. Houghton, and A. R. Boobis, *J. Chromatogr.*, *318*:393 (1985).

187. J. Hermansson and C. Von Bahr, *J. Chromatogr.*, *227*:113 (1982).

188. H. Weber, H. Spahn, E. Mutschler, and W. Mohrke, *J. Chromatogr.*, *307*:145 (1984).

189. I. Demian and F. Gripshover, *J. Chromatogr.*, *466*:415 (1989).

190. S. K. Chin, A. C. Hui, and K. M. Giacomini, *J. Chromatogr.*, *489*: 438 (1989).

191. R. Mehvar, *J. Pharm. Sci.*, *78*:1035 (1989).

192. M. J. Wilson, K. D. Ballard, and T. Walle, *J. Chromatogr.*, *431*: 222 (1988).

193. E. Wuis, E. W. J. Beneken Kolmer, L. E. C. Van Beijsterveldt, R. C. M. Burgers, T. B. Vreem, and E. Van Der Kleyn, *J. Chromatogr.*, *415*:419 (1987).

194. H. Spahn, D. Kraub, and E. Mutschler, *Pharm. Res.*, *5*:107 (1988).

195. H. Spahn, I. Spahn, G. Pflugmann, E. Mutshler, and L. Z. Benet, *J. Chromatogr.*, *433*:331 (1988).

196. A. J. Sedman and J. Gal, *J. Chromatogr.*, *278*:199 (1983).

197. K. J. Miller, J. Gall, and M. M. Ames, J. Chromatogr., *307*:335 (1984).

198. R. Shimizu, K. Ishii, N. Tsumagari, M. Tanigawa, M. Matsumoto, and I. T. Harrison, *J. Chromatogr.*, *253*:101 (1982).

199. C. Prakash, H. K. Jajoo, I. A. Blair, and R. F. Mayol, *J. Chromatogr.*, *493*:325 (1989).

200. C. R. Clark and J. M. Barksdale, *Anal. Chem.*, *56*:958 (1984).

201. J. Gal, *J. Chromatogr.*, *307*:220 (1984).

202. N. Nimura, Y. Kasahara, and T. Kinoshita, *J. Chromatogr.*, *213*: 327 (1981).

203. F. Jamali, R. Mehvar, C. Lemko, and O. Eradiri, *J. Pharm. Sci.*, *77*:963 (1988).

204. B. C. Sallustio, A. Abas, P. Hayball, Y. J. Purdie, and P. J. Meffin, *J. Chromatogr.*, *374*:329 (1986).

205. R. Mehvar and F. Jamali, *Pharm. Res.*, *5*:53 (1988).

206. H. F. Schmittenner, M. Fedorchuk, and D. J. Walter, *J. Chromatogr.*, *487*:197 (1989).

207. J. Turgeon, H. K. Kroemer, C. Prakash, I. A. Blair, and D. M. Roden, *J. Pharm. Sci.*, *79*:265 (1990).

208. S. Severini, F. Jamali, F. M. Pasutto, R. T. Coutts, and S. Gulam-husein, *J. Pharm. Sci.*, *79*:257 (1990).
209. J. Gal, D. M. Desai, and S. M. Lehnert, *Chirality*, *2*:43 (1990).
210. S. Pedrazzini, W. Z. Muciaccia, C. Sacchi, and A. Forgione, *J. Chromatogr.*, *415*:214 (1987).
211. B. W. Berry and F. Jamali, *Pharm. Res.*, *5*:123 (1988).
212. M. P. Knadler and S. D. Hall, *J. Chromatogr.*, *494*:173 (1984).
213. S. A. Matlin, A. Belenguer, R. G. Tyson, and A. N. Brookes, *J. High Res. Chromatogr.*, *10*:86 (1987).
214. D. F. Wu, M. M. Reidenberg, and D. E. Drayer, *J. Chromatogr.*, *433*:141 (1988).
215. E. D. J. Lee, K. M. Williams, G. G. Graham, R. O. Day, and G. D. Champion, *J. Pharm. Sci.*, *73*:1542 (1984).
216. M. R. Euerby, L. Z. Partridge, and P. Rajani, *J. Chromatogr.*, *447*: 392 (1988).
217. H. Nagashima, Y. Tanaka, H. Watanabe, R. Hayashi, and K. Kawada, *Chem. Pharm. Bull.*, *32*:251 (1984).
218. R. T. Foster and F. Jamali, *J. Chromatogr.*, *416*:388 (1987).
219. R. Mehvar, *J. Chromatogr.*, *493*:402 (1989).
220. M. Ahnoff, S. Chen, A. Green, and I. Grundevik, *J. Chromatogr.*, *506*:593 (1990).
221. G. Plugmann, H. Spahn, and E. Mutschler, *J. Chromatogr.*, *416*: 331 (1987).
222. Y. Fujimoto, K. Ishi, H. Nishi, N. Tsugumari, T. Kakimoto, and R. Shimizu, *J. Chromatogr.*, *402*:344 (1987).
223. R. Hasegawa, M. Kushiya, T. Komuro, and T. Kimura, *J. Chromatogr.*, *494*:381 (1989).
224. A. Siouffi, D. Colussi, F. Matfil, and J. P. Dubois, *J. Chromatogr.*, *414*:131 (1987).
225. T. Walle, D. D. Christ, U. K. Walle, and M. J. Wison, *J. Chromatogr.*, *341*:213 (1985).
226. Y. Gietl, H. Spahn, and E. Mutschler, *J. Chromatogr.*, *426*:305 (1988).
227. M. J. Wilson and T. Walle, *J. Chromatogr.*, *310*:333 (1984).
228. J. Hermansson and C. C. Von Bahr, *J. Chromatogr.*, *221*:109 (1980).
229. B. Silber, and S. Riegelman, *J. Pharmacol. Exp. Ther.*, *215*:643 (1980).
230. R. J. Guttrndorf, H. B. Kostenbauder, and P. J. Wedlund, *J. Chromatogr.*, *489*:333 (1989).
231. S. Laganiere, E. Kwong, and D. D. Shen, *J. Chromatogr.*, *488*:407 (1989).

232. H. Langguth, B. Padkowik, E. Stahl, and E. Mitschler, *J. Anal. Toxicol.*, *15*:209 (1991).
233. W. Linder, M. Rath, K. Stoschitzky, and G. Uray, *J. Chromatogr.*, *487*:375 (1989).
234. M. Christensen, A. J. Aasen, K. E. Sasmussen, and B. Salvesen, *J. Chromatogr.*, *491*:355 (1989).
235. E. Pianezzola, V. Bellotti, E. Fontana, E. Moro, J. Gal, and D. M. Desai, *J. Chromatogr.*, *495*:205 (1989).
236. M. Benedetti, V. Belloti, E. Pianezzola, E. Moro, P. Carminati, and P. Dostert, *J. Neural Transmission*, *77*:47 (1989).
237. G. Lovell and P. H. Corran, *J. Chromatogr.*, *525*:287 (1990).
238. K. Hoffmann, L. Renberg, and C. Baarhielm, *Eur. J. Drug Metab. Pharmacokinet.*, *9*:215 (1984).
239. R. A. Carr, R. T. Foster, D. Freitag, and F. M. Pasutto, *J. Chromatogr.*, *566*:155 (1991).
240. C. Banfield and M. Rowland, *J. Pharm. Sci.*, *72*:921 (1983).
241. E. Grushka and R. Lesham, The use of eluents containing metal cations in high performance liquid chromatography, *TRACS, 1*:95 (1981).
242. Y. Yuang and X. Zheng, *Chirality*, *1*:92 (1989).
243. K. Shimada, E. Haniuda, T. Oe, and T. Nambara, *J. Liquid Chromatogr.*, *10*:3161 (1987).
244. H. K. Lim, Master's Thesis, University of Saskatchewan, Saskatoon, Canada.
245. R. Wernicke, Separation of underivatized amino acid enantiomers by means of a chiral solvent-generated phase, *J. Chromatogr.*, *318*:117 (1985).
246. S. Weinstein and N. Grinberg, Enantiomeric separation of underivatized α-methyl-α-amino acids by high-performance liquid chromatography, *J. Chromatogr.*, *318*:117 (1985).
247. E. Busker, K. Gunther, and J. Martens, Application of chromatographic chiral stationary phases to pharmaceutical analysis, *J. Chromatogr.*, *350*:179 (1985).
248. E. Oelrich, H. Preusch, and E. Wilhelm, Separation of enantiomers by high-performance liquid chromatography using chiral eluents, *HRC&CC, J. High Res. Chromatogr. Chromatogr. Commun.*, *3*:269 (1980).
249. J. Florance, High-performance liquid chromatographic separation of peptides and amino acid stereoisomers, *J. Chromatogr.*, *414*:313 (1987).
250. V. A. Davankov and A. A. Kurganov, The role of achiral sorbent matrix in chiral recognition of amino acid enantiomers in ligand-exchange chromatography, *Chromatographia*, *17*:696 (1983).

251. C. Pettersson and G. Schill, Chiral resolution of amino alcohols by ion-pair chromatography, *Chromatographia, 16*:192 (1982).

252. C. Pettersson and H. W. Stuurman, Direct separation of enantiomer of ephedrine and some analogues by reversed-phase liquid chromatography using (+)-di-*n*-butyitartrate as the liquid stationary phase, *J. Chromatogr. Sci., 22*:441 (1984).

253. C. Pettersson and K. No, Chiral resolution of carboxylic and sulfonic acids by ion-pair chromatography, *J. Chromatogr., 282*:671 (1983).

254. C. Pettersson, Chromatographic separation of enantiomers of acids with quinine as chiral counter ion, *J. Chromatogr., 316*:553 (1984).

255. C. Pettersson, T. Arvidsson, A.-L. Karlsson, and I. Marle, Chromatographic resolution of enantiomers using albumin as complexing agent in the mobile phase, *J. Pharm. Biomed. Anal., 4*:221 (1986).

256. K. C. Cundy and P. A. Crook, Unexpected phenomenon in the high-performance liquid chromatographic analysis of racemic ^{14}C-labelled nicotine, separation of enantiomers in a totally achiral system, *J. Chromatogr., 281*:17 (1983).

257. R. Dappen, H. Arm, and V. R. Meyer, Applications and limitations of commercially available chiral stationary phases for high-performance liquid chromatography, *J. Chromatogr., 373*:1 (1986).

258. I. Wainer and M. Alembik, The enantiomeric resolution of biologically active molecules. In *Chromatographic Chiral Separations* (M. Zief and L. Crane, eds.), Marcel Dekker, Inc., New York, 1987.

259. H. G. Kicinski and A. Kettup (Columns available from either Serva, Daicel) *Fresenius Zf. Analf. Chem., 320*:51 (1985).

260. G. Gubitz, F. Juffmann, and E. Jellens, Direct separation of amino enantiomers by high-performance ligand exchange on chemically bonded chiral phases, *Chromatographia, 16*:103 (1982).

261. W. H. Pirkle, M. H. Hyun, A. Tsiporas, B. C. Hamper, and B. Banks, A rational approach to the design of highly effective chiral stationary phases for the liquid chromatographic separation of enantiomers, *J. Pharm. Biomed. Anal., 2*:173 (1984).

262. W. H. Pirkle, J. M. Finn, J. L. Schreiner, and B. L. Hamper, A widely useful chiral stationary phases for the high-performance liquid chromatography of enantiomers, *J. Am. Chem. Soc., 103*:3964 (1981).

263. W. H. Pirkle, D. W. House, and J. E. Finn, Broad spectrum resolution of optical isomers using chiral high-performance liquid chromatographic bonded phase, *J. Chromatogr., 192*:143 (1980).

264. I. W. Wainer, T. D. Doyle, and W. M. Adams, Liquid chromato-

graphic chiral stationary phases in pharmaceutical analysis: determination of trace amount of the (−)-enantiomer in (+)-amphetamine, *J. Pharm. Sci.*, 73:1162 (1984).

265. E. D. Lee, J. D. Henion, C. A. Brunner, I. W. Wainer, T. D. Doyle, and J. Gal, High-performance liquid chromatographic chiral stationary phase separation with filament on thermospray mass spectrometric identification of the enantiomer contaminant (S)-(+)methamphetamine, *Anal. Chem.*, 58:1349 (1986).

266. C. A. Domerson, L. G. Humber, N. A. Abraham, G. Schilling, R. R. Mantel, and C. Pace-Ascik, Resolution of etodalac and anti-inflammatory and prostaglandin synthetase inhibiting properties of the enantiomers, *J. Med. Chem.*, 26:1778 (1980).

267. W. H. Pirkle, H. Tsipouras, M. H. Hyan, D. J. Hart, and C.-S. Lee, Use of chiral stationary phases for the chromatographic determination of enantiomeric purity and absolute configuration of some β-lactams, *J. Chromatogr.*, 358:377 (1986).

268. H. Yuki, Y. Okamoto, and I. Okamoto, Resolution of racemic compounds by optically active poly (triphenylmethyl methacrylate), *J. Am. Chem. Soc.*, 102:6356 (1980).

269. Y. Okamoto and K. Hatada, Optically active poly (triphenymethyl methacrylate) as chiral stationary phases. In *Chromatographic Chiral Separation* (M. Zief and L. Crane, eds.), Marcel Dekker, Inc., New York, 1987.

270. H. Yamaguchi, Y. Itakura, and K. Kunihiro, Analysis of the stereoisomers of alpha-tocopheryl acetate by HPLC (high performance liquid chromatography), *Iyakuhin Kenkyu*, 15:536 (1984).

271. T. J. Ward and D. W. Armstrong, Cyclodextrin bonded phases. In *Chromatographic Chiral Separations* (M. Zief and L. Crane ed.), Marcel Dekker, Inc., New York, 1987.

272. M. Pawlowska, Enantiomer separations by normal HPLC systems with permethylated β-cyclodextrin dynamically coated on silica solid supports, *J. Liquid Chromatogr.*, 14:2273 (1991).

273. M. Pawlowska, Separating anatiomers by HPLC on silica dynamically coated with a chiral stationary phase of permethylated-cyclodextrin, *Chirality*, 3:136 (1991).

274. K. Cabrera and G. Schwinn, Cyclodextrins as movile phase additives in the HPLC separation of enantiomers, *Am. Lab.*, 22 (June 1990).

275. N. Thuaud, B. Sebille, A. Deratani, and G. Lelievre, Retention behavior and chiral recognition of β-cyclodextrin derivative polymer adsorbed on silica for warfarin, structurally related compounds and Dns-amino acids, *J. Chromatogr.*, 555:53 (1991).

276. Y. Okamoto, M. Kawashima, and K. Hatada, Optical resolution of racemic drugs by chiral HPLC on cellulose and amylose tris(phenyl-carbamate) derivative, *J. Liquid Chromatogr.*, *11*:2147 (1988).

277. S. R. Narayanan, Immobilized proteins as chromatographic supports for chiral resolution, *J. Pharm. Biomed. Anal.*, *10*:251 (1992).

278. I. Marle, A. Karlsson, and C. Pettersson, Separation of enantiomers using α-chymotrypsin-silica as a chiral stationary phase, *J. Chromatogr.*, *604*:185 (1992).

279. M. Enquist and J. Hermansson, Separation of the enantiomers of β-receptor blocking agents and other cationic drugs using a CHIRAL–AGP column: Binding properties and characterization of immobilized α_1-acid glycoprotein, *J. Chromatogr.*, *519*:285 (1990).

280. J. Haginaka, J. Wakai, K. Takahashi, H. Yasuda, and T. Katagi, Chiral separation of propranolol and its ester derivatives in ovomucoid-bonded silica: Influence of pH, ionic strenght and organic modifier on retention, enantioselectivity and enantiomeric elution order, *Chromatographia*, *29*:587 (1990).

281. L. Oliveros, C. Minguillon, B. Desmazieres, and P.-L. Desbene, Preparation and evaluation of chiral HPLC stationary phases of mixed character for the resolution of racemic compounds, *J. Chromatogr.*, *543*:277 (1991).

282. J. R. Kern, Chromaotgraphic separation of the optical isomers of naproxen, *J. Chromatogr.*, *543*:355 (1991).

283. J. Haginaka, Ch. Seyama, H. Yasuda, and K. Takahashi, Retention, enantioselectivity and enantiomeric elution order of propranolol and its ester derivatives on an alpha$_1$-acid glycoprotein-bonded column, *Chromatographia*, *33*:127 (1992).

284. J. Haginaka, C. Seyama, H. Yasuda, and K. Takahashi, Investigation of enantioselectivity and enantiomeric elution order of propranolol and its ester derivatives on an ovomucoid-bonded column, *J. Chromatogr.*, *598*:67 (1992).

285. K. M. Kirkland, K. L. Neilson, D. A. Mc Combs, and J. J. DeStefano, Optimized HPLC separations of racemic drugs using ovomucoid protein-based chiral column, *LC–GC*, *10*:322 (1992).

286. J. Vanggaard Andersen and S. H. Hansen, Simultaneous determination of (R)- and (S)-naproxen and (R)- and (S)-6-O-desmethyl-naproxen by HPLC on a Chiral-AGP column, *J. Chromatogr.*, *577*:362 (1992).

287. T. A. G. Noctor and I. W. Wainer, The use of displacement chromatography to alter retention and enantioselectivity on a human serum albumin-based HPLC chiral stationary phase: A mini review, *J. Liquid Chromatogr.*, *16*:783 (1993).

288. I. W. Wainer, S. A. Schill, and G. Schill, α-Acid glycoprotein chiral stationary phase, *LC–GC Mag.*, *4*:422 (1986).

289. C. Vandenbosch, D. Luc Massart, and W. Linder, Evaluation of six chiral stationart phases in LC for selectivity towards drug enantiomers, *J. Pharm. Biomed. Anal.*, *10*:895 (1992).

290. A. M. Dyas, The chiral chromatographic separation of β-adrenoceptor blocking drugs, *J. Pharm Biomed. Anal.*, *10*:383 (1992).

291. F. A. Maris, R. J. M. Veroort, and H. Hindriks, applicability of new chiral stationary phases in the separation of racemic pharmaceutical compounds by HPLC, *J. Chromatogr.*, *547*:45 (1991).

292. I. W. Wainer, Some observations on choosing an HPLC chiral stationary phase, *LC–GC*, *7*:378 (1989).

293. J. V. O'Connor, The use of peptide mapping for the detection of heterogeneity in recombinant DNA-derived proteins, *Biologicals*, *21*: 111–117 (1993).

294. G. Krystal, H. R. C. Pankratz, N. M. Farber, and J. E. Smart, Purification of human erythropoietin to homogeneity by a rapid five-step procedure, *Blood 67*:71–79 (1986).

295. K. Asahara, H. Y. Yamada, and S. Yoshida, Stability of human insulin in solutions containing sodium bisulfite, *Chem. Pharm. Bull.*, *39*:2662–2666 (1991).

296. P. S. Adams and R. F. Haines-Nutt, Analysis of bovine, porcine and human insulins in pharmaceutical dosage forms and drug delivery systems *J. Chromatogr.*, *351*:574 (1986).

297. E. Conova-Davis, I. P. Baldonado, J. A. Moore, C. G. Rudman, W. F. Bennett, and W. S. Hancock, Properties of a cleaved two-chain form of recombinant human growth hormone, *Int. J. Peptide Protein Res.*, *35*:17–24 (1990).

298. E. Watson and W. C. Kenney, High-performance size exclusion chromatography of recombinant derived proteins and aggregated species, *J. Chromatogr.*, *436*:289–298 (1988).

299. R. M. Riggin, C. J. Shaar, C. K. Dorulla, D. S. Lefeger, and D. J. Miner, High-performance-exclusion chromatographic determination of the potency of biosynthetic human growth products, *J. Chromatogr.*, *435*:307 (1988).

300. N. A. Farid, L. M. Atkins, G. W. Becker, A. Dinner, R. E. Heiney, and D. J. Miner, Liquid chromatographic control of the identity, purity and "potency" of biomolecules used as drugs, *J. Pharm. Biomed Anal.*, *7*:185 (1989)

301. T. P. Patel, R. B. Parekh, B. J. Moellering, and C. P. Prior, Different culture methods lead to differences in glycosylation of a murine IgG monoclonal antibody, *Biochem. J. 285*:839–845 (1992)

302. M. Weitzhandler, M. Hardy, M. S. Co, and Avdalovic, Analysis of carbohydrates on IgG preparations, *J. Pharm. Sci.*, *83*:1670 (1994).

303. C. L. Clogston, S. Hu, T. C. Boone, H. S. Lu, Glycosidase digestion electrophoresis and chromatographic analysis of recombinant human granulocyte colony-stimulating factor glycoforms produced in Chinese hamster ovary cells, *J. Chromatogr.*, *637*:55–62 (1993).

304. A. Riggin, J. R. Sportsmen, and F. E. Regnier, Immunochromatographic analysis of proteins: identification, characterization and purity determination, *J. Chromatogr.*, *632*:37–44 (1993).

305. Characterization of biotechnology pharmaceutical products, FDA Workshop Proceedings, Washington DC, 1995.

306. R. J. M. Vervoort, F. A. Maris, and H. Hindriks, Comparison of HPLC methods for the analysis of basic drugs, *J. Chromatogr.*, *623*: 207 (1992).

307. I. Jane, The separation of a wide range of drugs of abuse by HPLC, *J. Chromatogr.*, *111*:227 (1975).

308. B. A. Bildenmeyer, J. K. Del Rios, and J. Korpi, Separation of organic amine compounds on silica gel with reversed-phase eluents, *Anal. Chem.*, *54*:442 (1982).

309. H. Richardson and B. A. Bildenmeyer, Bare silica as a reversed-phase stationary phase: liquid chromatographic separation of antihistamines with buffered aqueous organic mobile phases, *J. Pharm. Sci.*, *73*:1480 (1984).

310. P. Helboe, Separation of catacholamines by high-performance liquid chromatography on dynamically modified silica, *J. Pharm. Biomed. Anal.*, *3*:293 (1985).

311. *The United States Pharmacopeia*, Twenty-First Revision, United States Pharmacopeia Convention, Inc., Rockville, MD, 1985, pp. 520, 1847.

312. H. Lingeman, H. A. Van Munster, J. H. Beyner, W. J. M. Underberg, and A. Hulshoff, HPLC analysis of basic compounds on non-modified silica gel and aluminum oxide with aqueous solvent mixtures, *J. Chromatogr.*, *352*:261 (1986).

313. *The United States Pharmacopeia*, Twenty-First Revision, United States Pharmacopeia Convention, Inc., Rockville, MD, 1985, p. 869.

314. *The United States Pharmacopeia*, Twenty-First Revision, United States Pharmacopeia Convention, Inc., Rockville, MD, 1985, pp. 913, 1760.

315. *The United States Pharmacopeia*, Twenty-First Revision, United States Pharmacopeia Convention, Inc., Rockville, MD, 1985, pp. 986, 1098.

316. R. Gill, M. D. Osselton, and R. M. Smith, International collabora-

tive study of the retention reproducibility of basic drugs in HPLC on a silica column with methanol–ammonium nitrate eluent, *J. Pharm. Biomed. Anal.*, *7*:447 (1989).

317. John A. Adamovics and K. Shields, Alternate strategies for analysis of pharmaceuticals. In *Chromatography of Pharmaceuticals* (S. Ahuje, ed.), American Chemical Society, Washington, DC. 1992, pp. 26–39.

318. S. H. Hansen, P. Helboe, and M. Thomsen, Dynamically modified silica — an alternative to reversed-phase high-performance liquid chromatography on chemically bonded phases, *J. Pharm. Biomed. Anal.*, *2*:167 (1984).

319. R. J. Flanagan, G. C. A. Storey, R. K. Bhamra, and I. Jane, HPLC analysis of basic drugs on silica columns using nonaqueous ionic eluents, *J. Chromatogr.*, *247*:15 (1982).

320. S. B. Mahato, N. P. Sahu, and S. K. Maitra, Simultaneous determination of chlorpheniramine and diphenydramine in cough syrups by reversed-phase ion-pair high-performance liquid chromatography, *J. Chromatogr.*, *351*:580 (1986).

321. R. J. Flanagan and I. Jane, HPLC analysis of basic drugs on silica columns using aqueous ionic eluents, *J. Chromatogr.*, *323*:173 (1985).

322. J. Adamovics, Determination of antibiotics and antimicrobial agents in human serum by direct injection onto silica liquid chromatographic columns, *J. Pharm. Biomed. Anal.*, *5*:267 (1987).

323. B. G. Cox and R. W. Stout, Study of the retention mechanism for basic compounds on silica under "pseudo-reversed-phase" conditions, *J. Chromatogr.*, *384*:315 (1987).

324. D. Helton, M. Ready, and D. Sacks, Advantages of a common column and common mobile phase system for steroid analysis by normal phase HPLC, *Pharmacopeial Forum*, 2794 (1983).

325. *The United States Pharmacopeia*, Twenty-First Revision, United States Pharmacopeia Convention, Inc., Rockville, MD, 1985, p. 439.

326. *The United States Pharmacopeia*, Twenty-First Revision, United States Pharmacopeia Convention, Inc., Rockville, MD, 1985, p. 111.

327. D. M. Pearce, Critique of a proposal to use a common column and common mobile phase for USP high-performance liquid chromatographic steroid analyses in the normal-phase mode, *Pharmacopeial Forum*, 4821 (1984).

328. P. Helboe, Separation of corticosteroids by high performance liquid chromatography on dyamically modified silica, *J. Chromatogr.*, *366*: 191 (1986).

329. F. F. Cantwell, Pre-column reactions to eliminate interferences in the

liquid chromatographic analysis of *p*-hydroxybenzoates on complex pharmaceuticals, *Anal. Chem.*, *48*:1854 (1976).

330. T. J. Franks and J. D. Stodola, A reverse phase HPLC assay for the determination of calcium pantothenate utilizing column switching, *J. Liq. Chromatogr.*, *7*:823 (1984).

331. R. A. Kenley, S. Chandry, and G. C. Visor, An automated column-switching HPLC method for analyzing active and excipient materials in both cream and ointment formulations, *Drug Dev. Ind. Pharm.*, *4*:1781 (1985).

332. E. J. Benjamin and D. L. Conley, On-line HPLC method for clean-up and analysis of hydrocortisone and sulconazole nitrate in a cream, *Int. J. Pharm.*, *13*:205 (1983).

333. E. Hillier, R. Cotter, and M. Andrews, HPLC analysis of topical creams using on-line sample clean-up, *Am. Lab.*, 67 (August 1985).

334. E. Erni and R. W. Frei, Two-dimensional column liquid chromatographic technique for resolution of complex mixture, *J. Chromatogr.*, *149*:561 (1978).

335. P. Cockaerts, E. Roets, and J. Hoogmartens, Analysis of a complex analgesic formulation by high performance liquid chromatography with column-switching, *J. Pharm. Biomed. Anal.*, *4*:367 (1986).

336. R. A. Kenley, S. Chandry, and G. A. Visor, Multidimensional column-switching liquid chromatographic method for dissolution testing of enprostil soft elastic gelatin capsules, *J. Pharm. Sci.*, *75*:999 (1986).

337. Y. Tapuhi, M. Miller, and B. L. Kanger, Practical considerations in the chiral separation of D,L-amino acids by reversed-phase liquid chromatography using metal chelate additives, *J. Chromatogr.*, *205*:325 (1981).

338. Th. Cachet, K. De Turck, E. Roets, and J. Hoofmartens, Quantitative analysis of erythromycin by reversed-phase liquid chromatography using column-switching, *J. Pharm. Biomed. Anal.*, *9*:547 (1991).

339. A. A. Petraukas and V. K. Svedas, Hydrophobicity of β-lactam antibiotics: Explanation and prediction of their behavior in various partitiong solvent systems and reversed-phase chromatography, *J. Chromatogr.*, *585*:3 (1991).

340. W. W. Wright, Use of liquid chromatography for the assay of antibiotics, *Pharmacopeial Forum*, *20*:8155 (1994).

341. J. Hoogmartens, Liquid chromatography for the quantitative analysis of antibiotics—some applications using poly(styrene-divinyl-benzene), *J. Pharm. Biomed. Anal.*,*10*:845 (1992).

342. W. A. Moats and L. Leskinen, Comparison of bonded, polymeric

and silica colunms for chromatography of some penicillins, *J. Chromatogr.*, *386*:79 (1987).

343. H. J. Rodriquez and A. J. Vanderweilen, Molecular weight determination of commercial heparin sodium USP and its sterile solutions, *J. Pharm. Sci.*, *68*:588 (1979).

344. K. Sreenivasan, P. D. Nair, and K. Rathinam, A GPC method for analysis of low molecular weight drugs, *J. Liquid Chromatogr.*, *7*: 2297 (1984).

345. K. R. Majors and E. L. Johnson, High-performance exclusion chromatography of low molecular weight additives, *J. Chromatogr.*, *67*: 17 (1978).

346. M. A. Quarry, R. L. Grob, and L. R. Snyder, Prediction of isocratic retention data from two or more gradient runs. Analysis of some associated errors, *Anal. Chem.*, *58*:907 (1986).

347. P. C. Rahn, M. Woodman, W. Beverung, and A. Heckendorf, Preparation liquid chromatography and its relationship in thin-layer chromatography, Waters Assoc., 1979.

348. T. Okumura, Application of thin-layer chromatography to high performance liquid chromatographic separation of steroidal hormones and cephalosporin antibiotics, *J. Liquid Chromatogr.*, *4*:1035 (1981).

349. M. Gazdag, G. Szepesi, K. Varsanyi-Riedl, Z. Vegh, and Z. Pap-Sziklay, Chromatography of bis-quaternary amino steroids, I. Separation or silica by thin-layer chromatography, *J. Chromatogr.*, *318*: 279 (1985).

350. N. Lammers, J. Zeeman, G. J. DeJong, and U. A. Brinkman, TLC as a pilot technique for reversed-phase HPLC, *Methodol. Surv. Biochem. Anal.*, *14*:101 (1984).

351. J. W. Dolan, Obtaining separations, Part I: A look at retention, *LC–GC*, *12*:368 (1994).

352. J. W. Dolan, Obtaining separations, Part II: Adjusting selectivity, *LC–GC*, *12*:446 (1994).

353. J. W. Dolan, Obtaining separations, Part III: Adjusting column conditions, *LC–GC*, *12*:520 (1994).

6
Capillary Electrophoresis

SHELLEY R. RABEL* and JOHN F. STOBAUGH
University of Kansas, Lawrence, Kansas

I. INTRODUCTION

In the development of therapeutic agents, the pharmaceutical industry essentially imposes on itself the highest standards currently available. Due to continuing progress in techniques, instruments and so on, current methods of pharmaceutical analysis must utilize the available advanced technologies, which are generally characterized by high sensitivity, selectivity, robustness, precision, accuracy, and speed. The multitude of high-performance liquid chromatographic (HPLC) instruments in the typical pharmaceutical laboratory attests to the dominance of this analytical technique in pharmaceutical analysis. However, the advent of automated commercial capillary electrophoresis instrumentation has resulted in a substantial increase in the number of capillary electrophoresis (CE) applications in the pharmaceutical industry over the last 5 years. Capillary electrophoresis has been utilized in the quantitation of drug-related impurities [1], stability studies [2], chiral analysis [3], stereoisomeric separations [4], and formulation analysis [5]. Continued interest in the research and development of biotechnology-derived products has promoted the widespread use of CE to monitor the synthetic and purification processes in addition to the analysis of these therapeutic entities in formulations [6].

Current affiliation: DuPont Merck Pharmaceutical Company, Wilmington, Delaware.

A review of capillary electrophoresis applications in pharmaceutical analysis was published in 1993, and the goal of this chapter is to provide an update on the various disciplines within the technique and includes selected applications. Recent developments in the areas of capillary technology, instrumentation, and detection will be reviewed here. Useful strategies for method development involving several classes of pharmaceuticals and biotechnology products will be addressed. The formats within capillary electrophoresis have evolved to such an extent that this chapter is not comprehensive in scope. Therefore, the reader will be directed to other reviews on the various aspects of capillary electrophoresis. Of particular interest to many separation scientists may be a special issue of an *Applied Biosystems Newsletter*, which addresses the future role of CE, method development in CE, and selected applications in the area of drug analysis and protein separations [7].

II. CAPILLARY ELECTROPHORESIS FORMATS

A. Capillary Zone Electrophoresis

Capillary electrophoresis performed in free solution is perhaps the most straightforward of CE formats. Selectivity in capillary zone electrophoresis (CZE) is based on differences in electrophoretic mobility of individual analytes, which depends on the mass-to-charge ratio and the solution conformation of the analyte. The simultaneous analysis of both positively and negatively charged species within a single analysis is possible due to electroosmosis, which is the "bulk flow" of solvent within the capillary. Electroosmotic activity is derived from the interaction of cationic buffer consituents with the negatively charged silanol groups at the inner surface of the capillary. The density of the electrical layer deceases exponentially with distance from the capillary wall, generating a zeta potential. The bulk movement of material within the capillary arises due to the migration of hydrated cations toward the cathode. This results in the "plug" flow profile, shown in Figure 6.1, as opposed to laminar flow in HPLC.

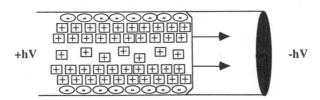

Figure 6.1 "Plug-flow" in capillary electrophoresis.

Jorgenson and Lukacs [8] have described a number of parameters, which may be helpful in characterizing CE separations. The linear velocity (v) of a given analyte in free-solution capillary electrophoresis may be represented by the following equation:

$$v = (\mu_{eo} + \mu_{el})E \qquad (1)$$

where μ_{eo} and μ_{el} represent the contributions from the electroosmotic flow coefficient and the electrophoretic mobility, respectively. The effective electric field (E) is equal to the applied voltage divided by the length of the capillary. As in HPLC, the peak efficiency may be evaluated by the number of the theoretical plates (N):

$$N = \frac{1}{2D} (\mu_{eo} + \mu_{el}) V \qquad (2)$$

The efficiency is therefore directly proportional to the the applied voltage (V) and inversely proportional to the diffusion coefficient of the analyte (D). The sum of contributions from electoosmotic flow (μ_{eo}) and the electrophoretic mobility (μ_{el}) result in the overall, or "apparent," electrophoretic mobility of the analyte:

$$\mu = \mu_{eo} + \mu_{el} \qquad (3)$$

The expression used to calculate resolution is shown in Equation (4):

$$Rs = 0.177(\mu_1 - \mu_2) \left(\frac{V}{D (\bar{\mu} + \mu_{eo})} \right)^{0.5} \qquad (4)$$

in which ($\mu_1 - \mu_2$) represents the difference in the overall electrophoretic mobilities of the two species and $\bar{\mu}$ is the average of the two electrophoretic mobilities. A more detailed discussion and interpretation of theoretical concepts in CZE may be found in the recent text by Camilleri [9].

The key operational parameter in free-solution capillary electrophoresis is the pH of the running buffer, as the electroosmotic flow and ionization of the analyte can be regulated by this variable. The role of buffers in capillary electrophoresis has been discussed in detail, with emphasis on buffer concentration, buffer type, and pH effects [10]. The effect of organic solvents on separation and migration behavior has been studied for dipeptides [11] and somatostatin analog peptides [12]. The order of migration as well as the selectivity may be manipulated by organic modifiers in the running buffer, however, in the case of the dipeptides, the separation deteriorated with greater than 15% acetonitrile. Furthermore, the type of organic solvent used for the separation of somatostatin analog peptides had a profound effect on the order of elution.

In the near future, CE will very likely serve as an orthogonal separation technique, which complements the results generated by HPLC. The use of such contrasting techniques to confirm results is particularly useful in the determination of peak homogeneity [13] and identification [14]. Although the quantitive capabilities of CE have been met with skepticism, strides toward improvement in this area are being made as workers become familiar with different facets of the technique and as the instrumentation has improved. A growing awareness of the need to normalize the peak area in order to quantitate drug-related impurities has improved the overall quality of quantitative CE methods reported in the literature in recent years [15]. Unlike HPLC, analytes do not migrate with uniform linear velocity. The migration rate of each analyte will differ according to the inherent electrophoretic mobility of the molecule; thus, slower migrating peaks will require more time to pass through the detector. Corrections for detection window residence time are made by dividing the peak area response by the migration time. The use of automated instrumentation generally results in enhanced reproducibility of injection and migration times. In addition, autosamplers contribute to the ease of analysis in routine stability studies [2]. For example, Qin et al. report on a stability-indicating assay for an inhibitor of angiotensin-converting enzyme [16], in which the precision of injection was 0.62%.

B. Micellar Electrokinetic Capillary Chromatography

Many pharmaceutical preparations contain multiple components with a wide array of physico-chemical properties. Although CZE is a very effective means of separation for ionic species, an additional selectivity factor is required to discriminate neutral analytes in CE. Terabe first introduced the concept of micellar electrokinetic capillary chromatography (MEKC) in which ionic surfactants were included in the running buffer at a concentration above the critical micelle concentration (CMC) [17]. Micelles, which have hydrophobic interiors and anionic exteriors, serve as a pseudostationary phase, which is pumped electrophoretically. Separations are based on the differential association of analytes with the micelle. Interactions between the analyte and micelles may be due to any one or a combination of the following: electrostatic interactions, hydrogen bonding, and/or hydrophobic interactions. The applicability of MEKC is limited in some cases to small molecules and peptides due to the physical size of macromolecules and their inability to fully partition into the interior of the micelle. However, MEKC was successful in the separation of a family of decapeptide antibiotics [18] utilizing a zwitterionic surfactant. The selectivity in MEKC may be manipulated by variables such as the type of surfactant, pH (in the case of ionic solutes), temperature, and various additives (organic, ion-pairing agent, etc.) [19]. Donato et al. examined the effects of pH, surfactant concentration and influence of organic modifer on the separation of some non-steroidal antiinflammatory drugs [20]. The same group also applied CE and MEKC to the direct analysis of the non-steroidal antiinflammatory drugs in a several pharmaceutical formulations without sample pretreatment [21]. A cationie surfactant (CTAB) was effective in the reverse electroosmotic flow (EOF) separation for both neutral and anionic antibiotics [5]. The levels of neomycin and hydrocortisone determined in otosporin ear drops were in good agreement with those reported on the label. Janini and Issaq have presented an overview of the different surfactants employed in MEKC and the effects of additives [22].

C. Electrokinetic Chromatography (Chiral Separations)

The importance of stereoselective separations in pharmaceutical analysis cannot be underestimated. Drug racemates often display unique characteristics with respect to therapeutic effectiveness and the potential to cause adverse side effects. Although the majority of pharmaceuticals have been marketed as racemic mixtures in the past, the current trend in the pharmaceutical industry is to develop new drug candidates as a pure isomer. The development of optically pure pharmaceuticals demands analytical methods to confirm enantiomeric purity throughout synthetic processes and stability-indicating assays capable of detecting small amounts of enantiomerically related impurities. Various analytical approaches have been utilized in the HPLC separations of enantiomers, including chiral mobile-phase additives [23], chiral stationary phases [24], or reaction with a chiral reagent to form diastereomers. Whereas the separation of diasteromerically related analytes based on differences in physico-chemical properties is routinely performed by HPLC techniques, the same separation in CE is not straightforward due to the absence of an interactive stationary phase. Therefore, the general approach to separating enantiomers by CE has been to add chiral selectors to the running electrolyte to form a pseudostationary phase. This direct approach to the separation of enantiomers is preferred because inaccurate results can arise following derivatization due to racemization or the presence of chiral impurites in the reagent itself. Recently, several excellent reviews have characterized the different groups of selectors and examined the different mechanisms of chiral recongnition [25,26]. Chiral selectors may be conveniently grouped into four major classes [25] based on host–guest complexation, chiral micellar solubilization, ligand exchange, and polymer-based recognition. The following discussion of chiral separations is limited to the more widely used techniques and recent applications thereof.

Cyclodextrins are the most commonly used chiral selector in guest–host complexation. These cyclic oligosaccharides consists of six to eight glucose units (α-, β-, and γ-cyclodextrin), which are linked by glycosidic bonds to form truncated cone-shaped structures. The chiral selectivity inherent in cyclodextrins originates from the chiral centers present in the individual glucose units. The inner cavity is hydrophobic in nature, whereas the outer rim is composed of secondary hydroxyl groups, which provide a degree of hydrophilicity. The propensity for an analyte to partition into the cavity of the host molecule is a function of its hydrophobicity, hydrogen-bonding capability, and the size and shape of the guest molecule. Investigations have been focused on the use of those modified cyclodextrins, which may have higher solubility than the neutral cyclodextrins [27] as well as those which may exhibit specific mechanisms of interaction between the host and guest molecules. Carboxymethylated, carboxyethylated, and succinylated β-cyclodextrins have been described for the separation of basic and neutral drugs [28]. The derivatized cyclodextrins are neutral at low pH; however, at pH values above 5, the carboxylic acid group is ionized and the cyclodextrins serve as chiral selectors for the separation of neutral enantiomers. Tait and co-workers [29] demonstrated the superiority of polyanionic β-cyclodextrin derivatives over neutral cyclodextrin selectors for the separation of racemic mixtures of ephedrine and pseudoephedrine. Numerous cyclodextrin derivatives varying in size and type of functionalities have been explored in the separation of basic drugs [3,30]. As shown in Figure 6.2, enantiomeric impurities for ephedrine and norephedrine have been quantitated at levels below 0.5% [30]. The larger cavity provided by γ-cyclodextrin was effective in the separation of a dansylated neuroactive drug and dansylated amino acids [31]. The dansyl-derivative of the N-methyl-D-aspartate antagonist provided adequate sensitivity such that 0.1% of the antipode was detectable [30]. Experimental parameters that influence resolution in chiral separations, such as the type and concentration of cyclodextrin, effect of organic solvent and other buffer additives, and pH, have been investigated in the separation of numerous enantiomers of pharmaceutical importance [4,32,33]. The addition of cyclodextrins in the running buffer has also been beneficial for the separation of the cis- and trans-epimers of pilocarpine in an opthalmic formulation [34]. Baseline separation of the two epimers and the primary hydrolysis product of pilocarpine was observed using a coated column in the presence of β-cyclodextrin. The

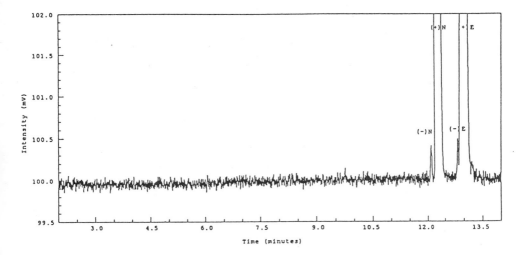

Figure 6.2 Separation of norepinephrine (N) and epinephrine (E) and the corresponding enantiomeric impurities present at 0.5%. (From Ref. 30.)

use of surface-modified capillaries in conjuction with cyclodextrins adds another dimension to the separation whereby efficiency may be increased without sacrificing enantioselectivity [35]. Nonionic polymers may be used to either dynamically or permanently modifiy the capillary surface. Significant increases in the separation efficiency, and thus resolution, of basic pharmaceutical enantiomers may result due to the diminished analyte–surface interactions. Li and Lloyd have recently reported electrochromatographic separations utilizing capillaries packed with a chiral stationary phase composed of β-cyclodextrin for the enantiomeric separations of hexobarbital and dansylated amino acids [36]. Cyclodextrins have not been exclusively used as chiral selectors; these modifiers have also been instrumental in achiral separations such as the separation of a series of sulphonamides in combination drug formulations [37].

Chiral micellar solubilization may involve the use of chiral surfactants, or a combination of achiral surfactants and a chiral selector. Terabe [26] and Bereuter [25] provide a comprehensive overview of applications involving chiral surfactants such as bile salts or synthetic amino acid surfactants. The use of cyclodextrins (CD) as the chiral selector in combination with MEKC was successful for the separation of neutral racemic nonsteroidal aromatase inhibitors and barbituates [38]. Further approaches to the separation of enantiomers utilizing a combination of CD–MEKC have been described in the review by Terabe [26].

Another class of chiral selectors gaining in popularity are those which are polymeric. Linear maltooligosaccharides were effective in the chiral discrimination of the enantiomeric pairs of nonsteroidal anti-inflammatory pharmaceuticals, coumarinic anticoagulant compounds, and diastereomers of cephalosporin antibiotics [39]. Proteins, such as bovine serum albumin (BSA), may be categorized as polymeric discriminators and have been utilized in running buffers as chiral selectors. Barker et al. have incorporated BSA as a chiral pseudostationary phase for the separation of (6R)- and (6S)-leucovorin [40]. Problems with respect to protein–wall interactions were encountered; however, these adsorptive processes were overcome by using PEG-coated capillaries. The addition of dextran to form a polymer seiving network allowed control of the mobility of the BSA, thereby allowing time for the mechanism of separation to manifest itself. The inclusion of dextran in the running buffer in addition to bovine serum albumin was effective in separating racemates of ibuprofen, leucovorin, and dansylated amino acids [41]. The addition of the enzymatic protein cellulase to the supporting electrolyte proved successful in the enantiomeric separation of several β-blockers [42]. In some instances, 2-propanol was required as a modifier to regulate the interactions between the analyte and the protein.

D. Sieving Separations

The growing interest in biopolymers as pharmaceutical entities has stimulated the need for rapid and readily automated methods of analysis for these complicated molecules. Whereas traditional slab–gel electrophoretic techniques are time-consuming and labor-intensive, capillary gel electrophoresis (CGE) provides rapid analysis times, the potential for automation, high-efficiency separations, and improved quantitation capabilities. The high resolving power of CGE is attributed to the anticonvective nature of the separation medium, which minimizes molecular diffusion contributions to band dispersion. Cross-linked polyacrylamide gels, in which the pore sizes can be controlled by the extent of cross-linking, have been widely used as sieving networks [43]. Several separation mechanisms may operate in CGE, depending on the separation conditions. In the absence of denaturants, the separation of native proteins is dependant on charge and size of the analyte. Incorporation of SDS and reducing agents for the separation of proteins results in protein complexation with SDS such that the charges on the protein are occluded. In this situation, the separation is based solely on the molecular weight [44]. This approach was effective in the analysis of recombinant bovine somatotropin in which the monomer, dimer, trimer, and tetramer forms could be resolved [45]. In addition to cross-linked polyacrylamide gels, applications utilizing linear polyacrylamide networks have been reported. Wu and Regnier employed linear polyacrylamide gels at concentrations of 4–12% for the separation of native proteins ranging from 20,000 to 45000 molecular weight [46]. Interestingly, separations were found to be dependent only on charge with no appreciable contribution from sieving mechansims. The use of linear polyacrylamide gels was also instrumental in conducting enzyme microassays in capillaries due to the ability to incubate the reaction within the capillary for several hours without band dispersion [46]. Several concerns have been expressed regarding the use of cross-linked polyacrylamide gels [47]. Bubble formation has been especially problematic and has been the source of irreproducible separations, erratic currents, and short column lifetimes. Paulus outlines precautions to be taken in the successful preparation and operation of polyacrylamide gel capillaries [47]. Column stability has been an issue and, in general, the need for coated capillaries to prevent the electroosmotic flow (EOF), and thus the movement of gel within the capillary, has been recognized. Nakatani produced a stable capillary coating through the use of SI–C bonds in order to stablize both cross-linked and linear polyacrylamide gels for the separation of oligonucleotides [48]. A comparison of coated and uncoated capillaries used in combination with non-cross-linked polyacrylamide gels revealed that the important factor in the stability of the gels was the viscosity [49]. In the presence of viscous gels, coated capillaries were not required to prevent EOF. In fact, the performance of the uncoated capillaries for the separation of high-molecular-weight proteins was superior to that of the coated capillaries. Although the use of linear polyacrylamide gels has extended the lifetime of the capillary, only 20–40 injections can normally be performed before capillary performance declines [50]. An additional problem is that polyacrylamide gels display significant ultraviolet (UV) absorbance below 214 nm. Efforts to avoid the aforementioned problems have been directed toward the use of replaceable sieving networks composed of liquid polymers of lower viscosity. UV transparent materials such as dextrans [50], poly(ethylene glycol), poly(ethylene oxide) [51], and methylcellulose have been utilized as sieving matrices. The use of a dextran gel as the sieving medium and a short length (7 cm) of capillary for separation resulted in resolution of SDS–protein complexes of carbonic anhydrase, ovalalbumin, bovine serum albumin, and phosphorylase b in under 2 min [52]. Figure 6.3 illustrates the excellent reproducibility and longevity of the columns when utilizing a replaceable linear polyacrylamide sieving medium [53]. A commercially available size separation kit including a hydrophilic linear polymer network (not polyacrylamide in origin) was successfully used in the separation of monomer–dimer forms of recombinant human ciliary neurotrophic factor as well as SDS–protein complexes ranging from 20,000–200,000 Da [54]. Parameters, such as column diameter

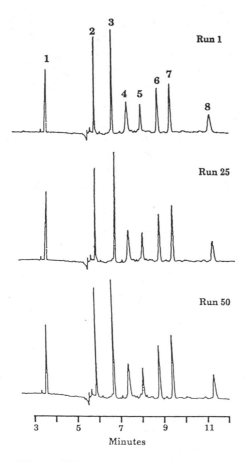

Figure 6.3 Reproducibility of SDS–protein separations utilizing a replaceable polymeric gel. The peaks correspond to mellitic acid (1), α-lactalbumin (2), carbonic anhydrase (3), ovalalbumin (4), bovine serum albumin (5), phosphorylase b (6), β-galactosidase (7), and myosin (8). (From Ref. 53.)

and length, and temperature, that may influence capillary SDS–gel separations, have been investigated in a nonacrylamide separation medium [55]. In a separate study, temperature effects on the sieving separation of SDS–protein complexes were examined for branched dextrans and poly(ethylene oxide) [51]. Capillary electrophoresis in a gel format has also been used extensively for the separation of oligonucleotides and DNA synthesis, as discussed in several reviews [56,57]. Guttman and co-workers have developed strategies for the prediction of migration behavior of oligonucleotides using polyacrylamide gels, and determined that both chain length and the base composition contribute to the migration patterns [58]. Additional studies probing the influence of pH on the migration of homooligomers have determined that pH plays a significant role and equilibration time can be minimized by preparing the gel at the same pH as the running electrolyte [59]. A recent area of interest in the treatment of disease states has been the concept of antisense therapy. Capillary gel electrophoresis has been an effective tool with which to characterize interactions of peptide nucleic acid antisense reagents to complementary oligonucleotides because the bound and unbound forms can be easily separated [60]. The anticonvective nature of the gel allowed the measurement of binding kinetics through multiple injections without zone dispersion.

E. Capillary Isoelectric Focusing

Capillary isoelectric focusing (CIEF) provides yet another dimension to capillary electrophoresis by introducing a separation mechansim based on differences in the isoelectric points (pI) of biomolecules [61]. Separations of proteins in CIEF rely on the formation of a pH gradient along the longitudinal axis of the capillary and the migration of analytes to the pH region equal to their pI, at which point migration of the neutral molecule ceases. Several steps, either performed independently or simultaneously, are involved in executing an analysis by CIEF. The sample is first dissolved in a mixture of carrier ampholytes, which will form the foundation of the pH gradient in the capillary. The choice of pH ranges will depend on the pI values of the analytes; however, the narrowest range which will sufficiently encompass the pIs results in the greater resolving power. A focusing step is required to establish the pH gradient and concurrently focus the analytes into discrete zones along the pH gradient. Once migration of the analytes ceases, the current will diminish. In order to detect the proteins, the individual zones must be mobilized, or eluted, past the detection window. This step may be performed electrophoretically by the addition of salt, hydrodynamically or electroosmostically. Chen and Wiktorowicz [62] utilized hydrodynamic mobilization in the presence of an electric field for the detection of proteins and related mutant forms. This mobilization procedure allowed the detection of the analytes while maintaining the pH gradient and minimized the distortion of the zones due to laminar flow.

Electroosmotic flow in CIEF is detrimental to the focusing process; thus, initial efforts were aimed at the development of coated capillaries to eliminate EOF. An additional concern, which also prompted the use of coated capillaries, was the propensity for proteinaceous material to adsorb onto the fused silica surface. Hjerten and Zhu proposed the use of hydrophilically modified surfaces in order to facilitate the formation of stable pH gradients [63,64]. However, the hydrolytic stability of these coatings was a problem, especially under the extreme alkaline conditions required in CIEF. Nelson has developed three new coatings for use in CIEF using alternative derivatization methods rather than typical siloxane chemistries [65]. An alternative approach to circumvent the problem of limited capillary lifetime and protein adsorption has been the addition of polymeric additives to the running buffer [66]. The inclusion of viscous polymers such as methylcellulose not only minimizes surface adsorption but, more importantly, allows simultaneous focusing and mobilization. Supression but not elimination of the EOF permits the formation of zones while electroosmotically pumping the analytes past the detector. Analysis times of proteins with methylcellulose and uncoated capillaries were improved by the addition of tetramethylethylenediamine (TEMED) to the sample and by reversal of the polarity [67]. The inclusion of TEMED effectively blocked the cathodic portion of the capillary and forced the focusing of analytes at the anodic region such that the length to the detection window could be decreased. The concentration of phosphoric acid (anolyte) was seen as a crucial factor in minimizing anodic drift and thus improve separations of acidic proteins [67]. Furthermore, the need for desalting of recombinant protein samples was essential in all CIEF methods due to the high currents generated during focusing and the resultant poor separations. Dialysis of monoclonal antibody isoform samples was necessary to remove salt prior to CIEF analysis [68]. The generation of joule heat in the presence of >10 mM salt concentrations may contribute to protein denaturation and/or precipitation [69]. The addition of nonionic detergents to the sample and ampholytes has reduced the risk of precipitation and improved the separation reproducibility of immunoglobulins [69]. Molteni and Thormann have examined various parameters that affect separations in CIEF with electroosmotic mobilization [70]. Factors such as capillary conditioning procedures, capillary dimensions, types and concentrations of additives, ampholytes, anolytes, catholytes, and the length of the sample zone were shown to influence the separation and resolution.

The performance of commerically available coated and uncoated capillaries utilizing electroosmotic mobilization has been assessed for model protein mixtures [71]. Although successful separations of basic and neutral proteins were achieved on uncoated capillaries, coated capillaries were required for the separation of acidic proteins in order to provide a constant EOF throughout the capillary. Yeo and Regnier [72] have reported novel coatings for CIEF, which are produced by dynamically modifying octadecylsilane-derivatized capillaries by adsorption of methylcellulose or surfactant PF-108. Again, the EOF was adequately suppressed such that focusing and mobilization steps could be combined. Under these conditions, hemoglobin variants differing by 0.03 pI units could be resolved. For a comprehensive overview of CIEF methods and applications to peptides/proteins and antibodies the reader is directed to a review by Mazzeo and Krull [73].

III. INSTRUMENTATION

The basic components of any capillary electrophoresis system (Fig. 6.4) are a high-voltage power supply, detector, capillary, and safety unit to shield the user from the high voltages associated with the technique. The availability of commercial CE systems, which have rendered improvements in injection and migration reproducibility as well as detection, has encouraged the use of this technique in pharmaceutical analysis. Altria recently reported values routinely obtained for relevant analytical figures of merit utilizing commercial instruments [74]. Limits of detection for pharmaceuticals were in the low microgram per milliliter range and the reproducibility, as determined by the relative standard deviation (RSD), of migration times was approximately 1%. Drug-related impurities could be detected at 0.02% (w/w), provided external calibrations were utilized for cases where the main drug peak was off-scale. The precision (RSD) with regard to peak area was less than 1–2% for analyses in which an internal standard was included.

A. Sample Introduction

The small dimensions associated with CE preclude the injection of large volumes. The sample may be introduced to the capillary either by a diplacement technique (i.e., pressure, vacuum, or siphoning) or via electrokinetic injection. The majority of commercial instruments apply a pressure differ-

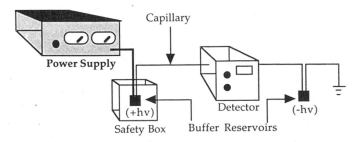

Figure 6.4 Components of a capillary electrophoresis apparatus.

ential for sample injection. Theoretical discussions concerning the attributes of different methods of injection may be found in the literature [9,75]. Various preconcentration strategies have been used to overcome the lack of sensitivity resulting from the injection of small volumes. One of the most simplistic methods to increase the sample loading in CE, is the use of "peak stacking" [76]. Sample stacking is achieved by hydrodynamically loading the sample, which may differ either in pH or buffer concentration from that of the running electrolyte. The higher electric field within the sample plug results in rapid migration of charged species toward the interface between the sample plug and the running buffer. Once the boundry is reached, migration of ions is retarded and sample stacking occurs. Alternatively, the use of discontinuous buffer systems (isotachophoretic preconcentration) has been effective in increasing the injection volume by at least a factor of 30 for the micropreparative separation of several peptides (Fig. 6.5) [77]. Isotachophoretic preconcentration techniques employ a leading and terminating electrolyte on either side of the injection plug. Isotachophoretic sample preconcentration was combined with capillary electrophoresis–mass spectrometry (CE–MS), which enhanced the limit of detection for proteins by a factor of 100 [78]. Another type of on-line preconcentration employs the principles of liquid chromatography by using commercially available capillaries, which are 1 mm in length and packed with a polymeric reverse-phase material. The application of this preconcentration method to the quantitative analysis of several pharmaceuticals improved the sensitivity by several orders of magnitude [79]. Yet another on-line preconcentration strategy takes advantage of the affinity of proteins for metal ions and thus utilizes a preconcentration capillary containing immobilized metal chelates to preconcentrate proteins in a dilute sample [80].

B. Capillary Technology

The fused silica capillaries used in CE are inexpensive and flexible due to the outer polyimide coating, and are available in inner diameters ranging from 10 to 300 μm. Fused silica is transparent to UV light, which allows the capillary serve as its own detection flow cell [81]. The success of CE is primarily due to the large surface-to-volume ratio of these capillaries, which allows the effective dissipation of heat generated by high voltages while retaining high-efficiency separations. However, this high surface-to-volume ratio also poses a dilemma in CE due to the high potential for solute–surface interations. Coulombic interactions with the capillary surface are especially problematic in the case of basic proteins/peptides, resulting in a loss of efficiency, and reproducibility, and in extreme cases elution may not occur. Various approaches, many of which will be discussed in more detail

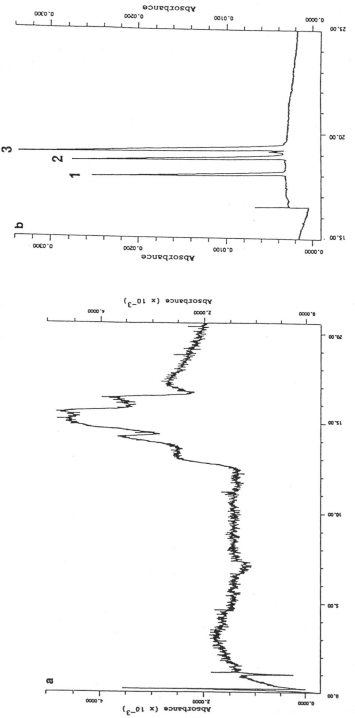

Figure 6.5 Separations of standard peptides performed under nonstacking conditions (left) and using an isotacophoretic preconcentration technique (right). The time for injection was 20 s for both analyses. (From Ref. 77.)

in future sections addressing method development, have been used to minimize adsorption of proteins to capillary surfaces. One of the more popular strategies has been to chemically modify the inner capillary surface to produce a nonionic, hydrophilic coating and, thus, sterically shield the silanol functionalities. In addition to preventing the adsorption of analytes, various workers advocate the use of surface coatings to reduce, or possibly eliminate, electroosmotic flow in CE [82]. In the absence of EOF, migration is solely dependent on the individual electrophoretic mobilites of the analytes; thus, an additional source of variation is removed. The elimination of EOF is also benefical in capillary isoelectric focusing as well as capillary gel electrophoresis. The presence of EOF in CIEF works against the establishment of a stable pH gradient, whereas the presence of EOF in gel-filled capillaries results in the extrusion of the gel [83].

Numerous approaches to the chemical modification of capillary surfaces, either by covalent or physical means, have been investigated and reviewed [81,83–85]. Table 6.1 summarizes the surface chemistries reported in the literature. The hydrolytic stability over a wide pH range and the reproducibility of chemically modified surfaces are of concern to those who produce and utilize these capillaries. Although polymeric coatings generally exhibit greater stability than the nonpolymeric counterparts, the issue of

Table 6.1 Capillary Coatings

Coating	Reference
Cross-linked, linear polyacrylamide (Si-C)	48
Linear polyacrylamide	63, 86
Poly(vinylpyrrolidinone)	87
Cross-linked polyacrylamide	88
Hydroxylated polyether functions	89
Polyethyleneimine (cross-linked)	90, 91
Epoxy coatings	92
C-8, C-18	93
Carbohydrates	94
Hydrogel polymers	95
Polyvinylmethylsiloxanediol	96
Polyethylene glycol	91, 97
Poly(methylglutamate)	82
Aryl pentafluoro group	98
Immobilized protein	99
C-18/surfactant additives	100

producing polymerized surface coatings reproducibily from batch to batch should be addressed. Schomburg points out several sources of difficulty in performing reproducible silanization reactions [83]. The number of silanol functionalities available for derivatization is unpredictable and dependent on the pretreatment of the capillary (i.e., etching, drying, and rinsing). Various silanization reagents react differently: The presence of water in trifunctional silanization reagents may result in polymerization, and by-products of silanization (such as HCl) may catalyze unwanted polymerizations. Although significant strides to prevent surface adsorption have been made through the use of modified capillaries, the separation efficiencies realized were far less than those predicted. Further research to probe the mechanism of interaction of proteins with various coatings [101] may provide insight into future chemistries that may result in higher theoretical plate counts.

C. Detection

The development of therapeutic entities with increased potency and the need to detect minute amounts of the active drug and related impurities exacerbates the issue of detector sensitivity in pharmaceutical analysis. In addition to the preconcentration methods previously discussed, alternative strategies used to overcome the lack of sensitivity in CE include improvements in optical design [102], the use of various capillary geometries, and alternative modes of detection. The following section gives a brief overview of the more commonly used detection methods in CE.

Spectrophotometric Methods

Although UV absorbance is the most widely used mode of detection in CE, it offers the least sensitivity due to the short optical path length. Efforts to solve this problem have resulted in new capillary geometries such as the bubble cell, z-cell, multireflection cell, and retangular capillaries. Attempts have been made to maximize the radial illumination efficiency and minimize the scattered light by positioning a sapphire or quartz ball lens next to the capillary [103]. Improvements in sensitivity associated with the various geometries along with additional enhancement techniques are summarized by Albin and co-workers [104]. Unlike HPLC, the use of low UV wavelengths (< 200 nm) is feasible in CE because there is no appreciable absorption of the running buffers. Most pharmaceuticals absorb more efficiently at lower wavelengths, which made the determination of a dimeric impurity of albuterol at 0.02% (w/w) possible [105]. Altria presents various methods, including wavelength switching, voltage gradients, and the use of wide-bore capillaries, to optimize sensitivity for the quantitation of selected pharmaceuticals [105].

Fluorescence offers the possibility for a 100-fold to 1000-fold improvement in detection sensitivity and improved selectivity as compared with UV detection; however, few solutes possess inherent fluorescence, thereby limiting the number of applications. Despite these limitations, numerous efforts continue in the area of laser-based fluorescence detection in CE [106]. The exceptional spatial coherence of the laser excitation source has proven ideal for use with capillary systems due to the ability to focus the beam onto the detection window of the capillary. Wu and Dovichi have described the components and parameters used to construct and operate LIF detectors [107]. The utility of LIF detection has been demonstrated in the detection of native proteins containing tryptophan and/or tyrosine residues. The 275.4-nm line of an argon-ion laser served as the excitation source for the detection of various biopharmaceuticals in dosage formulations, peptide mapping, and purity analysis at subnanomolar levels [108]. In another study, the feasiblity of a relatively inexpensive pulsed UV laser operating at 248 nm was evaluated for the detection of nanomolar levels of native peptides/proteins containing tryptophan residues [109]. Toulas and Hernandez describe a novel collinear CE–LIF design for the analysis of several drugs, peptides, amino acids and oligonucleotides, which permitted the detection limits ranging from 100 fM to 10 nM [110]. In cases where solutes lack a chromophore, precolumn or postcolumn derivatization may be possible [111]. In a recent review, Szulc and Krull [112] outline precolumn and postcolumn derivatization techniques and derivatization schemes for amino acids, peptides/proteins, DNA, and oligosaccharides. Yet another alternative for nonfluorescent analytes is the use of indirect detection methods [113]; however, the lack of selectivity of the method results in poor detection limits. In addition, sensitivity is generally compromised in the case of indirect techniques due to the the high background. Thermo-optical-based spectroscopy has demonstrated high sensitivity and is thus another means of detection used in CE [114]. This technique requires that analytes only absorb UV–visible radiation and, therefore, is applicable to a larger range of molecules than fluorescence detection.

Electrochemical Detection

Oxidative electrochemical detection has demonstrated high sensitivity and specificity for the detection of electroactive solutes. The success of electrochemical detection in CE depends on the ability to isolate the high electric fields associated with the method from the detector. Some of these detection schemes have been reviewed by Yik [115]. Several configurations utilize a fractured capillary covered with either with a porous glass joint [116] or more recently a on-column nafion joint [117]. Investigations into the use of chemically modified electrodes has resulted in increased selectivity of certain types of analytes such as thiols [118,119], glucose [119], and carbohydrates [120]. The electrochemical detection of peptides has been possible by on-column formation of a Cu(II)–peptide complex, which undergoes oxidation at a carbon fiber electrode [121]. Unlike amperometric detection, conductivity is nonspecific and serves as the basis for a universal detector for all solutes with typical detection limits in the micromolar range. Huang has described on-column [122] and end-column [123] conductivity detection schemes for CE.

Mass Spectrometry

Mass spectrometry is a valuble tool with which an abundancy of structural information may be obtained from a minute amount of material. Capillary electrophoresis may be interfaced with mass spectrometry by electrospray ionization [124–126] or continuous-flow, fast-atom bombardment methods [127,128]. Several reviews discuss applications of the interfacing techniques, and address the attributes and disadvantages associated with these methods [129,130]. Critical parameters involved in the optimization of CE-electrospray ionization mass spectrometry analysis have been reviewed as well [131].

III. METHODS DEVELOPMENT

A. Proteins and Peptides

The use of CE methods for routine quality control of synthetic or recombinant peptides–proteins necessitates optimization strategies for rapid method development. Ideally, the methods should be simple, fast, and robust. Because capillary electrophoresis in the zone format is the most simplistic, initial efforts should be directed toward the use of a simple buffer system [61]. The high efficiency and reproducibility in protein–peptide separations demands that interactions between the analyte and capillary wall be neglible. The use of low-pH buffers generally results in enhanced reproduciblity, and hence ruggedness, as slight variations in the capillary surface will have little impact on the already suppressed EOF.

Several strategies exist for optimizing protein–peptide separations on uncoated capillaries. Operation at low pH values discourages coulombic interactions between the analyte and capillary surface because the analyte will carry a positive charge and the silanol surface will be protonated

[87,132]. The other alternative is to utilize buffers of high pH above the pI of the peptide/protein such that the negative charge of the protein–peptide will repel the ionized silanol surface [133–136]. Although the use of extreme acidic or alkaline pH buffers has been successful in some instances, one should be aware of the potential for protein denaturation and/or aggregation and precipitation. Recently, Langenhuizen and Janssen [137] set forth to define a general optimization strategy for method development of pharmaceutical peptides. Four classes of pharmaceutical peptides, which demonstrated a large range of pI values, molecular masses and chain lengths, were used in the study. As expected, the most influential parameter was the pH of the buffer. The generalizations drawn from their results were that separations of basic to neutral peptides were optimal at low pH values, whereas acidic peptides behaved well at neutral pH. After an optimal pH region is established, the next variable to manipulate is the concentration of the buffer. Buffers of medium to high concentration (50–100 m*M*) were the most effective, acheiving good peak shape. The presence of high salt concentrations, which may be carried over from the purification process, most likely will be detrimental to the peak shape and thus efficiency [138], therefore some type of desalting is often required.

The separation may be refined by adjusting the temperature of the capillary, and the effects of temperature on the electrophoretic behavior of polyglycine peptides has been examined in detail [139]. Although the effect of temperature may differ for individual separations, the real emphasis should be placed on the careful control of temperature in order to obtain reproducible separations.

In situations where simple buffer systems are not adequate for separation, or analyte–wall interactions are still problematic, the use of additives in the running electrolyte may be effective yet simple alternative. High concentrations of alkali-metal salts, especially potassium salts, have met with some success in competing with cation-exchange sites on the silica surface [140]. The large currents associated with high buffer salt concentrations result in excessive joule heating, which forces the use of lower electric fields and longer analysis times. However, this may be avoided through the use of zwitterionic buffers, which are nonconducting [141]. Trialkylamine additives, which may also compete for cation-exchange sites at the silica surface, have been recommended for the analysis of therapeutic proteins [6]. Basic proteins are especially troublesome; however, the use of a buffer at a pH of 2.5 in combination with amine or amino sugar additives has been effective in the prevention of analyte surface adsorption [142]. Finally, in severe cases of protein–wall interaction, the use of modified capillaries as discussed in the section on instrumentation may be necessary. The increasing availability of commercially modified capillaries makes this alternative more feasible.

When all the options within capillary zone electrophoresis have been exhausted, alternative modes may be necessary. Micellar electrokinetic chromatography would be a possibility [143], followed by IEF, CGE, and, finally, CITP [61]. Regardless of the mode of CE employed, several precautionary tips may be of help in optimizing the robustness of the method [61]. As stated previously, maintenance of capillary temperature is crucial for reproducible separations. Minimization of solute–wall interactions will also increase the reproducibility of the migration times. Alteration of the solutions in the buffer resevoirs either through evaporative processes or changes in pH incurred during operation may also have detrimential effects on the ruggedness of the method. For example, frequent replenishment of the buffers at each electrode was important in maintaining a constant pH for the separation of proteins. Drift in the pH of the solutions was attributed to the oxidative and reductive processes occurring at the electrodes [144]. The ruggedness of methods may also be influenced by washing procedures; therefore, careful records on capillary pretreatment may be critical in transferring methods between laboratories.

B. Small Molecules

Again, for reasons of simplicity and the development of rugged methods, the use of free-zone capillary electrophoresis is preferred over other CE formats. Obviously, the type of analytes to be separated will dictate the format within capillary electrophoresis required. Separation of neutral analytes necessitates some type of pseudostationary phase, whereas chiral compounds require a chiral discriminator. In the case of ionizable analytes, manipulation of pH will affect the selectivity to the greatest extent. The initial choice of pH should encompass both the charged and uncharged form of the solute. Thus, knowledge of the pK_a values for the solutes contained in a mixture is paramount for rapid method development. One should note the importance of temperature control in maintaining the pH of the running electrolyte, especially in the case of temperature-sensitive buffers and the temperature dependence of analyte pK_a values. In addition to optimizing the pH of the buffer, parameters such as capillary dimensions, applied voltage, ionic strength, and temperature may be of consequence to the separation. The following discussion will address some of these variables; however, a detailed account of practical guidelines to follow for pharmaceutical analysis is given by McLaughlin et al. [145].

In choosing the ideal length and inner diameter of the capillary, there are often trade-offs between resolving power and the analysis time. In situations where many components in a complex mixture must be resolved, the use of a longer capillary may be required to obtain higher plate counts and resolution. The advantage of longer capillaries is the effective heat dissipation, which leads to high efficiencies. However, these efficiencies are obtained at the expense of longer analysis times. Conversely, the use of short capillaries are especially useful for method development in order to rapidly screen separation conditions. Furthermore, short capillaries often give adequate resolution of simple mixtures with fast analysis times. The choice of inner diameter will affect the mass loading and, thus, the concentration sensitivity for the method. Although improvements in sensitivity are observed with a larger inner diameter, the efficiency may be compromised due to the inability to remove heat. The applied voltage is another factor that will affect the the efficiency and resolution. Efficiency, resolution, and speed of analysis will improve with higher voltages, until the heat can no longer be effectively removed. The buffer type and ionic strength may also contribute to manipulation of selectivity, resolution, and peak shape. The use of high-ionic-strength buffers results in high theoretical plate counts and increased sensitivity due to peak stacking, again until the heat generated from the high currents cannot be eliminated.

The effects of temperature on a CE separation are severalfold. With increasing temperature, the viscosity of the running electrolyte decreases and analysis times are shorter. The high currents associated with elevated temperatures generates additional heat; thus, the efficiency and resolution may be altered. Changes in selectivity are often observed with different temperatures because solute mobilities are a function of diffusion coefficients, which are, in turn, dependent on temperature. Changes in selectivity may result from alteration of solute pK_a values with temperature changes.

The pH of the running buffer is the most important parameter to manipulate; however, a question arises regarding all of the other variables and the relative importance attached to each for a particular separation. Furthermore, it would be useful to know which initial experiments will render the most information. Khaledi and co-workers [146] outline two strategies for optimization of method development. The first of these is the sequential (or simplex) method in which the experiments are performed in series and the future latter experiments are based on results from the previous experiment. The other alternative is to design a set of experiments to be performed simultaneously. The use of an overlapping resolution mapping scheme has been successful in determining the optimum separation conditions for 12 amino acid derivatives as well as a mixture of seven antimalarial drugs from nine preliminary experiments [147]. The first step in the procedure was to determine the criteria for the separation (i.e., baseline resolution, analysis time); second, two variables were chosen based on the physico-chemical characteristics of the analytes. The nine mapping experiments were performed and the resolution under each condition was calculated to provide plots from which a grid could be constructed. The global optimum from the grid could then be defined in order to obtain the separation condition. Khaledi et al. [146] present a comprehensive overview of optimization strategies and introduce predictive models that lead to rapid method development for complex mixtures.

IV. CONCLUSION

The evolution of capillary electrophoresis has expanded the versatility of this method making possible the analysis of pharmaceuticals displaying a wide array of physico-chemical properties. A wealth of knowledge regarding the various modes of CE has been accumulated through the many applications to real problems, which have been reported in the literature. The largest impact leading to the widespread industrial use of CE has been the development of automated systems. As separation scientists gain experience from routine use of CE, idiosyncrasies leading to lack of reproducibility have been identified and addressed to improve the ruggedness of the method. Questions still remain regarding the lack of sensitivity of this microanalytical method; however, preconcentration techniques, new capillary geometries, and advancements in detector design have helped to overcome this problem to some degree. Although capillary electrophoresis was initially thought to be more of a qualitative rather than a quantitative technique, numerous advances in quantitative capabilities have been demonstrated for pharmaceutical applications; thus, one would expect to see expanding use in the pharmaceutical setting.

REFERENCES

1. K. D. Altria, *J. Chromatogr., 636*:125–132 (1993).
2. K. D. Altria and P. C. Connolly, *Chromatographia, 37*:176–178 (1993).
3. M. Heuermann and G. Blaschke, *J. Chromatogr., 648*:267–274 (1993).
4. T. E. Peterson, *J. Chromatogr., 630*:353–361 (1992).
5. M. T. Ackermans, F. M. Everaerts, and J. L. Beckers, *J. Chromatogr., 606*:229–235 (1992).
6. N. A. Guzman, J. Moschera, K. Iqbal, and A. W. Malick, *J. Chromatogr., 608*:197-204 (1992).

7. Capillary Electrophoresis The Future in Separations Technology, *Biosystems Reporter*, pp. 3–14 (September 1993).
8. J. W. Jorgenson and K. Lukacs, *Science, 222*:266–272 (1983).
9. P. Camilleri (ed.), *Capillary Electrophoresis: Theory and Practice*, CRC Press, Boca Raton, FL, 1993.
10. G. M. Janini and H. J. Issaq. In *Capillary Electrophoresis Technology* (N.A. Guzman, ed.), Marcel Dekker, Inc., New York, 1993, pp. 119–160.
11. G. M. Janini, K. C. Chan, J. A. Barnes, G. M. Muschik, and H. J. Issaq, *Chromatographia, 35*:497–502 (1993).
12. M. Idei, I. Mezo, Z. Vadasz, A. Horvath, I. Teplan, and G. Keri, *J. Liquid Chromatogr., 15*:3181–3192 (1992).
13. K. D. Altria and Y. K. Dave, *J. Chromatogr., 633*:221–225 (1993).
14. K. D. Altria and D. C. M. Luscombe, *J. Pharm. Biomed. Anal., 11*: 415–420 (1993).
15. K. D. Altria, *Chromatographia, 35*:177–182 (1993).
16. X. Z. Qin, D. P. Ip, and E. W. Tsai, *J. Chromatogr., 626*:251–258 (1992).
17. S. Terabe, K. Otsuka, K. Ichikawa, A. Tsuchiya, and T. Ando, *Anal. Chem., 56*:111–113 (1984).
18. H. K. Kristensen and S. H. Hansen, *J. Chromatogr., 628*:309–315 (1993).
19. S. Terabe. In *Capillary Electrophoresis Technology* (N.A. Guzman, ed.), Marcel Dekker, Inc., New York, 1993, pp. 65–88.
20. M. G. Donato, E. van den Eeckhout, W. van den Bossche, and P. Sandra, *J. Pharm. Biomed. Anal., 11*:197–201 (1993).
21. M. G. Donato, W. Baeyens, W. van den Bossche, and P. Sandra, *J. Pharm. Biomed. Anal., 12*:21–26 (1994).
22. G. M. Janini and H. J. Issaq, *J. Liquid Chromatogr., 15*:927–960 (1992).
23. A. Walhagen and L. Edholm, *Chromatographia, 32*: 215–223 (1991).
24. R. Kuhn and S. Hoffstetter-Kuhn, *Chromatographia, 34*:505–512 (1992).
25. T. L. Bereuter, *LC–GC, 12*:748–766 (1994).
26. S. Terabe, K. Otsuka, and H. Nishi, *J. Chromatogr., 666*:295–319 (1994).
27. T. Schmitt and H. Engelhardt, *J. High Resolut. Chromatogr., 16*: 525–529 (1993).
28. T. Schmitt and H. Engelhardt, *Chromatographia, 37*:475–481 (1993).
29. R. J. Tait, D. O. Thompson, V. J. Stella, and J. F. Stobaugh, *Anal. Chem., 66*:4013–4018 (1994).

30. M. W. F. Nielen, *Anal. Chem., 65*:885–893 (1993).
31. A. Werner, T. Nassauer, P. Kiechle, and F. Erni, *J. Chromatogr., 666*:375–379 (1994).
32. S. Pálmarsdóttir and L.-E. Edholm, *J. Chromatogr., 666*:337–350 (1994).
33. H. Nishi, Y. Kokusenya, T. Miyamoto, and T. Sato, *J. Chromatogr., 659*:449–457 (1994).
34. W. Baeyens, G. Weiss, G. Van Der Weken, and W. van den Bossche, *J. Chomatogr., 638*:319–326 (1993).
35. D. Belder and G. Schomburg, *J. Chromatogr., 666*:351–365 (1994).
36. S. Li and D. K. Lloyd, *J. Chromatogr., 666*:321–335 (1994).
37. C. L. Ng, H. K. Lee, and S. F. Y. Li, *J. Chromatogr., 632*:165–170 (1993).
38. E. Francotte, S. Cherkaoui, and M. Faupel, *Chirality, 5*:516–526 (1993).
39. A. D'Hulst and N. Verbeke, *J. Chromatogr., 608*:275–287 (1992).
40. G. E. Barker, P. Russo, and R. A. Hartwick, *Anal. Chem., 64*:3024–3028 (1992).
41. P. Sun, N. Wu, G. Barker, and R. A. Hartwick, *J. Chromatogr., 648*:475–480 (1993).
42. L. Valtcheva, J. Mohammad, G. Pettersson, and S. Hjerten, *J. Chromatogr., 638*:263–277 (1993).
43. B. L. Karger, A. S. Cohen, and A. Guttman, *J. Chromatogr. 492*: 585–614 (1989).
44. K. Weber and M. Osborn, *J. Biol. Chem., 244*:4406 (1969).
45. K. Tsuji, *J. Chromatogr., 652*:139–147 (1993).
46. D. Wu and F. E. Regnier, *Anal. Chem., 65*:2029–2035 (1993).
47. A. Paulus. In *Practical Capillary Electrophoresis* (D. Goodall, ed.), University Press Series, (in press).
48. M. Nakatani, A. Shibukawa, and T. Nakagawa, *Anal. Sci., 10*:1–4 (1994).
49. D. Wu and F. E. Regnier, *J. Chromatogr., 608*:349–356 (1992).
50. K. Ganzler, K. S. Greve, A. S. Cohen, B. L. Karger, A. Guttman, and N. C. Cooke, *Anal. Chem., 64*:2665–2671 (1992).
51. A. Guttman, J. Horvath, and N. Cooke, *Anal. Chem., 65*:199–203 (1993).
52. R. Lausch, T. Scheper, O. W. Reif, J. Schloesser, J. Fleischer, and R. Freitag, *J. Chromatogr., 654*:190–195 (1993).
53. W. E. Werner, D. M. Demorest, J. Stevens, and J. E. Wiktorowicz, *Anal. Biochem., 212*:253–258 (1993).
54. A. Guttman, J. A. Nolan, and N. Cooke, *J. Chromatogr., 632*:171–175 (1993).

55. K. J. Tsuji, *J. Chromatogr., 661*:257–264 (1994).
56. J. P. Landers, R. P. Oda, T. C. Spelsberg, J. A. Nolan, and K. J. Ulfelder, *BioTechniques, 14*:98–109 (1993).
57. A. S. Cohen, D. L. Smisek, and P. Keohavong, *Trends Anal. Chem. 12*:195–202 (1993).
58. A. Guttman, R. J. Nelson, and N. Cooke, *J. Chromatogr., 593*:297–303 (1992).
59. A. Guttman, A. Arai, and K. Magyar, *J. Chromatogr., 608*:175–179 (1992).
60. D. J. Rose, *Anal. Chem., 65*:3545–3549 (1993).
61. R. Weinberger, In *Practical Capillary Electrophoresis*, Academic Press, San Diego, CA, 1993.
62. S. M. Chen and J. E. Wiktorowicz, *Anal. Biochem., 206*:84–90 (1992).
63. S. Hjerten, *J. Chromatogr., 347*:191–198 (1985).
64. S. Hjerten and M.-D. Zhu, *J. Chromatogr., 346*:265–270 (1985).
65. T. J. Nelson, *J. Chromatogr., 623*:357–365 (1992).
66. J. R. Mazzeo and I. S. Krull, *Anal. Chem., 63*:2852–2857 (1991).
67. J. R. Mazzeo and I. S. Krull, *Chromatogr., 606*:291–296 (1992).
68. C. Silverman, M. Komar, K. Shields, G. Diegnan, and J. Adamovics, *J. Liquid Chromatogr., 15*:207–219 (1992).
69. M. Zhu, R. Rodriguez, and T. Wehr, *J. Chromatogr., 559*:479–788 (1991).
70. S. Molteni and W. Thormann, *J. Chromatogr., 638*:187–193 (1993).
71. J. R. Mazzeo, J. A. Martineau, and I. S. Krull, *Methods, 4*:205–212 (1992).
72. X. W. Yao, and F. E. Regnier, *J. Chromatogr., 632*:185–193 (1993).
73. J. R. Mazzeo and I. S. Krull. In:*Capillary Electrophoresis Technology* (N.A. Guzman, ed.), Marcel Dekker, Inc., New York, 1993, pp. 795–818.
74. K. D. Altria, *J. Chromatogr., 646*:245–257 (1993).
75. P. Jandik and G. Bonn. In *Capillary Electrophoresis of Small Molecules and Ions*, VCH Publishers, Inc., New York, 1993.
76. R.-L. Chien and D. S. Burgi, *Anal. Chem., 64*:1046–1050 (1992).
77. C. Schwer and F. Lottspeich, *J. Chromatogr., 623*:345–355 (1992).
78. T. J. Thompson, F. Foret, P. Vouros, and B. L. Karger, *Anal. Chem., 65*:900–906 (1993).
79. M. E. Swartz and M. Merion, *J. Chromatogr., 632*:209–213 (1993).
80. J. Cai and Z. El Rassi, *J. Liquid Chromatogr., 16*:2007–2024 (1993).
81. F. E. Regnier and D. Wu. In:*Capillary Electrophoresis Technology* (N.A. Guzman, ed.), Marcel Dekker, Inc., New York, 1993, pp. 287–309.

82. D. Bentrop, J. Kohr, and H. Engelhardt, *Chromatographia, 32*:171–178 (1991).
83. Schomburg. In *Capillary Electrophoresis Technology* (N.A. Guzman ed.), Marcel Dekker, Inc., New York, 1993, pp. 311–356.
84. C. Schoneich, S. K. Kwok, G. S. Wilson, S. R. Rabel, J F. Stobaugh, T. D. Williams, and D. G. Vander Velde, *Anal. Chem., 65*: 67R–84R (1993).
85. J. Kohr and H. Engelhardt. In *Capillary Electrophoresis Technology* (N.A. Guzman, ed.), Marcel Dekker, Inc., New York, 1993, pp. 357–381.
86. K. A. Cobb, V. Dolnik, and M. Novotny, *Anal. Chem., 62*:2478–2483 (1990).
87. R. M. McCormick *Anal. Chem., 60*:2322–2328 (1988).
88. M. Huang, W. P. Vorkink, and M. L. Lee, *J. Microcolumn Sep., 4*: 233–238 (1992).
89. W. Nashabeh and Z. El Rassi, *J. Chromatogr., 559*:367–383 (1991).
90. J. K. Towns and F. E. Regnier, *J. Chromatogr., 516*:69–78 (1990).
91. M. Huang, W. P. Vorkink, and M. L. Lee, *J. Microcolumn Sep., 4*: 135–143 (1992).
92. J. K. Towns, J. Bao, and F. E. Regnier, *J. Chromatogr., 599*:227–237 (1992).
93. A. M. Dougherty, C. L. Woolley, D. L. Williams, D. F. Swaile, R. O. Cole, and M. J. Sepaniak, *J. Liquid Chromatogr., 14*:907–921 (1991).
94. G. J. M. Bruin, R. Huisden, J. C. Kraak, and H. Poppe, *J. Chromatogr., 480*:339–349 (1989).
95. M. Huang and M. L. Lee, *J. Microcolumn Sep., 4*:491–496 (1993).
96. D. Schmalzing, C. A. Piggee, F. Foret, E. Carrilho, and B. L. Karger, *J. Chromatogr., 652*:149–159 (1993).
97. G. J. M. Bruin, J. P. Chang, R. H. Kuhlman, K. Zegers, J. C. Kraak, and H. Poppe, *J. Chromatogr., 471*:429–436 (1989).
98. S. A. Swedberg, *Anal. Biochem., 185*:51–56 (1990).
99. Y. F. Maa, K. J. Hyver, and S. A. Swedburg, *J. High Resolut. Chromatogr. Commun., 14*:65–67 (1991).
100. J. K. Towns and F. E. Regnier, *Anal. Chem., 63*:1126–1132 (1991).
101. Z. Zhao, A. Malik, and M. L. Lee, *Anal. Chem., 65*:2747–2752 (1993).
102. E. S. Yeung. In *Capillary Electrophoresis Technology* (N.A. Guzman, ed.), Marcel Dekker, Inc., New York, 1993, pp. 587–603.
103. G. J. M. Bruin, G. Stegeman, A. C. Van Asten, X. Xu, J. C. Kraak, and H. Poppe, *J. Chromatogr., 559*:163 (1991).
104. M. Albin, P. D. Grossman, and S. E. Moring, *Anal. Chem., 65*: 489A–497A (1993).

105. K. D. Altria, *LC–GC, 11*:438–440 (1993).
106. L. Hernandez, N. Joshi, P. Verdeguer, and N. A. Guzman. In *Capillary Electrophoresis Technology* (N.A. Guzman, ed.), Marcel Dekker, Inc., New York, 1993, pp. 605–614.
107. S. Wu and Dovichi, *J. Chromatogr., 480*:141–155 (1989).
108. T. T. Lee, S. J. Lillard, and E. S. Yeung, *Electrophoresis, 14*:429–438 (1993).
109. K. C. Chan, G. M. Janini, G. M. Muschik, and H. J. Issaq, *J. Liquid Chromatogr., 16*:1877–1890 (1993).
110. C. Toulas and L. Hernandez, *LC–GC, 10*:471–4722 (1992).
111. L. N. Amankwa, M. Albin, and W. G. Kuhr, *Trends in Anal.Chem., 11*:114–120 (1992).
112. M. E. Szulc and I. S. Krull, *J. Chromatogr., 659*:231–245 (1994).
113. E. S. Yeung and W. G. Kuhr, *Anal. Chem., 63*:275A–282A (1991).
114. J. M. Saz and J. C. Diez-Masa, *J. Liquid Chromatogr., 17*:499–520 (1994).
115. Y. F. Yik and S. F. Y. Li, *Trends Anal. Chem., 11*:325–332 (1992).
116. R. A. Wallingford and A. G. Ewing, *Anal. Chem., 59*:1762–1766 *1987).
117. T. J. O'Shea, R. D. Greenhagen, S. M. Lunte, C. E. Lunte, M. R. Smyth, D. M. Radzik, and N. Watanabe, *J. Chromatogr., 593*:305–312 (1992).
118. T. J. O'Shea and S. M. Lunte, *Anal. Chem., 65*:247–250 (1993).
119. T. J. O'Shea and S. M. Lunte, *Anal. Chem., 66*:307–311 (1994).
120. T. J. O'Shea, S. M. Lunte, and W. R. LaCourse, *Anal. Chem., 65*:948–951 (1993).
121. M. Deacon, T. J. O'Shea, S. M. Lunte, and M. R. Smyth, *J. Chromatogr., 652*:377–383 (1993).
122. X. Huang, T. K. J. Pang, M. J. Gordon, and R. N. Zare, *Anal. Chem., 59*:2747–2749 (1987).
123. X. Huang and R. N. Zare, *Anal. Chem., 63*:2193–2196 (1991).
124. R. D. Smith, N. T. Olivares, N. T. Nguyen, and H. R. Udseth, *Anal. Chem., 60*:436–441 (1988).
125. R. D. Smith, C. J. Barinaga, and H. R. Udseth, *Anal. Chem., 60*:1948–1952 (1988).
126. C. G. Edmonds, J. A. Loo, C. J. Barinaga, H. R. Udseth, and R. D. Smith, *J. Chromatogr., 474*:21–37 (1989).
127. M. A. Moseley, L. J. Deterding, K. B. Tomer, and J. W. Jorgenson, *Rapid Commun. Mass Spectrom., 3*:87–93 (1989).
128. M. A. Moseley, L. J. Deterding, K. B. Tomer, and J. W. Jorgenson, *J. Chromatogr., 516*:167–173 (1993).
129. R. D. Smith and H. R. Udseth. In *Capillary Electrophoresis Tech-*

nology (N.A. Guzman, ed.), Marcel Dekker, Inc., New York, 1993, pp. 525–567.

130. K. B. Tomer. In *Capillary Electrophoresis Technology* (N.A. Guzman, ed.), Marcel Dekker, Inc., New York, 1993, pp. 569–586.

131. M. A. Moseley, J. W. Jorgenson, J. Shabanowitz, D. F. Hunt, and K. B. Tomer, *J. Am. Soc. Mass Spectrom., 3*:289–300 (1992).

132. P. D. Grossman, J. C. Colburn, H. H. Lauer, R. G. Nielsen, R. M. Riggin, G. S. Sittampalam, and E. C. Rickard, *Anal. Chem., 61*: 1186–1194 (1989).

133. H. H. Lauer and D. McManigill, *Anal. Chem., 58*:166–170 (1986).

134. P. D. Grossman, J. C. Colburn, and H. H. Lauer, *Anal. Biochem., 179*:28–33 (1989).

135. M. Zhu, R. Rodriguez, D. Hansen, and T. Wehr, *J. Chromatogr., 516*:123–131 (1990).

136. F. T. A. Chen. *J. Chromatogr., 559*:445–453 (1991).

137. M. H. J. M. Langenhuizen and P. S. L. Janssen, *J. Chromatogr., 638*:311–318 (1993).

138. T. Satow, A. Machida, K. Funakushi, and R. L. Palmieri, *J. High Resolut. Chromatogr., 14*:276–279 (1991).

139. N. Chen, L. Wang, and Y. Zhang, *J. Chromatogr., 644*:175–182 (1993).

140. J. S. Green and J. W. Jorgenson, *J. Chromatogr., 478*:63–70 (1989).

141. M. M. Bushey and J. W. Jorgenson, *J. Chromatogr., 480*:301–310 (1989).

142. D. Corradini, A. Rhomberg, and C. Corradini, *J. Chromatogr., 661*:305–313 (1994).

143. M. A. Strege and A. L. Lagu, *J. Chromatogr., 630*:337–344 (1993).

144. M. A. Strege and A. L. Lagu, *J. Liquid Chromatogr., 16*:51–68 (1993).

145. G. M. McLaughlin, J. A. Nolan, J. L. Lindahl, R. H. Palmieri, K. W. Anderson, S. C. Morris, J. A. Morrison, and T. J. Bronzert, *J. Liquid Chromatogr., 15*:961–1021 (1992).

146. M. G. Khaledi, C. Quang, R. S. Sahota, J. K. Strasters, and S. C. Smith. In *Capillary Electrophoresis Technology* (N.A. Guzman, ed.), Marcel Dekker, Inc., New York, 1993, pp.

147. C. L. Ng, Y. L. Toh, S. F. Y. Li, and H. K. Lee, *J. Liquid Chromatogr., 16*:3653–3666 (1993).

7

Supercritical Fluid Chromatography of Bulk and Formulated Pharmaceuticals

JAMES T. STEWART *The University of Georgia, Athens, Georgia*

NIRDOSH K. JAGOTA *Bristol-Myers Squibb Pharmaceutical Research Institute, New Brunswick, New Jersey*

I. INTRODUCTION

Supercritical fluid chromatography (SFC) is a column chromatographic technique in which a supercritical fluid is used as a mobile phase. A supercritical fluid is a gas or liquid brought to a temperature and a pressure above its critical point. The first report of SFC dates back to 1962 when Kesper et al. [1] used supercritical fluid chlorofluorocarbons as a mobile phase for the separation of metal porphyrins. It was not until the early 1980s that an important breakthrough of the technique occurred. This was the introduction of capillary SFC and the availability of commercial instrumentation. These became major factors in the recent rise in popularity of SFC. According to the latest estimation, approximately 100 SFC articles are published in major journals every year.

In SFC, the mobile phase is delivered by a high-pressure pump. The sample is usually injected as a solution by means of a high-pressure injection valve. The column may either be a packed column, comparable to a high-performance liquid chromatographic (HPLC) column, or an open capillary column, comparable to a capillary gas chromatographic (GC) column, but with somewhat smaller internal diameters (50–100 μm). Detection is performed either on-line, (i.e., UV-VIS) or after the expansion of the fluid [i.e., flame ionization detection (FID) or mass spectrometry

239

(MS)]. The mobile phase is kept as a supercritical fluid by means of a restrictor until either on-line detection has been performed or just before the expansion into a gas phase detector. Figure 7.1 shows a line diagram for a typical SFC apparatus.

II. HARDWARE

A. Pumps

Three types of pump are used in SFC [2]:

1. Pneumatic amplifier pump: This pump is composed of two cylinders that are different in piston cross-sectional area. The piston cross-sectional area ratio between the two cylinders equals the pressure amplification factor from the low-pressure cylinder to the high-pressure cylinder, and also equals the flow rate attenuation factor from the high-flow-rate cylinder to the low-low-rate (high-pressure) cylinder. In practice, an area ratio of 5 : 10 is recommended for reasons such as safety, reliability of ultrahigh-pressure seals and connectors, fluid compressibility, and high-pressure cylinder volume.

2. Reciprocating piston pump: The reciprocating piston pump is a continuous-flow pump similar to an HPLC pump. Three major differences of a reciprocating pump from an liquid chromatographic (LC) pump are the addition of a pump cooling system, the requirement of a pulse dampener, and the greater minimization of postpump interface vol-

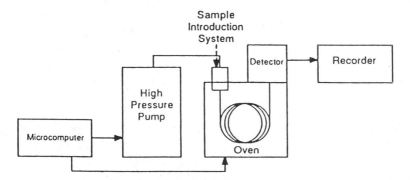

Figure 7.1 Schematic diagram of an SFC instrument. (Reprinted with permission from Ref. 2, *Analytical Supercritical Fluid Chromatography and Extraction*, Edited by M.L. Lee and K.G. Markides, 1990.)

ume. Commercially reciprocating pumps are available which have feed-back control to compensate for fluid compressibility, minimizing pressure ripple and, thus, producing reporducible results.

3. Syringe pump: This pump is widely employed for both open-tubular-column and packed-column SFC applications. Microprocessor control allows reproducible SFC fluid pressure or density programming.

B. Sample Introduction

Sample introduction is a major hardware problem for SFC. The sample solvent composition and the injection pressure and temperature can all affect sample introduction. The high solute diffusion and lower viscosity which favor supercritical fluids over liquid mobile phases can cause problems in injection. Back-diffusion can occur, causing broad solvent peaks and poor solute peak shape. There can also be a complex phase behavior as well as a solubility phenomenon taking place due to the fact that one may have combinations of supercritical fluid (neat or mixed with sample solvent), a subcritical liquified gas, sample solvents, and solute present simultaneously in the injector and column head [2]. All of these can contribute individually to reproducibility problems in SFC. Both dynamic and timed split modes are used for sample introduction in capillary SFC. Dynamic split injectors have a microvalve and splitter assembly. The amount of injection is based on the size of a fused silica restrictor. In the timed split mode, the SFC column is directly connected to the injection valve. High-speed pneumatics and electronics are used along with a standard injection valve and actuator. Rapid actuation of the valve from the load to the inject position and back occurs in milliseconds. In this mode, one can program the time of injection on a computer and thus control the amount of injection. In packed-column SFC, an injector similar to HPLC is used and whole loop is injected on the column. The valve is switched either manually or automatically through a remote injector port. The injection is done under pressure.

C. Oven

The chromatographic oven for SFC should meet the same requirements as a conventional GC oven. The oven is usually designed such that thermal gradients between any two points in the oven area where a column is placed are less than $\pm$ 0.1 °C.

D. Columns

Both capillary and packed columns are available for use. Much of the recent interest in SFC is based on the stability and high efficiencies of capillary columns. Because these columns are deactivated, the elution of analytes with unmodified carbon dioxide is possible. Some of the common stationary phases available as capillary columns are as follows:

1. SB-methyl-100: This is a 100% methylpolysiloxane stationary phase and is considered the least polar phase. Flexible siloxane bonds contribute to excellent diffusion characteristics and high efficiencies. This phase is often used as the first choice in the analysis of unknown samples in SFC because of the high resolution it delivers and the volatility-based elution order of solutes.
2. SB-phenyl-5: This stationary phase, which contains 5% phenyl substitution, has a slight amount of polarizability with increased capacity for polar solutes.
3. SB-phenyl-50: This stationary phase has 50% polarizable phenyl groups present on the silica surface and exhibits apparent dipoles of varying intensity depending on the local environment. This phase is moderately polar in a general sense and exhibits good selectivity for isomers containing polar functional groups, due to slight differences in the isomers' abilities to introduce dipoles within the stationary phase.
4. SB-biphenyl-30: This stationary phase has 30% biphenyl substitution which provides enhanced polarizability when compared to phenyl-substituted stationary phases. It is in the middle polarity range among the SFC stationary phases.
5. SB-cyanopropyl-25: This phase has 25% cyanopropyl substitution and often gives longer retention for polar solutes.
6. SB-cyanopropyl-50: This is the most polar stationary phase currently available in capillary SFC. It has 50% cyanopropyl substitution and is useful for polar analytes such as pharmaceuticals.
7. SB-smectic: This phase is a liquid polysiloxane. Its selectivity is based on solute size and shape. Solutes are separated on the basis of molecular geometry, with the length-to-breadth ratio determining elution order within an isomeric series.

One disadvantage of a capillary column is that the use of large injection volumes of sample solvent is not possible due to the low capacity of the column, thus resulting in difficulty in trace analysis. Another problem with capillary columns is that linear velocities of 10–20 times the optimum velocity are required in order to achieve a reasonable analysis time.

A column consisting of a deactivated silica-based stationary phase is used for the packed-column mode. A packed column allows larger volumes of sample solvent to be injected, thus improving sensitivity. Generally, the column dimensions are 1 × 100–250 mm and the particle size is 5 μm. Commercial SFC instruments are also available that will handle the classical 4.6 × 150-mm or 250-mm columns. With the introduction of electronically controlled variable restrictors to control the back pressure, the packed columns are becoming increasingly more popular. This feature allows the independent flow and pressure control of mobile phases, thus helping in rapid optimization of selectivities. Some of the commonly used packed columns are as follows:

1. Octadecylsilane (ODS): This is a polymer-coated silica packing with increased loadability of a porous silica support. Nonpolar compounds may be separated using the hydrophobic functionalities offered on ODS.
2. Cyanopropyl: It contains cyanopropyl packing and can be used in a normal-phase mode utilizing the strong dipole of cyano functional groups. It also offers high selectivity for phenols, alcohols, acids, and other polar compounds.
3. Si-60: It contains silica 60 Å and offers a large surface area compared to regular silica packing.
4. Polystyrene-divinylbenzene: This is a polymer-based column and gives better peak shape comparable to other columns such as octadecylsilane and cyanopropyl columns.

E. Detectors

Different types of detectors are used in conjunction with SFC. Some of the commonly used detectors are as follows:

(a) Flame ionization
(b) Ultraviolet
(c) Fluorescence
(d) Chemiluminescence
(e) Fourier transform–infrared
(f) Mass spectrometer
(g) Electrochemical
(h) Electron capture
(i) Thermionic ionization
(j) Light scattering
(k) Nitrogen–phosphorus

Other than chemiluminescence and electrochemical, all of the above detectors are commercially available.

F. Mobile Phases

Carbon dioxide is the most popular mobile phase for SFC because of its low critical temperature (31°C) and pressure (7.3 MPa). It is also inexpensive, nontoxic, nonflammable, and easily disposable. Other gases such as nitrous oxide and ammonia can also be used. Nitrous oxide is more polar than carbon dioxide, with ammonia being the most polar. Both nitrous oxide and ammonia are difficult to handle in the laboratory.

The solvating power of a supercritical fluid is manipulated by changing its density. An increase in density increases the solvating power of the supercritical fluid. In addition, the oven temperature can be varied to affect the selectivity of a supercritical fluid. The polarity of a supercritical fluid is altered by the addition of an organic modifier such as methanol or acetonitrile. Table 7.1 lists some of the commonly used organic modifiers added to carbon dioxide [2]. Supercritical fluids generally exhibit a high number of theoretical plates and selectivity when compared with HPLC or GC columns.

III.   APPLICATION OF SFC TO SELECTED BULK AND FORMULATED PHARMACEUTICALS

Jagota and Stewart [3] have reported the SFC analysis of diazepam and chlordiazepoxide and their related compounds. The separations were achieved using a 7-m × 50-μm capillary SB-cyanopropyl-50 column with a carbon dioxide mobile phase and FID detection. Typical analysis time for either mixture was within 30 min (see Figs. 7.2 and 7.3). A combination of oven temperature and pressure gradient was employed to give baseline resolution of the analytes. The limits of detection of the various compounds were in the 150–225-μg/ml range. The SFC method was applied to the assay of diazepam and chlordiazepoxide in their respective tablet dosage forms. Recoveries of 101–103% of the labeled amounts of drug were obtained from the assays. Accuracy and precision of spiked samples of both analytes were in the 0.4–4% and 1.8–4.0% ranges, respectively. A comparison of SFC to HPLC analysis of the drugs in their tablet dosage forms indicated that accuracy was comparable, but precision was better with HPLC (see Table 7.2). The SFC method was not sensitive enough to detect the related compounds of either diazepam or chlordiazepoxide at the 0.01–0.1% levels of the drug substance.

Table 7.1 Frequently Used Modifiers in SFC

Modifier	Temp. (°C)	Pressure (atm)	Molecular mass	Dielectric constant at 20°C	Polarity index[a]
Methanol	239.4	79.9	32.04	32.70	5.1
Ethanol	243.0	63.0	46.07	24.30	4.3
1-Propanol	263.5	51.0	60.10	20.33	4.0
2-Propanol	235.1	47.0	60.10	18.30	3.9
1-Hexanol	336.8	40.0	102.18	13.3	3.5
2-Methoxy ethanol	302	52.2	76.10	16.93	5.5
Tetrahydrofuran	267.0	51.2	72.11	7.58	4.0
1,4-Dioxane	314	51.4	88.11	2.25	4.8
Acetonitrile	275	47.7	41.05	37.50	5.8
Dichloromethane	237	60.0	84.93	8.93[b]	3.1
Chloroform	263.2	54.2	119.38	4.81	4.1
Propylene carbonate	352.0	–	102.09	69.0	6.1
N,N-dimethylacetamide	384	–	87.12	37.78[b]	6.5
Dimethyl sulfoxide	465.0	–	78.13	46.68	7.2
Formic acid	307	–	46.02	58.5[c]	–
Water	374.1	217.6	18.01	80.1	10.2
Carbon disulfide	279	78.0	76.13	2.64[c]	–

[a]Data from *Burdick and Jackson Solvent Guide*, 1984.
[b]25°C.
[c]Data from *Eastman Organic Chemical Bull., 47*: pp. 1–12 (1975).
Source: Reprinted from M.L. Lee and K.E. Markides. Ref. 2, with kind permission of M.L. Lee.

Jagota and Stewart [4] have also reported the SFC separation of various mixtures of chlordiazepoxide with clidinium bromide, amitriptyline hydrochloride, and the estrogens, estrone, equilin, *d*-equilenin, α-estradiol, and β-estradiol. The clidinium bromide-chlordiazepoxide mixture was separated on a 10-m × 50-μm ID (inner diameter) SB-biphenyl-30 capillary column within 25 min using pump and oven programs with FID detection as shown in Figure 7.4. The amitriptyline–chlordiazepoxide mixture was separated on a 7-m × 50-μm ID SB-cyanopropyl-50 capillary column using pump and oven programs and FID detection as shown in Figure 7.5. The retention times for amitriptyline and chlordiazepoxide were 20.3 and 13.8 min, respectively. The best SFC separation of the chlordiazepoxide–estrogens mix was achieved on a 10-m × 50-μm ID SB-cyanopropyl-50 capillary column using pump and oven programs with FID detection (see Fig. 7.6). The α-estradiol peak could not be baseline resolved from the nearby β-

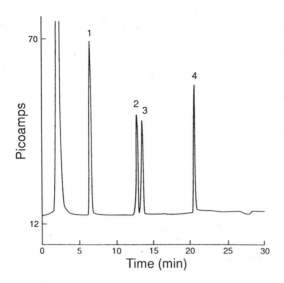

Figure 7.2 Typical SFC separation of MCAB (1), diazepam (2), ACMPC (3) and nordiazepam (4) on a SB-cyanopropyl-50 column using temperature and pressure gradients. Concentration of each compound was 1 mg/ml in methylene chloride. (From Ref. 3.)

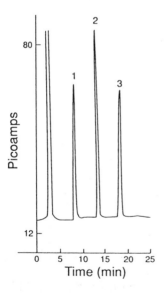

Figure 7.3 Typical SFC separation of ACB (1), chlordiazepoxide (2) and demoxepam (3) on a SB-cyanopropyl-50 column at 120°C oven temperature and a pressure gradient. Concentration of each compound was 1 mg/ml in methylene chloride. (From Ref. 3.)

Table 7.2 Comparison of Diazepam and Chlordiazepoxide Dosage Form Analysis Using SFC and HPLC

Compound	Labeled amount (mg)	Amount found by SFC (mg)	Amount found by HPLC (mg)
Diazepam[a]	10	10.29 ± 0.42[b] RSD, [c]4.08%	10.16 ± 0.12 RSD, 1.18%
Chlordiazepoxide[d]	10	10.13 ± 0.32 RSD, 3.16%	10.08 ± 0.04 RSD, 0.40%

[a]Valium tablet, 10 mg, Roche lot unknown.
[b]Mean ± standard deviation based on n = 3.
[c]RSD = relative standard deviation.
[d]Librium capsule, 10 mg, Roche lot 6159-02.
Source: Ref. 3.

estradiol peak. The retention times of the six analytes were in the 70–90-min range. It was interesting to note that the relative retention data obtained by these SFC methods for the various chlordiazepoxide mixtures was comparable to that obtained with HPLC methods, as shown in Table 7.3.

The SFC separation of selected estrogens (estrone, equilin, α-estradiol, β-estradiol, and d-equilenin) was achieved on a SB-cyanopropyl-50 capillary column using a carbon dioxide density gradient at an oven temperature of 73°C [5]. A typical analysis time on the 7-m column was 21

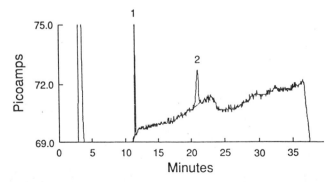

Figure 7.4 Typical SFC separation of clidinium bromide (1) and chlordiazepoxide (2) on a SB-biphenyl-30 column. (From Ref. 4.)

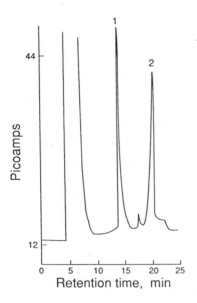

Figure 7.5 Typical SFC separation of chlordiazepoxide (1) and amitriptyline hydrochloride (2) on a SB-cyanopropyl-50 column. (From Ref. 4.)

min, as shown in Figure 7.7. Accuracy and precisions of the method were
in the 1–6% range. The SFC assay was applied to three commerical products: one in which the estrogens were conjugated, one in which they were
esterified, and a single entity dosage form containing only β-estradiol.
Based on a 20-nl injection and limited solubility of the estrogens in chloro-

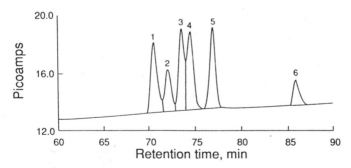

Figure 7.6 Typical SFC separation of chlordiazepoxide (1), estrone (2),
α-estradiol (3), β-estradiol (4), equilin (5), and d-equilenin (6) on a SB-
cyanopropyl-50 column. (From Ref. 4.)

Table 7.3 Comparison of SFC and
HPLC Relative Retention Data of
Chlordiazepoxide Mixtures

	Relative retention	
Mixture	SFC[a]	HPLC
1. Clidinium bromide	0.55	0.60[b]
Chlordiazepoxide	1.00	1.00
2. Chlordiazepoxide	0.68	0.71
Amitriptyline	1.00	1.00
hydrochloride		
3. Chlordiazepoxide	0.82	c
Estrone	0.83	0.85[d]
α-Estradiol	0.86	1.00
β-Estradiol	0.87	c
Equilin	0.90	0.75
d-Equilenin	1.00	0.69

[a]See chromatographic parameters, Experimental Section, Ref. 4.
[b]USP XXII method.
[c]No data were reported.
[d]Mobile phase for estrogens was 44 : 55 pH 2.5 phosphate buffer containing 0.1% TEA/ absolute methanol; octadecylsilane column (3.9 mm × 30 cm), flow rate 1.2 ml/min; electrochemical detection at + 600 mV versus Ag/AgCi.
Source: Ref. 4.

form, only the estrone levels could be determined in the conjugated or esterified dosage forms. The estrone levels found by SFC were close to labeled amounts expressed as total conjugated or esterfied estrogens. In the β-estradiol tablet, SFC analysis gave essentially 100% of the labeled amount. A comparison of SFC to HPLC assay of β-estradiol showed no statistical difference between the test results at the 95% confidence level.

The SFC separation of nonsteroidal anti-inflammatory drugs (NSAIDs) was investigated by Jagota and Stewart [6]. These compounds vary in chemical structure and functional group chemistry and provide a representative sample of acidic drugs to study. Using a 10-m × 50-μm ID

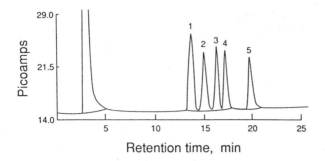

Figure 7.7 Typical SFC separation of estrone (1), equilin (2), α-estradiol (3), β-estradiol (4), and *d*-equilenin (5) on a SB-cyanopropyl-50 column. [Reprinted from Ref. 5, *J. Pharm. Biomed. Anal. 10*, Jagota and Stewart, "Supercritical Fluid Chromatography of Selected Oestrogens," p. 667 (1992) with kind permission from Elsevier Science Ltd., The Boulevard, Langford Lane, UK.]

SB-biphenyl-30 capillary column, many of the compounds were separated, but they could not all be separated in a single injection (see Figs. 7.8 and 7.9). Tailing factors were generally in the 1–1.5 range and the limits of detection (S/N = 3) were in the 75–250-μg/ml range. The SB-cyanopropyl-50 capillary column was also found to be suitable for SFC separation of the NSAIDs. Tailing factors were in the 1–1.8 range and the limits of detection (S/N = 3) were in the 80–400-μg/ml range. The biphenyl column was arbitrarily selected to demonstrate the applicability of the SFC assay to selected tablet or capsule dosage forms of ibuprofen, ketoprofen, and mefenamic acid. The amounts found were within 95–105% of the labeled amounts of each drug.

Jung and Schurig [7] have investigated enantiomeric separations using capillary SFC and an immobilized cyclodextrin stationary phase of film thickness 0.25 μm. The authors found no loss in selectivity and column performance after several months. Many NSAID enantiomers could be separated, but amine-containing drugs such as ephedrine or synephrine had to be derivatized with trifluoroacetic acid anhydride prior to assay (see Fig. 7.10). There was little or no enantioselectivity with beta blockers, oxazepam, lorazepam, fenoprofen, etololac, thalidomide, testosterone, and promethazine. Although the enantioselectivity was rather low in comparison to chiral HPLC, there are advantages of much higher column efficiencies and the high compatibility of SFC with FID and MS detection.

The analysis of controlled drugs using capillary and packed supercritical fluid chromatography interfaced with mass spectrometry was reported

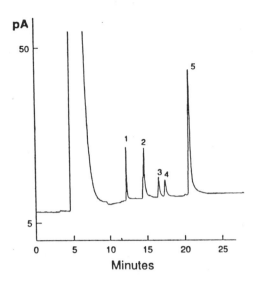

Figure 7.8 Typical SFC separation of ibuprofen (1), fenoprofen (2), naproxen (3), ketoprofen (4), and tolmetin (5) on a SB-cyanopropyl-50 column. [Reprinted from Ref. 6, *J. Chromatogr. 604*, 255 (1992) with kind permission of Elsevier Science Publishers, The Netherlands.]

by MacKay and Reed [8]. They emphasized the advantages of SFC versus HPLC such as identification of involatile adulterants and the promise of reproducible SFC–MS. A 10-m × 50-μm ID biphenylpolysiloxane capillary column and a 25-cm × 1-mm cyanobonded silica packed column were used in the study. The controlled drug sample consisted of a seizure of diamorphine, which was screened by HPLC and found to contain *N*-phenyl-2-naphthylamine, methaqualone, narcotine, papaverine, phenobarbitone, acetylcodeine, diamorphine, acetylmorphine, and an unknown compound which was not confirmed by GC–MS analysis. The authors mentioned that narcotine and papaverine could not be separated on a GC column but are easily separated by SFC because it enables solubility, polarity, and steric effects to influence a separation which would otherwise be governed by volatility.

The SFC analysis of mercaptopurine, trimethoprim, triprolidine, pseudoephedrine, permethrin, zidovudine, and trifluridine was reported by Mulcahey and Taylor [9]. The mobile phase was carbon dioxide containing methanol and the packed column was a cyanopropyl (250 × 4.6 mm ID). Figures 7.11 and 7.12 show separations of the compounds of interest. In order to elute trimethoprim from the column without severe tailing, tetra-

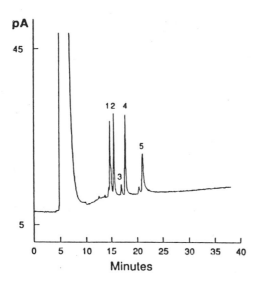

Figure 7.9 Typical SFC separation of flufenamic acid (1), mefenamic acid (2), acetylsalicylic acid (3), ketoprofen (4), and fenbufen (5) on a SB-cyanopropyl-50 column. [Reprinted from Ref. 6, *J. Chromatogr. 604*, 255 (1992) with kind permission of Elsevier Science Publishers, The Netherlands.]

butylammonium hydroxide was added to the methanol prior to adding to the carbon dioxide mobile phase. Upon addition of the base to the mobile phase, trimethoprim elutes easily. Pseudoephedrine and triprolidine would not elute from the cyanopropyl column with the tetrabutylammonium hydroxide containing mobile phase. Instead, these drugs had to be extracted using a supercritical fluid extraction experiment prior to SFC analysis. In this manner, SFC chromatograms of the two drugs could be obtained.

Crowther and Henion have reported the SFC assay of polar drugs on packed columns using mass spectrometric detection [10]. Their method was shown to be suitable for a multicomponent mixtures containing nonpolar and polar analytes. They mentioned that one of the advantages of SFC versus HPLC was the faster column reequilibration time. In this method, silica, amino, nitrile, and diol packed columns (20 cm × 2.1 mm ID) were used. Cocaine, codeine, caffeine, methocarbamol, phenylbutazone and oxyphenbutazone were separated on one or more of these columns and mass spectra were obtained.

Lee et al. [11] have reported the SFC separation of NSAIDs on a

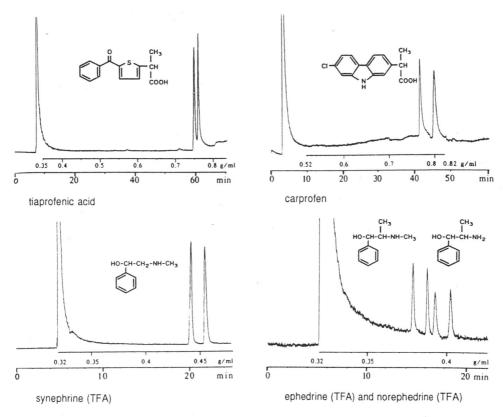

Figure 7.10 Separations of enantiomers by open-tubular SFC on a 5-m × 50-μm ID fused silica column coated with a 0.25-μm film of immobilized Chirasil-Dex. (Reprinted from Ref. 7 with kind permission of Dr. Alfred Huthig Verlag GmbH, Heidelberg.)

10-m × 50-μm ID SB-methyl capillary column with a carbon dioxide mobile phase and an oven temperature of 150°C. For some unexplained reasons, the interfacing of SFC and MS has produced reduced electron impact (EI) sensitivity compared to GC–MS. The use of chemical ionization (CI) helps to circumvent the problem but lacks the structural information available from EI. Charge exchange (CE) has been used to obtain EI-like mass spectra when combined with SFC. The authors showed that the CE technique blended with SFC–MS did indeed offer sensitivity approaching that of GC–MS.

Koski et al. [12] have developed an SFC separation of prostaglandins

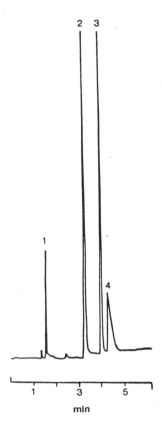

Figure 7.11 Separation of four of the compounds of interest employing a mobile-phase gradient. Elution order: (1) permethrin, (2) zidovudine, (3) trifluridine, (4) mercaptopurine. Flow = 2 ml/min CO_2, 150 μl/min methanol for 1.5 min, 150–450 μl/min in 2.0 min, hold at 450 μl/min column. Deltabond CN (250 × 4.6 mm ID). Oven temperature: 60°C; UV detection at 254 nm. (Reprinted from Ref. 9, with kind permission of Dr. Alfred Huthig Verlag, GmbH, Heidelerg.)

without derivatization using SFE coupled with open-tubular SFC. The compounds investigated were prostaglandin F2α and prostaglandin F1α and their esters and prostaglandin E2. Using a carbon dioxide mobile phase and FID, the prostaglandins were directly injected on-column or, if SFE were used, placed on the column by solute focusing. Eleven prostaglandins were efficiently separated within 35 min using a density program with constant

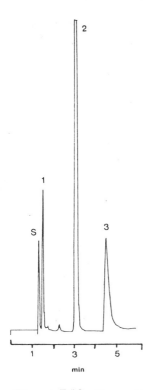

Figure 7.12 Separation of three compounds using tetrabutylammonium hydroxide in the mobile phase. Elution order: (1) permethrin, (2) zidovudine, (3) trimethoprim. Flow = 2 ml/min CO_2, 100 μl methanol–tetrabutylammonium hydroxide for 2.0 min, 100–500 μl/min methanol–tetrabutylammonium hydroxide for 2.0 min, 100–500 μl/min methanol–TBAOH in 4.0 min, hold at 500 μl/min. Column: Deltabond CN (250 × 4.6 mm ID). Oven temperature: 60°C; UV detection at 254 nm. (Reprinted from Ref. 9 with kind permission of Dr. Alfred Huthig Verlag GmbH, Heidelberg.)

oven temperature. With S/N = 3, the minimum detectable quantity was 9 ng for a typical compound such as the isopropyl ester of 15-proprionate PGF2α. When SFE was used, a 100-μl sample size could be placed in the extraction cell and this would produce an even lower minimum detectable quantity.

Miscellaneous steroids were separated on a packed phenyl column (Spherisorb, 10 cm × 4.6 mm ID) using UV detection at 254 nm and a carbon dioxide mobile phase with organic modifiers [13] (see Fig. 7.13). A light-scattering detector using either tungsten or a laser was used. It was found that some of the steroids that lacked a UV chromophore could be successfully detected with the light-scattering detector. The study showed that light-scattering detectors could be readily interfaced to SFC and will extend the range of detectors available for SFC.

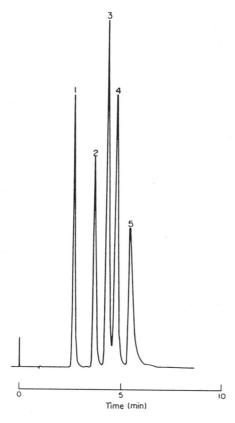

Figure 7.13 Typical chromatogram for steroids: (1) medroxyprogesterone acetate, 2.0 mg/ml; (2) cortisone acetate, 2.0 mg/ml; (3) methylprednisolone acetate, 2.0 mg/ml; (4) isoflupredone acetate, 2.0 mg/ml; (5) hydrocortisone, 1.6 mg/ml. [Reprinted from Ref. 13, *J. Pharm. Biomed. Anal. 8*, Loran and Cromie, "An Evaluation of the Use of Supercritical Fluid Chromatography with Light Scattering Detection for the Analysis of Steroids" (1990) with kind permission from Elsevier Science Ltd, The Boulevard, Langford Lane, UK.]

Lynam and Nicolas have evaluated chiral separations by HPLC versus SFC [14] The enantiomers studied were pharmaceutical synthetic precursors. Repeated injections of trans-stilbene oxide and carbobenzyloxy phenylalaninol were made and the chromatographic parameters Rs, N, and α were calculated daily. SFC gave superior enentiomeric resolution of peaks and there was a faster solvent equilibration. The columns were quite stable in both SFC and HPLC systems.

White et al. [15] have reported on the SFC determination of cyclosporin, ionic polyether antibiotics, and fat-soluble vitamins. Cyclosporin is a cyclic undecapeptide that has potent immunosuppressive activity and is particularly effective in the prevention of graft rejection after organ transplantation. A 9-m × 50-μm ID fused silica open-tubular column with a 0.20-μm film thickness of DB-5 was used for the cyclosporin. The column was heated at 150°C. The carbon dioxide mobile phase was held at 200 atm for 10 min, then subjected to a linear pressure program to 300 atm at 10 atm/min. The retention time of the cyclosporin was about 25 min. The antibiotics were organic sodium salts and belonged to a class called ionophores. A 10-m × 50-μm ID fused silica open-tubular column with a 0.2-μm DB-5 film thickness was used. The column was heated to 100°C and the carbon dioxide mobile phase held at 225 atm for 5 min and then a linear pressure program to 306 atm at 4 atm/min. The antibiotics were fully separated within 24 min. The vitamins were chromatographed on the same DB-5 stationary phase using 140°C column temperature and the carbon dioxide mobile phase at 150 atm for 10 min, then ramped to 200 atm at 5 atm/min. The following are the compounds and their retention times: vitamin K_3 menadione, 11 min; vitamin K_1, 32 min; vitamin E, 28 min; vitamin A, 20 min; and provitamin B, 34 min.

The SFC analysis of barbiturates has been reported by Smith and Sanagi [16]. The method utilzed packed columns of ODS bonded silica (200 × 3 mm ID) or polystyrene–divinylbenzene polymer (150 × 4.6 mm ID) with a carbon dioxide mobile phase and FID detection. The barbiturates were strongly adsorbed on the ODS column and were not observed to elute. The drugs did elute with reasonable retention times on the polymer column. Unfortunately, there was peak tailing, and adsorption remained a problem with all the barbiturates studied. The authors were prevented from adding methanol as a polar modifier to the mobile phase because the detection was by FID.·

Peytavin et al. [17] have reported on the chiral resolution of mefloquine, halofantrine, enpiroline, quinine, quinidine, chloroquine, and primaquine by subcritical fluid chromatography on a (S) naphthylurea column (250 × 4.6 mm ID). The mobile phase consisted of carbon dioxide, methanol, and triethylamine at a 3-ml/min flow rate. Except for primaquine and chloroquinine, the enantiomers were separated at a column temperature of 40–60°C and pressure below 15 MPa. A typical chromatogram of the separation of quinine and quinidine is shown in Figure 7.14. Carbon dioxide – methanol 0.1% triethylamine (98 : 2 v/v) was used to separate the halofantrine enantiomers, whereas a 80 : 20 v/v mix was more suitable for mefloquine, enpiroline, quinine, and quinidine enantiomers. The authors concluded that carbon dioxide entailed specific solvation of these antimalarial drugs and played an important role in chiral recognition on the (S) naphthylurea phase. Furthermore, methanol in the mobile phase affected capacity factors and preserved selectivity, whereas triethylamine was useful for efficiency optimization.

Berge and Deye studied the effect of column surface area on the retention of polar solutes [18]. They found that there was a linear relationship between retention and the surface area. 4-Hydroxy benzoic acid was used as a model acidic compound, and sulfamethazine, sulfanilamide, sulfisomidine, and sulfapyridine were used as the model basic compounds. The separations were carried out on a packed Nucleosil Diol column with a methanol-modified carbon dioxide as the mobile phase. The UV detector was used for the analysis. It was observed that 0.1% acetic acid for the acidic solutes and 0.1% isopropylamine for basic solutes was required in the methanol to achieve the separations. The efficiency was found to be similar for 100-, 300-, and 500-Å packing materials.

Karlsson et al. reported the supercritical fluid chromatography of methaqualone, cotinine, and reclopride, among other compounds, using capillary columns of different polarities [19]. Detection was either thermionic nitrogen–phosphorus or flame ionization. Supercritical nitrous oxide was used as the mobile phase. The detection limits obtained were in the range of 2–4 ppm and the precision was in the range of 3–12%.

Janicot et al. presented the separation of opium alkaloids using subcritical and supercritical fluid chromatography [20]. Carbon dioxide–methanol–triethylamine–water mixtures were used as the mobile phase with packed aminopropyl or bare silica columns. The influence of aminated polar modifiers such as methylamine, ethylamine, and triethylamine was studied. Figure 7.15 shows the separation of six opium alkaloids narcotine, papaverine, thebaine, codeine, cryptopine, and morphine on a Lichrosorb Si-60 column. The method gave comparable results with HPLC.

Steuer et al. demonstrated the use of supercritical fluid chromatography in the separation of enantiomers of 1,2 amino alcohols, namely pindolol, metoprolol, oxprenolol, propranolol, and DPT 201–106 using ion-pairing modifiers [21]. The mobile phase consisted of carbon dioxide mixed with acetonitrile containing triethylamine as a counterion and *N*-benzoxycarbonylglycyl-L-proline as a chiral counterion. They found that the ca-

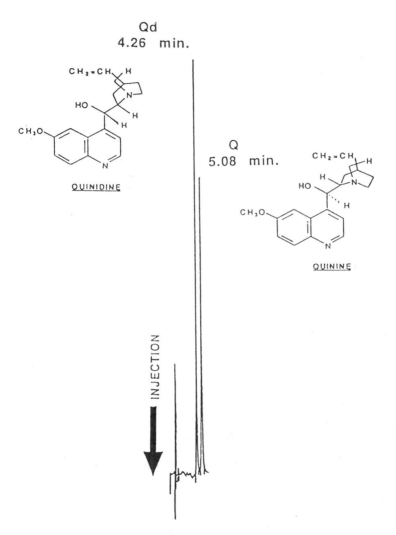

Figure 7.14 Chromatographic separation of diastereoisomers; quinine and quinidine; stationary phase; (S) naphthylurea 250 × 4.6 mm ID. Chromatographic conditions: CO_2 methanol, 0.1%, TEA (80 : 20), 3 ml/min, 50°C, 12.5 MPa, and λ = 230 nm. (Reprinted from Ref. 17 with kind permission of Wiley-Liss, Inc. New York.)

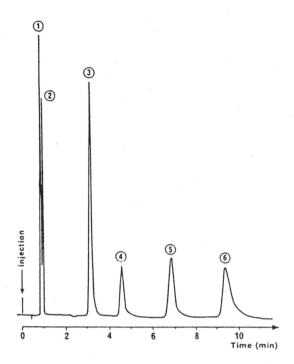

Figure 7.15 Separations of opium alkaloids on bare silica. Column = 23 × 0.46 cm ID; stationary phase, 5-μm LiChrosorb Si-60 silica; mobile phase, carbon dioxide–methanol–methylamine–water (83.37 : 16.2 : 0.15 : 0.23, w/w); Solutes: 1, narcotine; 2, papaverine; 3, thebaine; 4, codeine; 5, crytopine; 6, morphine. [Reprinted from Ref. 20, *J. Chromatogr. 437*, 351 (1988) with kind permission of Elsevier Science Publishers, The Netherlands.]

pacity factor increases with the concentration of counterion and decreases with the TEA concentration. The chiral selectivity was found to be larger at lower pressures, and increasing temperatures caused it to decrease. They also found that due to absence of isocratic conditions, H versus U curves for supercritical and subcritical conditions cannot be well described by the Van Deemter equation.

Perkins et al. reported the packed-column SFC of four veterinary antibiotics (levamisole, furazolidone, chloramphenicol, and lincomycin) using an amino column with methanol-modified carbon dioxide as the mobile phase [22]. A baseline separation of all four analytes was obtained in less than 4 min. The effect of several modifiers such as 2-methoxy-ethanol, propylene carbonate, 2-propanol, and dimethylformamide was also studied.

Matsumoto and co-workers demonstrated the use of atmospheric pressure chemical ionization mass spectrometry for the analysis of fat-soluble vitamins (vitamins K, E, and D_3 and vitamin A acetate) [23]. Carbon dioxide was used as the mobile phase.

Steuer et al. compared supercritical fluid chromatography with capillary zone electrophoresis (CZE) and high-performance liquid chromatography (HPLC) for its application in pharmaceutical analysis [24]. Efficiency, performance, sensitivity, optimization, sample preparation, ease of method development, technical capabilities, and orthogonality of the information were the parameters studied. They concluded that SFC is ideal for moderately polar compounds, such as excipients, for which mass detection is required.

Edder et al. reported the capillary supercritical fluid chromatography of basic drugs of abuse, namely nicotine, caffeine, methadone, cocaine, imipramine, codeine, diazepam, morphine, benzoylecgonine, papverine, narcotine, and strychnine [25]. They compared the separation of these drugs on DBS and DB wax columns. The chromatographic conditions included a carbon dioxide mobile phase and a flame-ionization detector. It was noted that on the DBS column, all peaks other than methadone and cocaine were separated. With the exception of benzoylecgonine and papaverine, all other peaks were separated on a DB wax column. A reproducibility of less than 5% was obtained with an internal standard method. The detection limits obtained were within 10–50 ppm on both the columns. A linearity of >0.99 was obtained for methadone, codeine, and morphine in the concentration range 10–1000 ppm.

Veuthey and Haerdi reported the separation of amphetamines using packed-column SFC [26]. The amphetamines were derivatized with 9-fluorenylmethyl chloroformate and chromatographed with a methanol or 2-propanol-modified carbon dioxide as the mobile phase. The separations were compared on bare silica and aminopropyl-bonded silica columns. Both columns gave comparable results and the separation of all five amphetamines (methylamphetamine, amphetamine, phenethylamine, ephedrine, and norephedrine) was achieved in less than 5 min. Both methanol and 2-propanol-modified carbon dioxide gave comparable results. It was observed that the modifier concentration had more effect on the solvating power than the mobile-phase density.

Later et al. analyzed steroids, antibiotics, and cannabinoids on a methylpolysiloxane (SE-33) capillary column using carbon dioxide as the mobile phase [27]. Figure 7.16 shows the separation of tetrahydrocannabinol and its six metabolites in less than 20 min.

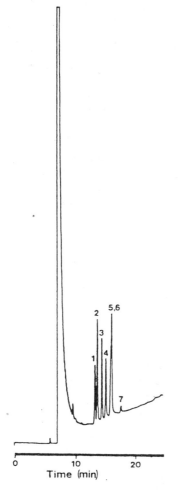

1. HEXAHYDROCANNABINOL
2. TETRAHYDROCANNABINOL
3. CANNABINOL
4. 3'-HYDROXY THC
5. 5'-HYDROXY THC
6. II-HYDROXY THC
7. II-NOR-9- CARBOXY THC

Figure 7.16 Capillary supercritical fluid chromatogram of tetrahydrocannabinol and six metabolites. Conditions: supercritical CO_2 at 120°C; 15-m × 50-μm ID SE-33 column; density programmed from 0.40 g/ml after a 7-min hold to 0.56 g/ml at 0.01 g/ml/min; FID at 280°C. (Reprinted from Ref. 27, *J. Chromatogr. Sci.* by permission of Preston Publications, a division of Preston Industries.)

Li and co-workers demonstrated the use of SFC in the analysis of panaxadiol and panaxatriol in ginseng, a famous traditional chinese medicine [28]. A capillary SB-cyanopropyl-50 column with a carbon dioxide mobile phase and flame-ionization detector was used for the analysis. Methyltestosterone was used as an internal standard for the quantitation. Figure 7.17 shows the SFC separation. The method was found to be linear (r > 0.999) in the range studied and the precision obtained was in the range 2.2–5.7%.

Smith and Sanagi reported the packed-column SFC of benzodiazepines (diazepam, lorazepam, lormetazepam, nordazepam, temazepam, estrazolam, chlordiazepoxide, triazolam, cloxazolam, ketazolam, and loprazolam) with methanol-modified carbon dioxide as the mobile phase [29]. The effect of methanol concentration on separation was studied on three columns: polystyrene–divinylbenzene, octadecylsilane, and cyanopropyl-bonded silica columns. They concluded that proportion of methanol has marked effect on the selectivity of compounds containing different functional groups.

Smith and Sanagi also studied the SFC of barbiturates (barbitone, butobarbitone, amylobarbitone, pentobarbitone, talbutal, quinalbarbitone, methohexitone, phenobarbitone, and heptabarbitone) using a packed polystyrene–divinylbenzene or octadecylsilane column with a methanol-modi-

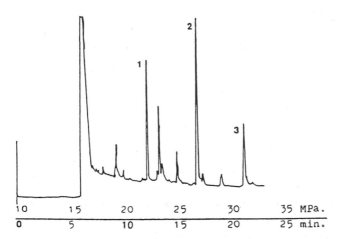

Figure 7.17 Capillary supercritical fluid chromatograms of ginseng extract. (1) Methyltestosterone (internal standard), (2) panaxadiol, (3) panaxatriol. [Reprinted from Ref. 28, *Biomedical Chrom.* (1992) with kind permission of John Wiley and Sons, Ltd., UK.]

fied carbon dioxide [30]. Again, they concluded that proportion of modifier has profound effect on the selectivity.

Steuer et al. used ion-pairing techniques in packed-column supercritical fluid chromatography [31]. They studied the combined effect of ion-pairing reagents and mobile-phase density on the selectivity for a wide range of pharmaceuticals such as pindolol, propranolol, bopindolol, isradipin, and spirapril. They concluded that the selectivity is more dependent on the density of supercritical fluids. They also demonstrated that the optimized separations were obtained faster by SFC than by HPLC. It was also observed that a diol phase with TBA and acetate in methanol as ion-pairing agents was the best choice for initial conditions for method development for both anionic and cationic compounds. Figure 7.18 shows the optimization of the method for the separation of bopindolol, its precursor, and benzoic acid as a degradation product within 30 min.

Scalia and Games developed a packed column SFC method for the analysis of free bile acids: cholic acid (CA), chenodeoxycholic acid (CDCA), deoxycholic acid (DCA), lithocholic acid (LCA), and ursodeoxycholic acid (UDCA) [32]. The baseline separation of all five bile acids was achieved on a packed phenyl column with a methanol-modified carbon dioxide in less than 4 min. The elution order showed a normal-phase mechanism because the solutes eluted in the order of increasing polarity following the number of hydroxyl groups on the steroid nucleus. The method was also applied to the assay of UDCA and CDCA in capsule and tablet formulations. The method was found to be linear in the range 1.5–7.5 ng/ml (r > 0.99, n = 6). The average recoveries (n = 10) for UDCA and CDCA were 100.2% with a RSD of 1.7% and 101.5% with a RSD of 2.2%, respectively. The reproducibility of the method was less than 1.5% (n = 10) for both UDCA and CDCA.

Perkins et al. reported the separation of mixtures of sulphonamides on packed silica and amino columns with a methanol-modified carbon dioxide along with a UV or mass spectrometer detector [33]. The MS was performed using both moving-belt and modified thermospray interfaces. Figure 7.19 shows the separation of sulphadoxine, sulphamethazine, sulphadimethoxine, sulphamerazine, sulphaquinoxaline, sulphachlorpyridazine, and sulphathiazole on an amino-bonded column. The mobile phase was methanol-modified carbon dioxide at a flow rate of 4 ml/min. The initial methanol concentration was 15% which was increased to 25% after 4 min and the temperature was increased to 90°C. The separation was compared on a silica column. Although all eight compounds were separated on a silica column, there was difference in the elution order. They also concluded that a reduction in methanol concentration enhanced the separation.

Berry et al. studied the SFC separation of xanthines, carbamates,

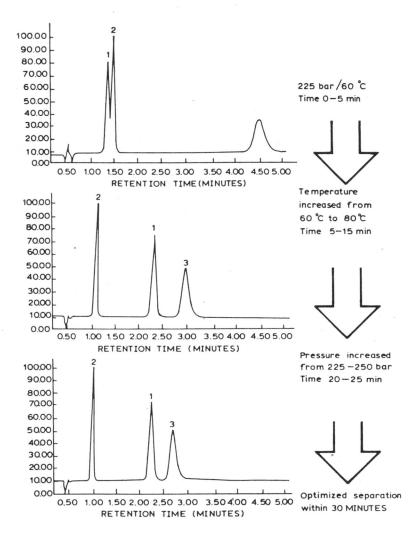

Figure 7.18 Optimization of the separation of bopindolol (2), its precursor (3), and benzoic acid (1) as a degradation product within 30 min. Conditions: carbon dioxide, 20% methanol, 20 mM TBA, 20 mM acetic acid; flow rate, 4 ml/min. Column: diol, 10 μm (100 mm × 4.6 mm ID). [Reprinted from Ref. 31, *J. Chromatogr. 500*, 469 (1990), with kind permission of Elsevier Science Publishers, The Netherlands.]

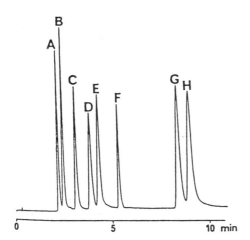

Figure 7.19 Ultraviolet trace (270 nm) obtained from SFC of a mixture of (A) sulphadoxine (1-SDX), (B) sulphamethazine (2-SMT), (C) sulpha-merazine (3-SMZ), (D) sulphadimethoxine (4-SDM), (E) sulphadiazine (5-SDZ), (F) sulphaquinoxaline (6-SQX), (G) sulphachlorpyridazine (7-SCP), and (H) sulphathiazole (8-STZ) on a 100 × 4.6-mm ID column packed with 5-μm amino-bonded Spherisorb. The mobile phase was initially carbon dioxide modified with 15% methanol at a flow rate of 4 ml/min. After 4 min, the concentration of methanol was increased to 25%. Column pressure, 361 bars; temperature, 90°C. [Reprinted from Ref. 33, *J. Chromatogr. 540*, 239 (1991) with kind permission of Elsevier Science Publishers, The Netherlands.]

sulphonamides, steroids, and ergot alkaloids [34]. A silica- or aminopro-pyl-bonded column with a methanol- or methoxy ethanol-modified carbon dioxide was used for the separations. SFC–MS was accomplished using a moving-belt HPLC–MS interfaced with a modified thermospray device.

Anton et al. reported the use of packed-column SFC for a stability-indicating assay of crotaniton cream and liquid dosage forms [35]. A silica or cyano column or both columns in tandem were used for the analysis. The methanol-modified carbon dioxide with or without pressure program-ming was used for the separation. UV, and evaporative light-scattering detectors were used in tandem. The method was found to be linear (r > 0.99) for bulk drug, bulk drug in cream, and bulk drug in lotion. The precision of the assay was less than 1%. The recoveries obtained were close to 99%.

Berger and Wilson presented the separation of 14 antipyschotic drugs (triflupromazine HCl, carphenazine maleate, methotrimeprazine, promazine HCl, perphenazine, chloroprothixine, deserpidine, thiothixene, reserpine, acetophenazine dimaleate, ethopropazine, promethazine HCl, propriomazine HCl, and triflupromazine HCl) using packed-column SFC [36]. A cyanopropyl column along with a tertiary mobile phase consisting of carbon dioxide, methanol, and isopropylamine and an UV detector were used for the separation. It was noted that none of the solutes eluted without the addition of isopropylamine in the mobile phase. They observed that change in temperature (over 30°C range) had much more effect on selectivity than the modifier concentration, which affected the retention time but not the selectivity. This may be due to the fact that a thick film of mobile-phase components is adsorbed on the packing and behaves as the part of the effective stationary phase. However, above 1% modifier concentration, the adsorbed film is essentially complete and changes little with increased modifier concentration or pressure. The small changes in temperature have a significant impact on both the thickness and composition of adsorbed film, thus affecting the selectivity. Detection limits obtained were as low as 125 ppb for a 5-μl injection.

Berger and Wilson also reported the separation of 10 antidepressants (amitriptyline, imipramine, nortriptyline, desipramine, protripyline, buclizine, benactyzine, hydroxyzine, perphenazine, and thioridazine) using a packed-column SFC with a tertiary mobile phase [37]. A Lichrosphere cyanopropyl column with a mobile phase consisting of supercritical fluid carbon dioxide with 10% modifier (methanol with 0.5% isopropylamine) was used for the separation. It was noted that solutes did not elute without the addition of isopropylamine. Detection limits obtained were as low as 88 ppb for a 5-μl injection.

Biermanns et al. reported the chiral resolution of β-blockers, including propranolol, metoprolol, and atenolol using packed-column supercritical fluid chromatography [38]. A Chiracel OD column with a mobile phase of 30% methanol with 0.5% isopropylamine in carbon dioxide was used for the separation. A baseline separation of isomers was obtained in less than 5 min at a mobile-phase flow rate of 2 ml/min. While keeping the column outlet pressure constant, the flow rate was increased to 4 ml/min and it was noted that, although the retention was reduced, the resolution remained the same. Both R- and S-propranolol gave linear responses from 0.25–2500 ppm with a correlation coefficient of >0.9999. The detection limit was approximately 250 ppb for a S/N ratio of 3. The reproducibility for both R- and S-propranolol was less than 1.5%. It was also noted that 0.09% R-propranolol can be quantitated in the presence of 2500 ppm of S-Propranolol.

A surfatron microwave-induced plasma was reported for the selective detection of catechols after SFC separation of the compounds as cylic boronate esters [39]. An available spectroscopic system was modified to utilize the 749.4-nm line of third-order boron emission. The separations were achieved on a 10 m × 50 μm ID SB-methyl-100 column using a 200-nl injection. Both supercritical-fluid-grade carbon dioxide and nitrous oxide were used as the mobile phase with separations of the esters within 10–15 min. Sensitivity of boron in catechol boronate was 25 pg/sec. Cortisone was also studied, but the small peak obtained demonstrated that the practical sensitivity of the microwave-induced plasma was too low for an application of real and practical importance.

Jung and Schurig have reported the chiral separations of several pharmaceuticals including NSAIDs, Norgestrel, and hexobarbital using supercritical fluid chromatography [40]. Separations were performed on a capillary-immobilized polysiloxane-B-cyclodextrin (Chirasil-Dex) column with carbon dioxide as the mobile phase. It was observed that separations were better at lower temperatures. In comparison with GC, SFC gave higher separation factors but similar resolution.

Heaton et al. have reported the supercritical fluid chromatography of taxicin I and taxicin II extracted by supercritical fluid extraction of Taxus baccata, the English yew tree [41]. They compared capillary- and packed-column SFC and concluded that packed-column SFC was better than capillary-column SFC for quantitative analysis of these compounds. Capillary SFC was done on either a biphenyl or carbowax column with unmodified carbon dioxide as the mobile phase. The packed-column SFC was performed on a nitrile column with a mobile phase consisting of a methanol gradient with carbon dioxide.

Gyllenhaal and Vessman have described the packed-column SFC of omeprazole and related compunds using an amino column with triethylamine- and methanol-modified carbon dioxide as the mobile phase [42]. Triethylamine (1%) was added to the methanol. The method was compared with an existing HPLC method, and it was concluded that SFC was about 2.5 times faster and gave a higher degree of selectivity than HPLC.

Masuda et al. have reported the supercritical fluid chromatography of retinol palmitate and tocopherol acetate using an octadecylsilyl silica gel column with ethanol-modified carbon dioide as the mobile phase [43]. Both retinol and tocophenol eluted within 5 min. The method was applied to the determination of these compounds in an ointment formulation. The method was found to be linear for both retinol and tocophenol from 0.5 to 2.5 μg/ml.

IV. CONCLUSIONS

Supercritical fluid chromatography offers high selectivity and efficiency. Method development is faster and easier than HPLC. The future of SFC lies in the use of both capillary and packed columns. Although capillary columns offer a higher number of theoretical plates for separation, they lack in sample loadability, thus affecting the sensitivity in terms of concentration. Packed columns, on the other hand, offer greater sample loadability and a reproducibility comparable to HPLC. SFC will emerge as a complimentary technique to HPLC and its use in applications such as chiral separation and purity analysis will increase. Many of these applications are just beginning to find their way into the scientific literature.

REFERENCES

1. E. Kesper, A.H. Corwin, and D.A. Turner *J. Organ Chem., 27*:700 (1962).
2. M.L. Lee and K.E. Markides (eds.), *Analytical Supercritical Fluid Chromatography and Extraction*, Chromatography Conferences, Inc., Provo, UT, 1990.
3. N.K. Jagota and J.T. Stewart, Analysis of diazepam and chlordiazepoxide and their related compounds using supercritical fluid chromatography, *J. Liquid Chromatogr., 15*:2429 (1992).
4. N.K. Jagota and J.T. Stewart, Separation of chlordiazepoxide and selected chlordiazepoxide mixtures using capillary SFC, *J. Liquid Chromatogr., 16*:291 (1993).
5. N.K. Jagota and J.T. Stewart, Supercritical fluid chromatography of selected oestrogens, *J. Pharm. Biomed. Anal., 10*:667 (1992).
6. N.K. Jagota and J.T. Stewart, Separation of non-steroidal anti-inflammatory agents using supercritical fluid chromatography, *J. Chromatogr., 604*:255 (1992).
7. M. Jung and V. Schurig, Extending the scope of enantiomer separation by capillary supercritical fluid chromatography on immobilized polysiloxane anchored permethyl-β-cyclodextrin (Chirasil-Dex), *J. High Resolut. Chromatogr., 16*:215 (1993).
8. G.A. MacKay and G. D. Reed, Application of capillary supercritical fluid chromatography, packed column supercritical fluid chromatography, and capillary supercritical fluid chromatography–mass spectrometry in the analysis of controlled drugs, *J. High Resolut. Chromatogr., 14*:537 (1991).
9. L.J. Mulcahey and L.T. Taylor, Gradient separation of pharmaceuticals employing supercritical carbon dixode and modifiers, *J. High Resolut. Chromatogr., 13*:393 (1990).
10. J.B. Crowther and J.D. Henion, Supercritical fluid chromatography of polar drugs using small particle packed columns with mass spectrometric detection, *Anal. Chem., 57*:2711 (1985).
11. E.D. Lee, S. Hsu, and J.D. Henion, Electron-ionization-like mass

spectra by capillary supercritical fluid chromatography–charge exchange mass spectrometry, *Anal. Chem., 60*:1990 (1988).

12.   I.J. Koski, B.A. Jansson, K.E. Markides and M.K. Lee, Analysis of prostaglandins in aqueous solution by supercritical fluid extraction and chromatography, *J. Pharm. Biomed. Anal., 9*:281 (1991).

13.   J.S. Loran and K.D. Cromie, An evaluation of the use of supercritical fluid chromatography with light scattering detection for the analysis of steroids, *J. Pharm. Biomed. Anal., 8*:607 (1990).

14.   K.G. Lynam and E.C. Nicolas, Chiral HPLC versus chiral SFC: evaluation of long term stability and selectivity of chiracel OD using various eluents, *J. Pharm. Biomed. Anal., 11*:1197 (1993).

15.   C.M. White, D.R. Gere, D. Boyer, F. Pacholec, and L.K Wong, Analysis of pharmaceuticals and other solutes of biochemical importance by supercritical fluid chromatography, *J. High Resolut. Chromatogr., 11*:94 (1988).

16.   R.M. Smith and M.M Sanagi, Application of packed column supercritical fluid chromatography to the analysis of barbituates, *J. Pharm. Biomed. Anal., 6*:837 (1988).

17.   G. Peytavin, F. Gimenez, B. Genissel, C. Gillotin, A. Baillet, I. Wainer, and R. Farnotti, Chiral resolution of some antimalarial agents by sub- and supercritical fluid chromatography on an (S) naphthylurea stationary phase, *Chirality, 5*:173 (1993).

18.   T.A. Berge and J. F. Deye, Correlation between column surface area and retention of polar solutes in packed-column supercritical fluid chromatography, *J. Chromatogr., 594*:291 (1992).

19.   L. Karlsson, L. Mathiasson, J. Akesson, and J.A. Jonsson, Quantitative aspects of the determination of compounds with wide varying polarity using capillary fluid chromatography, *J. Chromatogr., 557*: 99 (1991).

20.   J.L. Janicot, M. Caude, and R. Rosset, Separation of opium alkaloids by carbon dioxide sub- and supercritical fluid chromatography with packed columns; Application to the quantitative analysis of poppy straw extracts, *J. Chromatogr., 437*:351 (1988).

21.   W. Steuer, M. Schindler, G. Schill, and F. Erni, Supercritical fluid chromatography with ion-pairing modifiers; separation of enantiomeric 1,2-aminoalcohols as diastereomeric ion pairs, *J. Chromatogr., 447*:287 (1988).

22.   J.R. Perkins, D.E. Games, J.R. Startin, and J. Gilbert, Analysis of veterinary drugs using supercritical fluid chromatography and supercritical fluid chromatography–mass spectrometry, *J. Chromatogr., 540*:257 (1991).

23.   K. Matsumoto, S. Nagata, H. Hattori, and S. Tsuge, Development of directly coupled supercritical fluid chromatography with packed

capillary column–mass spectrometry with atmospheric pressure chemical ionization, *J. Chromatogr., 605*:87 (1992).

24. W. Steuer, I. Grant, and F. Erni, Comparison of high-performance liquid chromatography, Supercritical fluid chromatography and capillary zone electrophoresis in drug analysis, *J. Chromatogr., 507*:125 (1990).

25. P. Edder, W. Haerdi, I. Veuthey, and C. Staub, Quantitative capillary supercritical fluid chromatography and supercritical fluid extraction of basic drugs of abuse, *Chimia, 46*:141 (1992).

26. J. Veuthey and W. Haerdi, Separation of amphetamines by supercritical fluid chromatography, *J. Chromatogr., 515*:385 (1990).

27. D.W. Later, B.E. Richter, D.E. Knowles, and M.R. Andersen, Analysis of various classes of drug by capillary supercritical fluid chromatography, *J. Chromatogr. Sci., 24*:249 (1986).

28. Y. Li, X. Li, I. Hong, J. Liu, and M. Zhang, Determination of panaxadiol and panaxatriol in ginseng and its preparations by capillary supercritical fluid chromatography (SFC), *Biomed. Chromatogr., 6*:88 (1992).

29. R.M. Smith and M.M. Sanagi, Packed-column supercritical fluid chromatography of benzodiazepines, *J. Chromatogr., 483*:51 (1989).

30. R.M. Smith and M.M. Sanagi, Supercritical fluid chromatography of barbiturates, *J. Chromatogr., 481*:63 (1989).

31. W. Steuer, J. Baumann, and F. Erni, Separation of ionic drug substances by supercritical fluid chromatography, *J. Chromatogr., 500*:469 (1990).

32. S. Scalia and D.E. Games, Determination of free bile acids in pharmaceutical preparation by packed column supercritical fluid chromatography, *J. Pharm. Sci., 82*:44 (1993).

33. J.R. Perkins, D.E. Games, J.R. Startin, and J. Gilbert, Analysis of sulphonamides using supercritical fluid chromatography and supercritical fluid chromatography–mass spectrometry, *J. Chromatogr., 540*:239 (1991).

34. T.J. Berry, D.E. Games, and J.R. Perkins, Supercritical fluid chromatographic and supercritical fluid chromatographic–mass spectrometric studies of some polar compounds, *J. Chromatogr., 363*:147 (1986).

35. K. Anton, M. Bach, and A. Geiser, Supercritical fluid chromatography in the routine stability control of antipruritic preparations, *J. Chromatogr., 553*:71 (1991).

36. T.A. Berger and W.H. Wilson, Separation of drugs by packed column supercritical fluid chromatography 1. Phenothiazine antipsychotics, *J. Pharm. Sci., 83*:281 (1994).

37. T.A. Berger and W.H. Wilson, Separation of drugs by packed column

supercritical fluid chromatography 2. Antidepressants, *J. Pharm. Sci., 83*:287 (1994).

38. P. Biermanns, C. Miller, V. Lyon, and W. Wilson, Chiral resolution of β-blockers by packed-column supercritical fluid chromatography, *LC–GC Mag., 11*(10):744 (1993).

39. D. R. Luffer and M. V. Novotny, Capillary supercritical fluid chromatography and microwave-induced plasma detection of cyclic boronate esters of hydroxy compounds, *J. Microcolumn Sep., 3*:39–46 (1991).

40. M. Jung and V. Schurig, Extending the scope of enantiomer separation by capillary supercritical fluid chromatography on immobilized polysiloxane anchored permethyl-B-cyclodextrin (Chirasil-Dex), *J. High Resolut. Chromstogr., 16*:215 (1993).

41. D.M. Heaton, K.D. Bartle, C.M. Rayner, and A.A. Clifford, Application of supercritical fluid extraction and supercritical fluid chromatography to the production of taxanes as anti-cancer drugs, *J. High Resolut. Chromatogr., 16*:666 (1993).

42. O. Gyllenhaal and J. Vessman, Packed-column supercritical fluid chromatography of omeprazole and related compounds; selection of column support with triethylamine and methanol-modified carbon dioxide as the mobile phase, *J. Chromatogr., 628*:275 (1993).

43. M. Masuda, S. Koike, M. Handa, K. Sagara, and T. Mizutani, Application of supercritical fluid extraction and chromatography to assay fat-soluble vitamins in hydrophobic ointment, *Anal. Sci., 9*:29 (1993).

8
Applications

JOHN A. ADAMOVICS *Cytogen Corporation, Princeton,*
New Jersey

I. INTRODUCTION

The purpose of this chapter of the book is to provide the analyst with a supplementary up-to-date reference source that describes analytical procedures for detecting and characterizing drug substances and related impurities in a variety of matrices from plant extracts to drug product. This chapter is meant to complement the preceding chapters and to direct the reader to references on topics not discussed in the preceding chapters but considered relevant to the topic of this book.

The procedures described in this chapter vary from general methods as developed by chromatographic manufacturers to very exacting methods such as those used in stability-indicating assays. The substances have been arranged alphabetically according to their generic names. There are also general classifications such as methods for amino acids, analgesics, antibodies, antibiotics, antihistamines, barbiturates, colorants, opium alkaloids, prostaglandins, and sulphonamides. For more complete coverage of phytochemicals, several references are recommended [1–3] and for biotechnology-related products refer to Ref 4 and the most recent USP. Each procedure is subdivided into a description of the sample, mode of analysis, sample pretreatment steps, chromatographic sorbent, mobile phase, type of detection, pertinent comments, and reference.

Even though there are a number of citations to spectral data and physical data such as NMR, UV, IR, MS, mp, and so forth, the most comphrensive references are Clarke's [4] and Mills [5] and from the reference standard retailers (Sigma, Alltech Applied Science Labs, Quantimetrix, and USP).

References

1. N. G. Bisset (ed.), *Herbal Drugs and Phytochemicals*, CRC Press, Boca Raton, FL.
2. *Pharmacopoeia of the People's Republic of China* (English Edition, 1992), Complied by The Pharmacopoeia Commission of PRC, Beijing, China
3. *The Japanese Pharmacopoeia,* 10th ed., (Eng. ed. 1982), Yakuki Nippo, Ltd., Tokyo, 1981.
4. A. C, Moffat (ed.), *Clarke's Isolation and Identification of Drugs,* The Pharmaceutical Press, 1986.
5. T. M. Mills, W. N. Price, and J. C. Roberson (eds.), *Instrumental Data for Drug Analysis,* Vols. 1–5, Elsevier, New York.

II. ABBREVIATIONS

ACN	Acetonitrile
AcOH	Acetic acid
AOAC	Association of Offical Analytical Chemists
BP	British Pharmacopoeia
CE	Capillary electrophoresis
CI	Chemical ionization
CP	Chinese Pharmacopoeia
DAD	Photodiode array detection
DSC	Docusate sodium
DMAB	p-Dimethylaminobenzaldehyde
DSS	Decane sodium sulfonate
EC	Electrochemical detection
ECD	Electron-capture detector
EI	Electron impact mass spectrometry
EP	European Pharmacopoeia
EtOAc	Ethyl acetate
EtOH	Ethanol
FID	Flame-ionization detector

HPA	1-Heptane sulfonate
HSA	1-Hexane sulfonate
IPA	Isopropyl alcohol
ISTD	Internal standard
JP	Japanese Pharmacopoeia
LD	Limit of detection
MAb	Monoclonal antibody
MeOH	Methanol
MLL	Mean list length
MPPH	*p*-Methylphenytoin
MS	Mass spectrometer
MSTFA	*N*-methyl-*N*-(trifluorosily)trifluoracetamide
NH_4OH	Ammonium hydroxide
NPD	Nitrogen phosphorous detector
PO_4	Phosphate buffer
PrOH	Propanol
PRP	Poly(styrene)-divinylbenzene
PSA	1-Pentane sulfonate
REA	Radiative energy attenuation
RI	Refractive index detector
RIA	Radioimmunoassay
SFC	Supercritical fluid chromatography
SCX	Strong cation exchanger bonded to silica
SLS	Sodium lauryl sulfate
SPE	Solid-phase extraction
SRM	Selected reaction monitoring
TBA(H)	Tetrabutylammonium hydroxide
TCD	Thermal conductivity detector
TEA	Triethylamine
THF	Tetrahydrofuran
USP	*United States Pharmacopeia*

III. TABLE OF ANALYSIS

The following table is a listing of the chromatographic methods that have been used to test over 1,300 pharmaceuticals, their excipients, and impurities.

Drug	Mode	Sample pretreatment	Sorbent (temp.)	Mobile phase	Detection	Comments	Ref.
Acebutolol hydrochloride							
Bulk	TLC	—	Silica (KOH treated)	Ammonium hydroxide–methanol (1.5 : 100)	Char	GC [2], HPLC [3]. Also, TLC [267]; enantiomer resolution [703,707]	[1]
Acecainide							
Bulk	GC	—	USP-G2 (between 100 and 300°C)	—	FID	—	[2]
Acecarbromal							
Bulk	TLC	—	Silica	CH$_2$Cl$_2$–acetone (4 : 1)	Chlorinate	HPLC	[8]
Acedapsone							
Bulk	TLC	Prepare three solutions of 20 µl, 100 µl and 1 mg/ml in MeOH	Silica G	Toluene–acetone (2 : 1)	Dry plate in air, spray with 0.5% sodium nitrite in HCl, after a few minutes 0.1% n-(1-naphtyl) ethylenediamine HCl	No secondary spot with the 1-mg/ml solution is more intense than the principal spot obtained with the 200 µg/ml soln and no two spots are more intense than the spot obtained with 50 µg/ml soln, p. 385	[1138]
Acefylline piperazine							
Bulk	GC	—	USP-G2 (between 100 and 300°C)		FID	—	[2]
Acenocoumarin (See warfarin)							
Acepromazine maleate							
Bulk	TLC	Silica (spray 0.1M MeOH KOH)	NH$_4$OH–MeOH (1.5 : 100)	—	Acidified iodoplatinate	—	[1]
Bulk	HPLC	Dissolve in 0.05N HCl, 1 mg/ml → 5 ml/50ml H$_2$O	C-8, 0.4 × 15 cm	(H$_2$O–triethylamine (700 : 6)–CH$_3$CN (7 : 3)	280 nm	USP 23, p.15, assay; related compounds by TLC injection, assay, p. 16	[5]

Sample	Method	Stationary phase	Sample preparation	Mobile phase	Detection	Comments	Ref.
Acetaminophen (Paracetamol) Bulk	TLC	Silica gel	Transfer 1 g to a glass-stoppered, 15-ml centrifuge tube, add 5 ml ether, shake for 30 min, centrifuge for 1000 rpm for 15 min. Apply 200 μl. Alongside spot 10 μg of p-chloroacetanilide.	Hexane–acetone (75 : 25)	UV	Limit for p-chloroacetanilide, ≤0.001%, USP 23, p. 17	[5]
Capsules	HPLC	C_{18}, 0.39 × 30 cm	Weigh contents of not less than 20 capsules, calculate the average weight- dissolve in mobile phase to obtain final volume of 200 ml – filter (0.5 μm) discard first 10 ml.	H_2O–MeOH (3 : 1), 1.5 ml/min	243 nm	USP 23, p. 17, assay; oral solution, p. 18, suppositories, p. 19, after melting– add H_2O– wash with hexane– dilute with mobile phase; tablets	[5]
Capsules containing aspirin and caffeine	HPLC	C_{18}, 4.6 × 10 cm	Weigh the contents of at least 20 capsules, transfer accurately a weighed portion equivalent to 250 mg acetaminophen to a 100-ml volumetric flask, add 75 ml MeOH–AcOH (95 : 5), shake 30 min. Dilute to volume with MeOH–AcOH–transfer 2.0 ml to 3.0 ml internal std solution (benzoic acid in MeOH 6	H_2O–MeOH–AcOH (69 : 28 : 3), 2.0 ml/min, 45°C	275 nm	Assay, for salicylic acid use 302 nm and weigh equivalent to 250 mg aspirin–100 ml with AcOH–MeOH, p. 21; for aspirin tablets, sonicate powder with $CHCl_3$–MeOH–AcOH (78 : 20 : 2), 280 nm, benzoic acid ISTD, p. 20; TLC, GC, and HPLC review [11]	[5]

Drug	Mode	Sample pretreatment	Sorbent (temp.)	Mobile phase	Detection	Comments	Ref.
Capsules with codeine phosphate	HPLC	mg/ml)–add to 50 ml volumetric, dilute to volume with MeOH–AcOH. Remove 20 capsules, transfer equiv. of 550 mg acetaminophen to 250-ml vol. flask, dissolve in mobile phase– 15 ml dilute to 50 ml–filter, discard first few ml.	C_{18}, 3.9 x 30 cm, 1.5 ml/min	4.44 g DSC/L(MeOH–H_2O–THF–H_3PO_4 (600 : 360 : 40 : 1)	280 nm	USP 23, assay p. 24, similiar procedures for oral solution, oral suspension and tablets containing codeine, p. 26	[5]
Tablets with diphenhydramine citrate	HPLC	Shake powder with MeOH, add H_2O	C_8 (35°C)	H_2O–MeOH (3 : 2)	254 nm	USP 23, p. 27, ISTD–guaifenesin [2,12–17, 22, 73, 75, 141, 310, 649, 650, 1082, 1133]; stability [1104]; with pseudoephedrine tablets, p. 28	[5]
Acetanilide							
Bulk	GC	—	USP-G2 (100–300°C)	—	FID	—	[2]
Bulk	HPLC	—	Silica	MeOH–0.01M NH_4ClO_4 EtOAc	—	—	[3,10]
Bulk	TLC	—	Silica	EtOAc	Acidified $KMnO_4$	—	[1]
Acetazolamide							
Bulk	TLC	—	Silica	EtOAc	—	—	[1, 10, 18]
Bulk	HPLC	—	Silica	MeOH–0.01M NH_4ClO_4, pH 6.7	—	—	[2,8,10]
Tablets	HPLC	Sonicate powder (~100 mg) with 10 ml 0.5N NaOH, 5 min, dilute to 100 ml with H_2O–10	C_{18}, 4.6 x 25 cm, 2.0 ml/min	NaOAc (4.1 g/950 ml) H_2O, 20 ml MeOH, 30 ml ACN	254 nm	USP 23, p. 29, assay	[5]

Compound/Form	Method	Sample preparation	Column	Mobile phase	Detection	Notes	Ref.
Suspension	HPLC	Dilute in mobile phase [ml- +ISTD–sulfadiazine–100 ml with H_2O]	C_{18}	Phosphate buffer (pH 5.0)–MeOH (65:35)	254 nm	Stability-indicating; HPLC [109]	[103]
Acetic acid — In Bulk	GC	Dissolve in H_2O	— (150°C)	—	FID	BP, p. 395, ISTD–dioxane	[4]
Otic solution	GC	Dilute in MeOH	USP-G35	—	—	USP, ISTD–anisole	[5]
Acetohexamide — Bulk	HPLC	—	Silica	MeOH (1% NH_4OH)–CH_2Cl_2 (5:95)	254 nm	Also [112]	[10, 1105]
Bulk	TLC	—	Silica	EtOAc	—	—	[1]
Acetone — Benzethonium chloride, tincture	GC	Dilute with H_2O	USP-S3 (120°C)	—	FID	USP, ISTD-MeOH, resolves from EtOH	[5]
Neat	GC	—	USP-S4 (8°C/min)	—	FID	USP, assay, also [2], AOAC [19]; [399]	[5]
Acetophenazine maleate — Bulk	HPLC	—	Silica	MeOH (1% NH_4OH)–CH_2Cl_2 (95:5)	254 nm	—	[10]
Acetorphine Hydrochloride — Bulk	TLC	—	Silica (Spray 0.1M KOH, MeOH)	Ammonium hydroxide–methanol (1.5:100)	Acidified iodoplatinate	—	[1]
Bulk	HPLC	—	Silica	MeOH–0.01M NH_4ClO_4, pH 6.7	—	—	[3]
Acetylcarbromal — Bulk	GC	—	USP-G2 (100–300°C)	—	FID	—	[2]
Bulk	TLC	—	Silica (0.1 M KOH, MeOH)	MeOH 0.01M NH_4OH, pH 6.7	—	—	[1]
Acetylcholine — Bulk	HPLC	Dissolve in mobile phase	Silica	1% H_3PO_4	Conductivity	Unpublished	—
Ophthalmic Solution	HPLC	Dissolve in MeOH–HPA (1:90), pH 4.0 with AcOH	C_{18}	MeOH–HPA (1:90), pH 4.0 with AcOH	RI	USP 23, p. 33; TLC [1]; HPLC [20]; also, HPLC [111] resolved from choline, detected by postcolumn immobilized enzyme	[5]

Drug	Mode	Sample pretreatment	Sorbent (temp.)	Mobile phase	Detection	Comments	Ref.
Acetylcysteine							
Aqueous solutions	HPLC	Dilute in H_3PO_4, pH 2	C_{18} (60°C)	0.5% M/v aq. $NaClO_4$, pH 2.0	215 nm	ISTD–L-tyrosine	[1066]
Bulk and solution	HPLC	Dissolve in bisulfite soln. (1 in 2000)	C_{18}	KH_2PO_4 (6.5 g/l)	214 nm	USP 23, p. 34, ISTD–phenylalanine; inhalation solution with isoproterenol use 280 nm, p. 35	[5]
Acetyldigitoxin							
Bulk	TLC	—	Silica	Benzene–EtOH (7:3)	*p*-Anisaldehyde	HPLC [1138]	[8]
Acetyldihydrocodeine							
Bulk	HPLC	—	Silica, 30 cm × 4 mm, 10 μm	MeOH (1% NH_4OH)–CH_2Cl_2 (1:9)	254 nm	GC, UV, MS, NMR, and IR	[10, p. 24]
Acetylmethadol							
Bulk	HPLC	—	Silica, 30 cm × 4 mm, 10 μm	MeOH (1% NH_4OH)–CH_2Cl_2 (2:98)	254 nm	GC, UV, MS, NMR, and IR	[10, p. 30]
Acetylsalicylic acid							
Bulk	HPLC	—	C_{18}	CH_3CN–H_2O gradient	220 nm	GC, HPLC [113, 142, 143]	[8]
Tablets	HPLC	Extract with mobile phase	C_{18}	H_2O–CH_3CN–H_3PO_4 (76:24:0.5)	295 nm	Resolved from salicylic acid, caffeine, and acetaminophen	[28]
Acintrazole							
Bulk	TLC	—	Silica	NH_4OH–MeOH (1.5:100)	Acidified iodoplatinate	—	[1]
Acivicin							
Bulk	HPLC	Dissolve in mobile phase	C_{18}	H_2O–CH_3CN–ClO_4–HSA–DMCH (995 ml:5 ml:0.8 ml:1.88 g:0.62 g), pH 6.2	214 nm	—	[21]
Aconitine nitrate							
Bulk	TLC	—	Silica (0.1 M KOH, MeOH)	NH_4OH–MeOH (1.5:100)	Acidified iodoplatinate	—	[1]
Acriflavine							
Bulk	TLC	—	Silica (0.1 M KOH, MeOH)	NH_4OH–MeOH (1.5:100)	254 nm	—	[1]

	Method	Sample preparation	Stationary phase	Mobile phase	Detection	Notes	Ref.
Acrinol							
Ointment with zinc oxide	TLC	Shake with ether, acetic acid & H_2O, aq. layer used	Silica	Ether–EtOH–AcOH (40:10:1)	350 nm	JP, p. 1306 identification	[7]
Actinomycin							
Bulk	HPLC	—	C_{18}	CH_3CN–H_2O (1:1)	Electrometer	—	[782]
Acyclovir							
Bulk	HPLC	Dissolve 25 mg in 0.1 N NaOH–dilute to volume with H_2O, 50 ml	C_{18}, 4.2 × 30 cm	AcOH (1 in 1000 H_2O), 3 ml/min	254 nm	USP 23, p. 36, assay, limit for guanine and identification; HPLC, stability indicating, more stable in alkaline [128]	[5]
Adenine							
Bulk	GC	Head space	—	—	—	USP 23, volatile impurities [467]	[5]
Adiphenine hydrochloride							
Bulk	GC	—	USP-G2 (between 100 and 300°C)	—	FID	—	[3]
Bulk	HPLC	—	Silica	MeOH–0.01M NH_4ClO_4	Acidified iodoplatinate	—	[2]
Bulk	TLC	—	Silica (0.1 M KOH, MeOH)	NH_4OH–MeOH (1.5:100)		—	[1]
Adipiodone (Iodipamide)							
Bulk	TLC	Dissolve 1 mg in 0.08% MeOH–NaOH	Silica (HF_{254})	n-BuOH–AcOH–H_2O (4:1:5)	254 nm	CP, p. 409, identification & injection i.d.	[1138]
Adrenaline acid tartrate							
Bulk	TLC	Dissolve in H_2O, apply, spray with Na bicarbonate solution followed by acetic anhydride, heat	Silica	HCOOH–acetone–CH_2Cl_2 (0.5:50:50)	Ethylenediamine/ferric cyanide	EP, pt. II-7, pp. 254–3, noradrenaline also assayed; HPLC, resolved from lidocaine/bupivacaine [130]	[6]
Adrenaline Hydrochloride							
Bulk	TLC	—	Silica (0.1 M KOH, MeOH)	NH_4OH–MeOH (1.5:100)	Acidified $KMnO_4$	HPLC [23]	[1]
Injection	HPLC	Dissolve in mobile phase	C_{18}	MeOH–buffer (380:620); buffer = 7.8 g KH_2PO_4 & 0.14 g SDS, pH 3.8 with H_3PO_4/l with H_2O	280 nm	CP, assay p. 411; BP, vol. II, p. 586 [4] TLC, injection,	[1138]

Drug	Mode	Sample pretreatment	Sorbent (temp.)	Mobile phase	Detection	Comments	Ref.
Adriamycin (Doxorubicin)							
Ajmaline							
Bulk	HPLC	—	Silica	MeOH–0.01 M NH$_4$ClO$_4$	—	GC [11]	[3]
Bulk	TLC	Dissolve in CH$_3$Cl	Silica	CHCl$_3$–acetone–diethylamine (5:4:1)	Dragendorff	JP, p. 71, purity; [1,11]	[7]
Alanine							
Bulk	TLC	60% EtOH	Silica	n-BuOH–H$_2$O–AcOH (3:1:1)	Ninhydrin	—	[24]
Albenazole							
Bulk	TLC	Dissolve 50 mg in 3.0 ml AcOH, dilute to 50 ml with AcOH–1.0 ml to 100 ml with AcOH	Silica	CH$_3$Cl–AcOH–ether (6:1:1)	Short wavelength	USP 23, p. 37, purity; similiar procedure in CP [1138, p. 412]	[5]
Albuterol							
Bulk	TLC	Dissolve in MeOH at 0.1 mg/ml	Silica	MEK–IPA–EtoAc–H$_2$O–NH$_4$OH (50:45:35:18:3)	I$_2$	USP 23, purity, p. 38, same for sulfate salt and tablets, p. 39; HPLC for tablet assay, C$_{18}$ HSA (1.13 g/1.2 l) + 12 ml AcOH – MeOH (6:4); enantiomers [137]	[5]
Alclofenac							
Bulk	HPLC	—	C$_{18}$	IPA–CHOOH–0.1M KH$_2$PO$_4$ (13.61 g/L) (540:1:1000)	—	TLC and GC, p. 323	[25]
Alcometasone Dipropionate							
Bulk	HPLC	Dissolve in MeOH	C$_{18}$	0.05M KH$_2$PO$_4$–MeOH (1:2)	254 nm	USP 23, p. 40, ISTD–betamethasone, assay	[5]
Bulk	TLC	Dissolve in MeOH–CH$_2$Cl$_2$ (1:2)	Silica	CHCl$_3$–acetone (7:1)	350 nm	USP 23, p. 40	[5]
Cream	TLC/HPLC	Add MeOH, heat to 60°C until sample melts, remove and shake vigorously	Silica	CHCl$_3$–acetone (7:1)	350 nm		[5]

Drug	Method	Sample preparation	Layer	Solvent	Detection	Other	Ref.
Ointment	TLC/HPLC	Add 2,2,4-trimethylpentane vortex, add MeOH–H₂O (45:5), mix centrifuge, use lower layer until sample solidifies, place in ace–MeOH bath, centrifuge	Silica	CHCl₃–acetone (7:1)	350 nm	USP 23, p. 41	[5]
Alcuronium Chloride Bulk	TLC	—	Silica (0.1 M KOH)	NH₄OH–MeOH (1.5:100)	KMnO₄	HPLC [1105]	[1]
Alendronate Tablets	CE	Dissolve in solution consisting of 1.6 mM nitric acid–2mM CuSO₄, sonicate, filter	Silica, uncoated	16 mM HNO₃–2mM CuSO₄, 25 kV	240 nm	Compared to HPLC methods; HPLC—conductivity [139]; also [145]	[138]
Aletamine Bulk	TLC	—	Silica (0.1 M KOH)	NH₄OH–MeOH (1.5:100)	KMnO₄	—	[1]
Allobarbitone (Allobarbital) Bulk	TLC	—	Silica	S₁ = CHCl₃–acetone (8:2) S₂ = EtOAc S₃ = CHCl₃MeOH (9:1) S₄ = EtOAc–MeOH conc. NH₄OH (85:10:5)	—	hR$_{f1}$ = 50, hR$_{f2}$ = 66, hR$_{f3}$ = 56, hR$_{f4}$ = 31 also GC, SE-30 RI:1606; UV, MS, NMR, IR [9, p. 52]	[132]
Bulk	HPLC	—	Silica	CH₂Cl₂ (1% NH₄OH) (99:1)	254 nm	TLC, GC, p. 326 [25]	[9]
Allyprodine Bulk	HPLC	—	Silica	MeOH (1% NH₄OH)–CH₂Cl₂ (5:95)	254 nm	GC, UV, MS, NMR, & IR	[9, p. 56]

Drug	Mode	Sample pretreatment	Sorbent (temp.)	Mobile phase	Detection	Comments	Ref.
Allopurinol							
Bulk	TLC	Dissolve in 10% diethylamine	Cellulose	Butanol–5M NH$_4$OH (saturated)	254 nm	BP, p. 19, related substances, JP, p. 75 [7], for HPLC/TLC review [831], p. 326 [25]; GC [8], impurities [24]; CP [1138, p. 414]; USP 23, p. 45	[4]
Bulk	HPLC	—	Silica	CH$_2$Cl$_2$–MeOH (1% NH$_4$OH) (99:1)	254 nm	Tablets [26, 27]	[9]
Tablets	HPLC	Add 10 ml 0.1 N NaOH to 50 mg of allopurinol, shake, add H$_2$O to 50 ml volume	C$_{18}$, 4 × 30 cm, 1.5 ml/min	0.05 M KH$_2$PO$_4$	254 nm	USP 23, assay, p. 45	[5]
Allyoestrenol							
Tablet	GC	—	USP-G-1 (220°C)	—	—	ISTD–epiandrosterone	[419]
Aloxiprin							
Bulk	TLC	—	Silica	CHCl$_3$–acetone (4:1)	Various sprays	—	[1]
Alphadolone acetate							
Bulk	TLC	Dissolve in MeOH–CHCl$_3$ (1:1)	Silica	Toluene–EtOAc (1:1)	Ceric sulfate	BP, addendum 1983, p. 213	[4]
Alphaprodine							
Bulk	HPLC	—	Silica	MeOH–0.01M NH$_4$OH	—	HPLC [9], GC [6]	[3]
Alphaxalone							
Bulk	HPLC	Dissolve in IPA	C$_{18}$ (60°C)	IPA–H$_2$O (CO$_2$ free) (25:75)	205 nm	BP, addendum 1983, p. 213; TLC, p.330 [25]	[4]
Alprazolam							
Bulk	HPLC	—	C-18, Hypersil, 5 μm; flow rate 1.5 ml/min.	S$_1$ = MeOH, H$_2$O, PO$_4$ (0.1M) (55:25:20), pH: 7.25; S$_2$ = MeOH, H$_2$O, PO$_4$(0.1M) (7:1:2), pH: 7.67	—	Cap. Fact. 1: 4.70; Cap. Fact. 2: —	[135]

	Method	Sample preparation	Column	Mobile phase	Detection	Comments	Ref.
Bulk/tablets	HPLC		Silica, Spherisorb 5W, 2 ml/min.	S_3 = MeOH–perchloric acid (1L:100 μl), S_4 = MeOH, H_2O, TFA (997:2:1)	254 nm	Cap. Fact. 3: 2.04, Cap. Fact. 4: 3.38	[135]
	HPLC	Dissolve bulk in CH_3CN; shake powder with CH_3CN, centrifuge	Silica	n-BuOH–H_2O–AcOH–CH_3CN–$CHCl_3$ (50:20:0.5:850:80)	254 nm	USP 23, p. 46, ISTD-triazolam, also p. 331 [25], [9]	[5]
Alprenolol Bulk	HPLC	—	Silica	Methanolic 0.01M ammonium perchlorate, pH 6.7	—	GC and TLC, also [8]; [263]; TLC [1139]; HPLC enantiomers [703, 707, 919]	[3]
Alprostadil Bulk/injection	HPLC	Dissolve bulk in CH_2Cl_2, derivatize with 2-acetonapthone	Silica	t-Amyl alcohol–CH_2Cl_2–H_2O (7.5:1000:1)	254 nm	USP 23, p. 49, ISTD-methyltestosterone	[5]
Alteplase Bulk	HPLC	Dilute 1 mg in H_2O, dialyze–add dithiothreitol–dialyze against NH_4 carbonate-trypsin–24 hr	C_{18}, 4.6 × 10 cm, 1 ml/min	Mobile phase A—6.9 g KH_2PO_4/l, pH 2.85; mobile phase B—CH_3CN, gradient with 100% A-0-30%B at 0.33%/min–1%/min B until 60% B-10 min	214 nm	USP 23, p. 50, peptide mapping, 20 major peaks; purity by electrophoresis and single-chain content by size exclusion	[5]
Altiazide Tablets	HPLC	MeOH added to powder, centrifuged	C_{18}	CH_3CN–H_2O (1:1)	271/235 nm	ISTD–polythiazide	[29]
Alverine Bulk	HPLC	—	Silica	MeOH–0.01 M NH_4OH, pH 6.7	—	TLC [1]; GC [2]	[3]
Amantadine Bulk	HPLC	—	Silica	1% H_3PO_4	Conductivity	GC [8,10]; TLC [1,8]; HPLC [920]	—
Ambazone Bulk	TLC	—	Silica (0.1M KOH)	NH_4OH–MeOH (1.5:100)	Char	—	[1]

Drug	Mode	Sample pretreatment	Sorbent (temp.)	Mobile phase	Detection	Comments	Ref.
Ambenonium chloride							
Bulk	TLC	—	Silica (0.1M KOH)	NH_4OH–MeOH (1.5 : 100)	Char	HPLC [1105]	[1]
Ambucetamide							
Bulk	TLC	—	Silica (0.1M KOH)	NH_4OH–MeOH (1.5 : 100)	Char	—	[1]
Amcinonide							
Bulk	HPLC	Dissolve in mobile phase	Silica, 2.1 x 60 cm, 1 ml/min	IPA–CH_2Cl_2 (4 : 100)	254 nm	USP 23, assay, p. 74	[5]
Cream	HPLC	Partition between cyclohexane–$CHCl_3$ (1 : 1) and 10% H_2SO_4, use a Mg silicate column	Silica, 2.1 x 60 cm, 1 ml/min	IPA–CH_2Cl_2 (4 : 100)	254 nm		[5]
Ointment	HPLC	Dissolve ointment into beadlets in a centrifuge tube with MeOH, heat, cool, add mobile phase, filter	C_{18}, 4.6 × 25 cm, 1 ml/min	MeOH–H_2O (4 : 1)	254 nm	USP 23, assay, p. 74	[5]
Amdinocillin							
Bulk	HPLC	Dissolve in H_2O	C_{18}, 4.6 × 25 cm, 5 µm, 1 ml/min	Buffer–CH_3CN (85 : 15); buffer–KH_2PO_4, 1.36 g/l, pH 5.0	220 nm	USP 23, p. 75, assay	[5]
Ametazole							
Bulk	TLC	—	Silica (MeOH NH_4OH)	NH_4OH–MeOH (1.5 : 100)	Char	GC [2]	[2]
Amethocaine (Tetracaine hydrochloride)							
Eye drops	TLC	Dilute if necessary	Silica	n-Hexane–dibutyl ether–AcOH (80 : 16 : 4)	254 nm	BP, vol.II, p. 566, related substances, also [25]; HPLC [1105]	[4]
Amfepramone (Diethylpropion)							
Bulk	HPLC	—	Silica	MeOH (1% NH_4OH)–CH_2Cl_2 (2 : 98)	254 nm	UV, GC, MS, NMR, and IR	[9, p. 666]
Amiodarone							
Bulk	HPLC	—	Silica	MeOH–0.01M NH_4ClO_4, pH 6.7	—	TLC [1]; GC [2]; HPLC [449]	[3]

Compound	Method	Sample preparation	Column	Mobile phase	Detection	Notes	Ref.
Amidopyrine Bulk	HPLC	—	Silica	MeOH–0.01M NH₄OH, pH 6.7	—	GC [2]; TLC [1, 337]	[5]
Amikacin Bulk	TLC	Dissolve in H₂O	Silica	MeOH–CHCl₃–NH₄OH (60:30:25)	Ninhydrin	USP 23, p.76, identity	[5]
Bulk	HPLC	Dissolve in H₂O, derivatize with 2,4,6-trinitrobenzene–sulfonic acid	C₁₈, 4.6 × 25 cm, 5 μm, 30°C, 0.8 ml/min	Buffer–MeOH (28:72); buffer–KH₂PO₄, 2.7 g/l, pH 6.5	340 nm	USP 23, p.76, assay of sulfate salt and for injection	[5]
Bulk	TLC	Dissolve in H₂O	Silica gel H	CHCl₃–MeOH–NH₄OH–H₂O (1:4:2:1)	Ninhydrin	CP, p. 418, identity for bulk and injection	[1138]
Amiloride Bulk	TLC	Mix in MeOH–CHCl₃ (4:1)	Silica	THF–3 N NH₄OH (15:2)	350 nm	USP 23, p. 78, chromatographic purity; BP vol. II, p. 731 [4]	[5]
Tablets	HPLC	Sonicate powder with MeOH–0.1 N HCl (15:2), dilute with H₂O, sonicate, mix and filter	C₁₈, 3.9 × 30 cm, 1 ml/min	H₂O–MeOH–buffer (71:25:4); buffer–136 mg/ml KH₂PO₄, pH 3.0	286 nm	USP 23, p.78, assay, same procedure for tablets with hydrochlorothiazide [1, 8,46,140]; review [1046], HPLC [1105]	[5]
Amino acids Parenterals	HPLC	o-Phthalaldehyde	C₁₈	Gradient	Fluorescence	Common amino acids; enantiomers [142]; parenterals [144]	[1039]
Aminobenzoic acid Gel	HPLC	Dissolve sample and salicylic acid (ISTD) in MeOH	Phenyl (L11), 3.9 x 30 cm, 1 ml/min	MeOH–AcOH–H₂O (30:1:69)	280 nm	USP 23, p. 83, assay	[5]
Aminocaproic acid Bulk	HPLC	Dissolve in H₂O with methionine (ISTD)	C₁₈, 4.6 x 15 cm, 30°C, 0.7 ml/min	Buffer–KH₂PO₄ (10 g)–HPA (0.45 g)–MeOH (250 ml)	210 nm	USP 23, bulk and similar procedure for injection, p. 83	[9, p. 90]
Aminoglutethimide Bulk	HPLC	Dissolve in acetate buffer–MeOH (1:1); see mobile phase	C₁₈, 3.9 × 15 cm 40°C, 1.3 ml/min	Acetate buffer–MeOH (73:27); buffer—240 ml 0.1 N AcOH to 200 ml 0.1 N KOH, pH 5.0 dilute to 2 L	240 nm	USP 23, p. 85, assay, volatile impurities <467>, separate procedure for azoaminoglutethimide, same procedure for tablets, p. 86	[5]

Drug	Mode	Sample pretreatment	Sorbent (temp.)	Mobile phase	Detection	Comments	Ref.
Aminoglycoside Ear drops	CE	Dissolve in 100 mM CMA	Silica	0.01 M imidazole acetate, pH 5.0, 12 kV	214 nm	Resolved dihydrostreptomycin, lividomycin, amikacin, kanamycin, tobramycin, and sisomycin; HPLC–OPA derivatization [148]; HPLC–PAD [149–151]	[147]
Aminohippuric acid Bulk	TLC	—	Silica (MeOH–0.1 M NH$_4$OH)	NH$_4$OH–MeOH (1.5 : 100)	Char	—	[1]
Aminophenazone Bulk	TLC	—	Silica	S$_1$ = CHCl$_3$–acetone (8 : 2) S$_2$ = EtOAc S$_3$ = CHCl$_3$–MeOH (9 : 1) S$_4$ = EtOAc–MeOH conc. NH$_4$OH (85 : 10 : 5)	—	hR$_{f1}$ = 25, hR$_{f2}$ = 15, hR$_{f3}$ = 58, hR$_{f4}$ = 61; tablets GC [550]; HPLC, dosage forms [551]; TLC, stability [553]	[132]
Aminophylline Bulk	HPLC	Dissolve in H$_2$O–MeOH (4 : 1)	C$_{18}$, 3.9 × 15 cm, 1 ml/min	200 ml MeOH, 960 mg PSA/l, pH 2.9 AcOH	254 nm	USP 23, p. 87, assay, also for oral solution, p. 88	[5]
Aminopyrine Bulk	HPLC	—	Silica	CH$_2$Cl$_2$–MeOH (1% NH$_4$OH) (98 : 2)	254 nm	GC [10]	[9]
Aminosalicylic acid Bulk and tablets	HPLC	Shake with mobile phase, filter, use low-actinic flasks	C$_{18}$, 4.6 × 25 cm, 1.5 ml/min	MeOH–buffer (3 : 10); buffer—17.04 g K$_2$HPO$_4$, 16.56 g KH$_2$PO$_4$, 9.2 g TBAH/l;	254 nm	USP 23, assay, p. 92, also for tablets, limit for aminophenol also described, ISTD–	[5]
Tablets, granules, suspension	HPLC	Extracted with AcOH–H$_2$O (15 : 85), low-actinic flasks	μBondapak C$_{18}$	5 mM PICB5, pH 3.5 w/AcOH–MeOH (9 : 1)	230 nm	acetaminophen; TLC [1], GC [2], stability [1104]; suppositories [155,175]	[152]
Amiodarone hydrochloride Bulk and tablets	TLC	Dissolve in MeOH	Silica HF$_{254}$	CHCl$_3$–MeOH–CH$_3$OOH (80 : 15 : 5)	254 nm	CP, related substances, p. 421; [141]	[1138]

Compound/form	Technique	Sample preparation	Stationary phase	Mobile phase / conditions	Detection	Comments	Ref.
Amiphanazole Bulk	TLC	—	Silica (base treated)	NH$_4$OH–MeOH (1.5 : 100)	Char	—	[1]
Amitriptyline Bulk	TLC	—	Silica	S$_1$ = EtOAc–MeOH–conc. NH$_4$OH (85 : 10 : 4); S$_2$ = MeOH; S$_3$ = MeOH–1-butanol (3 : 2) – 0.1M NaBr (100 : 1.5); S$_4$ = MeOH–conc. NH$_4$OH (100 : 1.5); S$_5$ = cyclohexane–toluene–diethylamine (75 : 15 : 10); S$_6$ = CHCl$_3$–MeOH (9 : 1); S$_7$ = acetone	—	hR$_{f1}$ = 70, hR$_{f2}$ = 26, hR$_{f3}$ = 51, hR$_{f4}$ = 51, hR$_{f5}$ = 55, hR$_{f6}$ = 32, hR$_{f7}$ = 15, also, GC, SE-30/OV-1, RI: 2196	[132]
Bulk	TLC	Dissolve in methanol	Silica	Ammonium hydroxide–chloroform–methanol (1 : 135 : 15)	254 nm	USP 23, p. 93, volatile impurities <467>; impurities [925]; TLC [1, 8, 921]; GC [2,8,921]; and HPLC [3,8,168,922,1054]	[5]
Bulk	HPLC	—	Silica, 5 μm	Isooctane–diethylether–methanol–diethylamine–water (400 : 325 : 225 : 0.5 : 15); Flow rate: 2.0 ml/min	279 nm	k^1 = 0.93, RRT = 0.47 (diamorphine, 1.00, 49 min); embonate, BP, v. II, p. 686 and v.1, p. 29 [4], adsorption [177]	[133]
Injection and tablets	HPLC	Dilute with H$_2$O	C$_{18}$, 4 × 30 cm, 2 ml/min	KH$_2$PO$_4$, 11.04 g/l, pH 2.5	254 nm	USP 23, p. 93, assay; CP [p. 422, 1138], related substances, CHCl$_3$–toluene (1 : 1), silica	[5]
Amobarbital Bulk	TLC	Dissolve in ethanol	Silica	Ammonium hydroxide–chloroform–ethanol (5 : 80 : 15)	254 nm	EP, pt. II-5, p. 166-2, for GC, USP 23 [5] and [10, 924,1138]; stability [923]	[6]
Bulk	HPLC	—	Silica	CH$_2$Cl$_2$–MeOH (1% NH$_4$OH) (99 : 1)	254 nm	—	[9]

Drug	Mode	Sample pretreatment	Sorbent (temp.)	Mobile phase	Detection	Comments	Ref.
Amobarbitone (Amobarbital)							
Bulk	TLC	—	Silica	$S_1 = CHCl_3$–acetone (8 : 2); $S_2 = EtOAc$; $S_3 = CHCl_3$–MeOH (9 : 1); $S_4 = EtOAc$–MeOH conc. NH_4OH (85 : 10 : 5)	—	$hR_{f1} = 52$, $hR_{f2} = 65$, $hR_{f3} = 58$, $hR_{f4} = 40$; also GC, SE-30 OV-1, RI: 1718	[132]
Amodiaquine hydrochloride							
Bulk	TLC	Shake with $CHCl_3$ saturated with NH_4OH	Silica	$CHCl_3$ (sat. NH_4OH)–EtOH (dehydrated) (9 : 1)	254 nm	USP 23, p. 97, chromatographic purity, organic volatile impurities <467>	[5]
Amoxapine							
Bulk	HPLC	—	Silica	CH_2Cl_2–MeOH (1% NH_4OH) (95 : 5)	254 nm	GC [2,10], USP 23 TLC, p. 99 & tablet assay by HPLC, p.100; review [159]	[9]
Amoxicillin							
Bulk	GC	Add 1 N NaOH and cyclohexane, shake	—	—	FID	EP, pt. II-7, p. 260–3, ISTD–naphthalene [32,34,47,319,438]; stability [1104]; CE [176]	[6]
Capsules	HPLC		C_{18}	H_2O–MeOH–AcOH (74 : 25 : 1)	Amperometric		[33]
Capsules, tablets, suspension	TLC	Dissolve in 0.1 N HCl uae within 10 min	Silica	MeOH–pyridine–$CHCl_3$–H_2O (90:10:80:10)	Ninhydrin	USP 23, p. 100, identification, assay by HPLC, oral suspension & tablets with clavulanate, assay by HPLC, p. 103; also, BP, p. 31 [4]	[5]
Amphetamine sulfate							
Tablets	HPLC	Dissolution	C-18	Buffer–methanol (3 : 2); buffer — 1.1 g HPA, 25 ml 14% acetic acid to 575 ml water	254 nm	USP 23, p. 104; GC [2,8, 10,702,926,927]; tablets [336]; salts and optical isomers are DEA controlled; SCF [170]; HPLC [243, 251,335]	[5]

Sample	Technique	Sample preparation	Sorbent	Mobile phase	Detection	Notes	Ref
Bulk	HPLC	—	Silica	Methylene chloride–methanol (1% ammonium hydroxide) (9:1)	254 nm	Separation of enantiomers [339,343–346,703,928–934,1126]	[9]
Bulk	TLC	—	Silica	S_1 = EtOAc–MeOH–conc. NH_4OH (85:10:4); S_2 = MeOH; S_3 = MeOH–1-butanol (3:2)–0.1M NaBr; S_4 = MeOH–conc. NH_4OH (100:1.5); S_5 = cyclohexane–toluene–diethylamine (75:15:10); S_6 = $CHCl_3$–MeOH (9:1); S_7 = acetone	—	hR_{f1} = 44, hR_{f2} = 12, hR_{f3} = 75, hR_{f4} = 43, hR_{f5} = 15, hR_{f6} = 9, hR_{f7} = 18; also, GC, SE-30/OV-1, RI: 1123; also [51, 52], enantiomer resolution [57–59]; GC [264,273,484,485]; spectrophotofluorimetry [349] GC/MS [350, 713]; mp IR [348] and thermal analysis [353, 354]; TLC [488]; standard stability [678]; HPLC [231,481,567, 705, p. 218, 1138]; CE [759]; SPE (Varian); Toxilab, TLC [564]; HPLC [672]; GC [706, p. 156]; historical [652]; colorassay (Marquis test reagent kit) [1118]	[132]
Amphotercins Bulk, parenteral, and topical	HPLC	For bulk, weigh into volumetric and dissolve with DMSO	C_{18}	MeOH–PO_4 buffer (pH 2.6) (10:3)	313 nm	Method developed for potency of ampho A & B. PRP sorbent [776]	[116]
Ampicillin Capsules, oral suspensions	TLC	Dissolve in acetone–0.1 N HCl (4:1)	Silica	Acetone–toluene–H_2O–AcOH (650:100:100:25)	Ninhydrin	USP 23, pp. 106–9, identification, assay HPLC, p. 107, with probenecid capsules assay by HPLC, p. 111, with sulbactam, p. 113; with cloxacillin [171]; UV [174]	[5]

Drug	Mode	Sample pretreatment	Sorbent (temp.)	Mobile phase	Detection	Comments	Ref.
Amprolium Bulk	HPLC	Add 50 mg + ~100 ml (H$_2$O–MeOH–CH$_3$CN) (50 : 45 : 5)	Trimethylsilane, L13	Dissolve 1.10 g HPA in 500 ml H$_2$O + 12 ml AcOH + 2.0 ml TEA + 450 ml MeOH + 50 ml ACN, filter	254 nm	USP 23, assay, p. 114; also for oral solution and soluble powder	[5]
Amyl nitrate Neat	GC	—	USP-G2 (between 100 and 300°C)	—	FID	also, total nitrites by GC, USP 23, p. 115 [5]	[2]
Analgesics Bulk	HPLC	—	C$_{18}$	2-Propanol–formic acid–PO$_4$	240 nm	Forty drugs; haloethane [204]	[1044]
Capsules/tablets	HPLC	Grind, ultrasonicate with H$_2$O/CH$_3$CN, filter	Silica	5 mM PO$_4$, pH 2.6–CH$_3$CN (95 : 5), 1 ml/min	254 nm	Resolved aspirin, fenbufen, ibuprofen, indomethacin, ketoprofen, sulindac, naproxen, tolmetin	[178]
Androsterone Bulk	GC	—	USP-G2 (between 100 and 300°C)	—	FID	—	[2]
Anileridine Bulk	HPLC	—	Silica	Methylene chloride–methanol (1% ammonium hydroxide) (98 : 2)	254 nm	Also [3]; GC	[9]
Anisodamine hydrobromide Bulk	TLC	MeOH	Aluminum oxide	CHCl$_3$–dehydrated EtOH (95 : 5)	Dilute K, iodobismuthate TS-KI/I$_2$ (1 : 1)	CP, other alkaloids, isolated form root of *Scopolia tangutica* Maxim.	[1138]
Antazoline Bulk	HPLC	—	Silica	Methanolic 0.01 M ammonium perchlorate adjusted to pH 6.7	—	TLC [1]; GC [2]; HPLC [180,935,936,1061,1105]; GC [706, p. 164]; USP 23, p. 117, TLC, chromatographic stability	[3]

Substance / form	Technique	Sample preparation	Column	Mobile phase / conditions	Detection	Comments	Ref.
Anthralin Bulk	HPLC	Dissolve in CH$_2$Cl$_2$ followed by mobile phase	Silica	n-Hexane–CH$_2$Cl$_2$–AcOH (82:12:6)	354 nm	USP 23, assay, p. 118, use low-actinic glass, ISTD–o-nitroaniline, same procedure for cream & ointment	[5]
Antibiotics Formulations	HPLC	—	—	—	—	Review	[1142]
Bulk	CE	—	Silica (50 μm × 50 cm)	0.05M PO$_4$–0.1M borate, pH 7.06, 15 kV	PDA	Resolved pencillin-G, ampicillin, amoxicillin, chlorotetracycline, tylosin, HPLC β-lactam antibiotics [186], antibacterials [407], PRP [342], HPLC [601,602]; novobiocin [322,907]	[185]
Antibodies (IgG) Bulk	CE	Dilute in 10 mM HEPES–MES, pH 7.0, reduce with DTT	eCap SDS 200 coated capillary, 47 cm × 100 μm	14.1 kV, 30 min	214 nm	MW separations, paper [181,182]	[187]
Antihistamines Tablets	CE	Extract powder with MeOH, filter	Silica, 47 cm × 50 μm	10 mM SDS–0.05M borate–0.05M PO$_4$–10 mM TBA–10 mM 8, β-cyclodextrin	214 nm	Resolved pheniramine, doxylamine, metapyrilene, thonylamine, triprolidine, dimenhydrinate, cyclizine, promethazine, chlorpheniramine; HPLC [192,424,1063]	[193]
Antipyrine Bulk	HPLC	—	Silica	CH$_2$Cl$_2$–MeOH (1% NH$_4$OH) (98:2)	254 nm	GC [9]; ear drops [153,154]; USP 23, ordinary impurities, p. 142, with benzocaine solution & with phenylephrine, assay by HPLC, p. 125 [5]	[9]

Drug	Mode	Sample pretreatment	Sorbent (temp.)	Mobile phase	Detection	Comments	Ref.
Apomorphine hydrochloride							
Bulk	TLC	Dissolve in MeOH	Silica	CH_3CN–formic acid–EtOAc–CH_2Cl_2 (30 : 5 : 30 : 5)	NH_4 vapor	EP, pt. II-4, p.136–2, also, BP addendum 1983, p. 214 [4]	[6]
Bulk	HPLC	—	Silica	Methanolic 0.01M NH_4ClO_4, pH 6.7	—	TLC [1,157]; GC & HPLC [9]; also [156,653]; USP 23, ordinary impurities p. 127	[3]
Apraclonidine hydrochloride							
Bulk	HPLC	Dissolve in mobile phase	C_8, L7	CH_3CN–PO_4–buffer (6.8 ml H_3PO_4/2 l)–MeOH (56 : 40 : 4)	220 nm	USP 23, purity, p. 128, for ophthalmic solution TLC for ID and HPLC for assay, p.129	[5]
Aprindine							
Bulk	GC	—	USP-G2 (between 100 and 300°C)	—	FID	Also [194,1138,1143]	[2]
Aprobarbital (Aprobarbitone)							
Bulk	TLC	—	Silica	S_1 = $CHCl_3$–acetone (8 : 2) S_2 = EtOAc S_3 = $CHCl_3$–MeOH (9 : 1) S_4 = EtOAc–MeOH conc. NH_4OH (85 : 10 : 5)	—	hR_{f1} = 48, hR_{f2} = 65, hR_{f3} = 57, hR_{f4} = 37; also GC, SE-30 OV–1, RI: 1622	[132]
Bulk	HPLC	—	Silica	Methylene chloride–methanol (1% ammonium hydroxide) (98 : 2)	254 nm	GC [8,9]; TLC [8]; also p. 359 [25] and [166]; TLC [261]; GC [404]; GC, JW DB-17; p. 54 [705]	[9]
Apronal (Apronalide)							
Bulk	TLC	—	Silica	Chloroform–acetone (4 : 1)	Acidified potassium permanganate	—	[1]

	Method	Sample preparation	Column/Phase	Mobile phase	Detection	Notes	Ref.
Arginine Bulk, parenternal	HPLC	Dissolve in mobile phase	Silica	0.1% H_3PO_4	206 nm	Impurities TLC [24]; HPLC [554]; USP 23 organic volatile impurities p. 129 [5]	[30]
Ascorbic acid Bulk	TLC	Dissolve in abs. EtOH	Silica	MeOH–acetone–H_2O (20 : 4 : 3)	UV	Impurities; TLC, GC & HPLC review [49]; HPLC [64,196–198, 736,738,739–745] review [801]; stability [1104]; USP 23 organic volatile impurities p. 130 [5]; TLC [728]	[24]
Oral tonic	CE	Dissolve in 0.1M PO_4	Coated, 20 cm × 25 μm	0.1M PO_4, pH 5.0, 8 kV	265 nm		[195]
Aspartame Bulk	GC	Silylation	USP-G2 (200°C)	—	FID	USP 23, for determination 5-benzyl-3,6-dioxo 2-piperazine-acetic acid; HPLC [51,200–202,1124, 1144–1146]	[5]
Aspirin Bulk	TLC	—	—	—	—	USP 23, p. 1321, see procedure for propoxyphene; JP, p. 1200 [7]; [12,203,475,1074,1147]	[5]
Tablets	HPLC	Shake powder with $CHCl_3$–formic acid–ACN (EtOH free) (99 : 2 : 99)	C18	H_2O–CH_3CN–HPA (850 : 150 : 2 g) adjust pH to 3.4 with AcOH	280 nm	USP 23, pp. 133–136 also for capsules, extended & delay release tablets	
Atenolol Bulk	HPLC	—	Silica	Methanol–0.01M ammonium perchlorate, pH adjusted to 6.7	—	For pharmaceutical forms [52]; HPLC, GC and TLC [8,939,940,1148]; [188,190,191]; HPLC [272, 751, pp. 7,13]; HPLC review enantiomers [703,707]; CP TLC p-hydroxy-phenylacetamide p. 439 [1138]	[3]

Drug	Mode	Sample pretreatment	Sorbent (temp.)	Mobile phase	Detection	Comments	Ref.
Atropine Bulk	TLC	—	Silica	S_1 = EtOAc–MeOH–conc. NH$_4$OH (85:10:4) S_2 = MeOH S_3 = MeOH–1-butanol (3:2) – 0.1M NaBr S_4 = MeOH–conc. NH$_4$OH (100:1.5) S_5 = cyclohexane–toluene–diethylamine (75:15:10) S_6 = CHCl$_3$–MeOH (9:1) S_7 = acetone	—	hR_{f1}=25, hR_{f2}= 6, hR_{f3}=28, hR_{f4}=18, hR_{f5}= 6, hR_{f6}=3, hR_{f7}= 1; also GC, SE-30/OV-1, RI: 2199; enantiomers [199]; HPLC [205]	[132]
Bulk	TLC	Dissolve in methanol	Silica	Chloroform–acetone–diethylamine (5:4:1)	Iodoplatinate	USP 23, p. 145, impurities; p. 364 [25]	[5]
Tablets with Diphenoxylate Hydrochloride	GC	Add pH 2.8 buffer to powder, extract with chloroform, discard organic layer, adjust to pH 9.0 extract with chloroform	USP-G3 (230°C)	—	FID	USP 23, p. 535, ISTD–homatropine, assay; also [9, 159], enantiomers [57,1065]; stability [1104]	[5]
Bulk	GC	Dissolved in methanol	0.25-μm film chemically bonded 5% phenyl polydimethyl siloxane. Injector 275°C, column 60°C after inj. 40°C/min to 250°C, 30 min.	—	MS, scanned between m/z 33 and 500/sec. EI	Propoxyphene forms major artifacts in the injector zone. Benzoylecgonine degrades to ecgonine and benzoic acid. Atrophine forms dehydroatropine and O-(phenyl acetyl) tropine	—
Atropine sulfate Injection	TLC	Dissolve in EtOH	Silica	CHCl$_3$–DEA (9:1)	Iodoplatinate	JP, p. 297, also, EP, pt. II–2, p. 68-2 foreign alkaloids and decomposition products	[7]

Compound	Technique	Sample preparation	Stationary phase	Mobile phase	Detection	Notes	Ref.
Injection and tablets	GC	Render alkaline and extract with CH_2Cl_2	USP-G3 (225°C)	—	FID	[6], BP, v. II, p. 566 [4]; GC method USP 23, p. 537, BP, v. II, pp. 571, 583, & 736 [4], [8,9,53–56, 160–162,855,1105, 1149]	
Azacitidine (5-azacytidine) Bulk	HPLC	—	C-18	MeOH–PO_4 buffer (5:95)	UV	Cytarabine and degradants; [654,655,1150]; stability [1104]	[58]
Azacyclonal Bulk	HPLC	—	Silica	MeOH–0.01M NH_4ClO_4, pH 6.7	—	TLC [1]; GC [2]	[1]
Azaperone Injection	HPLC	Dissolve in MeOH (0.5 mg/ml) & dilute with ISTD 1:1 (benzophenone, 0.5 mg/ml)	C-18	CH_3CN–0.01M PO_4 buffer, adj. pH 7.8	243 nm	USP 23, assay, p. 149; chromatographic purity bulk, TLC, p. 149	[5]
Azapetidine Bulk	TLC	—	Silica (0.1M MeOH–KOH)	NH_4OH–MeOH (1.5:100)	Char	GC [2,11]	[1]
Azapropazone Bulk	TLC	—	Silica (above)	As above	Char	GC/HPLC [11,158]	[1]
Azatadine maleate Bulk	TLC	Dissolve in toluene–MeOH (1:1)	Silica	Toluene–2-propanol–diethylamine (10:10:1)	254 nm	USP 23, chromatographic purity, p. 150; tablets TLC ID, p.150	[5]
Azathioprine Bulk	TLC	Dissolve in dimethylamine	μ Cellulose	Butanol sat. 6N NH_4OH	254 nm/350 nm	USP 23, limit test, p. 151, same for injection, p. 152, volatile impurities <467>; JP, p. 42 [7]; BP, v. II, p. 737 and v. I, p. 42 [4]; p. 368 [25], [59]	[5]
Tablets	TLC	Shake powder with 6 N NH_4OH, filter	μ Cellulose	Butanol sat. 6N NH_4OH	254 nm/350 nm		[5]
Tablets	HPLC	Sonicate powder with NH_4OH–MeOH (1:25)	C-18	Buffer–MeOH (7:3), buffer–HPA 1.1 g in 700 ml H_2O	254 nm	USP 23, uniformity of dosage forms, p.151; [10,1104]	[5]

Drug	Mode	Sample pretreatment	Sorbent (temp.)	Mobile phase	Detection	Comments	Ref.
Azithromycin							
Bulk	HPLC	Dissolve in AcCN, dilute in mobile phase	Gamma alumina RP	5.8 g PO_4/2130 ml (30 ML H_2O + 870 ml CH_3CN, mix, adj pH 11.0 w/KOH	Amperometric +0.8 V	USP 23, assay, similiar procedure for capsules, p. 153	[5]
Azlocillin							
Bulk	HPLC	Dissolve in mobile phase	Silica	BtOH–AcOH–buffer–BtOAc (9:25:15:50)	I_2 vapor	USP 23, p. 154, ID, assay HPLC C-18, 210 nm	[5]
Aztreonam							
Bulk, parenteral	HPLC	Dissolve in mobile phase	Silica	0.1% H_3PO_4	206 nm	Also, USP 23, p. 156 diol L20; [60,61,210]; stability [1104]	[30]
Bacampicillin hydrochloride							
Bulk	HPLC	Dissolve in H_2O, 0.8 mg/ml, filter	C_{18}	pH 6.8, 0.02M PO_4 buffer–CH_3CN (1:1)	254 nm	USP 23, assay, p. 157, ID by TLC same for oral suspension & tablets, tablets assyed by HPLC	[5]
Bacitracin							
Bulk	TLC	Dissolve in 1% EDTA solution	Silica	Butanol–pyridine–EtOH–AcOH (60:15:10:6:5)	1% Triketohydrinene hydrate, heat	USP 23, ID, p. 158, also used for ID of other formulations; BP addendum 1983, p. 274 [4]; HPLC [71,169,220]	[5]
Baclofen							
Bulk	TLC	Dissolve in EtOH–AcOH (4:1)	Silica	Butanol–H_2O–AcOH (4:1:1)	Cl_2 vapor	USP 23, ID for bulk & tablets, p. 163; [8], [p. 369 [25]	[5]
Tablets	HPLC	Dissolve tablet powder in MeOH–AcOH (75:10 ml) dilute to 250 ml w/H_2O	C-18	0.3N AcOH–MeOH–0.36N PSA (55:44:2)	265 nm	USP 23, assay, ISTD benzoic acid, flow 0.6 ml/min, 3.9 x 30 cm, HPLC IP [213]; HPLC cyclodextrin [242]	[5]
Bamethan							
Bulk	HPLC	—	Silica	MeOH–0.01M NH_4ClO_4, pH 6.7	—	GC [2,8]; HPLC [8]; TLC [1]	[3]

Compound / Form	Method	Sample preparation	Stationary phase	Mobile phase	Detection	Notes	Ref.
Bamifylline							
Bulk	TLC	—	Silica (0.1M MeOH–KOH)	NH$_4$OH–MeOH (1.5 : 100)	Char	[8]	[1]
Suppositories	HPLC	MeOH added, shaken 15 min, filtered	C$_{18}$	MeOH–THF–PO$_4$ (60 : 4 : 40), 0.1M, pH 7.5	278 nm	Tablets and bulk also analyzed, resolved from impurities	[214]
Bamipine							
Bulk	TLC	—	C$_{18}$	MeOH–THF–PO$_4$ (60 : 4 : 40), 0.1M, pH 7.5	Char	GC [2]; also [11,214]	[1]
Cream	HPLC	Ultrasonic with MeOH, dilute with mobile phase	Silica	MeOH–CH$_2$Cl$_2$ (1 : 4), MeOH contains 0.02% NH$_4$OH	258 nm	Tablets, solutions & ointments	[74]
Various formulations	HPLC	Add EtOH (95%) for all formulations, for solids sonicate, let settle	C$_2$ (LiCrosorb)	NH$_4$OAc (15mM)–CH$_3$CN (35 : 65), pH 4.7, 0.9 ml/min	251 nm	Resolved from trimipramine, clomipramine, thioridazine, prochlorperazine, chlorprothixene, & haloperidol; HPLC [212,225,253,424]	
Barbital (Barbitone)							
Bulk	TLC	Dissolve in ethanol	Silica	Ammonium hydroxide–chloroform–ethanol (5 : 80 : 15)	254 nm	EP, pt. II–5, p. 170–2 [8], 372 [25]; HPLC and GC [9,167,632, 942]; general ref. [163–166]; stability [1104]	[6]
Bulk	TLC	—	Silica	S$_1$ = CHCl$_3$–acetone (8:2), S$_2$ = EtOAc, S$_3$ = CHCl$_3$–MeOH (9 : 1), S$_4$ = EtOAc–MeOH conc. NH$_4$OH (85 : 10 : 5)	—	hR$_{f1}$ = 41, hR$_{f2}$ = 61, hR$_{f3}$ = 57, hR$_{f4}$ = 33; also GC, SE-30 OV-1, RI: 1497	[132]

Drug	Mode	Sample pretreatment	Sorbent (temp.)	Mobile phase	Detection	Comments	Reference
Bulk	TLC	—	Silica	Chloroform–acetone (dry) (9 : 1)	Mercuric sulfate followed by diphenylcarbazone or potassium permanganate	[31], derivatization [65–67], reviews [68,69]; CE [691]; GC [528, 531]; GC/TLC [527]; GC/IR/MS [686]; color assay [1118]; [70–72,85, 106,107]	
Barbiturates							
Bulk	GC	—	25 m × 0.25 mm, Heliflex, AT-1, 0.2 μm, 160°C	H$_2$, flow rate 3.5 ml/min	FID	Resolved tetradecane, barbital, amobarbital, pentobarbital, secobarbital, hexobarbital, phenobarbital; resolved R,S hexo′/mepho-barbital [232], GC [250] HPLC [31,117,1113, 1117]	[230]
Beclamide							
Bulk	TLC	—	Silica (0.1M MeOH–KOH)	NH$_4$OH–MeOH (1.5 : 100)	Char	GC [2,8]; HPLC [8]	[1]
Beclomethansone dipropionate							
Bulk	HPLC	Dissolve in MeOH + ISTD (testosterone dipropionate)	C-18	CH$_3$CN–H$_2$O (3 : 2)	254 nm	USP 23, assay, p. 167; for cream and ointment, BP 1983, p. 268 v. II, p. 696; p. 374 [25]; CP assay, p. 444 [1138], same procedure for ointment HPLC [215]	[5]
Bulk	HPLC	Dissolve in mobile phase + ISTD (methyl testosterone)	C-18	MeOH–H$_2$O (74 : 26)	240 nm		[1138]
Bendroflumethiazide							
Dosage forms	HPLC	Dissolve in MeOH–buffer pH 2.0 (3 : 2)	Phenyl, 7 μm	MeOH–H$_2$O (2 : 3), 1.5 ml/min	270 nm	Resolved from degradants	[227]
Belladonna extract							
Bulk and tablets	GC	Dissolve in acid	USP-G3	—	FID	USP 23, assay, p. 167	[5]

Benorilate (Benorylate) Bulk	TLC	Dissolve in methanol–chloroform (1:9)	Silica	Methylene chloride–acetic acid–ether (80:5:15); dry, develop in a second mobile phase 2,2,4-trimethylpentane–ether–formic acid (45:45:10)	254 nm	BP, addendum 1983, p. 216; HPLC [945]; HPLC, GC, & TLC, p. 376 [937], [1138]	[4]
Benzalkonium Eye care	HPLC	SPE: C_{18}, THF–mobile phase (7:3)	Cyano, Zorbax	H_2O–THF–TEA (2500:1500:2), pH 3.0	215 nm	Column-switching, sample cleanup [218]	[216]
Benzethidine Bulk	HPLC	—	Silica	MeOH (1% NH_4OH)–CH_2Cl_2 (1:99)	254 nm	GC, MS, NMR, and IR	[9, p. 194]
Benzhexol Tablets	TLC	Shake powder with chloroform	Silica	Chloroform–methanol (9:1)	Iodoplatinate	BP, vol. 11, p.738; TLC [1]; GC [2]; HPLC [3, 922]	[4]
Benzocaine Bulk	TLC	—	Silica	S_1 = $CHCl_3$–acetone (8:2) S_2 = EtOAc S_3 = $CHCl_3$–MeOH (9:1) S_4 = EtOAc–MeOH conc. NH_4OH (85:10:5)	—	hR_{f1} = 56, hR_{f2} = 62, hR_{f3} = 63, hR_{f4} = 76; also GC, SE-30 OV-1, RI: 1555; topical formulation [229,386]; HPLC [249]; cream w/phenindamine [379]	[132]
Bulk	TLC	—	Silica	S_1 = EtOAc–MeOH–conc. NH_4OH (85:10:4) S_2 = MeOH S_3 = MeOH–1-butanol (3:2)–0.1M NaBr S_4 = MeOH–conc. NH_4OH (100:1.5) S_5 = cyclohexane-	—	hR_{f1} = 76, hR_{f2} = 84, hR_{f3} = 87, hR_{f4} = 67, hR_{f5} = 6, hR_{f6} = 57, hR_{f7} = 66; also, GC, SE-30/OV-1, RI: 1555	[132]

Drug	Mode	Sample pretreatment	Sorbent (temp.)	Mobile phase	Detection	Comments	Reference
Bulk	TLC	Dissolve with 40% alcohol	Silica	toluene–diethylamine (75 : 15 : 10) S_6 = CHCl₃–MeOH (9 : 1) S_7 = acetone Propanol–10% ammonium hydroxide (88 : 12)	UV	Impurities, GC [9]; HPLC [9,946]; p. 383 [937]; stability [923]; also, resolved from dextromethorphan and cetylpyridinium by HPLC [259]; GC [404,756]	[24]
Benzoylecgonine Bulk	GC	Dissolved in methanol	0.25-μm film chemically bonded 5% phenyl polydimethyl siloxane; injector 275°C, column 60°C after inj. 40°C/min to 250°C, 30 min	—	MS, scanned between m/z 33 and 500 per sec. EI	Propoxyphene forms major artifacts in the injector zone. Benzoylecgonine degrades to ecgonine and benzonic acid. Atrophine forms dehydroatrophine and O-(phenyl acetyl) atrophine; urine, HPLC [706, p. 159]	[143]
Benzphetamine Bulk	HPLC	—	Silica	Methanolic 0.01 M ammonium perchlorate pH adjusted to 6.7	—	TLC and GC, p. 385 [937] and [9]	[3]
Benztropine Injection	TLC	Evaporate, dissolve residue in acetone	Silica	Ethanol–13.5 M ammonium hydroxide	Iodoplatinate	BP, vol. II, p. 584 and p. 738; ToxiLab TLC [564], GC [404, 706, p. 156]	[4]

	Method	Sample preparation	Column	Mobile phase	Detection	Notes	Ref.
Injection	HPLC	Dilute with water	C-8	Acetonitrile –0.005 M buffer (65 : 35)	259 nm	USP 23, assay; p. 388 [937] and [9, 922] GC [404,706, p. 156], 388 [937] and [9, 922]	[5]
Tablets	HPLC	Shake powder with phosphoric acid–water–2-propanol (1 : 600 : 400), filter	C-8	Buffer–0.83 ml OSA/l dilute with water, adjust pH to 3.0 with phosphoric acid			
Benzydamine Cream	HPLC	Dissolve in CH_3CN–H_2O (1 : 1)	RP-select B (Merck)	$0.05M\ NH_4OAc$–CH_3CN, pH 7.0 (45 : 55)	230 nm	Bulk and gel, also assayed irradiated samples	[221]
Benzylmorphine Bulk	TLC	—	Silica	S_1 = EtOAc–MeOH–conc. NH_4OH (85 : 10 : 4) S_2 = MeOH S_3 = MeOH–1-butanol (3 : 2)–$0.1M\ NaBr$ S_4 = MeOH–conc. NH_4OH (100 : 1.5) S_5 = cyclohexane–toluene–diethylamine (75 : 15 : 10) S_6 = $CHCl_3$–MeOH (9 : 1) S_7 = acetone	—	hR_{f1} = 41, hR_{f2} = 20, hR_{f3} = 23, hR_{f4} = 41, hR_{f5} = 6, hR_{f6} = 23, hR_{f7} = 8; also, GC, SE-30/OV-1, RI: 3015	[132]
Bulk	HPLC	—	Silica	Methanolic 0.01 M ammonium perchlorate, pH adjusted to 6.7	—	GC [1]; TLC [2]	[3]
Betamethasone Tablets	HPLC	Derivatize with homochiral reagent	Silica	n-Hexane–CH_2Cl_2–IPA (100 : 100 : 4)	240 nm	Resolved epimers along with dexamethasone	[254]
Bezitramide Bulk	HPLC	—	Silica	Methanolic 0.01 M ammonium perchlorate pH adjusted to 6.7	—	TLC [1]	[3]

Drug	Mode	Sample pretreatment	Sorbent (temp.)	Mobile phase	Detection	Comments	Reference
Biotin							
Various formulations	HPLC	ISTD added in PO$_4$ buffer, filtered	C$_{18}$ (Tosoh)	50 mM PO$_4$–CH$_3$CN (9 : 1) pH 4.5, 25 mM 1-BSA	EX: 370 nm EM: 440 nm	Postcolumn derivatization, other methods discussed	[240]
Biperiden							
Bulk	TLC	Dissolve in chloroform	Silica (base treated)	Toluene–methanol (965 : 35)	Iodoplatinate	BP, p. 59; impurities [937], GC [8,10]; p. 396 [937]; [922]	[4]
Bisoprolol						HPLC [235]; degradation [238]	
Boldine							
Syrup	HPLC	Dilute with H$_2$O, Na carbonate added to saturation, extract with ether, centrifuge	PRP-1 Hamilton	H$_2$O–CH$_3$CN–MeOH–1M TEA (50 : 25 : 25 : 5)	304 nm	Tablets also assayed	[241]
Brallobarbital (Brallobarbitone)							
Bulk	TLC	—	Silica	S$_1$ = CHCl$_3$–acetone (8 : 2) S$_2$ = EtOAc S$_3$ = CHCl$_3$–MeOH (9 : 1) S$_4$ = EtOAc–MeOH–conc. NH$_4$OH (85 : 10 : 5)	—	hR$_{f1}$ = 52, hR$_{f2}$ = 69, hR$_{f3}$ = 57, hR$_{f4}$ = 29; also GC, SE-30 OV-1, RI: 1858	[132]
Bromazepam							
Bulk	TLC	—	Silica (spray 0.1 M methanolic potassium hydroxide)	Ammonium hydroxide–methanol (1.5 : 100)	Char	Also, p. 400 [937]; [1138]	[1]
Bulk	TLC	—	Silica	S$_1$ = CHCl$_3$–acetone (8 : 2) S$_2$ = EtOAc S$_3$ = CHCl$_3$–MeOH (9 : 1) S$_4$ = EtOAc–MeOH–conc. NH$_4$OH (85 : 10 : 5)	—	hR$_{f1}$ = 13, hR$_{f2}$ = 20, hR$_{f3}$ = 47, hR$_{f4}$ = 64; also, GC, SE-30 OV-1	[134]

Sample	Method	Sample prep	Stationary phase	Mobile phase / Solvent system	Detection	Comments	Ref.
Bulk	TLC	—	Silica	S_1 = EtOAc–MeOH–conc. NH_4OH (85:10:4) S_2 = MeOH S_3 = MeOH–1-butanol (3:2)–0.1M NaBr S_4 = MeOH–conc. NH_4OH (100:1.5) S_5 = cyclohexane–toluene–diethylamine (75:15:10) S_6 = CHCl$_3$–MeOH 9:1) S_7 = acetone	—	hR_{f1} = 64, hR_{f2} = 74, hR_{f3} = 69, hR_{f4} = 61, hR_{f5} = 12, hR_{f6} = 41, hR_{f7} = 53; also, GC, SE-30/OV-1, RI: 2663	[132]
Bulk	HPLC	—	C-18, Hypersil, 5 μm; Flow rate 1.5 ml/min. Silica, Spherisorb 5W, 2 ml/min	S_1 = MeOH, H$_2$O, PO$_4$ (0.1M) (55:25:20), pH 7.25 S_2 = MeOH, H$_2$O, PO$_4$ (0.1M)(7:1:2), pH : 7.67 S_3 = MeOH–perchloric acid (1L: 100 μl) S_4 = MeOH, H$_2$O, TFA (997:2:1)	—	Cap. Fact. 1: 3.91 Cap. Fact. 2: — Cap. Fact. 3: 0.07 Cap. Fact. 4: 0.13; also [699, 922]	[135] [135]
Brompheniramine							
Bulk	HPLC	—	Silica	Methylene chloride–methanol (1% ammonium hydroxide) (9:1)	254 nm	TLC, GC, & HPLC, p. 403 [937]; [1138]; impurities [925]; enantiomers [919]; HPLC [922]	[9]
Bulk	GC	—	DB-5	—	FID	Resolved 23 anti-histamines	[706, p.164]
Bromvaletone							
Tablets	HPLC	Extract w/CH$_3$CN, ISTD–carbamazepine	C$_{18}$ (Nucleosil)	CH$_3$CN–0.05M PO$_4$, pH 2.7 (2:3), 1.5 ml/min	220 nm	Other methods discussed	[255]
Buclizine							
Bulk	GC	—	3% OV-1 Silica	Methanolic 0.01M ammonium perchlorate, pH adjusted to 6.7	FID	UV, MS, NMR, and IR TLC [1]; GC [2], [706, p. 164]	[9, p. 262]
Bulk	HPLC	—			—		[3]

Drug	Mode	Sample pretreatment	Sorbent (temp.)	Mobile phase	Detection	Comments	Reference
Bufotenine							
Bulk	TLC	—	Silica	S_1 = EtOAc–MeOH–conc. NH_4OH (85 : 10 : 4) S_2 = MeOH S_3 = MeOH–1-butanol (3 : 2)–0.1M NaBr S_4 = MeOH–conc. NH_4OH (100 : 1.5) S_5 = cyclohexane–toluene–diethylamine (75 : 15 : 10) S_6 = $CHCl_3$–MeOH (9 : 1) S_7 = acetone	—	hR_{f1} = 33, hR_{f2} = 10, hR_{f3} = 34, hR_{f4} = 35, R_{f5} = 0, R_{f6} = 1, R_{f7} = 1; also, GC. SE-30/OV-1, RI: 2030; UV, MS, HPLC, NMR, and IR [9, p. 266]	[132]
Bulk	HPLC	—	Silica	Methanolic 0.01M ammonium perchlorate, pH adjusted to 6.7	—	TLC [1]; GC [2]; some trade and other names: 3-(β-dimethylamino-ethyl)-5-hydroxyindole; 3-(2-dimethylamino-ethyl)-5-indolol; *N*, *N*-dimethylserotonin; 5-hydroxy-*N*, *N*-dimethyltryptamine; mappine	[3]
Bumetamide							
Tablet	HPLC	Eluent–acetone (9.5 : 0.5)	RP-8 (Varian)	MeOH–H_2O–CH_3CN (60 : 40 : 0.5), 1 ml/min	231 nm	Other methods described	[236]
Bupivacine hydrochloride							
Bulk	TLC	Dissolve in isopropylamine–chloroform (1 : 99)	Silica	Hexanes–isopropylamine (97 : 3)	Char	USP 23 purity p. 226, and BP, vol. II, p. 586 [4]; [1,1138]	[5]
Injection	HPLC	Dilute in methanol	C-18	Acetonitrile–buffer (65 : 35); buffer–mono and dibasic sodium phosphate (1.95 and 2.48 g/l, pH 6.8)	263 nm	USP 23, p. 226, ISTD–dibutyl phthlate; [8,922]; silica [1,10]; GC [2,10,8]; enantiomers [919]; [569,937]	[5]

	Method	Sample preparation	Stationary phase	Mobile phase	Detection	Notes	Ref.
Buprenorphine Tablets	HPLC	Add base, extract with ether, evaporate to dryness, dissolve in methanol	C-18	0.05M PSA–methanol–acetonitrile, pH 2.0 (30 : 55 : 15)	290 nm	Applicable to injection formulations; TLC [1]; HPLC [3,234]	[947]
Butalbital Bulk	TLC	—	Silica	S_1 = CHCl$_3$–acetone (8 : 2) S_2 = EtOAc S_3 = CHCl$_3$–MeOH (9 : 1) S_4 = EtOAc–MeOH–conc. NH$_4$OH (85 : 10 : 5)	—	hR_{f1} = 54, hR_{f2} = 67, hR_{f3} = 57, hR_{f4} = 44; also GC, SE-30/OV-1, RI: 1668	[132]
Tablets with aspirin	HPLC	Acetonitrile–formic acid (400 : 4) with powder, filter	C-18	Water–acetonitrile–phosphoric acid (3100 : 725 : 4)	214 nm	USP 23, p. 234; TLC & GC: p. 414 [937,1138]; HPLC [9, 948]	[5]
Bulk	TLC	(Same procedure as for butabarbital bulk)	—	—	—	USP 23, p. 232; also, pK and spectral data [569,937]; GC [228]; CE [260]	[5]
Butethal Bulk	HPLC	—	Silica	Methylene chloride–methanol (1% ammonium hydroxide) (99 : 1)	254 nm	Also GC; HPLC [938]	[9]
Butorphanol Bulk	GC	MeOH	USP-G3 (250°C)	—	FID	USP 23, p. 240	[5]
Butoxyethyl nicotinate Bulk	TLC	—	Silica (spray 0.1M methanolic KOH)	NH$_4$OH–MeOH (1.5 : 100)	Acidified iodoplatinate GC [2]		[1]
Butriptyline Bulk	HPLC	—	Silica	MeOH–0.01M ammonium perchlorate, pH adjusted to 6.7	—	TLC [1]; GC [2]	[3]

Drug	Mode	Sample pretreatment	Sorbent (temp.)	Mobile phase	Detection	Comments	Reference
Butyl aminobenzoate (Butamben)							
Bulk	GC	—	USP-G2 (between 100 and 300°C)	—	FID	—	[2]
Caffeine							
Bulk	TLC	—	Silica	S_1 = EtOAc–MeOH–conc. NH_4OH (85:10:) S_2 = MeOH S_3 = MeOH–1-butanol (3:2)–0.1M NaBr S_4 = MeOH–conc. NH_4OH (100:1.5) S_5 = clohexane–toluene–diethylamine (75:15:0) S_6 = $CHCl_3$–MeOH (9:1) S_7 = acetone	—	hR_{f1} = 50, hR_{f2} = 60, hR_{f3} = 55, hR_{f4} = 52, hR_{f5} = 3, hR_{f6} = 58, hR_{f7} = 25; also, GC, SE-30/OV-1, RI: 1810; HPLC [949–952]; with paracetamol [260]; with K sorbate in neonatal oral solution [305]; HPLC review [394,420]; TLC [421]	[132]
Bulk	TLC	Dissolve in methanol–chloroform (2:3)	Silica	Ammonium hydroxide–acetone–chloroform	254 nm	ED, pt. II–7, p. 267–2, [1]	[6]
Capsules	TLC	(See propoxyphene hydochloride)				USP 23, p. 242; HPLC [119]	[5]
Bulk	TLC	—	Silica	S_1 = EtOAc–MeOH–conc. NH_4OH (85:10:4) S_2 = MeOH S_3 = MeOH–1-butanol (3:2)–0.1M NaBr S_4 = MeOH–conc. NH_4OH (100:1.5) S_5 = cyclohexane–toluene–diethylamine (75:15:10) S_6 = $CHCl_3$–MeOH (9:1) S_7 = acetone	—	hR_{f1} = 50, hR_{f2} = 60, hR_{f3} = 55, hR_{f4} = 52, hR_{f5} = 3, hR_{f6} = 58, hR_{f7} = 25; also, GC, SE-30/OV-1, RI: 1810	[132]

Sample	Technique	Extraction	Column	Mobile phase	λ	Notes	Ref.
Bulk	HPLC	—	Silica 5 μm	Isooctane–diethyl-ether–methanol–diethylamine–water (400:325:225:0.5:15); flow rate: 2.0 ml/min	279 nm	k^1 = 1.61, RRT = 0.64, (diamorphien, 1.00, 49 min)	[133]
Bulk	HPLC	—	Silica	Methylene chloride–methanol (1% ammonium hydroxide (98:2)	254 nm	[3,751], GC [2,9, 953], review [954]	[9]
Bulk	HPLC	—	C-18	Acetonitrile–buffer (156:344), buffer: 4.8 g phoshoric acid–6.66 g potassium dihydrogen phosphate/l, pH 2.30	220 nm	GC [404]; ToxiLab TLC [564, 656]; HPLC [705, p. 185]; pK and spectral data [937]	[5]
Calcifediol Bulk	HPLC	EtOAc	Silica	Heptane–CH_2Cl_2–EtOAc–heptane (water sat.) (6:3:5:6)	254 nm	USP, p. 243	[5]
Camazepam Bulk	HPLC	—	C-18	Acetonitrile–buffer (156:344), buffer: 4.8 g phoshoric acid–6.66 g potassium dihydrogen phosphate/l, pH 2.30	220 nm	Also GC & TLC, GC [2]; midazolam [862]; ketazolam [879]	[11]
Bulk	HPLC	—	C-18 HS Perkin–Elmer, 3 μm	S_1 = MeOH–H_2O (7:3) S_2 = 5 mM PO_4(pH 6)–MeOH–CH_3CN (57:17:26), flow 1.5 ml/min	254 nm	Resolved 19 benzo-diazepines, RRT_1: 1.403 RRT_2: 1.085 (azinphos-methyl, 1.00)	[134]
Bulk	HPLC	—	C-18, Hypersil, 5 μm flow rate 1.5 ml/min	S_1 = MeOH, H_2O, PO_4 (0.1M) (55:25:20), pH 7.25 S_2 = MeOH, H_2O, PO4 (0.1M) (7:1:2), pH 7.67	—	Cap. Fact 1: 3.61 Cap. Fact 2: 2.58	[135]

Drug	Mode	Sample pretreatment	Sorbent (temp.)	Mobile phase	Detection	Comments	Reference
Bulk	HPLC	—	Silica, Spherisorb 5W, 2 ml/min	S_3 = MeOH–perchloric acid (IL: 100 µl) S_4 = MeOH, H_2O, TFA (997 : 2 : 1)	—	Cap. Fact. 3 : 0.08 Cap. Fact. 4 : 0.14, also [108–112]	[135]
Calcitonin Aerosol	HPLC	Derivatized with fluorescamine	G2000 SW XL Supelco	0.1M PO_4 buffer, pH 7.5–CH_3CN–MeOH (6 : 3 : 1)	EX: 370 nm EM: 420 nm	Autosampler	[320]
Camphor Ointments	GC	Disperse in $CHCl_3$, 50°C, filtered, ISTD added	1.95% OV-210–1.5% OV-17, Supelco 130°C	—	FID	Resolved from menthol, methyl salicylate and thymol; 10-camphorsulphonic acid, HPLC [384]	[378]
Capreomycin sulfate Bulk	Paper	Dissolve in citrate buffer	—	n-Propanol–water–acetic acid–TEA (75 : 33 : 8 : 8)	UV	USP 23, p. 262	[5]
Caprolactam Parenteral	HPLC	Add carbonate, ISTD, extract with $CHCl_3$, dry, dissolve in IPA	Silica, Supelcosil 5 µm, 250 x 8 cm	n-Hexane–IPA (9 : 1), 1.5 ml/min	210 nm	Plasticizers identified, semipreparative	[318]
Captopril Bulk	TLC	—	Silica	Benzene–AcOH (75 : 25)	I_2	GC & HPLC; HPLC p.1134 [96]; stability [1104]; USP 23, p. 262	[510]
Carbamazepine Bulk	TLC	Dissolve in chloroform	Silica	Toluene–methanol (95 : 5)	Chromate	BP, p. 78, also JP, p. 78 [7]; [1]; impurities [925]	[4]
Bulk	HPLC	—	Silica	Methylene chloride – methanol (1% ammonium hydroxide) (98 : 2)	254 nm	Also p. 428 [937], [11]; GC [9]; HPLC [955]; USP 23, p. 266, organic volatiles GC [275]; GC/MS [705, p. 55]; Toxi Lab TLC [564]; CE [723]; HPLC [569]	[9]

	Technique	Sample preparation	Phase	Mobile phase	Detection	Notes	Ref.
Carbenicillin Tablets	TLC	Triturate powder with acetone–EtOAc–pyridine–AcOH–water (200:100:25:1.5:75)	Silica	Acetone–pyridine–water–EtOAc–AcOH (400:25:75:300:2)	I_2	USP 23, p. 260	[5]
Carbetidine	—	—	—	—	—	Would be expected to chromatographically migrate on silica HPLC, CH_3OH (1%NH_4OH)–CH_2Cl_2 (1:9), similar to piperdolate	[10]
Carbidopa Tablets with levodopa	TLC	Add MeOH–0.05 N HCl, mix filter	Silica	n-Butanol–CH_3Cl–AcOH–acetone–water (40:40:40:60:35)	Spray triketohydrindene	USP 23; [8]	[5]
Bulk	HPLC	Dissolve in mobile phase	C_{18}	$0.05M\ PO_4$–EtOH (95:5), pH 2.7	280 nm	USP 23, resolved methyldopa and 3-O-methylcarbidopa	[5]
Carboprost tromethamine Bulk/injection	HPLC	Partition between CH_2Cl_2/citric acid buffer, derivatize with bromo-2-acetonapthone	Silica	Water–1,3-butanediol–CH_2Cl_2 (0.5:7:992)	254 nm	USP 23, ISTD–guaifensin, resolved from impurities 15 R-epimer and 5-trans epimer	[132]
Carbromal Bulk	TLC		Silica	S_1 = CHCl₃–acetone (8:2); S_2 = EtOAc; S_3 = CHCl₃–MeOH (9:1); S_4 = EtOAc–MeOH–conc. NH_4OH (85:10:5)	—	hR_{f1} = 53, hR_{f2} = 56, hR_{f3} = 64, hR_{f4} = 73; also GC, SE-30/OV-1, RI: 1513	[132]
Bulk	TLC	Dissolve in chloroform	Silica	Chloroform–ether (3:1)	254 nm	BP, p. 32; [1,8]	[4]

Drug	Mode	Sample pretreatment	Sorbent (temp.)	Mobile phase	Detection	Comments	Reference
Bulk	HPLC	—	Silica	Methylene chloride–methanol (1% ammonium hydroxide) (99:1)	RI	Also GC [9]	[9]
Carboplatin Infusion fluids	HPLC	—	—	—	—	See cisplatin for details; stability [419]; [334]	[5]
Carisoprodol Tablets with aspirin	HPLC	Dissolution	C-18 (30°C)	Methanol–1–2% acetic acid (51:49)	RI	USP 23, p. 276; GC [2]	[5]
Tablets with aspirin and codeine	HPLC	Dissolution	C-18 (30°C)	Acetonitrile–buffer (45:55); buffer–2.2 g DSC and ammonium nitrate 0.8 g/550 ml water	254 nm	USP 23	[5]
Bulk	HPLC	—	Silica	Methylene chloride–methanol (1% ammonium hydroxide (98:2)	RI	TLC [1,11]; ToxiLab [564]	[9]
Carnitine Tablets	HPLC	Precolumn derivatization with 9-anthryldiazo-methane	Silica	MeOH–5% aq. SDS–H_3PO_4 (990:10:1)	250 nm or EX: 365 nm, EM: 412 nm	Recovery from crude drugs, vitamins, & antacids	[409]
Cathine ((+)-norpseudoephedrine) Bulk	HPLC	—	Silica, 30 cm × 4 mm	MeOH (1% NH_4OH)–CH_2Cl_2 (1:9)	254 nm	GC, UV, MS, NMR, and IR	[9, p. 1668]
Bulk	TLC	—	Silica (spray 0.1 M methanolic potassium hydroxide)	Ammonium hydroxide–methanol (1.5:100)	Acidified potassium permanganate	GC [2]; HPLC [3]	[1]
Cefaclor Bulk	HPLC	PO_4 buffer, pH 2.5	C_{18} (YMC)	50 mM PO_4, pH 4.0–CH_3CN gradient, 2.25–45%	220 nm	Resolved process-related impurities; USP 23 [5]	[316]

Compound	Method	Sample preparation	Stationary phase	Mobile phase	Detection	Comments	Ref.
Cefadroxil Bulk	HPLC	Mobile phase–30% 0.02M NaOS, ultrasonicate, centrifuge	PIRP–S, Hamilton and others, 50 °C	CH_3CN–0.02M OSA–0.2M H_3PO_4–H_2O (10 : 5 : 20 : 5), 1 ml/min	254 nm	Resolved related substances; USP 23; HPLC [97]	[272]
Cefazolin Bulk	HPLC	Dissolve in pH 7, PO_4 buffer	C_{18}	CH_3CN–buffer (1 : 9); buffer– PO_4, 0.9 g citric acid, 1.3 g/l H_2O	254 nm	USP 23, p. 284; [1092]; stability [262]	[5]
Ceftazidine Injectable	HPLC	H_2O	C_{18} (Ultrasphere)	0.01M NH_4OAc–MeOH (89 : 11), 1.5 ml/min	254 nm	Stability eye drops [417]	[385]
Cephalexin Capsules	TLC	Mix powder with H_2O, filter	Same procedure as USP cefaclor	—	—	USP 23; impurity profile [258]	[5]
Capsules	HPLC	Mix powder with mobile phase	Silica	0.1% H_3PO_4–MeOH (95 : 5)	220 nm	Unpublished, see below	[761]
Cephalosporin C Bulk	HPLC	—	Silica	2% aq. acid adipic acid	250 nm	[1128]	
Cephradine Bulk	HPLC	Dissolve in H_2O	SCX	1 N AcOH–sodium sulfate, 24.2 g/l, pH 4.7 (1 : 199)	254 nm	USP 23, ITD– sulfanilamide, TLC [815]; cephalexin & *m*-hydroxycephalexin are degradants, see cephalexin; CE [265]; cephalosporins [396,411]; high-MW impurities [397]	[5]
Chloralbetamine	—	—	—	—	—	MS, NMR, and IR	[9, p. 392]
Chlorambucil Tablet with melphalan	HPLC	Powder extracted with H_2O–CH_3CN (1 : 1), filtered, diluted with mobile phase	C_{18}	0.1M PO_4–1 mM Na_3–EDTA–CH_3CN (7 : 3, pH 4.0), 1.5 ml/min	PAD, 0.9V/0.95V	Methods reviewed	[332]

Drug	Mode	Sample pretreatment	Sorbent (temp.)	Mobile phase	Detection	Comments	Reference
Chloralhydrate							
Bulk	GC	—	3% OV-1,4' × 4", 80°C	—	FID	MS, NMR, and IR	[9, p. 394]
Chloramphenicol							
Topical	HPLC	MeOH, ISTD, mobile phase	μBondapak, C_{18} w/ precolumn	CH_3CN–H_2O (35:65)	280/240 nm	Methods reviewed; HPLC [412]; [6,121,122,136, 145,229,784,785]	[386]
Chlorcyclizine							
Bulk	TLC	Dissolve in methanol	Silica	Chloroform–methanol–13.5M ammonium hydroxide (90:8:2)	Iodoplatinate	BP, p. 98; HPLC [3, 10]	[4]
Chlordiazepoxide							
Bulk	TLC	—	Silica	S_1 = $CHCl_3$–acetone (8:2) S_2 = EtOAc S_3 = $CHCl_3$–MeOH (9:1) S_4 = EtOAc–MeOH–conc. NH_4OH (85:10:5)	—	hR_{f1} = 10, hR_{f2} = 11, hR_{f3} = 53, hR_{f4} = 52; also GC, SE-30/OV-1, RI: 2797	[132]
Bulk	TLC	—	Silica	S_1 = EtOAc–MeOH–conc. NH_4OH (85:10:4) S_2 = MeOH S_3 = MeOH–1-butanol (3:2)–0.1M NaBr S_4 = MeOH–conc. NH_4OH (100:1:5) S_5 = cyclohexane–toluene–diethylamine (75:15:10) S_6 = $CHCl_3$–MeOH (9:1) S_7 = acetone	—	hR_{f1} = 52, hR_{f2} = 76, hR_{f3} = 77, hR_{f4} = 62, hR_{f5} = 2, hR_{f6} = 50, hR_{f7} = 22; also GC, SE-30/OV-1, RI: 2797	[132]

Sample	Method	Sample preparation	Column	Mobile phase / Solvent	Detection	Comments	Ref.
Bulk	HPLC	—	C-18 HS, Perkin-Elmer, 3 μm	S_1 = MeOH, H₂O (7:3); S_2 = 5 mM PO₄ (pH 6)–MeOH–CH₃CN (57:17:26) Flow: 1.5 ml/min	254 nm	Resolved 19 benzodiazepines, RRT₁: 1.008; RRT₂: 0.462 (azinphosmethyl, 1.00)	[134]
Bulk and tablets	TLC	Dissolve in acetone	Silica	Ethyl acetate	Char	USP 23, p. 343, related sustances, TLC review [956] also, spray with sodium nitrite reagent, also, BP, vol. II, p. 526 and p. 745 [4] and EP vol. III, p. 183 [6]; [957,958]; stability [1104]	[5]
Bulk	HPLC	—	C-18, Hypersil, 5 μm; Flow rate: 1.5 ml/min	S_1 = MeOH, H₂O, PO₄ (0.1M) (55:25:20), pH: 7.25; S_2 = MeOH, H₂O, PO₄ (0.1M)(7:1:2), pH: 7.67	—	Cap. Fact 1: 6.41 Cap. Fact 2: —	[135]
Bulk	HPLC	—	Silica Spherisorb 5W, 2 ml/min	S_3 = MeOH–Perchloric acid (1L: 100 μL); S_4 = MeOH, H₂O, TFA (997:2:1)		Cap. Fact. 3: 2.27 Cap. Fact. 4: 5.9 HPLC–diode array [118, 699]; [769]; pK and spectral data [937]; tablets [422]	[135]
Chlorocresol Ointments	HPLC	Extract with hexane	C₁₈, ODS-2 Perkin Elmer	CH₃CN–H₂O (1:1), 1 ml/min	275 nm	Compared to colorimetry	[377]
Chloroquine Bulk	HPLC	—	Silica	Methylene chloride methanol (1% ammonium hydroxide) (4:1)	254 nm	TLC [1,11], GC [2, 9, 11]; HPLC [3,415,418]; review [959]; enantiomers [388]	[9]
Chlorothiazide Bulk	HPLC	Dissolve in CH₃CN & mobile phase	C₁₈	0.1M PO₄, buffer–CH₃CN (9:1)	254 nm	USP 23, EP, pt. II-9, p. 385 [6]; TLC/HPLC, p. 454 [25]; [9]; photodecomposition [408]	[5]

Drug	Mode	Sample pretreatment	Sorbent (temp.)	Mobile phase	Detection	Comments	Reference
Chlorpheniramine							
Elixir	TLC	Add water, render alkaline, extract with chloroform	Silica	Ethyl acetate–methanol–1M acetic acid (5:3:2)	Dragendroff	BP, vol. II, pp. 552, related substances, also, JP, p. 1206 [7], EP, pt. II-9, p. 386 [6]; impurities [925]; GC [2, 706, p. 164, 960]; HPLC [3,9,420,961–965,922]; enantiomers [919]; ToxiLab TLC [267, 564]	[4]
Chlorproamide							
Bulk	HPLC	—	C$_{18}$	MeOH–NaOAc–THF (5:4:1), 0.1M AcO; pH 4.8, 1.2 ml/min	EC +200 mV, +550 mV	TLC [1,4,8]; GC [2,8], HPLC [3,10,123, 124,1120]	[405]
Formulations - tolbutamide	HPLC	Sonicate powder in MeOH, filter	C$_{18}$ Hypersil	MeOH–0.01M PO$_4$, pH 3.0 (1:1), 1 ml/min	228 or 230 nm	Other methods reviewed	[330]
Chlorpromazine							
Tablets	TLC	Mix powder with methanol–diethyl-amine (95:5), filter	Silica	Cyclohexane–acetone–diethylamine (8:1:1)	254 nm	BP, vol. II, p. 552, p. 591, p. 723, and p. 748, also EP, vol. III, p. 188; TLC [1,11]; GC [2,9,11]; HPLC [3, 9, 922];TLC [224]; GC [264]; HPLC [567, 700]; GC/MS [705, p. 55]	[4]
Chlorprothixene							
Bulk	TLC	—	Silica	S$_1$ = EtOAc-MeOH–conc. NH$_4$OH (85:10:4); S$_2$ = MeOH; S$_3$ = MeOH-1-butanol (3:2)–0.1M NaBr; S$_4$ = MeOH-conc. NH$_4$OH (100:1.5)	—	hR$_{f1}$ = 74, hR$_{f2}$ = 34, hR$_{f3}$ = 51, hR$_{f4}$ = 56, hR$_{f5}$ = 51, hR$_{f6}$ = 50, hR$_{f7}$ = 22; also GC, SE-30/OV-1, RI: 2797; HPLC [700]; USP 23 [5], TLC [1,8]; GC [2,8,9]; HPLC [3,9]	[132]

S5 = cyclohexane–toluene–diethylamine (75:15:10)
S6 = CHCl3–MeOH (9:1)
S7 = acetone

Compound	Method	Sample preparation	Stationary phase	Mobile phase	Detection	Comments	Ref.
Chlortetracycline hydrochloride							
Bulk	TLC	MeOH	Kieselguhr with glycerol, EDTA	CHCl3–acetone–EtOAc (2:1:1), use lower layer with EDTA, pH 7.0	HN4 vapor, 350 nm	BP, p.108, EP, pt. II–5, p. 173 [6]; TLC [1]	[4]
Bulk/related substances	HPLC	—	PRP-1	2.5–65% 2-methyl-2-propanol, 5% 1.0 M ClO4–H2O, 1 ml/min	254 nm	Other methods reviewed, collaborative study; ointment [308]; HPLC [311,314,315]	[307]
Chlorzoxazone							
Bulk	TLC	Dissolve in methanol	Silica	Hexane–dioxane (63:37)	254 nm	USP 23, p. 358 [5], TLC [1,8]; GC [2,8,9]; HPLC [3,9]; impurities	[5]
Tablets with acetaminophen	HPLC	Add methanol to the powder, mix, and filter	C-18	Methanol–water linear gradient	280 nm	USP 23, p. 364; ISTD–phenacetin; HPLC [8,9]; GC [2,8]	[5]
Ciclopiroxolamine							
Topical	HPLC	Alkylation to form 1-alkyloxypirydone	C18—Delta Pak, 15 cm x 330 μm	CH3CN–H2O (1:1), 10 μl/min	300 nm	—	[389b]
Cimetidine							
Bulk and tablets	HPLC	Dissolve in mobile-phase buffer	C-18	Acetonitrile–buffer (16:84); buffer–acetic acid–0.2% sodium acetate (95:5)	228 nm	USP 23; [3,8,9]; TLC [1]; also, TLC [267]; ToxiLab TLC [564]; pK and spectral data [937]	[5]
Cinnarizine							
Bulk	TLC	—	Silica	S1 = EtOAc–MeOH–conc. NH4OH (85:10:4) S2 = MeOH S3 = MeOH–1-butanol (3:2)–0.1 M NaBr	—	hRf1 = 85, hRf2 = 79, hRf3 = 87, hRf4 = 76, hRf5 = 51, hRf6 = 78, hRf7 = 66; also GC, SE-30/OV-1, RI: 3065	[132]

Drug	Mode	Sample pretreatment	Sorbent (temp.)	Mobile phase	Detection	Comments	Reference
				S_4 = MeOH–conc. NH$_4$OH (100 : 1.5) S_5 = cyclohexane–toluene–diethylamine (75 : 15 : 10) S_6 = CHCl$_3$–MeOH (9 : 1) S_7 = acetone			
Bulk	HPLC	—	Silica	Methanolic 0.01 M ammonium perchlorate, pH 6.7	—	TLC [1,8]; GC [1,8,706 p. 164]	[3]
Cisplatin Bulk	HPLC	—	Alumina	Varied phosphate conc.. 0.01M–0.1M	205/220 nm	Stability, hydrolysis; [1104]	[395]
Clarithromycin Residues	HPLC	Swab with EtOH–H$_2$O (1 : 1)	C$_{18}$, Nucleosil	CH$_3$CN–H$_2$O (400 : 600), 16.3 g/l Na$_2$OAc, pH 6.6, 1.7 ml/min	EC	Methods reviewed; HPLC [392,393]	[389a]
Clavulanic acid Amoxicillin tablet, suspension	HPLC	H$_2$O, filter	C$_{18}$	CH$_3$OH–PO$_4$, pH 6–H$_2$O (15 : 1 : 84), 1 ml/min	235 nm	Methods reviewed; determination of clvan-2-carboxylate [410]	[319]
Clemastine Bulk and tablets	TLC	Dissolve in methanol	Silica	Diisopropyl alcohol–formic acid–water (70 : 20 : 5)	Spray potassium permangenate	USP 23, p. 385, identification, also JP, p. 161 [4]; [1,8]	[5]
Tablets	HPLC	Mix powder with methanol–water (1 : 1), centrifuge, filter supernatant layer	C-8	Methanol–buffer (83 : 17); buffer a) 9.47 g dipotassium hydrogen phosphate, b) 9.08 g potassium dihydrogen phosphate, dilute each to 3l, mix a–b (612 : 388)	220 nm	USP 23, assay; [3,10]; GC [3,8,10]	[5]

Compound	Method	Sample prep	Column	Mobile phase	Detection	Notes	Ref.
Clidinium Capsules	HPLC	H_2O sonicate, centrifuge	ODS-3, Partisil	CH_3CN–0.3M PO_4 (32 : 68), pH 4.3	325 nm	Resolved impurities and chlordiazepoxide; USP 23; TLC [1,8,24]; HPLC [8,1105,1109]	[277]
Clindamycin Topical	HPLC	—	C_{18}, Novapak	CH_3CN–acetate (27 : 73), pH 5.8, 0.01M HSA	600–625 nm	Postcolumn chemiluminescence with Ru(bpy)$_3^{+3}$; USP 23 [5]; BP [4]; TLC [1]; GC [10,771]; stability [1104]	[380]
Cliquinol Bulk	HPLC	MeOH–CH_3CN (1 : 1)	μBondapak, C-18	MeOH–H_2O–H_3PO_4 (800 : 298.8 : 1.2)	254 nm	ISTD-metronidazole; USP 23 [5]; BP [4]	[387]
Clobazam Bulk	TLC	—	Silica	S_1 = $CHCl_3$–acetone (8 : 2) S_2 = EtOAc S_3 = $CHCl_3$–MeOH (9 : 1) S_4 = EtOAc–MeOH–conc. NH_4OH (85 : 10 : 5)	—	hR_{f1} = 53, hR_{f2} = 49, hR_{f3} = 70, hR_{f4} = 74; also GC, SE-30/OV–1, RI: 2694	[132]
Bulk	HPLC	—	C-18 HS, Perkin-Elmer, 3 μm	S_1 = MeOH, H_2O (7 : 3) S_2 = 5 mM PO_4 (pH 6)–MeOH–CH_3CN (57 : 17 : 26) Flow: 1.5 ml/min	254 nm	Resolved 19 benzodiazepines, RRT$_1$: 0.755, RRT$_2$: 0.582 (azinphosmethyl, 1.00)	[134]
Bulk	HPLC	—	C-18, Hypersil, 5 μm; Flow rate 1.5 ml/min	S_1 = MeOH, H_2O, PO_4 (0.1M) (55 : 25 : 20), pH 7.25 S_2 = MeOH, H_2O, PO_4 (0.1M) (7 : 1 : 2), pH 7.67	—	Cap. Fact. 1: 3.91 Cap. Fact. 2:—	[135]
Bulk	HPLC	—	Silica, Spherisorb 5W, 2 ml/min	S_3 = MeOH–perchloric acid (1L: 100 μl) S_4 = MeOH, H_2O, TFA (997 : 2 : 1)		Cap. Fact. 3: 0.07 Cap. Fact. 4: 0.13	[135]

Drug	Mode	Sample pretreatment	Sorbent (temp.)	Mobile phase	Detection	Comments	Reference
Bulk	TLC	—	Silica (spray 0.1M methanolic potassium hydroxide)	Ammonium hydroxide–methanol (1.5 : 100)	Char	Various other visualization reagents can be used	[1]
Clobetasone-17 butyrate Ointment	HPLC	Shake with MeOH, 60°C, add ISTD	C_{18}, Novapak	H_2O–MeOH (7 : 3)	235 nm	ISTD–clobetasone propionate Ion-pair RPHPLC [270]	[321]
Clocinizine							
Clomethiazole (See chlormethiazole)							
Clomipramine Bulk	TLC	—	Silica	S_1 =EtOAc–MeOH–conc. NH_4OH (85 : 10 : 4) S_2 = MeOH S_3 = MeOH–1-butanol (3 : 2) – 0.1M NaBr S_4 = MeOH–conc. NH_4OH (100 : 1.5) S_5 = cyclohexane-toluene–diethylamine (75 : 15 : 10) S_6 = $CHCl_3$–MeOH (9 : 1) S_7 = acetone		hR_{f1} = 72, hR_{f2} = 26, hR_{f3} = 54, hR_{f4} = 51, hR_{f5} = 54, hR_{f6} = 34, hR_{f7} = 18; also GC, SE-30/OV-1, RI: 2406	[132]
Clomipramine hydrochloride Bulk/capsules	TLC	Dissolve in methanol	Silica	Acetone–13.5M ammonium hydroxide–ethyl acetate (5 : 1 : 15)	Spray dichromate	BP, addendum 1983, p. 222 and p. 265; TLC [1,8]; GC [2,8]; HPLC [3,8]	[4]
Clonazepam Bulk	TLC	—	Silica	S_1 = $CHCl_3$–acetone (8 : 2) S_2 = EtOAc S_3 = $CHCl_3$–MeOH (9 : 1) S_4 = EtOAc–MeOH–conc. NH_4OH (85 : 10 : 5)	—	hR_{f1} = 35, hR_{f2} = 45, hR_{f3} = 56, hR_{f4} = 68; also GC, SE-30/ OV-1, RI: 2885	[132]

Sample	Method	Sample preparation	Stationary phase	Mobile phase	Detection	Notes/Results	Ref.
Bulk	TLC	—	Silica	S_1 = EtOAc–MeOH–conc. NH$_4$OH (85:10:4); S_2 = MeOH; S_3 = MeOH–1-butanol (3:2)–0.1M NaBr; S_4 = MeOH–conc. NH$_4$OH (100:1.5); S_5 = cyclohexane–toluene–diethylamine (75:15:10); S_6 = CHCl$_3$–MeOH (9:1); S_7 = acetone	—	hR_{f1} = 68, hR_{f2} = 85, hR_{f3} = 87, hR_{f4} = 72, hR_{f5} = 0, hR_{f6} = 53, hR_{f7} = 61; also GC, SE-30/OV-1, RI: 2885	[132]
Bulk	HPLC	—	C-18, HS, Perkin-Elmer, 3 μm	S_1 = MeOH, H$_2$O (7:3); S_2 = 5 mM PO$_4$ (pH 6)–MeOH–CH$_3$CN (57:17:26) Flow: 1.5 ml/min	254 nm	Resolved 19 benzodiazepines, RRT$_1$:; 0.684 RRT$_2$: 0.406 (azinphosmethyl, 1.00)	[134]
Bulk	HPLC	—	C-18, Hypersil, 5 μm; flow rate 1.5 ml/min.	S_1 = MeOH, H$_2$O, PO$_4$ (0.1M) (55:25:20), pH: 7.25; S_2 = MeOH, H$_2$O, PO$_4$ (0.1M) (7:1:2), pH: 7.67	—	Cap. Fact. 1: 2.85; Cap. Fact. 2: —	[135]
	HPLC	—	Silica, Spherisorb 5W, 2 ml/min.	S_3 = MeOH–perchloric acid (1L:100 μl); S_4 = MeOH, H$_2$O, TFA (997:2:1)		Cap. Fact. 3: 0.25; Cap. Fact. 4: 0.29	[135]
Bulk and tablets	TLC	Dissolve in acetone, for tablets, filter	Silica	Ethyl acetate–carbon tetrachloride (1:1)	Sprays, see ref.	USP 23, p. 401, related substances	[5]
Tablets	HPLC	Dissolve in dimethylformamide	C-18	Water–methanol–acetonitrile (4:3:3)	254 nm	USP 23, p. 402, ISTD-o-dichlorobenzene, assay; GC & HPLC [8,9]; p. 480 [937]; HPLC [922] pK and spectral data [937]; GC/TLC [758]	[5]

Drug	Mode	Sample pretreatment	Sorbent (temp.)	Mobile phase	Detection	Comments	Reference
Clonidine Bulk	TLC	Dissolve in methanol	Silica	Toluene–dioxane–dehydrated alcohol–ammonium hydroxide (10 : 8 : 2 : 1)	Spray sodium hypochlorite and potassium iodide	USP 23, p. 403, chromatographic purity; [1, 8]; pK and spectral data [937]; GC [2,8,10]; review [159, p.109]	[5]
Clorazepate Bulk	HPLC	—	C-18, HS, Perkin-Elmer, 3 μm	S_1 = MeOH, H_2O (7 : 3) S_2 = 5 mM PO_4 (pH 6)–MeOH–CH_3CN (57 : 17 : 26) Flow: 1.5 ml/min	254 nm	Resolved 19 benzo-diazepines, RRT_1; 1.032, RRT_2; 0.598 (azinphosmethyl, 1.00)	[134]
Bulk	HPLC	—	Amino	Acetonitrile	254 nm	[966], HPLC [699]	[9]
Clostebol Bulk	GC	—	DB-1 J&W, 30 m × 0.25 mm, 0.1 m 180°–320°C (10°C/min); 320°C for 4 min		FID	Resolved 17 anabolic steroids; HPLC, [749]; review of GC/MS analysis of anabolic steroids [1119]	[706, p. 168]
Clotrimazole Ointments	HPLC	Sonicate with MeOH	C$_{18}$, Nucleosil	MeOH–H_2O (9 : 1), 1 ml/min	258 nm	ISTD–chrysene	[413]
Cloxacillin Syrup	HPLC	Dilute in H_2O, add ISTD	C$_{18}$, μBondapak	MeOH–AcOH (6 : 4), 1.5 ml/min	254 nm	Bulk, capsules, and injection formulations; LC–MS [317]; HPLC [173]	[301]
Cloxazolam Bulk	HPLC	—	C-18, Hypersil, 5 μm flow rate 1.5 ml/min.	S_1 = MeOH, H_2O, PO_4 (0.1M) (55 : 25 : 20), pH 7.25 S_2 = MeOH, H_2O, PO_4 (0.1M) (7 : 1 : 2), pH 7.67	—	Cap. Fact. 1: 14.27 Cap. Fact. 2: 2.89	[135]

Sample	Technique		Stationary phase	Mobile phase / conditions	Detection	Results / Notes	Ref.
Bulk	HPLC	—	Silica, Spherisorb 5W, 2 ml/min.	S_3 = MeOH–perchloric acid (1L: 100 μl), S_4 = MeOH, H$_2$O, TFA (997:2:1)	—	Cap. Fact. 3: 2.24, Cap. Fact. 4: 5.92; GC/TLC [758]	[135]
Cocaine Bulk	TLC	—	Silica	S_1 = EtOAc–MeOH–conc. NH$_4$OH (85:10:4), S_2 = MeOH, S_3 = MeOH–1-butanol (3:2) – 0.1M NaBr, S_4 = MeOH–conc. NH$_4$OH (100:1.5), S_5 = cyclohexane–toluene–diethylamine (75:15:10), S_6 = CHCl$_3$–MeOH (9:1), S_7 = acetone	—	hR_{f1} = 77, hR_{f2} = 35, hR_{f3} = 30, hR_{f4} = 65, hR_{f5} = 47, hR_{f6} = 47, hR_{f7} = 54; also, GC, SE-30/OV-1, RI: 2187	[132]
Bulk	HPLC	—	Silica, 5 μm	Isooctane–diethyl-ether–methanol–diethylamine–water (400:325:225:0.5:15), flow rate: 2.0 ml/min	279 nm	k^1 = 0.71, RRT = 0.47 (diamorphine, 1.00, 49 min)	[133]
Bulk	TLC	—	Silica	Acetonitrile	—	Diastereoisomers, resolved pseudoallocaine	
Bulk	GC	—	CP-Sil 8 (10-m)	—	NPD	Resolved from methaqualone	[35]
Bulk	HPLC	—	Silica	Methylene chloride–methanol (1% ammonium hydroxide (99:1)	254 nm	TLC [1,8]; GC [2,8,9]; HPLC [1,8,105,669, 1130]; review [967]; enantiomers [233, 919]; history [923]; stability [244,278]; ToxiLab TLC [564]; studied on 38 TLC systems [256]; GC [289,333]; TLC [273,331,404,466,709]	[9]

Drug	Mode	Sample pretreatment	Sorbent (temp.)	Mobile phase	Detection	Comments	Reference
Codeine							
Bulk	TLC	—	Silica	S_1 = EtOAc–MeOH–conc. NH$_4$OH (85 : 10 : 4) S_2 = MeOH S_3 = MeOH–1-butanol (3 : 2)–0.1M NaBr S_4 = MeOH–conc. NH$_4$OH (100 : 1.5) S_5 = cyclohexane–toluene–diethylamine (75 : 15 : 10) S_6 = CHCl$_3$–MeOH (9 : 1) S_7 = acetone	—	hR_{f1} = 35, hR_{f2} = 20, hR_{f3} = 22, hR_{f4} = 33, hR_{f5} = 6, hR_{f6} = 18, hR_{f7} = 3; also GC, SE-30/OV-1, RI: 2376; semisynthetic codeine base HPLC [271]; dissolution [383]	[132]
Bulk	HPLC	—	Silica, 5 μm	Isooctane–diethyl-ether–methanol–diethylamine–water (400 : 325 : 225 : 0.5 : 15), flow rate: 2.0 ml/min	279 nm	k^1 = 6.36, RRT = 1.57 (diamorphine, 1.00, 49 min)	[133]
Bulk	TLC	Dissolve in ethanol	(Same procedure as for codeine phosphate hemihydrate)		—	EP, pt. II-6, p. 76	[6]
Elixir Bulk	TLC HPLC	(See terpin hydrate) —	Silica	Methylene chloride–methanol (1% ammonium hydroxide) (95 : 5)	254 nm	USP 23, p. 418 Also p. 490 [937]; TLC, GC, & HPLC [8]; HPLC [13,119]	[5] [9]
Cough syrup	HPLC	Dilute with mobile phase	C-18	Methanol–0.05M potassium dihydrogen (13 : 87)	254 nm	Also [172,1105]	[152]
Bulk	TLC	Dissolve in 0.01N hydrochloric acid–methanol (4 : 1)	Silica	Ammonium hydroxide–cyclohexane–ethanol (6 : 30 : 72)	Dragendorff	EP, pt. II-6, p. 74–3; DAD [118]; HPLC [567,705, p. 215]; GC [264,273,300,404,709];	[6]

Compound	Method	Sample preparation	Column	Mobile phase	Detection	Other methods	Ref.
Colorants Bulk	HPLC	Dissolve in H$_2$O, 0.15%, use within 15 min	SAX	0.01M Na borate gradient with 0.1M Na perchlorate	254 nm	CE [280]; TLC [256,290,656,682]; ToxiLab TLC [564]; GC/MS [670]; spectral and pK data [937]; SCF [252]; general [286]	[764]
Cortalcerone Crude	HPLC	Extracted from mycelium	Separon SGX NH$_2$	CH$_3$CN–H$_2$O (7 : 3), pH 5.0 with 0.1M PO$_4$, 1.5 ml/min	195 nm	—	[414]
Cortisone acetate Bulk	TLC	Dissolve in MeOH–CH$_3$Cl (1 : 9)	Silica	CHCl$_3$–H$_2$O–MeOH (423 : 1 : 77)	Tetrazolium blue	JP, p.181; EP, pt.II-8, p. 321 [4]	[4]
Bulk	HPLC	Dissolve in THF	Silica	Butylchloride–butyl chloride (H$_2$O sat.)–THF–MeOH–AcOH (95 : 95 : 14 : 7 : 6)	254 nm	USP 23, p. 429; dosage forms [347]; [4, 24, 25,820,821,827–830]	[5]
Crotamiton Cream	HPLC	Add MeOH, shake, sonicate, filter	C$_{18}$	CH$_3$CN–MeOH (3 : 2)	254 nm	USP 23; TLC [1]; GC [2]	[5]
Cyanocobalamin (Vitamin B$_{12}$) Multivitamin	HPLC	SPE	C$_{18}$, μBondapak, 40°C	15–50% MeOH gradient in 0.02M PO$_4$, 15 min, 1.5 ml/min	550 nm	Alternative to microbiological methods; CE [730]; HPLC [416,727,730]; TLC [728]; review [726]	[376]
Cyclizine Bulk	HPLC	—	Silica, 10 μm	MeOH (1% NH$_4$OH) – CH$_2$Cl$_2$ (2 : 98)	254 nm	UV, GC, MS, NMR, and IR	[9, p. 532]

Drug	Mode	Sample pretreatment	Sorbent (temp.)	Mobile phase	Detection	Comments	Reference
Bulk	TLC	—	Silica	S_1 = EtOAc–MeOH–conc. NH$_4$OH (85:10:4) S_2 = MeOH S_3 = MeOH–1-butanol (3:2) – 0.1M NaBr S_4 = MeOH–conc. NH$_4$OH (100:1.5) S_5 = cyclohexane–toluene–diethylamine (75:15:10) S_6 = CHCl$_3$–MeOH (9:1) S_7 = acetone	—	hR$_{f1}$ = 68, hR$_{f2}$ = 40, hR$_{f3}$ = 52, hR$_{f4}$ = 57, hR$_{f5}$ = 49, hR$_{f6}$ = 41, hR$_{f7}$ = 16; also GC, SE-30/OV-1, RI: 2020	[132]
Bulk/injection	TLC	Dissolve in methanol	Silica	Chloroform–methanol–13.5M ammonium	Iodine vapor	BP, addedum 1983, p. 223, p. 277, and p. 301; GC [2]; HPLC [3]; TLC impurities [925]	[4]
Bulk	GC	—	DB-5	—	FID	Resolved 23 antihistamines	[706, p.164]
Cyclobarbital (Cyclobarbitone)							
Bulk	TLC	—	Silica	S_1 = CHCl$_3$–acetone (8:2) S_2 = EtOAc S_3 = CHCl$_3$–MeOH (9:1) S_4 = EtOAc–MeOH–conc. NH$_4$OH (85:10:5)	—	hR$_{f1}$ = 50, hR$_{f2}$ = 64, hR$_{f3}$ = 58, hR$_{f4}$ = 35; also GC, SE-30/OV-1, RI: 1963	[132]
Bulk	TLC	Dissolve in water, heat to 60°C	Silica	Ammonium hydroxide–ethanol–chloroform (5:15:80)	254 nm	EP, pt. II–5, p. 175–2; p. 498 [925]	[6]
Bulk	HPLC	—	Silica	Methylene chloride–methanol (1% ammonium hydroxide) (98:2)	254 nm	TLC [8]; GC [9,8]; HPLC [8,968]	[9]

Cyclopentolate Bulk	TLC	Dissolve in methanol	Silica	Cyclohexane–diethyl amine (95 : 5)	254 nm	USP 23, p. 439, identification, also BP, p. 132, p. 567, and p. 754 [4]; TLC [1]; impurities [925]; GC [2]; HPLC [3]; enantiomers [919]	[5]
Cyclophosphamide Tablets	HPLC	Shake with H_2O, filter	C_{18}	H_2O–CH_3CN (7 : 3)	195 nm	USP 23, p.441; [2,8,837–840]; stability [1104]	[5]
Cyclosporine Bulk	HPLC	CH_3CN–H_2O (1 : 1)	C_{18}	H_2O–CH_3CN–t butyl-methyl ether–H_3PO_4 (520 : 430 : 50 : 1) CH_3CN–H_2O (1 : 1)	210 nm	USP 23, p. 444, column at 80°C, 1.2 ml/min	[5]
Fermentation	HPLC	—	C_8, Supelco, 60°C	—	202 nm	Other methods discussed	[303]
Cyproheptadine Bulk	HPLC	—	Silica, 10 μm	MeOH (1% NH_4OH)–CH_2Cl_2 (5 : 95)	254 nm	UV, MS, GC, NMR, and IR	[9, p. 574]
Bulk	TLC	Dissolve in chloroform	Silica	Chloroform–methanol (9 : 1)	Char, UV	BP, p. 134; GC [2,8,10]; HPLC [3,8,10,922]; impurities [925]	[4]
Cytokine, recombinant (leukocyte A, interferon) Injectable	CE	0.05M PO_4, pH 7, 500 μg/ml human serum albumin, 50 μg/ml n-acetyl tryptophan	Silica, 57 cm × 75 μm, 10 kV	0.05M Na tetraborate buffer, pH 8.3–0.025N $LiCl_3$	200 nm	Resolved from interleukin-1α	[381]
Dapsone Bulk	TLC	Dissolve in methanol	Silica	Toluene–acetone (2 : 1)	Sodium nitrite solution	EP, pt. II–2, p. 77, related substances; impurities [925]	[6]
Daunorubicin Injection	HPLC	Dissolve in mobile phase	C_{18}	H_2O–CH_3CN (62 : 38)	254 nm	USP 23, p. 454; [841–843]; derivative [435]	[5]
Demeclocycline Capsules	HPLC	Powder, sonicate with mp, cenrifuge	PRP, 60°C	2-methyl-2-propanol–0.2M PO_4 (pH 9.0)–0.02M TBA (pH	254 nm	Resolved impurities, methods reviewed, ointment analyzed	[453]

Drug	Mode	Sample pretreatment	Sorbent (temp.)	Mobile phase	Detection	Comments	Reference
Denopamine Bulk/tablets	HPLC	GITC derivative	ODS-2, Inertsil	$0.05M$ PO_4, pH 3.0–MeOH (44 : 56) 9.0)–0.01M EDTA (pH 9.0)water (8 : 10 : 15 : 10 : 57)	250 nm	Enantiomers resolved	[459]
Desipramine Bulk	TLC	—	Silica	S_1 = EtOAc–MeOH– conc. NH_4OH (85 : 10 : 4) S_2 = MeOH S_3 = MeOH–1-butanol (3 : 2)–0.1M NaBr S_4 = MeOH–conc. NH_4OH (100 : 1.5) S_5 = cyclohexane– toluene–diethylamine (75 : 15 : 10) S_6 = $CHCl_3$–MeOH (9 : 1) S_7 = acetone	—	hR_{f1} = 41, hR_{f2} = 7, hR_{f3} = 71, hR_{f4} = 26, hR_{f5} = 20, hR_{f6} = 11, hR_{f7} = 3; also, GC, SE-30/OV-1, RI: 2242	[132]
Tablets	TLC	Shake powder with methanol, filter	Silica	Toluene–methanol (95 : 5)	Potassium dichromate	BP, vol. II, p. 756, also EP vol. III, Supplement, p. 99 [6]	[4]
Bulk	HPLC	—	Silica	Methylene chloride–methanol (1% ammonium) hydroxide (4 : 1)	254 nm	TLC [1,8]; GC [2,8,9]; HPLC [3,8,858,922]; ToxiLab [564]; GC [404,706, p. 163]; pK and spectral data [937]	[9]
Desmethyldiazepam Bulk	HPLC	—	C-18, HS, Perkin-Elmer, 3 μm	S_1 = MeOH, H_2O (7 : 3) S_2 = 5 mM PO_4 (pH 6)–MeOH–CH_3CN (57 : 17 : 26) Flow: 1.5 ml/min	254 nm	Resolved 19 benzo-diazepines, RRT_1; 1.040, RRT_2: 0.589 (azinphosmethyl, 1.00); CE [691]	[134]

Compound	Technique	Sample prep	Stationary phase	Mobile phase	Detection	Comments	Ref.
Desomorphine Bulk	HPLC	—	Silica, 10 μm	MeOH (1% NH$_4$OH)–CH$_2$Cl$_2$ (1:4)	254 nm	GC, UV, MS, NMR, IR	[9, p. 622]
Dexamethasone Bulk	HPLC	MeOH	C$_8$	H$_2$O–CH$_3$CN (3:2)	254 nm	USP 23, numerous TLC and HPLC methods for wide array of formulations; AOAC methods 988.26/0.27; compatability [442]; HPLC [458,832–836, 859]	[5]
Dexbrompheniramine Tablets	TLC, GC& HPLC	—	—	—	—	Compendial evaluation	[969]
Dexchlorpheniramine Bulk	TLC	Ethyl acetate	Silica	Isopropanol–10% ammonium hydroxide (7:3)	UV	Impurities	[925]
Dextromethorphan (Racemethorphan) Bulk	TLC	Dissolve in methanol	Silica	Ammonium hydroxide–methylene chloride–methanol–toluene–ethyl acetate (2:10:13:55:20)	Dragendorff	EP, pt. II-6, p. 20–2. Also BP, addendum 1983, p. 224 [4]; [1,8,925]	[6]
Bulk	HPLC	Dissolve in water	C-18	Buffer–water (7:3), pH 3.4	280 nm	USP 23, p. 482; resolved from bamipine and phenylephrine [424]	[5]
Syrup	HPLC	Dilute with water	C-18	Buffer–0.007M DCS–0.007M ammonium nitrate	—	—	
Bulk	HPLC	—	Silica	Methylene chloride–methanol (1% ammonium hydroxide) (9:1)	254 nm	Also [3]; GC [2,8]	[9]

Drug	Mode	Sample pretreatment	Sorbent (temp.)	Mobile phase	Detection	Comments	Reference
Dextromoramide							
Bulk	TLC	—	Silica	S_1 = EtOAc–MeOH–conc. NH_4OH (85 : 10 : 4) S_2 = MeOH S_3 = MeOH–1-butanol (3 : 2)–0.1M NaBr S_4 = MeOH–conc. NH_4OH (100 : 1.5) S_5 = cyclohexane–toluene–diethylamine (75 : 15 : 10) S_6 = $CHCl_3$–MeOH (9 : 1) S_7 = acetone	—	hR_{f1} = 78, hR_{f2} = 71, hR_{f3} = 78, hR_{f4} = 76, hR_{f5} = 40, hR_{f6} = 71, hR_{f7} = 61; also GC, SE-30/OV-1, RI: 2940	[132]
Tablets	TLC	Mix powder with methanol, centrifuge	Silica	Methanol	Dragendorff	BP, addendum 1983, p. 300; [8]	[4]
Dextropropoxyplene (Propoxyphene)							
Bulk	HPLC	—	Silica	Methanolic 0.01M ammonium perchlorate, pH adjusted to 6.7	—	TLC [1]; GC [2]; HPLC [970]; impurities [430]	[3]
Bulk	TLC	—	Silica	S_1 = EtOAc–MeOH–conc. NH_4OH (85 : 10 : 4) S_2 = MeOH S_3 = MeOH–1-butanol (3 : 2)–0.1M NaBr S_4 = MeOH–conc. NH_4OH (100 : 1.5) S_5 = cyclohexane–toluene–diethylamine (75 : 15 : 10) S_6 = $CHCl_3$–MeOH (9 : 1) S_7 = acetone	—	hR_{f1} = 80, hR_{f2} = 50, hR_{f3} = 63, hR_{f4} = 68, hR_{f5} = 59, hR_{f6} = 55, hR_{f7} = 54; also GC, SE-30/OV-1, RI: 2188; also GC [264]; GC, TMS derivative [705, p. 55]; HPLC [567]	[132]

Compound	Method	Detection	Stationary phase	Mobile phase		Results / Notes	Ref.
Diamorphine (Heroin) Bulk	HPLC	—	Silica	Methanolic 0.01M ammonium perchlorate, pH adjusted to 6.7	—	TLC [1]; GC [2]; aq. stability [437]; studies on 38 TLC systems [256]	[3]
Diazepam Bulk	TLC	—	Silica	S_1 = $CHCl_3$–acetone (8:2) S_2 = EtOAc S_3 = $CHCl_3$–MeOH (9:1) S_4 = EtOAc–MeOH–conc. NH_4OH (85:10:5)	—	hR_{f1} = 58, hR_{f2} = 48, hR_{f3} = 72, hR_{f4} = 76; also GC, SE-30/OV-1, RI: 2425	[132]
Bulk	TLC	—	Silica	S_1 = EtOAc–MeOH–conc. NH_4OH (85:10:4) S_2 = MeOH S_3 = MeOH–1-butanol (3:2)–0.1M NaBr S_4 = MeOH–conc. NH_4OH (100:1.5) S_5 = cyclohexane–toluene–diethylamine (75:15:10) S_6 = $CHCl_3$–MeOH (9:1)	—	hR_{f1} = 76, hR_{f2} = 82, hR_{f3} = 85, hR_{f4} = 75, hR_{f5} = 23, hR_{f6} = 73, hR_{f7} = 59; also GC, SE-30/OV-1, RI: 2425; GC/HPLC [454]	[132]
Bulk	HPLC	—	C-18, HS, Perkin-Elmer, 3 μm	S_7 = acetone S_1 = MeOH, H_2O (7:3) S_2 = 5 mM PO_4 (pH 6)–MeOH–CH_3CN (57:17:26) Flow: 1.5 ml/min	254 nm	Resolved 19 benzo-diazepines, RRT$_1$: 1.198, RRT$_2$: 0.879 (azinphosmethyl, 1.00)	[134]
Bulk	HPLC	—	C-18, Hypersil, 5 μm; flow rate 1.5 ml/min.	S_1 = MeOH, H_2O, PO_4 (0.1M) (55:25:20), pH 7.25 S_2 = MeOH, H_2O, PO_4 (0.1M) (7:1:2), pH 7.67	—	Cap. Fact. 1: 9.47 Cap. Fact. 2: 2.13	[135]

Drug	Mode	Sample pretreatment	Sorbent (temp.)	Mobile phase	Detection	Comments	Reference
	HPLC		Silica, Spherisorb 5W, 2 ml/min.	S$_3$ = MeOH–perchloric acid (1L: 100 μl)		Cap. Fact. 3: 1.75	[135]
				S$_4$ = MeOH, H$_2$O, TFA (997 : 2 : 1)		Cap. Fact. 4: 3.59	
Bulk	TLC	Dissolve in acetone	Silica	Heptane–ethyl acetate (1 : 1)	254 nm	USP 23, p. 489, identification and related compounds, TLC and GC review [971]; also BP, vol. II, p. 529 [4] and EP, pt. II, p. 22 [8]	[5]
Capsules	HPLC	Shake powder with methanol	C$_{18}$	Methanol–water (65 : 35)	254 nm	USP 23, ISTD– sulfanilamide; [3,8,9]; GC [2,8,9]; with otilonium [432]; acid degradants [441]	[5]
Extended-release capsules	HPLC	Same as above plus centrifugation and filteration				For extended–release capsules, ISTD– ethylparaben. Injection, ISTD– tolualdehyde; for tablets, ISTD–	
Injection	HPLC	Dilute in methanol				ethylparaben; HPLC [922]; stability [923]; tablets TLC [426] GC [264, 273]; CE [691,723]; TLC [268, 276]; ToxiLab TLC [564]; TLC/GC [758]; HPLC [146,406,556, 567, 569,699,858]	
Tablets	HPLC	Same as for capsules				GC/MS [705, p.55]; pK and spectral data [937]; color assay for ketobenzodiazepine derivatives [1118];	

Compound	Technique	Sample preparation	Sorbent	Mobile phase	Detection	Notes / references	Ref.
Dibenzepin Bulk	HPLC	—	Silica	Methanolic 0.01M–ammonium perchlorate, pH adjusted to 6.7	—	TLC [1, 8]; GC [2,8]; HPLC [8]	[3]
Diclofenac Bulk	TLC	—	Silica	S_2 = CHCl$_3$–acetone (8:2); S_2 = EtOAc; S_3 = CHCl$_3$–MeOH (9:1); S_4 = EtOAc–MeOH–conc. NH$_4$OH (85:10:5)	—	hR_{f1} = 25, hR_{f2} = 40, hR_{f3} = 47, hR_{f4} = 13; also GC, SE-30/OV-1, RI: 2771	[132]
Bulk	TLC	—	Silica	Chloroform–ethanol (4:1)	Chromic acid	GC & HPLC, p. 533 [937]; HPLC [427,431,444,972]; TLC [434]	[1]
Dicloxacillin Capsules with amoxycillin	HPLC	Extract powder with H$_2$O, filter	C$_8$	MeOH–0.02M Acetate (pH 5.0) (1:1), 1 ml/min	230 nm	ISTD–pencillin; HPLC cyano [171]; capsules [456]	[438]
Diethylpropion (Amfepramone) Bulk	HPLC	—	Silica	Methanolic 0.01M ammonium perchlorate, pH adjusted to 6.7	—	TLC [1,9]; GC [2,9]; HPLC [9]; stability [923]; ToxiLab [564]	[3]
Digitalis Powder	TLC	Mix, shake and boil with EtOH–lead acetate soln., decant supernatant, extract with CHCl$_3$, centrifuge, evaporate, dissolve in MeOH–CHCl$_3$ (1:1)	Silica	EtOAc–MeOH–H$_2$O (30:4:3)	Chloramine T	USP 23, p. 512; TLC [440]; HPLC [844–847]	[5]
Digitoxin Bulk	TLC	MeOH–DMF–CHCl$_3$ (2:2:1)	Silica	Cyclohexane–AcOH–acetone (49:49:2)	Chloramine T	USP 23, JP, p. 201 [7], EP, pt. II–2, p.78 [6];	[5]

Drug	Mode	Sample pretreatment	Sorbent (temp.)	Mobile phase	Detection	Comments	Reference
Dihydralazine							
Bulk	TLC	—	Silica (sprayed with 0.1*M* MeOH–KOH)	NH$_4$OH–MeOH (1.5 : 100)	Char	TLC [25]; impurities [24]; GC [2,8]; HPLC [848,849]	[1]
						HPLC [8,443]	
Dihydrocodeine							
Bulk	TLC	—	Silica	S$_1$ = EtOAc–MeOH–conc. NH$_4$OH (85 : 10 : 4)	—	hR$_{f1}$ = 27, hR$_{f2}$ = 11, hR$_{f3}$ = 19, hR$_{f4}$ = 26, hR$_{f5}$ = 8, hR$_{f6}$ = 13, hR$_{f7}$ = 2; also GC, SE-30/OV-1, RI: 2365	[132]
				S$_2$ = MeOH			
				S$_3$ = MeOH–1-butanol (3 : 2)–0.1*M* NaBr			
				S$_4$ = MeOH–conc. NH$_4$OH (100 : 1.5)			
				S$_5$ = cyclohexane–toluene–diethylamine (75 : 15 : 10)			
				S$_6$ = CHCl$_3$–MeOH (9 : 1)			
				S$_7$ = acetone			
Bulk	HPLC	—	Silica	Methylene chloride–methanol (1%)	254 nm	Also GC [2,8]; TLC & HPLC [1, 8]; ToxiLab TLC [564]; GC, TMS derivative [705, p. 55]	[9]
Dihydromorphine							
Bulk	HPLC	—	Silica	Methylene chloride–methanol (1% ammonium hydroxide) (4 : 1)	254 nm	Also [3]; TLC [1]; GC [2,9]	[9]
Diltiazem							
Bulk	HPLC	Derivatize with S (-)-*N*-1-(2-naphthyl)-sufonyl)–2–pyrrolidine–carbonyl chloride	Silica	Methylene chloride–ethyl acetate (100 : 9)	254 nm	Resolution of enantiomers; [455,703,922]; tablets [428]	[973]

Compound / Form	Method	Treatment	Column	Mobile phase	Detection	Notes / other methods	Ref.
Dimethindene Bulk	HPLC	—	Silica	Methylene chloride–methanol (1% ammonium hydroxide) (4 : 1)	254 nm	TLC [1]; GC [2,10]; HPLC [3,922]; enantiomers [919]	[10]
Diphenoxylate hydrochloride Tablets with atropine sulfate	HPLC	Dissolution	C-8	Methanol–potassium phosphate, dibasic 0.05M (73 : 27)	210 nm	USP 23, TLC, BP, p. 162 [4]; GC [2,974]; HPLC [3,975]	[5]
Diphenyldramine Tablets	HPLC	Powder extracted with diluent	C_8	CH_3CN–5 mM HSA–AcOH (70 : 30 : 1), 1.8 ml/min	258 nm	Peak purity, methods reviewed; LC–MS [425]; HPLC–diode array [118]; ToxiLab TLC [564]	[423]
Diphenylpyraline hydrochloride Bulk	GC	Dissolve in water render alkaline, extract with chloroform	—	—	FID	BP, p. 162, ISTD–dibenzyl	[4]
Bulk	HPLC	—	Silica	Methylene chloride–methanol (1% ammonium hydroxide) (9 : 1)	254 nm	TLC [1]; GC [2,10]; HPLC [3]	[10]
Dipipanone hydrochloride Tablets	TLC	(See TLC procedure for cyclizine)				BP, addendum 1983, p. 300 and vol. II, p. 604; TLC [1]; GC [2]; HPLC [3,567]	[4]
Dipivefrin Ophthalmic formulations	HPLC	No treatment	C_{18}, μBondapak	CH_3CN–1% DDS–AcOH (51 : 46 : 3), 2 ml/min	254 nm	MS, preparative TLC of degradants	[451]
Dirithromycin Bulk	HPLC	—	C_{18}, Hypersil	CH_3CN–5-mM PO_4, pH 7.5 (44 : 19 : 37)	205 nm	Purity, related substances	[439]

Drug	Mode	Sample pretreatment	Sorbent (temp.)	Mobile phase	Detection	Comments	Reference
Disopyramide phosphate							
Bulk	TLC	Dissolve in methanol	Silica	Toluene–ammonium hydoxide–ethanol (dehydrated) (170 : 2 : 28)	Spray bismuth iodide solution	USP 23, p. 541, identification,	[5]
Capsules	TLC	Mix with methanol and filter				chromatographic purity, also, BP, p. 164 and p. 530 [4], [1,8]; GC [2,8]; HPLC [3,11,976]; enantiomers [919]; CE [691]	
Docusate							
Soft gelatin capsules	HPLC	H_2O–MeOH, filter	C_{22} Chromega-bond	aq. $0.005M$ PIC A–CH_3CN (3 : 7)	214 nm	Resolved from progester-one (ISTD), methods reviewed	[452]
Doretinel							
Alcohol gel	HPLC	SPE, C_{18}	OSD-2, Spherisorb	MeOH–THF–$0.025M$ PO_4 (5 : 1 : 4), 1.5 ml/min	300 nm	Retinoid stability	[457]
Dothiepin							
Tablet	CE	Extract powder with CH_3CN–H_2O (3 : 2), filter	Fused silica, 72 cm x 50 μm	$50\ mM\ PO_4$ w/$10\ mM$ β-cyclodextrin-1-propanol (9 : 1)	220 nm	Resolved cis, trans isomers, isomers [450]	[429]
Doxepin							
Bulk	TLC	—	Silica	S_1 = EtOAc–MeOH–conc. NH_4OH (85 : 10 : 4) S_2 = MeOH S_3 = MeOH–1-butanol (3 : 2)–$0.1M$ NaBr S_4 = MeOH–conc. NH_4OH (100 : 1.5) S_5 = cyclohexane–toluene–diethylamine (75 : 15 : 10) S_6 = $CHCl_3$–MeOH (9 : 1) S_7 = acetone	—	hR_{f1} = 65, hR_{f2} = 24, hR_{f3} = 45, hR_{f4} = 51, hR_{f5} = 52, hR_{f6} = 37, hR_{f7} = 13; also GC, SE-30/OV-1, RI: 2217; E/Z [436]	[132]

Substance	Technique	Sample solvent	Stationary phase	Mobile phase	Detection	Remarks	Ref.
Bulk	HPLC		—	Silica	Methylene chloride–methanol (1% ammonium hydroxide) (98:2)	254 nm	Also [3]; GC [2,8,9, 977,978]; TLC [1,8]; ToxiLab TLC [564]; GC [706, p. 163]; HPLC [858]; resolved isomers, related compounds, volatile impurities, capsules [433]	[9]
Doxorubicin (Adriamycin) Bulk/injection	HPLC	Mobile phase	C_{18}	H_2O–CH_3CN (adj. pH 2) (69:31)	254 nm	ISTD–2-naphthalene sulfonic acid, USP 23; TLC [1]; HPLC [461,850–854,856, 857,860]	[5]
Doxycycline Bulk	TLC	MeOH	Silica spray with 10% EDTA, pH 9.0, Machery–Nagel	CH_2Cl_2–MeOH–H_2O (59:35:6)	280 nm	Methods reviewed, correlated to HPLC; CE [445]; USP 23 [5]; BP [4]; EP, pt. II–7 [6]	[460]
Dronabinol						Names: Soft gelatin capsule (-)-delta-9-(trans) tetrahydrocannabinol; [129]	
Droperidol Bulk	TLC	—	—	Acetone	Dragendroff	Review; TLC [1,8]; GC [2]; HPLC [3,8]	[979]
Bulk	TLC	—	—	S_1 = EtOAc–MeOH–conc. NH_4OH (85:10:4) S_2 = MeOH S_3 = MeOH–1-butanol (3:2)–0.1M NaBr S_4 = MeOH–conc. NH_4OH (100:1.5)	—	hR_{f1} = 59, hR_{f2} = 71, hR_{f3} = 73, hR_{f4} = 67, hR_{f5} = 2, hR_{f6} = 48, hR_{f7} = 36; also GC, SE-30/OV-1, RI: 3430	[132]

Drug	Mode	Sample pretreatment	Sorbent (temp.)	Mobile phase	Detection	Comments	Reference
Egonine Bulk	HPLC	—	Silica	Methanolic 0.01M ammonium perchlorate, pH adjusted to 6.7	—	TLC [1]	[3]
Bulk	GC	TMS derivative	DB-5, J&W, 20 m × 1.8 mm, 0.4 μM, 200°C (1 min)—300°C at 10°C/min	—	MS	Resolved from cocaine, ecgonine methyester, and benzoylecgonine	[706, p.1]
Econazole Powder	HPLC	Extract with MeOH	C_{18}, Nova-Pak	MeOH–THF–TEA (0.1M, pH 7.0) (70:12:18)	230 nm	Methods reviewed, creams, resolved from miconazole	[463]
Enalapril Tablets	CE	H_2O	Silica, 57 cm × 75 μm	50 mM PO_4, pH 8.25, 16.1 kV	200 nm	Stability indicating; HPLC [471]	[470]
Ephedrine Bulk	TLC	—	Silica	S_1 = EtOAc–MeOH–conc. NH_4OH (85:10:4) S_2 = MeOH S_3 = MeOH–1-butanol (3:2)–0.1M NaBr S_4 = MeOH–conc. NH_4OH (100:1.5) S_5 = cyclohexane–toluene–diethylamine (75:15:10) S_6 = $CHCl_3$–MeOH (9:1) S_7 = acetone	—	hR_{f1} = 27, hR_{f2} = 10, hR_{f3} = 64, hR_{f4} = 30, hR_{f5} = 5, hR_{f6} = 5, hR_{f7} = 1, also GC, SE-30/OV-1, RI: 1363	[132]

Herbal extract	CE	1 g Ephedra herba extracted 50% EtOH (15 ml) RT, 30 min, centrifuged, filtered ISTD–benzyltriethyl ammonium chloride	Fused silica, 60 cm × 75 μm ID, 28 kV	0.02M isoleucine–0.005M barium hydroxide (pH 10)	185 nm	Resolved: ISTD: 4 min; methylpseudoephedrine: ~5 min; pseudoephedrine: 6.2 min; ephedrine: 6.8 min; methylephedrine: 7.8 min; norpseudoephedrine: 8.1 min; norephedrine: 8.8 min HPLC studies [567,569]; studied on 38 TLC systems [256,257]; HPLC enantiomers [274,703]; enantiomers [340]; GC/MS [351]; GC [404]; pK and spectral data [937]	[131]
Ephedrine hydrochloride Oily formulations	HPLC	ISTD, acetic acid and chloroform, vortex, inject. aq.	C-8	Acetonitrile–0.1M dihydrogen phosphate	214 nm	Also USP 23, p. 589 [5]; GC [2,8,9,279–282, 980–982]; HPLC [3,8,9,984–986]; TLC [1,925]; enant. [464,919]	[323]
Epicillin Bulk	HPLC		Cyano Nucleosil	0.05M (pH 7) PO4–MeOH (7 : 3. 6)	220 nm	Resolved from ampicillin, oxacillin, flucloxacillin, dicloxacillin	[171]
Epinephrine Bulk	CE	Dilute in 0.01M HCl	Silica, 50 cm × 75 μm	10mM Tris–18mM Me, β-CD, pH 2.4, 15 kV	206 nm	Resolved d/l forms; HPLC [500]	[463]
Ergocalciferol (Vitamin D$_2$) Bulk	TLC	Dissolve in CHCl$_3$	Silica	CHCl$_3$–MeOH (9 : 1)	Spray 0.1% bromocresol purple	USP 23, p. 602; [863]	[5]
Ergotamine Bulk	TLC	—	Silica	S$_1$ = EtOAc–MeOH–conc. NH$_4$OH (85 : 10 : 4) S$_2$ = MeOH S$_3$ = MeOH–1-butanol (3 : 2)– 0.1M NaBr	—	$hR_{f1} = 42$, $hR_{f2} = 67$, $hR_{f3} = 64$, $hR_{f4} = 63$, $R_{f5} = 1$, $hR_{f6} = 34$, $hR_{f7} = 22$; also GC, SE-30/OV-1, RI: 2366	[132]

Drug	Mode	Sample pretreatment	Sorbent (temp.)	Mobile phase	Detection	Comments	Reference
				S₄ = MeOH–conc. NH₄OH (100 : 1.5) S₅ = cyclohexane–toluene–diethylamine (75 : 15 : 10) S₆ = CHCl₃–MeOH (9 : 1) S7 = acetone			
Bulk	TLC	Dissolve in methanol–chloroform (1 : 9)	Silica	Dimethylformamide–ether–chloroform–ethanol (dehydrated) (15 : 70 : 10 : 5)	Spray DMAB	USP 23, p. 606, related alkaloids prior to development place each spot over ammonia; also ED, pt. II–6, p. 224 [6] and BP, vol. II, p. 765 [4]; TLC [1]; GC [2]; HPLC [3,986,987]; GC [988]; stability [923]	[5]
Erthomycin Bulk	HPLC	MeOH–0.2M PO₄ (pH 7.0) (1 : 1)	RPR-1, Hamilton, 70°C	CH₃CN–2-methyl-2-propanol–0.2M PO₄ (pH 9.0)–H₂O (3 : 16.5 : 5 : 75.5)	215 nm	Resolved from impurities; HPLC [474,476,477,489]; [4,6,865,867–871,874–877]	[472]
Estradiol valerate Bulk	HPLC	MeOH, sonicate, refrigerate, 4 hr	C₈, Zorbax	MeOH–H₂O (85 : 15), 1 ml/min	222 nm	Resolved from norethisterone;	[482]
Estazolam Bulk	HPLC	—	C-18, Hypersil, 5 μm; Flow rate 1.5 ml/min Silica, Spherisorb 5W, 2 ml/min.	S₁ = MeOH, H₂O, PO₄ (0.1M) (55 : 25 : 20), pH 7.25 S₂ = MeOH, H₂O, PO₄ (0.1M) (7 : 1 : 2), pH: 7.67 S₃ = MeOH–perchloric acid (1L : 100 μl) S₄ = MeOH, H₂O, TFA (997 : 2 : 1)	—	Cap. Fact. 1: 3.80 Cap. Fact. 2: — Cap. Fact. 3: 1.47 Cap. Fact. 4: 1.68	[135] [135]

Compound	Technique	Sample preparation	Column/Phase	Mobile phase/Solvent	Detection	Comments	Ref.
Estrogen, conjugated Bulk/tablets	HPLC	ISTD–ethyl paraben, powder, glass beads, mobile phase	C_8, Zorbax	PO_4 (6.8 g/730 ml)–CH_3CN (730 : 270), 1 ml/min	210 nm	Method review, resolve equilin sulfate, estrone sulfate along with other conjugates; history [478]; fluorodensitometry [480]; USP 23 [5]; HPLC [880–882]; [483,878]; USP 23 [5]; impurities [24]	[479]
Ethambutol Bulk	HPLC	Perbenzoyl derivative	D-phenyl glycine (Pirkle column)	n-Hexane–IPA (75 : 25), 1 ml/min	235 nm	Resolved stereoisomers; [468,469]; BP [4]; JP [7]; GC review [884]	[465]
Ethanol Cough Syrup	HPLC	—	Fast analysis column (bio Rad)	—	RI	See USP 23 [5] for GC methods; [661]	[462]
Ethinamate Bulk	TLC	—	Silica	S_1 = $CHCl_3$–acetone (8 : 2) S_2 = EtOAc S_3 = $CHCl_3$–MeOH (9 : 1) S_4 = EtOAc–MeOH–conc. NH_4OH (85 : 10 : 5)	—	hR_{f1} = 49, hR_{f2} = 59, hR_{f3} = 58, hR_{f4} = 72; also GC, SE-30/OV-1, RI: 1363	[132]
Bulk	TLC	Dissolve in acetone	Silica	Methanol–chloroform–acetic acid (15 : 8 : 1)	Potassium permanganate	Impurities; TLC, GC, & HPLC [1,2,8,9]	[24]
Ethionamide Tablets	TLC	Shake powder with methanol, centrifuge	Silica	Chloroform–methanol (9 : 1)	254 nm	BP, vol. II, p. 768, addendum 1983, p. 230 and EP, pt. II–4, p. 141; [1]; GC [2] impurities [925]	[4]
Ethisterone Bulk	TLC	Dissolve in EtOH–$CHCl_3$ (1 : 3)	Silica	$CHCl_3$–MeOH (95 : 5)	Char	EP, pt. II–4, p. 142; BP [4]; HPLC [887]	[6]

Drug	Mode	Sample pretreatment	Sorbent (temp.)	Mobile phase	Detection	Comments	Reference
Ethosuximide							
Bulk	TLC	—	Silica	S_1 = CHCl$_3$–acetone (8 : 2) S_2 = EtOAc S_3 = CHCl$_3$–MeOH (9 : 1) S_4 = EtOAc–MeOH–conc. NH$_4$OH (85 : 10 : 5)	—	hR$_{f1}$ = 50, hR$_{f2}$ = 56, hR$_{f3}$ = 59, hR$_{f4}$ = 58; also GC, SE-30/OV-1, RI : 1206	[132]
Bulk	GC	Dissolve in chloroform	USP-G5 (140°C), Silica	—	FID	USP 23, p. 642	[5]
Bulk	HPLC	—	Silica	Methylene chloride–methanol (1% ammonium hydroxide) (99 : 1)	254 nm	impurities; BP, vol. II, p. 555, assay [4] TLC [1,8]; GC [2,8]; HPLC [8]	[9]
Bulk	GC	—	Cyclodex-B, 30 m × 0.25 mm, 0.2 μm, 160°C	—	FID, 300°C, N$_2$, 30 ml/min		[706, p.158]
Ethybenztropine (Benzhydryl ether)						Expected to migrate, chromatographically by HPLC similar to *N*-ethylamphetamine	
***N*-Ethylamphetamine**							
Bulk	HPLC	—	Silica, 10 μm, 30 cm × 4 mm	MeOH (1% NH$_4$OH)–CH$_2$Cl$_2$ (2 : 8)	254 nm	GC described, also TLC [267]; *N*-ethylamphetamine, *N*, *N*-dimethylamphetamine (also known as *N*,*N*-alpha-trimethylbenzeneethanamine: *N*,*N*,alpha-trimethylphenethyl-amine)	[10]
Ethylenediaminetetracetic acid (Edetic acid)							
Bulk	HPLC	Add cupric nitrate	C$_8$	TBH (10 ml 25% MeOH)–MeOH–H$_2$O (10 : 90 : 900), pH 7.5	254 nm	Disodium form USP 23, p. 571; opthalmic HPLC [478]	[5]

Drug/Form	Technique	Sample preparation	Column/Phase	Mobile phase	Detection	Comments	Ref.
Ethylestrenol Bulk	GC	—	3% OV-1, Chromosorb, 280°C, 4" × 1/4"	N_2, 32 ml/min	FID	MS, NMR, IR described; NMR [773]	[10]
Ethyl loflazepate Bulk/tablet	HPLC	Powder, add MeOH, sonicate, add mobile phase	MCH-10, Micropak	CH_3CN–H_2O (1 : 1), 1 ml/min	230 nm	TLC also evaluated; benzodiazepine; HPLC [774] GC–MS, urine [778]	[485]
Etorphine Bulk	HPLC	—	Silica	Methanolic 0.01M ammonium perchlorate, pH adjusted to 6.7	—	TLC [1]	[3]
Etoxeridine (Carbetidine) Famotidine Parenterals	HPLC	Mobile phase	Silica	MeOH–H_2O–0.05M PO_4 (10 : 4 : 16)	254 nm	Resolved from parabens & degradants; stability [494,504]; LC–MS [503]; USP 23, p. 652 [5]	[888]
Fencamfamin(e) Bulk	HPLC	—	Silica	Methanolic 0.01 M ammonium perchlorate, pH adjusted to 6.7	—	TLC [1,8]; GC [2,8]	[3]
Fenethylline Bulk	HPLC	—	Silica	Methylene chloride–methanol (1% ammonium hydroxide) (98 : 2)	254 nm	GC [2,9]; HPLC, p. 613 [937]; [(±)cis-4-methylaminorex (±)cis-4,5-dihydro-4-methyl-5-phenyl-2-oxazolamine)]	[9]
Fenfluramine Tablets	TLC	Shake powder with cholorform, filter	Silica	Methanol–13.5 M ammonium hydroxide	Dragendorff	BP, vol. II, p. GC assay, p. 770; TLC, GC, & HPLC, p. 614 [937]; HPLC & GC [9,264]; CE [501]	[4]

Drug	Mode	Sample pretreatment	Sorbent (temp.)	Mobile phase	Detection	Comments	Reference
Fenoterol Tablets	CE	Sonicate powder in H_2O	Silica, 50 cm × 75 μm	0.01M Tris, pH 5.0, 12.5 kV	214 nm	Compared to isotachophoresis & HPLC, review, resolved salbutamol/ terbutaline	[491]
Fenproporex Bulk	GC	—	USP-G2 (100–300°C)	—	FID	TLC, GC, & HPLC [8]	[2]
Fentanyl Bulk	TLC	—	Silica	S_1 = EtOAc–MeOH–conc. NH_4OH (85:10:4) S_2 = MeOH S_3 = MeOH–1-butanol (3:2)–0.1M NaBr S_4 = MeOH–conc. NH_4OH (100:1.5) S_5 = cyclohexane–toluene–diethylamine (75:15:10) S_6 = $CHCl_3$–MeOH (9:1) S_7 = acetone	—	hR_{f1} = 75, hR_{f2} = 69, hR_{f3} = 77, hR_{f4} = 70, hR_{f5} = 45, hR_{f6} = 74, hR_{f7} = 58; also GC, SE-30/OV-1, RI: 2650	[132]
Bulk	HPLC	—	Silica, 30 cm × 4 mm, 10 μm	MeOH (1% NH_4OH)– CH_2Cl_2 (2:98)	254 nm	UV, MS, NMR, IR, & GC data	[10, p. 928]
Injection	GC	Dissolve in water, render alkaline, extract with chloroform	USP-G2 (240°C)	—	FID	USP 23, p. 654, [ISTD–papaverine; TLC [1,8]; GC [2,8,9]; HPLC [3,8,9]; GC [243,705, p. 57, 706, p. 159]; pK and spectral data [937]; homologs [189]	[5]
Bulk	HPLC	—	Silica, 30 cm × 4 mm, 10 μm	MeOH (1% NH_4OH)– CH_2Cl_2 (2:98)	254 nm	UV, MS, NMR, IR, & GC data	[11, p. 930]

	Method	Sample preparation	Column	Mobile phase / Solvent	Detection	Comments	Ref.
Flucytosine Capsules	HPLC	Extract powder with H_2O, filter	C_{18}, Pharmex	$0.05M$ PO_4, pH 4.5, 1 ml/min	300 nm	ISTD-5-aminouracil, stability; review [499]; USP 23; [889,890]	[493]
Fludiazepam						Structurally similar to flunitrazepam expected to have similar chromatographic behavior	
Flumcinol Bulk	HPLC	—	C_{18}, Lichrosorb	$MeOH$–H_2O (7:3)	PDA, 210 nm	Review, impurities, also chiral separation, compared to GC	[492]
Flunitrazepam Bulk	TLC	—	Silica	S_1 = $CHCl_3$–acetone (8:2) S_2 = EtOAc S_3 = $CHCl_3$–MeOH (9:1) S_4 = EtOAc–MeOH–conc. NH_4OH (85:10:5)	—	hR_{f1} = 54, hR_{f2} = 48, hR_{f3} = 72, hR_{f4} = 76; also GC, SE-30/OV-1, RI: 2645	[132]
Bulk	TLC	—	Silica	S_1 = EtOAc–MeOH–conc. NH_4OH (85:10:4) S_2 = MeOH S_3 = MeOH–1-butanol (3:2)–$0.1M$ NaBr S_4 = MeOH–conc. NH_4OH (100:1.5) S_5 = cyclohexane–toluene–diethylamine (75:15:10) S_6 = $CHCl_3$–MeOH (9:1) S_7 = acetone	—	hR_{f1} = 76, hR_{f2} = 79, hR_{f3} = 82, hR_{f4} = 63, hR_{f5} = 10, hR_{f6} = 72, hR_{f7} = 63; also GC, SE-30/OV-1, RI: 2645	[132]
Bulk	HPLC	—	C-18, HS, Perkin-	S_1 = MeOH, H_2O (7:3)	254 nm	Resolved 19 benzo-diazepines, RRT_1 :	[134]

346

Adamovics

Drug	Mode	Sample pretreatment	Sorbent (temp.)	Mobile phase	Detection	Comments	Reference
			Elmer, 3μm	S_2 = 5mM PO_4 (pH6)–MeOH–CH_3CN (57:17:26) Flow: 1.5 ml/min		0.725, RRT_2: 0.510(azinphosmethyl, 1.00)	[135]
Bulk	HPLC	—	C-18, Hypersil, 5μm, flow rate 1.5 ml/min.	S_1 = MeOH, H_2O, PO_4 (0.1M)(55:25:20), pH 7.25	—	Cap. Fact. 1: 3.10	[135]
				S_2 = MeOH, H_2O, PO_4 (0.1M)(7:1:2), pH 7.67		Cap. Fact. 2: —	
			Silica, Spherisorb 5W, 2 ml/min.	S_3 = MeOH–perchloric acid (1L: 100 μl)		Cap. Fact. 3: 0.34	
				S_4 = MeOH, H_2O, TFA (997:2:1)		Cap. Fact. 4: 0.38	
Bulk	TLC	—	Silica (spray 0.1M methanolic potassium hydroxide)	Ammonium hydroxide–methanol (1.5:100)	Various reagents	TLC [8]; GC [2,8]; HPLC [8]	[1]
Fluocinolone acetonide Cream	TLC	Partition between H_2O–$CHCl_3$, centrifuge, discard aq. layer	Silica	$CHCl_3$–DEA (2:1)	Char	USP 23, p. 667— numerous formulations evaluated, also HPLC	[5]
Fluorouracil Bulk	HPLC	—	Silica	CH_2Cl_2–MeOH (1% NH_4OH) (95:5)	254 nm	USP 23 [5]; JP [7]; BP [4]; TLC review [891]; stability [1104]	[10]
Fluoxetine Capsules	HPLC	Mix powder with 0.1 N HCl, filter	C_{18} μBondapak	10 mM PO_4, 0.2% AcOH, 2 ml/min	234 nm	Stability indicating	[498]
Fluoxymesterone Bulk	TLC	—	Silica	S_1 = $CHCl_3$–acetone (8:2); S_2 = EtOAc; S_3 = $CHCl_3$–MeOH (9:1)	—	hR_{f1} = 3, hR_{f2} = 3, hR_{f3} = 41, hR_{f4} = 72; also GC, SE-30/OV-1, RI: 2785	[132]

	Technique	Column/system	Solvent system	Detection	Comments	Ref.
			S_4 = EtOAc–MeOH–conc. NH_4OH (85:10:5)			
Fluparoxan Bulk	CE	In electolyte	(10 mM borax, 10 mM Tris, 150mM β cyclodextrin, 6M urea)–IPA (8:2), pH 2.5, 16 kV	214 nm	1% limit of quantitation	[491]
Fluphenazine Bulk	HPLC	—	Silica	CH_2Cl_2–MeOH (1% NH_4OH) (4:1)	254 nm	TLC [1,8,24]; GC [2,8,9]; HPLC [3,1105]; USP 23 [5]; BP [4]: review [892]	[9]
Flurazepam Bulk	TLC	—	Silica	S_1 = EtOAc–MeOH–conc. NH_4OH (85:10:4) S_2 = MeOH S_3 = MeOH–1-butanol (3:2)–0.1M NaBr S_4 = MeOH–conc. NH_4OH (100:1.5) S_5 = cyclohexane–toluene–diethylamine (75:15:10) S_6 = $CHCl_3$–MeOH (9:1) S_7 = acetone	—	hR_{f1} = 72, hR_{f2} = 52, hR_{f3} = 45, hR_{f4} = 62, hR_{f5} = 30, hR_{f6} = 48, hR_{f7} = 40; also GC, SE-30/OV-1, RI: 2785	[132]
Bulk	HPLC	—	C-18, HS, Perkin-Elmer, 3 µm	S_1 = MeOH, H_2O (7:3) S_2 = 5mM PO_4 (pH6)–MeOH–CH_3CN (57:17:26) Flow: 1.5 ml/min	254 nm	Resolved 19 benzo-diazepines, RRT$_1$: 2.050, RRT$_2$: 1.329 (azinphosmethyl, 1.00)	[134]
Bulk	HPLC	—	C-18, Hypersil, 5 µm, flow rate 1.5	S_1 = MeOH, H_2O, PO_4 (0.1M) (55:25:20), pH 7.25	—	Cap. Fact. 1: — Cap. Fact. 2: 3.11	[135]

Drug	Mode	Sample pretreatment	Sorbent (temp.)	Mobile phase	Detection	Comments	Reference
			ml/min. Silica, Spherisorb 5W, 2 ml/min.	S_2 = MeOH, H_2O, PO_4 (0.1M) (7:1:2), pH 7.67; S_3 = MeOH–perchloric acid (1L: 100 μl); S_4 = MeOH, H_2O, TFA (997:2:1)		Cap. Fact. 3: 6.10 Cap. Fact. 4: 13.41	[135]
Bulk	TLC	Dissolve in methanol	Silica	Ammonium hydroxide–ethyl acetate (1:200)	254 nm	USP 23, p. 687, & capsules, ID by TLC and GC review [989] and BP, addendum 1983, p. 235 and p. 266 [4]; [1,8]	[5]
Bulk	HPLC	—	Silica	Methylene chloride–methanol (1% ammonium hydroxide) (98:2)	254 nm	GC [2,8,264]; HPLC [8,922]; GC [273,404]; TLC [276]; CE [723]; [496]	[9]
Folic acid Bulk	TLC	MeOH–NH₄OH (9:2)	Silica	EtOH–propanol–NH₄OH (3:1:1)	350 nm	EP, pt. II–2, p. 67–2 & BP, p. 202 [4]	[6]
Fosinopril Bulk	HPLC	Mobile phase	Silica	CH_3CN–H_2O–H_3PO_4 (4000:15:2)	205 nm	Related impurities	[497]
Fursemide Bulk	HPLC	Mobile phase	C₈, Nucleosil	0.02M PO_4–IPA–cetrimide (780:300; 2.5), pH 7.0, 1 ml/min	250 nm	Review, related impurities, also TLC; HPLC–EC [505]	[495]
Gentamicin Cream/eye drops	TLC	Disperse with CHCl₃, extract with H₂O	Silica	CHCl₃–MeOH–13.5M NH₄OH (1:1:1), lower layer	Ninhydrin	BP, vol. II, p. 547 & 610	[4]
Glutathione Pharmaceutical prep.	HPLC	H₂O, filtration	C₁₈, Spherisorb ODS-2	5 mM octylamine orthophosphate, pH 6.4, 1.5 ml/min	230 nm	Review, resolved oxidized glutathione; HPLC [506,507]	[507]

Sample	Technique	Sample preparation	Stationary phase	Mobile phase	Detection	Notes	Ref.
Glutethimide Bulk	TLC	—	Silica	S_1 = CHCl$_3$-acetone (8:2) S_2 = EtOAc S_3 = CHCl$_3$-MeOH (9:1) S_4 = EtOAc-MeOH-conc. NH$_4$OH (85:10:5)	—	hR_{f1} = 63, hR_{f2} = 62, hR_{f3} = 70, hR_{f4} = 78; also GC, SE-30/OV-1, RI: 1836	[132]
Bulk	TLC	Dissolve in methanol	Silica	Ethyl acetate–water (95:5)	Chlorine gas and potassium iodide	BP, vol. II, p. 211 and p. 774, review [990]; HPLC, USP 23 [5]; impurities [925]	[4]
Bulk	HPLC	—	Silica	Methylene chloride–methanol (1% ammonium hydroxide) (99:1)	254 nm	TLC, GC, & HPLC, p. 642 [937], [8,991]; GC [9,264,273,404,705,992]; HPLC [993]	[9]
Guaiphenesin Capsules/elixir	HPLC	H$_2$O, filter	C$_{18}$	10 mM PO$_4$–MeOH–CH$_3$CN (8:2:1), pH 5.5, 1.2 ml/min	245 nm	Methods reviewed, resolved from theophylline, compared to 3rd derivative UV; GC [2]	[508]
Guaizulene Ophthalmic	HPLC	Borate buffer (1.71 g boric acid/0.265 g borate) H$_2$O/100 ml, extract with light petroleum, residue dissolved in EtOH	Silica	n-Hexane–EtOAc (98:2), 1 ml/min	600 nm	—	[509]
Granulocyte colony stimulating factor Bulk	CE	Sialic acids removed with neuramindase	Silica, 100 cm × 75 μm	pH 8.0, 50 mM PO$_4$–50 mM borate, 30 kV	214 nm	Glycoforms resolved; peptide mapping [518]	[510]
Halazepam Bulk	HPLC	—	C-18, Hypersil, 5 μm, flow rate 1.5 ml/min.	S_1 = MeOH, H$_2$O, PO$_4$ (0.1M) (55:25:20), pH 7.25 S_2 = MeOH, H$_2$O, PO$_4$ (0.1M) (7:1:2), pH 7.67	—	Cap. Fact. 1: 16.46 Cap. Fact. 2: 2.82 Cap. Fact. 3: 1.05	[135] [135]

Drug	Mode	Sample pretreatment	Sorbent (temp.)	Mobile phase	Detection	Comments	Reference
Tablets	HPLC	Extract powder with methanol, centrifuge	Silica, Spherisorb 5W, 2 ml/min. C-18	S_5 = MeOH–perchloric acid (1L: 100 μl) S_4 = MeOH, H$_2$O, TFA (997:2:1) Water–acetonitrile (3:1)	254 nm	Cap. Fact. 4: 1.34 GC & HPLC [10,699]	[994]
Haloperidol Bulk	TLC		Silica	S_1 = EtOAc–MeOH–conc. NH$_4$OH (85:10:4) S_2 = MeOH S_3 = MeOH–1-butanol (3:2)–0.1M NaBr S_4 = MeOH–conc. NH$_4$OH (100:1.5) S_5 = cyclohexane–toluene–diethylamine (75:15:10) S_6 = CHCl$_3$–MeOH (9:1) S_7 = acetone		hR_{f1} = 74, hR_{f2} = 51, hR_{f3} = 75, hR_{f4} = 67, hR_{f5} = 10, hR_{f6} = 27, hR_{f7} = 33; also GC, SE-30/OV-1, RI: 2942	[132]
Bulk/solution	TLC	Dilute with methanol	Silica	Chloroform–methanol–13.5M ammonium hydroxide (92:8:1)	Dragendorff	BP, addendum 1983, p. 297, vol. II, p. 612 and p. 776: [1,8,925]	[4]
Bulk	HPLC	—	Silica	Methylene chloride–methanol (1% ammonium hydroxide) (95:5)	254 nm	GC [2,8]; HPLC [8,922,995,996]; TLC [564]	[3]
Heparin						CE [516]	
Heroin (Diamorphine/ diacetylmorphine) Bulk	CE	—	Fused silica; ID—75 μm, L—90 cm	75 mM SDS–6 mM Na$_2$B$_4$O$_7$–10 mM Na$_2$HPO$_4$ (pH 9.1), 20 kV, cathode detector side	195 nm	Fraction A: methaqualone. Fraction B: morphine, heroin, codeine, methaqualone, 6-acetyl morphine.	[130]

Substance	Method	Column / Phase	Solvent system / Conditions	Detection	Remarks	Ref.
				—	Fraction C: benzoyl-ecgonine, morphine, 6-acetyl morphine, codeine. Detection limit 100 ng/ml [748]; impurities [746]; history [244]; CE [265]; SCF [266]; GC [273,292–300,404,705 p. 55]; GC–IR [284]; general [280–283,285]; HPLC [81,84–92,105, 306,569]; IR [325]; TLC [656]; GC–MS [326–328]; poppy seed [747], see morphine; stability [517]	[132]
Hexobarbital (Hexobarbitone) Bulk	TLC	Silica	S_1 = CHCl$_3$–acetone (8 : 2); S_2 = EtOAc; S_3 = CHCl$_3$–MeOH (9 : 1); S_4 = EtOAc–MeOH–conc. NH$_4$OH (85 : 10 : 5)	—	$hR_{f1} = 65$, $hR_{f2} = 65$, $hR_{f3} = 69$, $hR_{f4} = 51$; also GC, SE-30/OV-1, RI: 1857 [697,705, p. 53]; GC, p. 54 [705]; EP, pt. II-5, p. 183-2; TLC, GC, HPLC p. 656 [937], [8,9,126,924, 938]; enantiomers [919]; GC [275,404]; bulk also by TLC, see cyclobarbital [6]	[604]
Human growth hormone, (HCH, somatotrophin) Dosage	HPLC	C-4, Vydac, 4.6 × 250 nm, 5 μm	29% 1-PrOH–0.5M phosphate, pH 6.5, 45°C, 1.0 ml/min	Fluorescence	(Somatotropin) column conditioned by injecting 20 × 10 μl of 5 mg/ml protein, resolved from N-methionyl recombinant human growth	[724]

Drug	Mode	Sample pretreatment	Sorbent (temp.)	Mobile phase	Detection	Comments	Reference
						hormone; also [760]; HPLC [767]; somatropin for injection [918]	
Human normal immunoglobulins							
Bulk	HPLC	Dilute with pH 7 buffer	Agarose	pH 7.0 buffer	280 nm	EP, pt. II–8, p. 338–4	[6]
Hydralazine							
Bulk	TLC	Dissolve with 0.1M methanolic hydrochloric acid	Silica	Ethyl acetate–13.5M ammonium hydroxide–hexane (2 : 2 : 8) upper layer	DMAB	BP, addendum 1983, p. 237; [1,999]	[4]
Bulk	HPLC	Dissolve in mobile phase	SCX	Acetonitrile–0.5 M potassium phosphate, dibasic adjust to pH 3.5 (55 : 45)	280 nm	USP 23, ISTD–methylparaben, assay, GC [1000]; HPLC [9,515,1001,1002]; stability [511]	[5]
Hydrochlorothiazide							
Bulk	HPLC	—	Silica	Methylene chloride–methanol (1% ammonium hydroxide) (9 : 1)	254 nm	HPLC [8,218,937,997, 922]; stability [923];USP 23, for TLC, EP, pt. II–9, p. 394 [6]; tablets [5]; also see methyldopa and timolol maleate; [515]	[9]
Hydrocodone							
Bulk	TLC	—	Silica	S_1 = EtOAc–MeOH–conc. NH_4OH (85 : 10 :4) S_2 = MeOH S_3 = MeOH–1-butanol (3 : 2)–0.1M NaBr S_4 = MeOH–conc. NH_4OH (100 : 1.5)		$hR_{f_1} = 31$, $hR_{f_2} = 11$, $hR_{f_3} = 13$, $hR_{f_4} = 25$, $hR_{f_5} = 4$, $hR_{f_6} = 20$, $hR_{f_7} = 4$; also, GC, SE-30/OV-1, RI: 2440	[132]

Form	Technique	Sorbent	Solvent systems	Detection	Applications / Remarks	Ref.
Bulk	HPLC	—	S_5 = cyclohexane–toluene–diethylamine (75:15:10) S_6 = $CHCl_3$–MeOH (9:1) S_7 = acetone			
		Silica	Methylene chloride–methanol (1% ammonium hydroxide) (95:5)	254 nm	TLC [1,8]; GC [2,8,9,848,849]; HPLC [3,8,996,998]	[9]
Hydrocortisone Bulk	HPLC	C_{18}, adsorbosphere	MeOH–H_2O (7:3)	254 nm	SCF creams, ointments [512]; hydrocortisone 17-butyrate [514]; cream SPE [513]; USP 23, p. 759, numerous methods and formulations [5]; impurities [24]; BP [4]; JP [6]; stability [1104]; [893–903]	[513]
Hydromorphone Bulk	TLC	Silica	S_1 = EtOAc–MeOH–conc. NH_4OH (85:10:4) S_2 = MeOH S_3 = MeOH–1-butanol (3:2)–0.1M NaBr S_4 = MeOH–conc. NH_4OH (100:1.5) S_5 = cyclohexane–toluene–diethylamine (75:15:10) S_6 = $CHCl_3$–MeOH (9:1) S_7 = acetone	—	hR_{f1} = 18, hR_{f2} = 12, hR_{f3} = 14, hR_{f4} = 23, hR_{f5} = 3, hR_{f6} = 9, hR_{f7} = 2; also GC, SE-30/OV-1, RI: 2467	[132]

Drug	Mode	Sample pretreatment	Sorbent (temp.)	Mobile phase	Detection	Comments	Reference
Bulk	HPLC	—	Silica	Methylene chloride–methanol (1% ammonium hydroxide) (4:1)	254 nm	TLC [1,8,564]; GC [2,8,9]; HPLC [3,8,9]; IR, UV, MS, NMR, and IR [10, p. 1126]; GC, TMS derivative [705, p. 55]	[9]
Hydroxyzine Bulk	TLC	—	Silica	S_1 = EtOAc–MeOH–conc. NH$_4$OH (85:10:4) S_2 = MeOH S_3 = MeOH–1-butanol (3:2)–0.1 M NaBr S_4 = MeOH–conc. NH$_4$OH (100:1.5) S_5 = cyclohexane–toluene–diethylamine (75:15:10) S_6 = CHCl$_3$–MeOH (9:1) S_7 = acetone	—	hR_{f1} = 53, hR_{f2} = 56, hR_{f3} = 65, hR_{f4} = 68, hR_{f5} = 9, hR_{f6} = 54, hR_{f7} = 19; also GC, SE-30/OV-1, RI: 2049	[132]
Bulk	TLC	Dissolve in a mixture of sodium hydroxide solution–acetone (1:1)	Silica	Toluene–ethanol–ammonium hydroxide (150:95:1)	Iodoplatinate	JP, p. 281	[7]
Syrup	TLC	Dilute with methanol	Silica	Toluene–ethanol–ammonium hydroxide (150:95:1)	Iodoplatinate	USP 23, identification; [1]; TLC [564]; also tablets	[5]
Syrup	HPLC	Add 0.1 N sodium hydroxide, extract methanol–water, filter through sodium sulfate, evaporate, dissolve with methanol	SCX	Methanol–0.05 M potassium dihydrogen phosphate (6:4)	232 nm	USP 23;[3,8,9,1002, 1003]; GC [2]; HPLC [1004]	[5]
Capsules	HPLC	Dissolution	SCX	Methanol–0.05 M potassium dihydrogen phosphate (6:4)	232 nm	USP 23 [5]	[5]

Sample	Method	Sample preparation	Column	Mobile phase	Detection	Notes	Ref.
Capsules	HPLC	Mix with methanol, filter	—	Buffer–methanol (1:1), buffer 7 g/l potasssium dihydrogen phosphate, pH 4.4	232 nm	USP 23 assay; GC [404,706, p. 164]	[5]
Hyoscine Bulk	TLC	Dissolve in methanol	Silica	Ammonium hydroxide–methanol–acetone–chloroform (2:10:30:50)	Dragendorff	EP, pt. II–7, p. 106–2, foreign alkaloids, also BP, vol. II, p. 615 and 777 [4]; [1]; GC [2]	[6]
Hyoscyamine Bulk	TLC GC	Dissolve in methanol	Silica USP–G2 (225°C)	Chloroform–acetone —	Iodoplatinate FID	USP 23 USP 23 ISTD–homatropine hydrobromide, assay; GC [2]; HPLC [3]; TLC, root cultures [223]	[5] [5]
Tablets, sulfate elixir, injection, oral solution, and tablets		Dissolve powder in a buffered pH 9.0 solution, extract with methylene chloride					
Ibogaine Bulk	HPLC	—	Silica	Methylene chloride–methanol (1% ammonium hydroxide) (1:1)	254 nm	TLC [1]; GC [2,9]; HPLC [3]	[9]
Ibuprofen Bulk	HPLC	—	Cyclobond 1	CH_3CN–0.01% TEA (3:2), pH 4.0, 0.6 ml/min	254 nm	Resolved d,l isomers; USP 23, p. 786; BP [4]; enantiomers [521, 909,910]; [911,917]; ointment [520] with methods review	[519]
Ibutilide Saline	HPLC	Derivatize with 1-naphthylisocynate, C₁₈, SPE	Pirkle column	MeOH–0.05% TFA–0.05% TEA	230 nm	Resolved enantiomers	[522]
Iloprost 5% Dextrose	HPLC	CHCl₃ extract dried, redissolve in MeOH	C₁₈, Hypersil	$0.02 M\ PO_4$ (pH 3.0)–MeOH–CH_3CN (456:144:400), 1.8 ml/min	207 nm	ISTD–2-naphthoic acid	[523]

Drug	Mode	Sample pretreatment	Sorbent (temp.)	Mobile phase	Detection	Comments	Reference
Imipramine Bulk	TLC	—	Silica	S_1 = EtOAc–MeOH–conc. NH_4OH (85 : 10 : 4) S_2 = MeOH S_3 = MeOH–1-butanol (3 : 2)–$0.1M$ NaBr S_4 = MeOH–conc. NH_4OH (100 : 1.5) S_5 = cyclohexane–toluene–diethylamine (75 : 15 : 10) S_6 = $CHCl_3$–MeOH (9 : 1) S_7 = acetone	—	hR_{f1} = 67, hR_{f2} = 21, hR_{f3} = 47, hR_{f4} = 48, hR_{f5} = 49, hR_{f6} = 23, hR_{f7} = 13; also GC, SE-30/OV-1, RI: 2223	[132]
Bulk	TLC	Dissolve in ethanol	Silica	Ethyl acetate–water–hydrochloric acid–acetic acid (11 : 1 : 1 : 7)	Potassium dichromate solution	JP, p. 287 and BP, vol. II, p. 238 and p. 799 GC [404, 706, p. 163, 978]	[7]
Various dosage forms	HPLC	—	Silica	Methylene chloride–methanol–water–diethylamine (850 : 150 : 1 : 0.25)	251 nm	TLC and GC review; TLC, GC, & HPLC, p. 679 [937], [8]; HPLC [9,567,569,698,978, 922]	[1088]
Indapamide Bulk	HPLC	MeOH, mobile phase	C_{18}, Lichrosorb	MeOH–aq. 1% AcOH with TEA (0.2%) (1 : 1)	250 nm	Method review, stability indicating; [1,8]	[524]
Indobufen Dosage Forms	HPLC	—	Chiracel OD	Hexane–IPA–formic acid (160 : 40;1), 1.5 ml/min	270 nm	Various chiral sorbents evaluated along with C_{18}	[525]
Indomethacin Bulk	TLC	—	Silica	S_1 = $CHCl_3$–acetone (8 : 2) S_2 = EtOAc S_3 = $CHCl_3$–MeOH (9 : 1)	—	hR_{f1} = 16, hR_{f2} = 20, hR_{f3} = 38, hR_{f4} = 6; also GC, SE-30/OV-1, RI: 2685	[132]

Substance	Method	Sample preparation	Column	Mobile phase	Detection	Remarks	Ref.
Bulk	TLC	Dissolve in methanol	Silica coated with phosphate buffer	S_4 = EtOAc–MeOH–conc. NH$_4$OH (85 : 10 : 5); Ether–light petroleum (7 : 3)	254 nm	EP, pt. II–2, p. 92; [1,8,925], p. 239, BP [4]	[5]
Insulin Injectable	HPLC	—	C$_4$ Macrosphere	Gradient A : 0.3% TFA in H$_2$O B : 0.3% TFA in 95% CH$_3$CN/5% H$_2$O; 25% B to 50% B in 30 min, 1 ml/min	280 nm	Resolved bovine and porcine sourced insulin; HPLC [535,983,1021, 1092]; human insulin-like growth factor [537]	[534]
Iodine Tablets	HPLC	H$_2$O, sonicate, filter	C$_{18}$ μBonda-pak	10 mM trimethylphenyl ammonium bromide	231 nm	Dry kelp	[526]
Iodochlorohydroxquin Cream/ointment	HPLC	THF, sonicate, acetylated	Amino Zorbax 35°C	1-chlorobutane–THF–AcOH–MeOH (97.4 : 2.0 : 0.5 : 0.1), 1 ml/min	244 nm	Stability indicating	[529]
Iododoxorubicin Bulk	HPLC	H$_2$O–CH$_3$CN (58 : 42) with 1 g/l SLS, pH 2.0, 1.4 ml/min	TMS Zorbax	254 nm	254 nm	Stability indicating	[530]
Isomethadone Bulk	HPLC	—	Silica, 30 cm × 4 mm, 10 μm	MeOH(1% NH$_4$OH)–CH$_2$Cl$_2$ (2 : 98)	254 nm	GC, UV, MS, NMR, and IR	[10, p. 1226]
Isopilocarpine Bulk	HPLC	Water	Cyano Spherisorb, 25°C	Aq. 0.1% TEA (pH 2.5)	220 nm	Resolved from pilocarpine, pilocarpic acid, and isopilocarpic acid	[532]
Isosorbide Bulk	HPLC	H$_2$O	C$_{18}$ Hypersil	H$_2$O–MeOH (4 : 1)	210 nm	Methods reviewed; also TLC and DSC; HPLC	[533]

Drug	Mode	Sample pretreatment	Sorbent (temp.)	Mobile phase	Detection	Comments	Reference
						[536,1094–1096,1098,1100]; USP 23 [5]	
Ketamine Bulk	TLC	Dissolve in methanol	Silica	Benzene–methanol–ammonium hydroxide (80:20:1)	Dragendorff	USP 23, foreign amines; TLC [1,8,564,925]; GC [2,8,9]; HPLC [8,9,272,569,861]; enantiomer [541,919]	[5]
Ketazolam Bulk	HPLC	—	C-18, Hypersil 5 μm, flow rate 1.5 ml/min	S_1 = MeOH, H_2O, PO_4 (0.1M) (55:25:20), pH 7.25 S_2 = MeOH, H_2O, PO_4 (0.1M) (7:1:2), pH 7.67	—	Cap. Fact. 1: 13.02 Cap. Fact. 2: 2.29	[135]
			Silica, Spherisorb 5W, 2 ml/min.	S_3 = MeOH–perchloric acid (1L: 100 μl) S_4 = MeOH, H_2O, TFA (997:2:1)		Cap. Fact. 3: 0.07 Cap. Fact. 4: 0.12	[135]
Bulk	TLC	—	Silica (spray 0.1M methanolic potassium hydroxide)	Ammonium hydroxide–methanol (1.5:100)	Dragendorff	GC [2]; HPLC [937,1005]	[1]
Ketobemidone Bulk	HPLC	—	Silica	Methanolic 0.01M ammonium perchlorate, pH 6.7	—	TLC [1]; GC [2]	[3]
Ketoconazole Bulk	TLC	Dissolve in $CHCl_3$	Silica	Hexane–MeOH–EtOAc–H_2O–AcOH (42:15:40:2:1)	I_2	USP 23	[5]
Ketoprofen Gel	HPLC	MeOH, 50°C, 10 min, shaker, centrifuge	C_{18}, Hypertsil, guard column	$CH_3CH_2PO_4$ (pH 3.0) (2:3)	254 nm	Resolved from parabens; [1,1115,1116]	[542]

	Technique	H₂O	Stationary phase	Mobile phase	Detection	Notes	Ref.
Ketorlac Ophthalmic with thimerosal	HPLC	H₂O	C₈, Whatman, RAC II	MeOH–10mM acetate (pH 4.5)–THF (30 : 67 : 3)	254 nm	Compared to colorimetric techniques	[539]
Khellin Fruits/tablets	HPLC	Extract with CHCl₃–MeOH (1 : 1)	C₁₈, μBonda-pak	H₂O–MeOH–CH₃CN (49 : 49 : 2), 1.5 ml/min	250 nm	Methods reviewed	[538]
Labetalol Bulk	HPLC	—	Silica	Methanolic 0.01M ammonium perchlorate, pH 6.7	—	GC [2] and for enantiomers [1006]; TLC & HPLC [8]; HPLC [922]; HPLC, enantiomers [707]; HPLC [569]	[3]
Levomoramide Bulk	TLC	—	Silica (spray 0.1M methanolic KOH)	NH₄OH–MeOH (1.5-100)	Acidified potassium permanganate	GC [2], procedure described for morinamide (morphazinamide)	[1]
Levorphanol Bulk	TLC	—	Silica	S_1 = EtOAc–MeOH–conc. NH₄OH (85 : 10 : 4) S_2 = MeOH S_3 = MeOH–1-butanol (3 : 2)–0.1M NaBr S_4 = MeOH–conc. NH₄OH (100 : 1.5) S_5 = cyclohexane–toluene–diethylamine (75 : 15 : 10) S_6 = CHCl₃–MeOH (9 : 1) S_7 = acetone	—	hR_{f1} = 41, hR_{f2} = 11, hR_{f3} = 49, hR_{f4} = 35, hR_{f5} = 15, hR_{f6} = 8, hR_{f7} = 4; also GC, SE-30/OV-1, I: 2232	[132]
Bulk	HPLC	—	Silica	Methylene chloride–methanol (1% ammonium OH) (1 : 1)	254 nm	TLC [1,8]; GC [2,8,9]; HPLC [3,8]; GC, TMS derivatives [705, p. 55]	[9]

Drug	Mode	Sample pretreatment	Sorbent (temp.)	Mobile phase	Detection	Comments	Reference
Lidocaine (Lignocaine) Injection with epinephrine and solution	HPLC	Dilute with mobile phase	C-18	Buffer–acetonitrile (4:1) Buffer–Acetic acid (50 ml)–water (930 ml) pH 3.4 with 1 N sodium hydroxide	254 nm	USP 23; TLC [1,8]; GC [2,8,9,1008]; HPLC [3,8,9,922]; review [1007]	[5]
Injection with dextrose	HPLC	Dilute with methanol	C-18	Buffer–acetonitrile (45:55); pH 7 buffer–Mono- and dihydrogen phosphate	261 nm	USP 23, assay; TLC resolved from diazepam and promethazinium maleate [226]; GC [404,714]; HPLC [406,569,671]; ToxiLab, TLC [564]	[5]
Loperamide Bulk	TLC	Dissolve in chloroform	Silica	Chloroform–methanol–formic acid (85:10:5) For capsules: methanol–buffer (95:5)	Iodine Vapor	USP 23, identification and purity; TLC [1,8,925]; GC [8]; HPLC [8,10,543, 922,1008]	[5]
Capsules	TLC	Shake contents with methanol, filter		Buffer–water–acetic acid (3:1), adjust to pH 4.7 with 1 N sodium hydroxide			
Loprazolam Bulk	HPLC	—	C-18, Hypersil, 5 μm flow rate 1.5 ml/min; Silica, Spherisorb 5W, 2 ml/min	S_1 = MeOH, H_2O, PO_4 (0.1M) (55:25:20), pH 7.25; S_2 = MeOH, H_2O, PO_4 (0.1M) (7:1:2), pH 7.67; S_3 = MeOH–perchloric acid (1L:100 μl); S_4 = MeOH, H_2O, TFA (997:2:1)	—	Cap. Fact. 1: 6.09; Cap. Fact. 2: —; Cap. Fact. 3: 20.9; Cap. Fact. 4: 50.7	[135] [135]

Lorazepam		Method		Stationary phase	Solvent systems	Detection	Results	Ref.
Bulk		TLC	—	Silica	S_1 = CHCl$_3$–acetone (8:2) S_2 = EtOAc S_3 = CHCl$_3$–MeOH (9:1) S_4 = EtOAc–MeOH conc. NH$_4$OH (85:10:5)	—	hR_{f1} = 23, hR_{f2} = 39, hR_{f3} = 41, hR_{f4} = 45; also GC, SE-30/OV-1, RI: 2402	[132]
Bulk		TLC	—	Silica	S_1 = EtOAc–MeOH–conc. NH$_4$OH (85:10:4) S_2 = MeOH S_3 = MeOH–1-butanol (3:2)–0.1M NaBr S_4 = MeOH–conc. NH$_4$OH (100:1.5) S_5 = cyclohexane–toluene–diethylamine (75:15:10) S_6 = CHCl$_3$–MeOH (9:1) S_7 = acetone	—	hR_{f1} = 45, hR_{f2} = 82, hR_{f3} = 82, hR_{f4} = 52, hR_{f5} = 1, hR_{f6} = 36, hR_{f7} = 28; also GC, SE-30/OV-1, RI: 2402	[132]
Bulk		HPLC	—	C-18, HS, Perkin-Elmer, 3 μm	S_1 = MeOH, H$_2$O (7:3) S_2 = 5 mM PO$_4$ (pH 6)–MeOH–CH$_3$CN (57:17:26) Flow: 1.5 ml/min	254 nm	Resolved 19 benzodiazepines, RRT$_1$: 0.772, RRT$_2$: 0.417	[134]
Bulk		HPLC	—	C-18, Hypersil, 5 μm flow rate 1.5 ml/min	S_1 = MeOH, H$_2$O, PO4 (0.1M) (55:25:20), pH 7.25 S_2 = MeOH, H$_2$O, PO4 (0.1M) (7:1:2), pH 7.67	—	Cap. Fact. 1: 4.60 Cap. Fact. 2: —	[135]
				Silica, Spherisorb 5W, 2 ml/min.	S_3 = MeOH–perchloric acid (1L: 100 μl) S_4 = MeOH, H$_2$O, TFA (997:2:1)		Cap. Fact. 3: 0.11 Cap. Fact. 4: 0.15	[135]

Drug	Mode	Sample pretreatment	Sorbent (temp.)	Mobile phase	Detection	Comments	Reference
Bulk	TLC	Dissolve in acetone	Silica	Chloroform–methanol (10 : 1)	254 nm	BP, addendum 1983, p. 242	[4]
Bulk	HPLC	—	Silica	Methylene chloride–methanol (1% ammonium hydroxide) (95 : 5)	254 nm	GC [2,8]; HPLC [3,699,922,1114]: CE [691]; enantiomers [540]; TLC/GC [758]	[9]
Lormetazepam							
Bulk	HPLC	—	C-18, HS, Perkin-Elmer, 3 μm	S_1 = MeOH, H_2O (7 : 3) S_2 = 5 mM PO_4 (pH 6)–MeOH–CH_3CN (57 : 17 : 26) Flow: 1.5 ml/min	254 nm	Resolved from 19 benzodiazepines, RRT_1; 0.902, RRT_2: 0.644 (azinphosmethyl, 1.00)	[134]
Bulk	HPLC	—	C-18, Hypersil,5 μm flow rate1.5 ml/min	S_1 = MeOH, H_2O, PO_4 0.1M) (55 : 25 : 20), pH 7.25 S_2 = MeOH, H_2O, PO_4 (0.1M) (7 : 1 : 2), pH 7.67	—	Cap. Fact. 1: 6.39 Cap. Fact. 2:—	[135]
			Silica, Spherisorb 5W, 2 ml/min.	S_3 = MeOH–perchloric acid (1L: 100 μl) S_4 = MeOH, H_2O, TFA (997 : 2 : 1)		Cap. Fact. 3: 0.10 Cap. Fact. 4: 0.14; resolution enantiomers [793]	[135]
Lysergic acid diethylamide							
Bulk	TLC	—	Silica	S_1 = EtOAc–MeOH–conc. NH_4OH (85 : 10 : 4) S_2 = MeOH S_3 = MeOH–1-butanol (3 : 2)–0.1M NaBr S_4 = MeOH–conc. NH_4OH (100 : 1.5) S_5 = Cyclohexane–toluene–diethylamine (75 : 15 : 10) S_6 = $CHCl_3$–MeOH (9 : 1) S_7 = acetone	—	hR_{f1} = 59, hR_{f2} = 60, hR_{f3} = 59, hR_{f4} = 60, hR_{f5} = 3, hR_{f6} = 39, hR_{f7} = 18; also GC, SE-30/OV-1, RI: 3445; see [25–27] Thermally unstable, GC not generally applicable [43,61–64,101–103, 105]	[132]

Sample	Method	Preparation	Stationary phase	Mobile phase / Solvent	Detection	Notes	Ref.
Bulk	HPLC	—	Silica, 30 cm × 4 mm, 10 μm	MeOH (1% NH$_4$OH)–CH$_2$Cl$_2$ (5 : 95)	254 nm	GC, UV, MS, NMR, and IR; analysis [355–357, paper/spectrophoto. TLC [358,360–366]; GC/MS [514, 725]; MS [359,368]; HPLC [119,367,569]; Erlich test [1118]	[10, p. 1290]
Maprotiline							
Bulk	TLC	—	Silica	S_1 = EtOAc–MeOH–conc. NH$_4$OH (85 : 10 : 4) S_2 = MeOH S_3 = MeOH–1-butanol (3 : 2)–0.1M NaBr S_4 = MeOH–conc. NH$_4$OH (100 : 1.5) S_5 = cyclohexane–toluene–diethylamine (75 : 15 : 10) S_6 = CHCl$_3$–MeOH (9 : 1) S_7 = acetone	—	hR_{f1} = 35, hR_{f2} = 6, hR_{f3} = 71, hR_{f4} = 15, hR_{f5} = 17, hR_{f6} = 5, hR_{f7} = 2; also GC, SE-30/OV-1, RI: 2356	[132]
Bulk	TLC	Dissolve in methanol	Silica	2-Butanol–ethyl acetate–2 N ammonium hydroxide (6 : 3 : 1)	Expose to hydrochloric acid followed by high-intensity UV	USP 23, chromatographic purity; TLC [1,8]; GC [2,8,10]; review [700, 706 p. 163,1009]	[5]
Mazindol							
Bulk	TLC	Dissolve in methanol–chloroform (1 : 9)	Silica	Chloroform–ethanol–ammonium hydroxide (80 : 20 : 1)	254 nm	USP 23 chromatographic purity; [1,8]	[5]
Tablets	HPLC	Shake powder with methanol, filter	Cyano	Methanol–0.01M potassium monohydrogen phosphate (1 : 4)	—	USP 23, ISTD–amitripyline, assay; HPLC [3,9]	[5]
Tablets	HPLC	Dissolution	C-8	Acetonitrile–buffer (3 : 2) Buffer–Mono and dihydrogen phosphate	271 nm	USP 23; GC [2,8,9]	[5]

Drug	Mode	Sample pretreatment	Sorbent (temp.)	Mobile phase	Detection	Comments	Reference
Mebutamate							
Bulk	TLC	—	Silica	S_1 = CHCl$_3$–acetone (8 : 2) S_2 = EtOAc S_3 = CHCl$_3$–MeOH (9 : 1) S_4 = EtOAc–MeOH–conc. NH$_4$OH (85 : 10 : 5)	—	hR_{f1} = 10, hR_{f2} = 35, hR_{f3} = 35, hR_{f4} = 60; also GC, SE-30/ OV-1, RI: 1889	[132]
Bulk	TLC	—	Silica	Chloroform–acetone (4 : 1)	Various sprays	GC [2]	[1]
Mecloqualone							
Bulk	HPLC	—	Silica	Methylene chloride–methanol (1% ammonium hydroxide) (99 : 1)	254 nm	GC [9]	[9]
Medazepam							
Bulk	TLC	—	Silica	S_1 = EtOAc–MeOH–conc. NH$_4$OH (85 : 10 : 4) S_2 = MeOH S_3 = MeOH–1-butanol (3 : 2)–0.1M NaBr S_4 = MeOH–conc. NH$_4$OH (100 : 1.5) S_5 = cyclohexane–toluene–diethylamine (75 : 15 : 10) S_6 = CHCl$_3$–MeOH (9 : 1) S_7 = acetone	—	hR_{f1} = 78, hR_{f2} = 79, hR_{f3} = 83, hR_{f4} = 67, hR_{f5} = 40, hR_{f6} = 74, hR_{f7} = 62; also GC, SE/OV-1, RI: 2226	[132]
Bulk	HPLC	—	C-18, HS, Perkin-Elmer, 3 Fm	S_1 = MeOH, H$_2$O (7 : 3) S_2 = 5 mM PO$_4$ (pH 6)– MeOH–CH$_3$CN (57 : 17 : 26) Flow: 1.5 ml/min	254 nm	Resolved 19 benzodiazepines, RRT$_1$; 3.104, RRT$_2$: 0.393	[134]

Substance	Method	Sample prep	Column/conditions	Mobile phase	Detection	Data	Ref.
Bulk	HPLC	—	C-18, Hypersil, 5 Fm, flow rate 1.5 ml/min	S_1 = MeOH, H_2O, PO_4 (0.1M) (55:25:20), pH 7.25; S_2 = MeOH, H_2O, PO_4 (0.1M) (7:1:2), pH 7.67; S_3 = MeOH–perchloric acid (IL: 100 Fl) -; S_4 = MeOH, H_2O, TFA (997:2:1)	—	Cap. Fact. 1: —; Cap. Fact. 2: 6.44	[135]
			Silica, Spherisorb 5W, 2 ml/min			Cap. Fact. 3: 3.66; Cap. Fact. 4: 9.50	[135]
Bulk	TLC	Dissolve in methanol	Silica	Cyclohexane–acetone–ammonium hydroxide (60:40:1)	254 nm	JP, p. 425; TLC [1,8]; GC [2,8,978]; HPLC [3,406,569,699,922]	[7]
Mefenamic acid Bulk	TLC	—	Silica	S_1 = $CHCl_3$–acetone (8:2); S_2 = EtOAc; S_3 = $CHCl_3$–MeOH (9:1); S_4 = EtOAc–MeOH–conc. NH_4OH (85:10:5)	—	hR_{f1} = 41, hR_{f2} = 54, hR_{f3} = 54, hR_{f4} = 14; also GC, SE-30/OV-1, RI: 2201	[132]
Bulk	TLC	Dissolve in methanol–chloroform (1:3)	Silica	Isobutanol–ammonium hydroxide (3:1)	254 nm	JP, p. 57 and BP, p. 273 [4]; TLC [1,8]; GC [8,10]; HPLC [8,10]	[7]
Mefonorex Bulk	HPLC	—	Silica, 30 cm × 4 mm, 10 Fm	MeOH (1% NH_4OH)–CH_2Cl_2 (2:98)	254 nm	GC, UV, MS, NMR, and IR	[10, p. 1332]
Meperidine Bulk	GC	H_2O	USP-G3 (190°C)	—	FID	SP 23 chromatographic purity; HPLC [1010]; stability [922]; GC [264,273,404]; GC/MS [705, p. 55]; ToxiLab TLC [564]	[5]
Meprobamate Bulk	TLC	—	Silica	S_1 = $CHCl_3$–acetone (8:2)	—	hR_{f1} = 9, hR_{f2} = 34, hR_{f3} = 32, hR_{f4} = 60; also	[132]

Drug	Mode	Sample pretreatment	Sorbent (temp.)	Mobile phase	Detection	Comments	Reference
Bulk	TLC	Dissolve in ethanol	Silica	S_2 = EtOAc S_3 = CHCl₃–MeOH (9 : 1) S_4 = EtOAc–MeOH–conc. NH₄OH (85 : 10 : 5) Acetone–toluene (1 : 1)	Vanillin acid	GC, SE-30/OV-1, RI: 1796 EP, vol. III, p. 278; TLC [1,8,925]; GC [2,8,9,264,404,1011]; HPLC [8,1012,1013]; GC, p. 54 [705]; ToxiLab TLC [564]	[6]
Mepyramine (Pyrilamine)							
Bulk	TLC	Dissolve in water, render alkaline, extract with chloroform	Silica	Ethyl acetate–$3M$ acetic acid–methanol (5 : 2 : 3)	254 nm	BP, p. 278, and vol. II, p. 556, also, EP, pt. II–7, p. 278–2; TLC [1]; GC [2]; HPLC [3]	[4]
Bulk	HPLC	—	Silica	Methylene chloride–methanol (1% ammonium hydroxide) (95 : 5)	254 nm	TLC [24]; GC [9]; HPLC [569,967]	[9]
Mescaline							
Bulk	TLC	—	Silica	S_1 = EtOAc–MeOH–conc. NH₄OH (85 : 10 : 4) S_2 = MeOH S_3 = MeOH–1-butanol (3 : 2)–0.1M NaBr S_4 = MeOH–conc. NH₄OH (100 : 1.5) S_5 = cyclohexane–toluene–diethylamine (75 : 15 : 10) S_6 = CHCl₃–MeOH (9 : 1) S_7 = acetone	—	hR_{f1} = 24, hR_{f2} = 6, hR_{f3} = 63, hR_{f4} = 20, hR_{f5} = 4, hR_{f6} = 10, hR_{f7} = 12; also GC, SE-30/OV-1, RI: 1688; TLC [1,8]; GC [2,8,9]; HPLC [3,8,119]	[132]

Compound	Method	Sample preparation	Stationary phase	Mobile phase	Detection	Comments	Ref.
Plant	HPLC	Fresh plant lyophilized, defatted with ether and extracted with MeOH–NH$_3$, 33% (99:0), containing 150 mg/l methoxamine hydrochloride ISTD	ODS-1, Spherisorb, 150 × 4.6 mm, 3 Fm	CH$_3$CN–H$_2$O (108–892) containing 5.0 ml of phosphoric acid (85% and 0.28 ml of hexylamine per liter; flow rate 1.0 ml/min	DAD	Structure assignments confirmed by GC/MS, resolved from anhalamine, anhalonidine, isopellotine, methoxamine, *N*-methyl mescaline, pellotine, and anhalorine	[706, p. 168]
Mesterolone Bulk	GC	—	DB-1, J&W, 30 m × 0.25 mm, 0.1 Fm	180–320EC (10EC/min), 320EC for 4 min	FID, N2, 30 ml/min	Resolved 17 anabolic steroids; review [1119]; GC–MS review of anabolic steroids [1119]; HPLC, FT–IR, & MS [1141]	
Metacycline Bulk	HPLC	—	PRP-1, 60EC	2-Methyl-2-propanol–PO$_4$ buffer (pH 9.0, 2*M*)–EDTA (pH 9.0,10 m*M*)–H$_2$O (2.5:10:77.5)	254 nm	TLC and HPLC comparison	[544]
Methadone Bulk	TLC	—	Silica	S$_1$ = EtOAc–MeOH–conc. NH$_4$OH (85:10:4) S$_2$ = MeOH S$_3$ = MeOH–1-butanol (3:2)–0.1*M* NaBr S$_4$ = MeOH–conc. NH$_4$OH (100:1.5) S$_5$ = cyclohexane–toluene–diethylamine (75:15:10) S$_6$ = CHCl$_3$–MeOH (9:1) S$_7$ = acetone	—	hR$_{f1}$ = 77, hR$_{f2}$ = 16, hR$_{f3}$ = 60, hR$_{f4}$ = 48, hR$_{f5}$ = 61, hR$_{f6}$ = 20, hR$_{f7}$ = 27; also GC, SE-30/OV-1, RI: 2148	[132]

Drug	Mode	Sample pretreatment	Sorbent (temp.)	Mobile phase	Detection	Comments	Reference
Bulk	HPLC	—	Silica, 5 Fm	Isooctane–diethyl-ether–methanol–diethylamine–water (400 : 325 : 225 : 0.5 : 15); flow rate: 2.0 ml/min	279 nm	k^1 = 1.00, RRT = 0.49 (diamorphine, 1.00, 49 min)	[133]
Injection	GC	Make alkaline and extract with methylene chloride	USP-G2 (170EC)	—	FID	USP 23, ISTD–procaine, assay, procedure tablets, p. 649; GC [2,8,9,60, 1015,1017,1095]	[5]
Oral concentrate	HPLC	Dilute with water	C-18	Gradient using formic acid–water–ammonium hydroxide to acetonitrile	280 nm	USP 23, assay, procedure for oral solution, p. 650; [3,9,1016]; enantiomer [919]; USP 23 [5], identification, also BP, p. 280 [4], TLC, GC, and HPLC review [1014,1016], EP, vol. II, p. 282, pt. II–9, p. 408 [6]; [1,8]; impurities [925]	[5]
Methamphetamine							
Bulk	HPLC	—	Silica, 5 Fm	Isooctane–diethyl-ether–methanol–diethylamine–water (400 : 325 : 225 : 0.5 : 15); flow rate: 2.0 ml/min	279 nm	k^1 = 3.50, RRT = 1.12 (diamorphine, 1.00, 49 min)	[133]
Bulk	HPLC	—	Silica	Methylene chloride–methanol (1% ammonium hydroxide) (4 : 1)	254 nm	TLC [8]; GC [8,9]; Simon test reagent kit [1118] HPLC [119,567,672, 705, p. 218; GC [264, 273,404]; CE [759];	[9]

Compound	Method	Sample preparation	Stationary phase / column	Mobile phase	Detection	Comments	Ref.
Methandienone Bulk	TLC	Dissolve in ethanol	Silica	Cyclohexane–ethyl acetate (1:1)	Vanillin acid	ToxiLab TLC [564]; optical isomers [339]; GC/MS [352,712,703]; duration of detectability [662]; impurities [114,116]; BP, p. 282; TLC, p. 744 [937]; GC [2]	[4]
Methandriol Bulk	GC	—	3% OV-1, Chromosorb WHP, 80/100 mesh, 4N × 1/4", N_2, 32 ml/min	—	FID	MS, NMR, and IR	[10, p. 1384]
Methandrostenolone (Methandienone) Bulk	HPLC	—	C-18, Accubond, J&W, 4.6 mm × 25 cm, 5 Fm	Time CH₃CN H₂O / 0 45 55 / 8 45 55 / 17 90 10 / 21 90 10 / 1.5 ml/min	254 nm	Resolved from 8 other anabolic steroids, also SPE [706, p. 322]; GC [706, p. 168]; HPLC [806]; review [1119]; BP, p. 282 [4]; GC [2]	[706, p. 273]
Methaqualone Bulk	TLC	Dissolve in methanol–chloroform (1:1)	Silica	Ether (water sat.)	DMAB	BP, p. 282	[4]
Bulk	TLC	Dissolve in methanol	Silica	Chloroform–acetic acid–cyclohexane (5:1:4)	254 nm spray fluorescin solution	USP 23, identification; impurities [925]; TLC [1,8]; GC [1,8,9]; HPLC [3,8,978]	[5]
						HPLC–diode array [118]; and HPLC [751, p. 13]; ToxiLab TLC [564]; GC [264,273,404]	[130]
Bulk	TLC	—	Silica	S_1 = EtOAc–MeOH–conc. NH_4OH (85:10:4), S_2 = MeOH	—	$hR_{f1} = 78$, $hR_{f2} = 78$, $hR_{f3} = 84$, $hR_{f4} = 70$, $hR_{f5} = 37$, $hR_{f6} = 80$, $hR_{f7} = 56$, also GC, SE-30/OV-1,	[132]

Drug	Mode	Sample pretreatment	Sorbent (temp.)	Mobile phase	Detection	Comments	Reference
				S_3 = MeOH–1-butanol (3 : 2) – 0.1M NaBr S_4 = MeOH–conc. NH_4OH (100 : 1.5) S_5 = cyclohexane–toluene–diethylamine (75 : 15 : 10) S_6 = $CHCl_3$–MeOH (9 : 1) S_7 = acetone		RI: 2125; also CE [265, 280]; SCF [266]	
Methixene							
Bulk	HPLC	—	Silica	Methylene chloride–methanol (1% ammonium hydroxide (9 : 1)	254 nm	TLC [1]; GC [2,10]; HPLC [3]	[10]
Methocarbamol							
Bulk	HPLC	Dissolve in methanol	C-18	Buffer–methanol (75 : 25)	274 nm	USP 23, ISTD–caffeine; TLC [1,8,925]; GC [9,1017]; HPLC, p. 698 [564,754,018]; [8,9]	[5]
Injection	HPLC	Dilute with mobile phase		Buffer–Potassium dihydrogen phosphate (6.8 g/l, pH 4.5)			
Tablets	HPLC	Mix powder with mobile phase					
Methohexital (Methohexitone)							
Bulk	TLC	—	Silica	S_1 = $CHCl_3$–acetone (8 : 2) S_2 = EtOAc S_3 = $CHCl_3$–MeOH (9 : 1) S_4 = EtOAc–MeOH–conc. NH_4OH (85 : 10 : 5) —	—	hR_{f1} = 73, hR_{f2} = 72, hR_{f3} = 71, hR_{f4} = 58; also GC, SE-30/ OV-1, RI: 1766	[132]
Injection	GC	Reconstitute, add hydrochloric acid, extract with chloroform	USP-G10	—	FID	USP 23, ISTD–apobarbital, assay; TLC [1,8,925]; GC [8,9]; HPLC [8,9]; p. 753 [937]	[5]

	Technique	Derivative	Sorbent	Conditions	Detection	Notes	Ref
Methoxyamphetamine							
Bulk	TLC	—	Silica	S_1 = EtOAc–MeOH–conc. NH$_4$OH (85:10:4); S_2 = MeOH; S_3 = MeOH–1-butanol (3:2)–0.1M NaBr; S_4 = MeOH–conc. NH$_4$OH (100:1.5); S_5 = cyclohexane–toluene–diethylamine (75:15:10); S_6 = CHCl$_3$–MeOH (9:1); S_7 = acetone	—	hR_{f1} = 44, hR_{f2} = 11, hR_{f3} = 74, hR_{f4} = 73, hR_{f5} = 36, hR_{f6} = 77, hR_{f7} = 69; also, GC, SE-30/OV-1, RI: 1412; 4-ethoxyamphetamine, some trade or other names: 4-methoxy-α-methylphenethylamine; paramethoxy amphetamine; PMA	[132]
Methoxyphenamine							
Bulk	HPLC	—	Silica	Methanolic 0.01M ammonium perchlorate, pH 6.7	—	TLC [925]; GC [2]; ToxiLab TLC [564]; [338,341]	[3]
Methylenedioxyamphetamine							
Bulk	TLC	—	Silica	S_1 = EtOAc–MeOH–conc. NH$_4$OH (85:10:4); S_2 = MeOH; S_3 = MeOH–1-butanol (3:2)–0.1M NaBr; S_4 = MeOH–conc. NH$_4$OH (100:1.5); S_5 = cyclohexane–toluene–diethylamine (75:15:10); S_6 = CHCl$_3$–MeOH (9:1); S_7 = acetone	—	3,4-Methylenedioxy amphetamine, hR_{f1} = 45, hR_{f2} = 11, hR_{f3} = 76, hR_{f4} = 39, hR_{f5} = 17, hR_{f6} = 12, hR_{f7} = 17; also GC, SE-30/OV-1, RI: 1472; also [822,823,825,826]	[132]
Bulk	HPLC	—	Silica	MeOH (1% NH$_4$OH)–CH$_2$Cl$_2$ (1:4)	254 nm	GC, UV, MS, NMR, IR	[10, p. 1460]
Bulk	GC	TMS derivative	J&W, DB-5, 20 m × 0.18 mm	—	—	p. 56; HPLC, GC, UV, MS, NMR, IR [10, p. 1460]	[705]

Drug	Mode	Sample pretreatment	Sorbent (temp.)	Mobile phase	Detection	Comments	Reference
Methyldopa Oral suspension	HPLC	Dilute with 0.1 N H_2SO_4	C_{18}	PO_4 (6.8 g/l; pH 3.5)	280 nm	USP 23, p. 994, assay, also applicable to methyldopa–glucose reaction product; [1121, 1123,1125,1127]; [4,6]; methods review [545]	[5]
Methylephedrine Bulk	HPLC	—	Silica	Methanolic 0.01M ammonium perchlorate, pH 6.7	—	TLC [1]; GC [2,264]	[3]
Herbal extract	CE	1 g Ephedra herba extracted 50% EtOH (15 ml) RT, 30 min, centrifuged, filtered. ISTD–benzyltriethyl-ammonium chloride	Fused silica, 60 cm × 75 μm ID, 28 kV	0.02M isoleucine–0.005M barium hydroxide (pH 10)	185 nm	Resolved ISTD: 4 min, methylpseudo-ephedrine: ~5 min, pseudoephedrine: 6.2 min, ephedrine: 6.8 min, methyl-ephedrine: 7.8 min, norpseudoephedrine: 8.1 min, norephedrine 8.8 min	[131]
Methylpentynol Bulk	TLC	—	Silica	$CHCl_3$–acetone (4 : 1)	$KMnO_4$	GC [2]	[1]
Methylphenidate Bulk	TLC	—	Silica	S_1 = EtOAc–MeOH–conc. NH_4OH (85 : 10 : 4) S_2 = MeOH S_3 = MeOH–1-butanol (3 : 2)–0.1M NaBr S_4 = MeOH–conc. NH_4OH (100 : 1.5) S_5 = cyclohexane–toluene–diethylamine (75 : 15 : 10)	—	hR_{f1} = 70, hR_{f2} = 42, hR_{f3} = 70, hR_{f4} = 57, hR_{f5} = 34, hR_{f6} = 34, hR_{f7} = 23; also GC, SE-30/OV-1, RI: 1737	[132]

Sample	Technique	Preparation	Stationary phase	Mobile phase	Detection	Comments	Ref.
Bulk	TLC	Dissolved in methanol		S₆ = CHCl₃–MeOH (9:1), S₇ = acetone, Chloroform–methanol–ammonium hydroxide (190:10:10)	Dragendorff	USP 23 limit test erythro R and S isomers; TLC [1]; GC [2,8,1020]; HPLC [3,9,1019]; stability [923] Enantiomer resolution [703]	[5]
Methylphenobarbital (Mephobarbital) Bulk	TLC	(Same procedure as for cyclobarbital)				EP, pt. II–5, p. 189–2; [1,8]; GC [2,9,404]; HPLC p. 771 [937], [8]	[6]
Methyltestosterone Bulk	TLC	Dissolve with chloroform–ethanol (1:1)	Silica (activate at 105°C, 1 hr)	Toluene–ethyl acetate (1:1)	Extract, elute measure at 241 nm	USP 23 assay, also BP, vol. II, p. 788 [4], EP, pt. II–9, p. 410 [6]; GC & HPLC [9]; impurities [925]; resolved 9 anabolic steroids. Also SPE [706, p. 322]; GC [706, p. 168]; GC/MS [852]; HPTLC of anabolic compounds [222]; seven eluents; review [1119]	[5]
Bulk	HPLC	—	C-18, Accubond, J&W, 4.6 mm x 25 cm, 5 μm	Time / CH₃CN / H₂O: 0 / 45 / 55; 8 / 45 / 55; 17 / 90 / 10; 21 / 90 / 10	254 nm		[706, p. 273]
Methyprylon(e) capsules and tablets	HPLC	Dissolution	C-18	Methanol–water (3:2)	280 nm	USP 23, ISTD–ethyl paraben; TLC [1,8]; GC [2,8,9]; HPLC [8,9]	[5]
Metipranolol Solution	HPLC	H₂O	C₁₈	CH₃CN–H₂O–buffer (175:315:10); buffer: PIC B6	254 nm	Degradants resolved	[382]
Metolazone Bulk	HPLC	—	Silica	Methylene chloride–methanol (1% ammonium hydroxide) (98:2)	254 nm	TLC [1]; HPLC, p. 779 [937], [8,922]	[10]

Drug	Mode	Sample pretreatment	Sorbent (temp.)	Mobile phase	Detection	Comments	Reference
Metoprolol							
Bulk	TLC	Dissolve in chloroform	Silica	Chloroform, equilibrate chamber with ammonium hydroxide	Place plate in with potassium permanganate and hydrochloric acid, spray potassium iodide	USP 23, chromatographic purity; [1,8,1022]; HPLC [922] TLC [564]	[5]
Tablets	HPLC	Mix powder and methanol, heat, sonicate, centrifuge, dilute with mobile phase, filter	C-18	PSA (961 mg)—sodium acetate (82 mg) in 550 ml methanol, 470 ml water, and 0.57 ml acetic acid	254 nm	USP 23, assay, ISTD-oxprenolol; GC [2,8]; HPLC [3,8,10,272, 703,707,751,919]	[5]
Metronidazole							
Bulk	TLC	Dissolve in acetone	Silica	Chloroform–ethanol (dehydrated)–diethylamine– water (80 : 10 : 10 : 1)	Spray titanium trichloride	USP 23, for tablets, BP, vol. II, p. 790 [4], for GC and TLC; TLC [1,8]	[5]
Injection	HPLC	Dissolve in methanol	C-18	Buffer–methanol (93 : 7), buffer–0.68 g potassium dihydrogen phosphate, pH 4.0 in 930 ml water	320 nm	USP 23, assay; GC [2,8]; HPLC [8,9,922,1024, 1025]	[5]
Metyrapone B							
Bulk	TLC	MeOH	Silica	CHC_3–MeOH (48 : 3)	KI	USP 23, BP, p. 294 & 356 [4]; TLC [1,8]; GC [2,8,10]; HPLC [8,10]	[5]
Mexiletine							
Bulk	TLC	Dissolve in methanol	Silica	Chloroform–methanol–18M ammonium hydroxide (85 : 14 : 1)	Ninhydrin	BP, addendum 1983, p. 243; TLC [1]; GC [2]; HPLC [3]; enantiomer resolution [703]	[4]
Capsules	TLC	Shake contents with methanol, filter					
Mianserin							
Bulk	TLC	Dissolve in methanol– 13.5M ammonium hydroxide	Silica	Methylene chloride– methanol (9 : 1)	Iodine vapor	BP, addendum 1983, p. 244, for GC assay, p. 303; TLC [1,8]; GC [2, 8]; HPLC [3,922]	[4]

Compound	Method	Sample prep	Stationary phase	Mobile phase	Detection	Notes	Ref.
Miconazole Bulk/injection	TLC	CHCl$_3$	Silica	n-Hexane–CHCl$_3$–MeOH–NH$_4$OH (60 : 30 : 10 : 1)	I$_2$	USP 23; purity, assay, identity	[5]
Minocycline Bulk	HPLC	Mobile phase	Silica	0.2 NH$_4$ oxalate–DMF–0.1M EDTA (550 : 250 : 200), adj. pH 6.2 with TBA	280 nm	USP 23; HPLC [10,546, 547]; review [759]	[5]
Minoxidil Bulk	HPLC	Dissolve in mobile phase	C-18	Methanol–water–acetic acid (7 : 3 : 1), add 3.0 g DSC/l, adjust to pH 3.0	254 nm	USP 23; CE [548]; TLC [1,552]; HPLC [10, 549,569,922]	[5]
Monensins A and B Fermentation broth	HPLC	Dry aliquot, dissolve in MeOH	C$_{18}$ (Separon, Czech), 30°C	MeOH–H$_2$O (88 : 12), 0.8 ml/min	RI/210 nm	MS detection	[555]
Moperone Bulk	HPLC	—	Silica, 30 cm × 4 mm, 10 μm	MeOH (1% NH$_4$OH)–CH$_2$Cl$_2$ (2 : 98)	254 nm	GC, UV, MS, NMR, and IR [10, p. 1550]; GC [870,872]	[10]
Morazone Bulk	HPLC	—	Silica	MeOH–0.01M NH$_4$ClO$_4$, pH 6.7	—	TLC [1,8]; GC [2,8]; HPLC [8]	[3]
Morphine Bulk	TLC	—	Silica	S$_1$ = EtOAc–MeOH–conc. NH$_4$OH (85 : 10 : 4) S$_2$ = MeOH S$_3$ = MeOH–1-butanol (3 : 2)–0.1M NaBr S$_4$ = MeOH–conc. NH$_4$OH (100 : 1.5) S$_5$ = cyclohexane–toluene–diethylamine (75 : 15 : 10) S$_6$ = CHCl$_3$–MeOH (9 : 1) S$_7$ = acetone	—	hR$_{f1}$ = 20, hR$_{f2}$ = 18, hR$_{f3}$ = 23, hR$_{f4}$ = 37, hR$_{f5}$ = 0, hR$_{f6}$ = 9, hR$_{f7}$ = 1; also GC, SE-30/OV-1, RI: 2454	[132]

Drug	Mode	Sample pretreatment	Sorbent (temp.)	Mobile phase	Detection	Comments	Reference
Bulk	HPLC	—	Silica	Isooctane–diethyl-ether–methanol–diethylamine–water (400 : 325 : 225 : 0.5 : 15) flow rate: 2.0 ml/min	279 nm	k^1 = 7.86, RRT = 2.16 (diamorphine, 1.00, 49 min) Bulk: TLC [6, EP, pt. II–2, p. 97], HPLC [9]; for injection with pseudomorphine [556]; with hydromorphone/bupivacaine [559]	[133]
Illicit	TLC	—	Silica (saturated with 0.1 N NaOH)	2-dimensional development Cyclohexane–toleune–diethyamine (75 : 15 : 10) Chloroform–methanol (9 : 1)	—	Compared 26 TLC systems, selectivity of 100 drugs examined, 2-D development time 2–3 hr	[18]
Illicit	HPLC	—	Silica (Porasil T), 30 cm	Chloroform–methanol	210–400 nm	Resolved from diamorphine and 6-acetylmorphine in 25 min. Also [77–92]; TLC [1,4,8,19,35–42, 115,1026] HPTLC analysis by dansyl derivatization [247]; as opium in plant extract by TLC [565, p. 78]; ToxiLab TLC [564]; poppy straw extract SFC [252]; SFC [279]; studied on 38 TLC systems [256, 287–291,656,682]; GC [2,8,9,50,264,300,302, 304,404,705, p. 55,709]; CE [265,280]; HPLC [3,8,93,94,119, 120,269,286,309,312, 313,323,324,490,567]	[76]

Compound	Method	Sample prep	Column	Mobile phase	Detection	Notes	Ref
Morphine-*N*-oxide Bulk	HPLC	—	Silica, 30 cm × 4 mm, 10 μm	MeOH (1% NH₄OH)–CH₂Cl₂ (1:4)	254 nm	GC, UV, MS, IR	[10, p. 1556]
Moxlactam disodium Injection	HPLC	H₂O	C₁₈	0.01*M* NH₄OAc–MeOH (19:1)	254 nm	USP 23, assay, resolve R/S isomers	[5]
Myrophine						Benzylmorphine myristic acid ester, recommend HPLC procedure described above for morphine-*N*-oxide.	
Nabilone Bulk	TLC	CHCl₃	Silica	Benzene–EtOAc	Visual	Another name for nabilone: (±)-*trans*-3-(1,1-dimethylheptyl)–6,6a,7,8,10,10a-hexahydro-1-hydroxy-6,6-dimethyl-9H–dibenzo [b,d] pyran-9-one	[446]
Nadolol Bulk	TLC	CHCl₃–MeOH (1:1)	Silica	Acetone–CHCl₃–NH₄OH (8:1:1)	254 nm	USP 23; TLC [1,1131]; GC [2,8,1131]; HPLC [562,573] enantiomer [561,1065]	[5]
Nafcillin Bulk	TLC	—	Silica	CH₂Cl₂–MeOH (1% NH₄OH) (9:1)	254 nm	—	[10]
Naloxone Bulk	TLC	Dissolve in water, pH 5.5–9.0, extract with CHCl₃	Silica (activated)	MeOH-ammoniated butanol, butanol prepared by shaking 100 ml butanol with 60 ml NH₄OH, discard 1% lower layer	Ferric chloride	USP 23; [2,3,10,958, 1133,1134]	[5]
Nandrolone (19-Nortestosterone) Injection sites	HPLC	Extract with tetrahydrofuran	Lichrosorb, C-18, 5 μm	CH₃CN–H₂O gradient	UV	Resolved from estradiol, testosterone,	[611]

Drug	Mode	Sample pretreatment	Sorbent (temp.)	Mobile phase	Detection	Comments	Reference
Bulk	TLC	Dissolve in acetone	Silica	n-Heptane–acetone (3:1)	Char	nortestosterone; also HPLC [706, p. 273]; GC [307, p. 168]; SPE [706, p. 322]; GC/MS [852]; review [1119] USP 23, identification of injection, BP, vol. II, p. 641 [4]	[5]
Naproxen Bulk	TLC	MeOH	Silica	Toluene–THF–AcOH (30:3:1)	254 nm	USP 23, [1,2,4,8,9,1099, 1105]; enantiomers [1065]; [568,570,571, 572,574]	[5]
Narcotine (Noscapine)						Also, SCF [266]; plant extract by TLC [565], p. 67; TLC [682]	
Natamycin Bulk	TLC	—	Silica	CHCl₃–MeOH–H₂O (6:2:1)	Ninhydrin	Degradants	[447]
Neomycin Elixir	TLC	H₂O	Silica	3.85% NH₄OAc	Heat, spray sodium hypochlorite	BP, addendum 1983, p. 271; EP, pt. II–5, p.197; USP 23 [5]; [580,619,1135–1137, 1139,1140,1152]	[4]
Niacin (Nicotinic acid) Nicodeine Bulk	HPLC	—	Silica, 30 cm × 4 mm, 10 cm	MeOH (1% NH₄OH)–CH₂Cl₂ (5:95)	254 nm	UV, MS, NMR, and IR	[10]
Nicomorphine Bulk	HPLC	—		MeOH (1% NH₄OH)–CH₂Cl₂ (5:95)	254 nm	UV, MS, NMR, and IR	[10, p. 1606]
Nicotine Bulk	TLC	—	Silica	S_1 = EtOAc–MeOH–conc. NH₄OH (85:10:4)	—	$hR_{f1} = 61$, $hR_{f2} = 39$, $hR_{f3} = 22$, $hR_{f4} = 54$, $hR_{f5} = 39$, $hR_{f6} = 35$,	[132]

Substance	Method	Extraction	Stationary phase	Mobile phase	Detection	Notes / other methods	Ref
(continued)				S_2 = MeOH; S_3 = MeOH–1-butanol (3:2)–0.1M NaBr; S_4 = MeOH-conc. NH$_4$OH (100:1.5); S_5 = cyclohexane–toluene–diethylamine (75:15:10); S_6 = CHCl$_3$–MeOH (9:1); S_7 = acetone		hR_{f7} = 13; also GC, SE-30/OV-1, RI: 1348	
Bulk	HPLC	—	Silica	Methylene chloride–methanol (1% ammonium hydroxide) (9:1)	254 nm	TLC [1,564]; GC [2,9,264]; HPLC [3,9,578,1027]; in plant extract by TLC [565]	[9]
Nicotinic acid / Bulk	TLC	—	Silica (KOH treated)	NH$_4$OH–MeOH (1.5:100)	—	GC [2,4,1153–1155]	[1]
Nifedipine / Bulk	TLC	CHCl$_3$	Silica	Diisopropyl ether	254 nm	USP 23; TLC [1,8]; GC [2,8,10,1099]; HPLC [3,8,10,563,575,584,807]	[5]
Nikethamide / Bulk	TLC	MeOH	Silica	Propanol–CHCl$_3$	254 nm	EP, pt. II–6, p. 233–2; TLC [1]; GC [1,10,19,45,1099]; HPLC [10,1099]	[6]
Nimetazepam / Bulk	HPLC	—	C-18, Hypersil, 5 μm; Flow rate 1.5 ml/min	S_1 = MeOH, H$_2$O, PO$_4$ (0.1M) (55:25:20), pH 7.25; S_2 = MeOH, H$_2$O, PO$_4$ (0.1M) (7:1:2), pH 7.67	—	Cap. Fact. 1: 3.62 / Cap. Fact. 2: —	[135]
			Silica, Spherisorb 5W, 2 ml/min.	S_3 = MeOH–perchloric acid (1L: 100 μl); S_4 = MeOH, H$_2$O, TFA (997:2:1)		Cap. Fact. 3: 1.12 / Cap. Fact. 4: 1.56	[135]

Drug	Mode	Sample pretreatment	Sorbent (temp.)	Mobile phase	Detection	Comments	Reference
Nitrazepam							
Bulk	TLC	—	Silica	S_1 = CHCl$_3$–acetone (8 : 2) S_2 = EtOAc S_3 = CHCl$_3$–MeOH (9 : 1) S_4 = EtOAc–MeOH–conc. NH$_4$OH (85 : 10 : 5)	—	hR_{f1} = 35, hR_{f2} = 45, hR_{f3} = 55, hR_{f4} = 61; also GC, SE-30/OV-1, RI: 2750	[132]
Bulk	TLC	—	Silica	S_1 = EtOAc–MeOH–conc. NH$_4$OH (85 : 10 : 4) S_2 = MeOH S_3 = MeOH–1-butanol (3 : 2) – 0.1M NaBr S_4 = MeOH–conc. NH$_4$OH (100 : 1.5) S_5 = cyclohexane–toluene–diethylamine (75 : 15 : 10) S_6 = CHCl$_3$–MeOH (9 : 1) S_7 = acetone	—	hR_{f1} = 61, hR_{f2} = 84, hR_{f3} = 86, hR_{f4} = 68, hR_{f5} = 0, hR_{f6} = 36, hR_{f7} = 55; also GC, SE-30/OV-1, RI: 2750	[132]
Bulk	HPLC	—	C-18, HS, Perkin-Elmer, 3 μm	S_1 = MeOH, H$_2$O (7 : 3) S_2 = 5 mM PO$_4$ (pH 6)–MeOH–CH$_3$CN (57 : 17 : 26) Flow: 1.5 ml/min	254 nm	Resolved 19 benzo-diazepines, RRT$_1$: 0.703, RRT$_2$: 0.380 (Azinphosmethyl, 1.00)	[134]
Bulk	HPLC	—	C-18, Hypersil, 5 μm, Flow Rate 1.5 ml/min.	S_1 = MeOH, H$_2$O, PO$_4$ (0.1M) (55 : 25 : 20), pH 7.25 S_2 = MeOH, H$_2$O, PO$_4$ (0.1M) (7 : 1 : 2), pH 7.67	—	Cap. Fact. 1: 3.00 Cap. Fact. 2: —	[135]
			Silica,	S_3 = MeOH–Perchloric		Cap. Fact. 3: 0.99	[135]

Compound	Method	Sample prep	Stationary phase	Mobile phase	Detection	References	Ref.
			Spherisorb 5W, 2 ml/min	acid (IL: 100 µL) S_4 = MeOH, H_2O, TFA (997 : 2 : 1)		Cap. Fact. 4: 1.78	
Bulk	TLC	Dissolve in methanol–chloroform (1 : 1)	Silica	Nitromethane–ethyl acetate (85 : 15)	254 nm	EP, supplement, vol. III, p. 132, for capsules; BP, addendum 1983, p. 267, for tablets; vol. II, p. 793 [4], EP, pt. II–9, p. 415 [6]; TLC [1, 1028]; GC [2,9,1027]; HPLC [3,9,1029, 1027]; synthesis [558]; TLC/GC [758]; HPLC [567,699,762]; CE [723]	[6]
Nitrendipine Bulk	HPLC	MeOH	C_{18}, Novapak	THF–CH_3CN–buffer (5 : 35 : 60); Buffer: 0.05M NH_4OH, 0.05M PO_4 (pH 5.0), 1.5 ml/min	235 nm	Related compounds; HPLC [577]	[576]
Nitrofurantoin Bulk	TLC	DMF	Silica	Nitromethane–methanol (9 : 1)	Heat, UV	EP, pt. II–2, p.101; HPLC [10,904,1099]; [4,24, 1104]; USP 23 [5]	[6]
Nitroglycerin Diluted	TLC	Acetone, filter	Silica	Toluene–EtOAc–AcOH (16 : 4 : 1)	Spray 5% diphenylamine, 340 nm	USP 23; stability [1104]; [1,473,646,905]	[5]
Nomifensine Bulk	TLC	—	Silica (KOH treated)	Ammonium hydroxide–methanol (1.5 : 100)	—	GC [2]; HPLC [3,922]	[1]
Nordiazepam Bulk	TLC	—	Silica	S_1 = EtOAc–MeOH–conc. NH_4OH (85 : 0 : 4) S_2 = MeOH S_3 = MeOH–1-butanol (3 : 2)–0.1M NaBr	—	$hR_{f1} = 69$, $hR_{f2} = 82$, $hR_{f3} = 83$, $hR_{f4} = 62$, $hR_{f5} = 4$, $hR_{f6} = 55$, $hR_{f7} = 60$; also, GC, SE-30/OV-1, RI: 2496	[132]

Drug	Mode	Sample pretreatment	Sorbent (temp.)	Mobile phase	Detection	Comments	Reference
				S_4 = MeOH–conc. NH₄OH (100 : 1.5) S_5 = cyclohexane–toluene–diethylamine (75 : 15 : 10) S_6 = CHCl₃–MeOH (9 : 1) S_7 = acetone			
Bulk	HPLC	—	C-18, Hypersil, 5 µm; Flow Rate 1.5 ml/min	S_1 = MeOH, H₂O, PO₄ (0.1M) (55 : 25 : 20), pH 7.25	—	Cap. Fact. 1: 8.00	[135]
				S_2 = MeOH, H₂O, PO₄ (0.1M) (7 : 1 : 2), pH 7.67		Cap. Fact. 2: —	
			Silica, Spherisorb 5W, 2 ml/min	S_3 = MeOH–perchloric acid (1L: 100 µl)		Cap. Fact. 3: 1.36	
				S_4 = MeOH, H₂O, TFA (997 : 2 : 1)		Cap. Fact. 4: 3.18	[135]
Bulk	TLC	—	Silica (KOH treated)	Ammonium hydroxide–methanol (1.5 : 100)	Dragendorff	GC [2,1027]; HPLC [3, 1027]; ToxiLab TLC [564]; HPLC [699,858]; CE [723]; TLC/GC [758]	[1]
Norephedrine Injection	HPLC	AcOH	C₁₈	1.1 g HPES/800 ml, 200 MeOH, pH 3.0	280 nm	USP 23, p. 1097: HPLC [579]; GC [264]; HPLC photodiode array [685]	[5]
Norethanrolone Tablets	HPLC	—	C-18	Methanol–water (7 : 3)	240 nm	TLC, p. 822 [937]; GC [1030]	[1031]
Bulk	HPLC	—	C-18, Accabond, J&W, 4.6 mm × 25 cm, 5 µm	Time CH₃CN H₂O 0 45 55 8 45 55 17 90 10 21 90 10	254 nm	Resolved 9 anabolic steroids. Also SPE [706, p. 322]; HPLC, MS, FT–IR [1141]	[706, p. 273]
Bulk	GC	—	DB-1, J&W, 30 m × 0.25 mm,	—	FID, N₂ at 30 ml/min	Resolved 17 anabolic steroids. Also, SPE	[706, p.168]

Substance	Method	Sample preparation	Stationary phase / conditions	Mobile phase	Detection	Notes	Ref.
			0.1 μm, 180–320°C (10°C/min), 320°C for 4 min				[706, p. 322]; review [1119]
Norethisterone Tablets	HPLC	Powder in MeOH, centrifuge, mobile phase	Cyclobond-1	0.05M PO$_4$, pH 7–MeOH (6:4), 0.5 ml/min	280 nm	Resolved from ethinylestradiol and norgestrel	[483]
Normethadone Bulk	TLC	—	Silica (KOH treated)	Ammonium hydroxide–methanol (1.5:100)	Acidified iodoplatinate	TLC [8]; GC [2,8,1027]; HPLC [1027]	[1]
Normorphine Bulk	HPLC	—	Silica	Methylene chloride–methanol (1% ammonium hydroxide) (1:1)	254 nm	TLC [1]; GC [2,9]; resolved from norcodeine, codeine, morphine and it O-glucuronide conjugates by HPLC–EC [239]	[9]
Norpipanone Bulk	TLC	—	Silica (KOH treated)	Ammonium hydroxide–methanol (1.5:100)	Dragendorff	GC [2]	[1]
Nortriptyline Bulk	TLC	—	Silica	S$_1$ = EtOAc–MeOH–conc. NH$_4$OH S$_2$ = MeOH S$_3$ = MeOH–1-butanol (3:2)–0.1M NaBr S$_4$ = MeOH–conc. NH$_4$OH (100:1.5) S$_5$ = cyclohexane–toluene–diethylamine (75:15:10) S$_6$ = CHCl$_3$–MeOH (9:1) S$_7$ = acetone	—	hR$_{f1}$ = 45, hR$_{f2}$ = 9, hR$_{f3}$ = 71, hR$_{f4}$ = 34, hR$_{f5}$ = 27, hR$_{f6}$ = 16, hR$_{f7}$ = 4; also GC, SE-30/OV-1, RI: 2210	[132]

Drug	Mode	Sample pretreatment	Sorbent (temp.)	Mobile phase	Detection	Comments	Reference
Bulk	HPLC	—	Silica	Methylene chloride–methanol (1% ammonium hydroxide) (9 : 1)	254 nm	TLC and GC [8,1027]; review [1032]; HPLC [922]	[9]
Noscapine (Narcotine)							
Bulk	TLC	—	Silica	S_1 = EtOAc–MeOH–conc. NH$_4$OH (85 : 10 : 4) S_2 = MeOH S_3 = MeOH–1-butanol (3 : 2)–0.1M NaBr S_4 = MeOH–conc. NH$_4$OH (100 : 1.5) S_5 = cyclohexane–toluene–diethylamine (75 : 15 : 10) S_6 = CHCl$_3$–MeOH (9 : 1) S_7 = acetone	—	hR_{f1} = 78, hR_{f2} = 72, hR_{f3} = 75, hR_{f4} = 64, hR_{f5} = 22, hR_{f6} = 74, hR_{f7} = 64; also GC, SE-30/OV-1, RI: 3120	[132]
Bulk	HPLC	—	Silica, 5 μm	Isooctane–diethyl-ether–methanol–diethylamine–water (400 : 325 : 225 : 0.5 : 15); flow rate: 2.0 ml/min	279 nm	k^1 = 0.37, RRT = 0.34 (diamorphine, 1.00, 49 min)	[133]
Bulk	HPLC	—	Silica	Methylene chloride–methanol (1% ammonium hydroxide) (98 : 2)	254 nm	TLC [1,8,925,1033]; GC [2,9,8,1034]; HPLC [3,8,1025,1035]; studied on 38 TLC systems [256]	[9]
Nystatin Ointment with cliquinol	TLC	Add acetone, warm on steam bath, filter with pledget glass wool	Silica	Benzene–MeOH (9 : 1)	254 nm	USP 23, p. 1113; HPLC [10,581,776,908]; stability [1104]	[5]
Opipramol Bulk	TLC	—	Silica (KOH treated)	Ammonium hydroxide–methanol (1.5 : 100)	Acidified iodoplatinate	TLC [8]; GC [8]; HPLC [3,8,922]	[1]

	Technique	Sample prep	Stationary phase	Mobile phase	Detection	Notes	Ref.
Bulk	HPLC	—	Cyano	Acetonitrile–dioxane–1% ammonium acetate (8:1:1)	254 nm	—	[1036]
Opium alkaloids Cough mixtures	HPLC	H_2O	Silica	MeOH–H_2O–0.2M PO_4 (pH 7.0), (35:60:5), 2.5 mM CTMA, 1 ml/min	254 nm	Resolved morphine, codeine, thebaine, papaverine, noscapine; GC [329]	[583]
Orphenadrine citrate Bulk	TLC	Dissolve in methanol	Silica	Methanol–ammonium hydroxide (100:1)	254 nm	USP 23, chromatographic purity; TLC [1,8,564]; GC [2,8,9]; HPLC [3,8,9,923,1037]	[5]
Oxandrolone Bulk	TLC	Dissolve in chloroform	Silica	Chloroform–methanol (19:1)	Char	USP 23, identification; impurities [925]	[5]
Tablets	GC	Mix powder with chloroform	USP-G1 (250°C)	—	FID	USP 23, assay	[5]
Oxazepam Bulk	TLC	—	Silica	S_1 = $CHCl_3$–acetone (8:2) S_2 = EtOAc S_3 = $CHCl_3$–MeOH (9:1) S_4 = EtOAc–MeOH–conc. NH_4OH (85:10:5)	—	hR_{f1} = 22, hR_{f2} = 37, hR_{f3} = 42, hR_{f4} = 47; also GC, SE-30/OV-1, RI: 2336; [879]	[132]
Bulk	TLC	—	Silica	S_1 = EtOAc–MeOH–conc. NH_4OH (85:10:4) S_2 = MeOH S_3 = MeOH–1-butanol (3:2)–0.1M NaBr S_4 = MeOH–conc. NH_4OH (100:1.5) S_5 = cyclohexane–toluene–diethylamine (75:15:10)	—	hR_{f1} = 47, hR_{f2} = 81, hR_{f3} = 82, hR_{f4} = 56, hR_{f5} = 0, hR_{f6} = 40, hR_{f7} = 51; also GC, SE-30/OV-1, RI: 2336	[132]

Drug	Mode	Sample pretreatment	Sorbent (temp.)	Mobile phase	Detection	Comments	Reference
Bulk	HPLC	—	C-18, HS, Perkin–Elmer, 3 μm	S_6 = $CHCl_3$–MeOH (9 : 1) S_7 = acetone S_1 = MeOH, H_2O (7 : 3) S_2 = 5 mM PO_4 (pH 6)–MeOH–CH_3CN (57 : 17 : 26) Flow: 1.5 ml/min	254 nm	Resolved 19 benzo-diazepines, RRT_1: 0.802, RRT_2: 0.390 (azinphosmethyl, 1.00)	[134]
Bulk	HPLC	—	C-18, Hypersil, 5 μm; flow rate1.5 ml/min	S_1 = MeOH, H_2O, PO_4 (0.1M) (55 : 25 : 20), pH 7.25	—	Cap. Fact. 1: 4.62	[135]
				S_2 = MeOH, H_2O,PO_4 (0.1M) (7 : 1 : 2), pH 7.67		Cap. Fact. 2: —	[135]
			Silica, Spherisorb 5W, 2 ml/min.	S_3 = MeOH–perchloric acid (1L: 100 μl) S_4 = MeOH, H_2O,TFA (997 : 2 : 1)		Cap. Fact. 3: 0.47	
						Cap. Fact. 4: 0.55	
Tablets	TLC	Mix powder with acetone, centrifuge	(Same procedure as for lorazepam)			BP, addendum 1983, p. 267; TLC [1,8]; stability [923];	[4]
Tablets	HPLC	Dissolution	C-8	Methanol–water–acetic acid (60 : 40 : 1)	232 nm	USP 23; TLC and GC review [1038]; GC [2,8,9,1039]; HPLC [2,8,9,1040]; HPLC, enantiomers [406,751, p.15]; HPLC [569,697,699, 858]	[5]
Oxprenolol Bulk	TLC	Dissolve in mobile phase	Silica	Chloroform–methanol (9 : 1)	Potassium hexacyanoferrate	BP, p. 320; TLC [1,8,267]; GC [2,8]; HPLC [3,8,582,707, 1027,922]; enantiomer [919]; stability [592]	[5]

Compound/Form	Method	Sample preparation	Stationary phase	Mobile phase/Conditions	Detection	Notes	Ref.
Oxycodone Bulk	TLC	—	Silica	S_1 = EtOAc–MeOH–conc. NH$_4$OH (85:10:4); S_2 = MeOH; S_3 = MeOH–1-butanol (3:2)–0.1 M NaBr; S_4 = MeOH–conc. NH$_4$OH (100:1.5); S_5 = cyclohexane–toluene–diethylamine (75:15:10); S_6 = CHCl$_3$–MeOH (9:1); S_7 = acetone	—	hR_{f1} = 60, hR_{f2} = 30, hR_{f3} = 33, hR_{f4} = 50, hR_{f5} = 23, hR_{f6} = 51, hR_{f7} = 39; also GC, SE-30/OV-1, RI: 2524	[132]
Bulk	HPLC	—	Silica	Methylene chloride–methanol (1% ammonium hydroxide) (98:2)	254 nm	TLC [1,8,564]; GC [2,8,9,404]; HPLC [3,8]; GC, TMS derivatives [705, p. 55]	[9]
Oxymetholone Bulk	GC	—	DB-1, J&W, 30 m × 0.25 mm, 0.1 μm, 180–320°C (10°C/min), 320°C/4 min	—	FID, N$_2$ at 30 ml/min	Resolved 17 anabolic steroids, review [1119]	[706, p.168]
Tablets	TLC	Dissolve in ethanol–chloroform (1:1)	Silica	Ethanol (absolute)–toluene (2:98)	Vanillin acid	BP, vol. II, p. 321 and p. 797	[4]
Oxymorphone Bulk	TLC	—	Silica	S_1 = EtOAc–MeOH–conc. NH$_4$OH (85:10:4); S_2 = MeOH; S_3 = MeOH–1-butanol (3:2)–0.1 M NaBr; S_4 = MeOH–conc. NH$_4$OH (100:1.5); S_5 = cyclohexane–toluene–diethylamine (75:15:10)	—	hR_{f1} = 33, hR_{f2} = 26, hR_{f3} = 36, hR_{f4} = 48, hR_{f5} = 10, hR_{f6} = 37, hR_{f7} = 30; also GC, SE-30/OV-1, RI: 2538	[132]

Drug	Mode	Sample pretreatment	Sorbent (temp.)	Mobile phase	Detection	Comments	Reference
Suppositories	HPLC	Extract with 0.1 N hydrochloric acid and chloroform, discard chloroform layer	SAX	S_6 = CHCl$_3$–MeOH (9:1) S_7 = acetone 0.05 M sodium borate, pH 9.1	254 nm	USP 23, ISTD–procaine hydrochloride, assay; GC & HPLC [9]; TLC [1]; GC [2]; HPLC [3]	[5]
Oxyphenbutazone Bulk	TLC	—	Silica	S_1 = CHCl$_3$–acetone (8:2) S_2 = EtOAc S_3 = CHCl$_3$–MeOH (9:1) S_4 = EtOAc–MeOH–conc. NH$_4$OH (85:10:5)	—	hR_{f1} = 52, hR_{f2} = 62, hR_{f3} = 57, hR_{f4} = 7; also GC, SE-30/OV-1, RI: 1630	[132]
Tablets	TLC	Shake powder with ethanol (absolute), centrifuge, expose plate to carbon dioxide, 2 min	Silica	Chloroform–acetic acid (4:1) with 0.02% butylated hydroxytoluene	254 nm	BP, vol. II, p. 798, for eye ointment, p. 572, EP, pt. II–9, p. 418 [6]; TLC [1,8,1041]; GC [2,8]; HPLC [8,9,1042]	[4]
Oxyphencylimine Bulk	TLC	0.01 N methanolic HCl	Silica	CHCl$_3$–MeOH (65:27)	Blue tetrazolium	USP 23; TLC [1]; GC [2,8,9]; HPLC [3]	[5]
Oxytetracycline hydrochloride Bulk	TLC	Same procedure as for chloro-tetracycline	—	—	—	EP, pt. II–5, p. 198; TLC [1]; HPLC [1156]; stability [1104]	[4]
Oxytocin	—	—	—	—	—	HPLC [1093,1132]	
Pancuronium Bulk	HPLC	—	Silica	CH$_3$CN–H$_2$O (96:4) containing ClO$_4$	213 nm	Resolved from vecuronium and pipercuronium; BP [4]; TLC [1,1157]	[605]

	Technique	Sample preparation	Stationary phase	Mobile phase	Detection	Notes	Ref.
Papaverine							
Bulk	TLC	—	Silica	S_1 = EtOAc–MeOH–conc. NH$_4$OH (85:10:4) S_2 = MeOH S_3 = MeOH–1-butanol (3:2)–0.1 M NaBr S_4 = MeOH–conc. NH$_4$OH (100:1.5) S_5 = cyclohexane–toluene–diethylamine (75:15:10) S_6 = CHCl$_3$–MeOH (9:1) S_7 = acetone	—	hR$_{f1}$ = 69, hR$_{f2}$ = 74, hR$_{f3}$ = 74, hR$_{f4}$ = 61, hR$_{f5}$ = 8, hR$_{f6}$ = 65, hR$_{f7}$ = 47; also GC, SE-30/OV-1, RI: 2825	[132]
Bulk	HPLC	—	Silica, 5 μm	Isooctane–diethyl-ether–methanol–diethylamine–water (400:325:225:0.5:15); flow rate: 2.0 ml/min	279 nm	k' = 1.33, RRT = 0.58 (diamorphine, 1.00, 49 min)	[133]
Powder with ethyl aminobenzoate and scopolia extract	TLC	Add water, shake, filter, add 1 N hydrochloric acid, extract with chloroform, discard the chloroform, render alkaline, extract with ether	Silica	Chloroform–methanol–acetone–ammonium hydroxide (75:15:10:2)	Dragendorff	JP, p. 1216; also poppy straw SCF [252]; in plant extract, TLC [565], p. 82; TLC-densitometry [656]; ToxiLab TLC [564]; SCF [266]; GC [404]; HPLC [209,567]	[7]
Bulk	TLC	Dissolve in ethanol–water (1:1)	Silica	Diethylamine–ethyl acetate–toluene (1:2:7)	254 nm	EP, pt. II–2, p. 102; TLC [1,8]; GC [2,8,9]; HPLC [3,8,9]	[6]
Paracetamol (Acetaminophen)							
Bulk	TLC	Add peroxide–free ether to the powder, shake, centrifuge, dilute with methanol	Silica	Chloroform–acetone–toluene (65:25:10)	254 nm	EP, pt. II, p. 49, also, BP, p. 326 [4]; TLC [1,8]; GC [2,8]; HPLC [8,75,593,1043,1045,1047,1048]	[6]
Parahexyl						Some trade or other names: 3-hexyl-1-hydroxy-7,8,9,10-	

Drug	Mode	Sample pretreatment	Sorbent (temp.)	Mobile phase	Detection	Comments	Reference
						tetrahydro-6,6,9-trimethyl-6H-dibenzo[b,d]pyran; synhexyl; synthesis [883]; see cannabinoids	
Paraldehyde Bulk	GC	—	USP-G2	—	FID	Also p. 729 [1018], [9]	[2]
Parabens	—	—	—	—	—	[915,916]	—
Pecazine (Mepazine) Bulk	TLC	—	Silica	NH_4OH–MeOH (1.5 : 100)	Acidified iodoplatinate	GC [2]; HPLC [3]	[1]
Pemoline Bulk	HPLC	—	Silica	Methanolic 0.01M ammonium perchlorate, pH 6.7	—	TLC [1]; GC [2,9]; HPLC [8,1049]	[3]
Penfluridol Bulk	TLC	—	Silica	S_1 = EtOAc–MeOH–conc. NH_4OH (85 : 10 : 4) S_2 = MeOH S_3 = MeOH–1-butanol (3 : 2)–0.1M NaBr S_4 = MeOH–conc. NH_4OH (100 : 1.5) S_5 = cyclohexane–toluene–diethylamine (75 : 15 : 10) S_6 = $CHCl_3$–MeOH (9 : 1) S_7 = acetone	—	hR_{f1} = 83, hR_{f2} = 67, hR_{f3} = 89, hR_{f4} = 76, hR_{f5} = 17, hR_{f6} = 60, hR_{f7} = 60; also GC, SE-30/OV-1, RI: 3380	[132]
Bulk	TLC	—	Silica (KOH treated)	Ammonium hydroxide–methanol (1.5 : 100)	Dragendroff	TLC [8]; GC [2,8,1027]; HPLC [1027]	[1]
Penicillamine Tablets	HPLC	Dissolution	C_{18}	0.01M PO_4, pH 7	254 nm	USP 23; [624,1158,1159]	[5]

Compound	Technique	Sample preparation	Stationary phase	Mobile phase	Detection	Remarks	Ref.
Penicillin G (procaine) Bulk	TLC	Dissolve in acetone–0.1M citric acid–0.1M sodium citrate (2:1:1)	Silica	Toluene–dioxane–AcOH (90:25:4)	254 nm/350 nm starch solution and DMAB	USP 23; [46,48,73]	[5]
Pentaerythritol tetranitrate Formulations	HPLC	Mobile phase	C_{18}	H_2O–CH_3CN (35:65)	230 nm	Collaborative study; [8]	[585]
Pentazocine Bulk	TLC	—	Silica	S_1 = EtOAc–MeOH–conc. NH_4OH (85:10:4) S_2 = MeOH S_3 = MeOH–1-butanol (3:2)–0.1M NaBr S_4 = MeOH–conc. NH_4OH (100:1.5) S_5 = cyclohexane–toluene–diethylamine (75:15:10) S_6 = $CHCl_3$–MeOH (9:1) S_7 = acetone	—	hR_{f1} = 72, hR_{f2} = 33, hR_{f3} = 72, hR_{f4} = 61, hR_{f5} = 15, hR_{f6} = 12, hR_{f7} = 28; also GC, SE-30/OV-1, RI: 2275	[132]
Tablets with aspirin	TLC	Dissolve in methanol–chloroform (1:1)	Silica	Formic acid–ethyl acetate–methanol (90:5:5)	254 nm	USP 23, identification, with naloxone tablets, for lactate counterion, BP, vol. II, p. 647 and p. 801 [4]; TLC [1,8, 1050]; GC [2,8,9,264, 564]; HPLC [3,8,9, 1051]; enantiomer [919]	[5]
Pentobarbital (Pentobarbitone) Bulk	TLC	—	Silica	S_1 = $CHCl_3$–acetone (8:2) S_2 = EtOAc S_3 = $CHCl_3$–MeOH (9:1) S_4 = EtOAc–MeOH–conc. NH_4OH (85:10:5)	—	hR_{f1} = 55, hR_{f2} = 66, hR_{f3} = 59, hR_{f4} = 44; also GC, SE-30/OV-1, RI: 1740	[132]

Drug	Mode	Sample pretreatment	Sorbent (temp.)	Mobile phase	Detection	Comments	Reference
Elixir	TLC	Dilute with ethanol	Silica	2-Propanol–acetone–chloroform–ammonium hydroxide (9:2:2:4)	254 nm	USP 23, identification; also EP, pt. II–5, p. 200 and pt. II–9, p. 419 [6]; impurities [925]; TLC [1,8]; GC [2,9,824,968, 1052]; HPLC [8,9]	[5]
Pentoxifylline							
Bulk	HPLC	—	Silica, 30 cm × 4 mm, 10 μm	MeOH (1% NH$_4$OH)–CH$_2$Cl$_2$ (1:99)	254 nm	UV, MS, NMR, IR	[10, p. 1760]
Pentoxyverine (Carbetapentane)							
Bulk	TLC	—	Silica (spray with 0.1M methanolic KOH)	NH$_4$OH–MeOH (1.5:100)	Dragendroff	GC [2]	[1]
Perphenazine							
Injection and Oral Solution	TLC	Dilute with methanol	Silica	Acetone–ammonium hydroxide (200:1)	Iodoplatinate	USP 23, identification, also BP, vol. II, p. 802. [4]; TLC [1,8,1053]; GC [2,9]; HPLC [3,9,700,1054]	[5]
Syrup	TLC	Add water, render alkaline, extract with chloroform	Silica	"			
Tablets	TLC	Mix powder with chloroform, filter	Silica	"			
Pethidine (Meperidine)							
Bulk	TLC	—	Silica	S$_1$ = EtOAc–MeOH–conc. NH$_4$OH (85:10:4) S$_2$ = MeOH S$_3$ = MeOH–1-butanol (3:2)–0.1M NaBr S$_4$ = MeOH–conc. NH$_4$OH (100:1.5) S$_5$ = cyclohexane–toluene–diethylamine (75:15:10) S$_6$ = CHCl$_3$–MeOH (9:1) S$_7$ = acetone	—	hR$_{f1}$ = 62, hR$_{f2}$ = 34, hR$_{f3}$ = 40, hR$_{f4}$ = 52, hR$_{f5}$ = 37, hR$_{f6}$ = 34, hR$_{f7}$ = 11; also GC, SE-30/OV-1, RI: 1751	[132]

Bulk	TLC	Dissolve in water, render alkaline, extract with ether	Coat kieselguhr with phenoxyethanol	Diethylamine–light petroleum–phenoxyethanol (1 : 100 : 8), use supernatant	Dichlorofluorescein	EP, vol. II, p. 318 and vol. II–9, p. 419 [6]; BP, p. 335 [4]; TLCC [1,8]; GC [2,8,9,264]; HPLC [3,8,9,922]	[6]
Phenacetin Powder with aspirin and caffeine	TLC	Mix with MeOH, filter	Silica	CHCl$_3$–acetone–NH$_4$OH (45 : 5 : 1)	Dragendorff	JP, p. 122 and p. 1226; EP, pt. II–6, p. 241–2 [6]; TLC [1,8]; GC [2,8,1099]; HPLC [8,9,928,929]	[7]
Phenazocine Bulk	HPLC	—	Silica	Methylene chloride–methanol (1% ammonium hydroxide) (95 : 5)	254 nm	TLC [1,267]; GC [2,9]	[9]
Phenazone (Antipyrin) Bulk	TLC	—	Silica (KOH treated)	Ammonium hydroxide–methanol (1.5 : 100)	Acidified iodoplatinate	TLC [8]; GC [2,8]; HPLC [3,8]	[1]
Phencyclidine Bulk	HPLC	—	Silica, 30 cm × 4 mm, 10 μm	MeOH (1% NH$_4$OH)–CH$_2$Cl$_2$ (5 : 95)	254 nm	UV, MS, NMR, IR, ethyl analog [10, p. 1782]; 4-hydroxyl analog [10, p. 1784]; morpholine analog [10, p. 1786]; pyrrolidine analog [10, p. 1788]; color test [1118]; [400–403]	[10, p. 1780]
Phenethylamine (Phenylethylamine) Bulk	HPLC	—	Silica, 30 cm × 4 mm, 10 μm	MeOH (1% NH$_4$OH–CH$_2$Cl$_2$) (1 : 4)	254 nm	UV, MS, NMR, IR; also TLC [267,564]; GC [896,899]; HPLC [897,898,900]	[10, p.1794]
Pheniramine Bulk	TLC	Dissolve in methanol	Silica	Cyclohexane–chloroform–diethylamine (5 : 4 : 1)	Dragendorff	BP, p. 339; GC & HPLC [9]; TLC [1]; GC [2,9, 264,706, p. 164]; HPLC [3,922]; enantiomers [919]	[4]

Drug	Mode	Sample pretreatment	Sorbent (temp.)	Mobile phase	Detection	Comments	Reference
Bulk	TLC	—	Silica	S_1 = EtOAc–MeOH–conc. NH_4OH (85 : 10 : 4) S_2 = MeOH S_3 = MeOH–1-butanol (3 : 2)–0.1M NaBr S_4 = MeOH–conc. NH_4OH (100 : 1.5) S_5 = cyclohexane–toluene–diethylamine (75 : 15 : 10) S_6 = $CHCl_3$–MeOH (9 : 1) S_7 = acetone	—	hR_{f1} = 46, hR_{f2} = 34, hR_{f3} = 45, hR_{f4} = 50, hR_{f5} = 14, hR_{f6} = 21, hR_{f7} = 14; also GC, SE-30/OV-1, RI: 1431; TLC [9, 564]; GC [2,9,264]	[132]
Phenobarbital Tablets	HPLC	Mix powder with methanol–buffer (1 : 1)	C-18	Methanol–buffer (2 : 5), buffer–6.6 g sodium acetate–3 ml acetic acid/L water, pH 4.5	254 nm	USP 21, p. 817, ISTD–caffeine, assay; also EP, pt. II–5, p. 201 [6]; TLC [1,8]; GC [2,8,9]; HPLC [8,9,1055,1056]; stability [923] General methods; SCF [266]; GC, p. 54 [705]; GC [275]; CE [691]; [406,569]	[5]
Elixir	HPLC	Add hydrochloric acid, extract with chloroform					
Phenobarbitone Bulk	TLC	—	Silica	S_1 = $CHCl_3$–acetone (8 : 2) S_2 = EtOAc S_3 = $CHCl_3$–MeOH (9 : 1) S_4 = EtOAc–MeOH–conc. NH_4OH (85 : 10 : 5)	—	hR_{f1} = 47, hR_{f2} = 65, hR_{f3} = 56, hR_{f4} = 29, also GC, SE-30/OV-1, RI: 1957	[132]
Phenomorphane Bulk	HPLC	—	Silica, 30 cm × 4 mm, 10 μm	MeOH (1% NH_4OH–CH_2Cl_2) (1 : 9)	254 nm	UV, MS, NMR, IR	[10, p. 1808]

Compound	Method	Sample prep	Stationary phase	Mobile phase	Detection	Other methods	Ref.
Phenoperidine Bulk	HPLC	—	Silica	Methanolic 0.01M ammonium perchlorate, pH 6.7	—	TLC [1]; GC [2,1027]	[3]
Phenothiazine Bulk	HPLC	—	Silica	CH_2Cl_2–CH_3CN (4:1)	254 nm	GC, p. 738 [19]; [10, 639]; HPLC [587]	[10]
Phenprobamate Bulk	TLC	—	Silica (KOH treated)	Ammonium hydroxide–methanol (1.5:100)	—	TLC [8]; GC [2,8]	[1]
Phentermine Bulk	TLC	—	Silica	S_1 = EtOAc–MeOH–conc. NH_4OH (85:10:4); S_2 = MeOH; S_3 = MeOH–1-butanol (3:2) – 0.1M NaBr; S_4 = MeOH–conc. NH_4OH (100:1.5); S_5 = cyclohexane–toluene–diethylamine (75:15:10); S_6 = $CHCl_3$–MeOH (9:1); S_7 = acetone	—	hR_{f1} = 48, hR_{f2} = 11, hR_{f3} = 78, hR_{f4} = 46, hR_{f5} = 26, hR_{f6} = 31, hR_{f7} = 15; also GC, SE-30/OV-1, RI: 147	[132]
Capsules and tablets	HPLC	Add 0.04M phosphoric acid, filter	C-18	Buffer–methanol (3:2), buffer–HPA, 1.1 g/575 ml water, 25 ml diluted acetic acid	254 nm	USP 23, assay; TLC [1, 264,564]; GC [1,9]; HPLC [3,9]	[5]
Bulk	GC	—	3% OV-1, 4' × 1/4", 32 ml/min nitrogen	—	FID	MS, NMR, IR	[10, p. 1824]
Bulk	GC	TMS	DB-5	—	FID	Amphetamine precursor	[706, p. 156]
Phenylbutazone Bulk	TLC	—	Silica	S_1 = $CHCl_3$–acetone (8:2)	—	hR_{f1} = 78, hR_{f2} = 68, hR_{f3} = 76, hR_{f4} = 64;	[132]

Drug	Mode	Sample pretreatment	Sorbent (temp.)	Mobile phase	Detection	Comments	Reference
				S_2 = EtOAc S_3 = CHCl$_3$–MeOH (9:1) S_4 = EtOAc–MeOH—conc. NH$_4$OH (85:10:5)			
1-Phenylcyclohexylamine Bulk	HPLC	—	Silica, 30 cm × 4 mm	MeOH (1% NH$_4$OH)–CH$_2$Cl$_2$ (1:4)	254 nm	also GC, SE-30/OV-1, RI: 2365HPLC [406]	[10, p. 1828]
Phenylpropanolamine Bulk	HPLC	—	Silica	Methylene chloride–methanol (1% ammonium hydroxide) (4:1)	254 nm	TLC [1,925]; GC [2,9, 588,603,1057–1060]; HPLC [3, 1062–1065, 1067]; stability [1066]	[9]
Phenytoin Injection	HPLC	Dissolve in mobile phase	C-18	Methanol–water (55:45)	254 nm	USP 23, assay; TLC [1,8, 925]; GC [2,8]; HPLC [8,1068,922,1069] GC [273]; CE [691]	[5]
Pholcodine Bulk	HPLC	—	Silica, 30 cm × 4 mm, 10 μm	MeOH (1% NH$_4$OH–CH$_2$Cl$_2$) (1:9)	254 nm	UV, MS, NMR, IR; also HPLC [567]; cough mixtures [596]	[10]
Physostigmine Bulk	TLC	—	Silica	S_1 = EtOAc–MeOH–conc. NH$_4$OH (85:10:4) S_2 = MeOH S_3 = MeOH–1-butanol (3:2)–0.1M NaBr S_4 = MeOH–conc.–NH$_4$OH (100:1.5) S_5 = cyclohexane–toluene–diethylamine (75:15:10) S_6 = CHCl$_3$–MeOH (9:1) S_7 = acetone	—	hR_{f1} = 57, hR_{f2} = 40, hR_{f3} = 38, hR_{f4} = 55, hR_{f5} = 12, hR_{f6} = 36, hR_{f7} = 18; also GC, SE-30/OV-1, RI: 1804	[132]

Compound	Method	Sample preparation	Stationary phase	Mobile phase	Detection	References	Ref.
Bulk	TLC	Dissolve in ethanol	Silica	Ammonium hydroxide-isopropanol-cyclohexane (2:23:100)	Dragendorff	EP, pt. II, p. 286; TLC [1,8]; USP 23 [5]	[6]
Phytomenadione (Vitamin K$_1$) Tablets	TLC	Disperse powder with H$_2$O, extract with 2,2,4-trimethylpentane	Silica	Cyclohexane–ether–MeOH (50:20:1)	254 nm	BP, vol. II, p. 807; GC [2]; review [796]	[4]
Pilocarpine Ophthalmic solution	HPLC	Dilute in MeOH	Silica	n-Hexane-2% NH$_4$OH-in IPA (3:7)	220 nm	USP 23; TLC [1,24,572]; GC [2]; HPLC [19,272,569,570,589, 597,608,710,871,942]	[5]
Piminodine Bulk	TLC	—	Silica (KOH treated)	Ammonium hydroxide-methanol (1.5:100)	Acidified potassium permanganate	HPLC [3]; GC [1027]	[1]
Pinazepam Bulk	HPLC	—	C-18, Hypersil, 5 μm, flow rate 1.5 ml/min	S$_1$ = MeOH, H$_2$O, PO$_4$ (0.1M) (55:25:20), pH 7.25	—	Cap. Fact. 1: 10.96	[135]
				S$_2$ = MeOH, H$_2$O, PO$_4$ (0.1M) (7:1:2), pH 7.67		Cap. Fact. 2: 2.13	
			Silica, pherisorb 5W, 2 ml/min	S$_3$ = MeOH–perchloric acid (1L: 100 μl)		Cap. Fact. 3: 1.32	[135]
				S$_4$ = MeOH, H$_2$O, TFA (997:2:1)		Cap. Fact. 4: 2.19; also [906,907]	
Pindolol Bulk	HPLC	—	Silica	Methanolic 0.01M ammonium perchlorate, pH 6.7	—	TLC [1,8]; GC [2,8, 1027]; HPLC [8,607, 922,1027]; enantiomer [706,707,919]	[3]
Piperacillin Bulk	HPLC	Dissolve in mobile phase	C$_{18}$	MeOH–0.2M PO$_4$– TBA (450:547:3), pH 5.5	254 nm	USP 23, [941]	[5]

Drug	Mode	Sample pretreatment	Sorbent (temp.)	Mobile phase	Detection	Comments	Reference
Piritramide							
Bulk	TLC	—	Silica (KOH treated)	Ammonium hydroxide–methanol (1.5 : 100)	Dragendorff	TLC [8]; GC [8]; HPLC [3,8,1099]	[1]
Piroxicam							
Bulk	TLC	—	Silica	Chloroform–acetone (4 : 1)	—	TLC [8]; HPLC [5,591]; review [1070]	[1]
Polymyxin B Sulfate							
Bulk	TLC	Dissolve in H₂O–HCl (1 : 1), heat at 135°C for 5 hr	Silica	Water–phenol (25 : 75)	Ninhydrin	EP, pt. II–5, p. 203–2 and BP, addenum 1983, p. 274; HPLC [944]	[6]
Practolol							
Bulk	TLC	Dissolve in methanol	Silica	Methanol–13.5 *M* ammonium hydroxide (99 : 1)	254 nm	BP, p. 361; TLC [1]; HPLC [3] Enantiomer resolution by HPLC [703, 707]	[4]
Prazepam							
Bulk	TLC	—	Silica	S₁ = EtOAc–MeOH–conc. NH₄OH (85 : 10 : 4) S₂ = MeOH S₃ = MeOH–1-butanol (3 : 2)–0.1*M* NaBr S₄ = MeOH–conc. NH₄OH (100 : 1.5) S₅ = cyclohexane–toluene–diethylamine (75 : 15 : 10) S₆ = CHCl₃–MeOH (9 : 1) S₇ = acetone	—	hR_{T1} = 81, hR_{T2} = 85, hR_{T3} = 89, hR_{T4} = 65, hR_{T5} = 36, hR_{T6} = 74, hR_{T7} = 63; also, GC, SE-30/OV-1, RI: 2641	[132]
Bulk	HPLC	—	C-18 HS, Perkin-Elmer, 3 μm	S₁ = MeOH–H₂O (7 : 3) S₂ = 5m*M* PO₄ (pH 6)–MeOH–CH₃CN (57 : 17 : 26) Flow: 1.5 ml/min	254 nm	Resolved 19 benzodiazepines, RRT₁: 1.979, RRT₂: 2.130 (azinphosmethyl, 1.00)	[134]

Sample	Method	Sample preparation	Stationary phase	Mobile phase	Detection	Standard/notes	Ref.
Bulk	HPLC	—	C-18, Hypersil, 5 μm, flow rate 1.5 ml/min	S_1 = MeOH, H₂O, PO₄ (0.1M) (55:25:20), pH 7.25; S_2 = MeOH, H₂O, PO₄ (0.1M) (7:1:2), pH 7.67; S_3 = MeOH–perchloric acid (1L: 100 μl); S_4 = MeOH, H₂O,TFA (997:2:1)	—	Cap. Fact. 1: —; Cap. Fact. 2: 4.20; Cap. Fact. 3: 1.49; Cap. Fact. 4: 2.71	[135] [135]
			Silica, Spherisorb 5W, 2 ml/min				
Bulk	TLC	Dissolve in acetone	Silica	Heptane–ethyl acetate (1:1)	254 nm	USP 23	[5]
Capsules	TLC	Shake contents with acetone, filter	Silica				
Tablets	TLC	Shake powder with chloroform, filter	Silica				
Capsules and tablets	HPLC	Add water to contents, followed by acetic acid and methanol, mix, filter	C-18	Methanol–water–acetic acid	254 nm	USP 23; TLC [1,8]; GC [2,8,9]; HPLC [8,699, 1027]; TLC/GC [758]	[5]
Prazosin Capsules	TLC	Shake contents with chloroform–methanol (1:1), centrifuge	Silica	Ethyl acetate–diethylamine (19:1)	254 nm	USP 23, identification; TLC [1,8]	[5]
Bulk	HPLC	Dissolve in methanol, dilute with methanol–water (7:3)	Silica	Methanol–water–acetic acid–diethylamine (700:300:10:0.2)	254 nm	USP 23, assay; HPLC [3,8,19,922,1071]	[5]
Capsules	HPLC	Shake contents with 0.08% hydrochloric acid–methanol (3:2), filter	Silica				
Prednisolone Bulk	HPLC	Dissolve in water saturated CHCl₃–THF	Silica	Butyl chloride–butyl chloride (water saturated)–THF–MeOH–AcOH (59:95:14:7:6)	254 nm	USP 23, BP [4]; HPLC [9,369–373]	[5]

Drug	Mode	Sample pretreatment	Sorbent (temp.)	Mobile phase	Detection	Comments	Reference
Prednisone							
Bulk	HPLC	Dilute in H_2O	C_{18}	H_2O–MeOH–THF (688 : 62 : 250)	254 nm	USP 23, BP [4]; [1,24,25,374]	[5]
Prilocaine							
Injection	HPLC	Dilute with water	Cyano	Buffer–methanol acetonitrile–acetic acid (950 : 10 : 20 : 20); buffer–OSA (2.16 g) and EDTA (37 mg) in 950 ml water	254 nm	USP 23, assay, BP, p. 364 [4]; GC [10]; TLC [267]	[5]
Primidone							
Tablets	GC	Add EtOH to powder, boil 1 hr	USP-G3 (260°C)	—	FID	USP 23, BP [4]; [1,2,8,9, 375]	[5]
Procainamide							
Bulk and capsules	TLC	Dissolve in methanol	Silica	Methanol–ammonium hydroxide–ethyl acetate (2 : 1 : 22)	254 nm	USP 23, identification, also BP, vol. II, p. 655 [4]; TLC [1,11,24]; GC [2,8,9]; HPLC [3,9,1072]	[5]
Injection	TLC	Dilute with water	Silica		254 nm		
Bulk	HPLC	Dissolve in water	C-18	Described as (3 in 20) and triethylamine (1 in 1000) in 0.02M aqueous sodium acetate, pH 4.5	254 nm	USP 23, ISTD–procaine hydrochloride; HPLC [705, p. 185]; CE [691]; ToxiLab TLC [564]	[5]
Injection	HPLC	Dilute with water	C-18				
Prochlorperazine							
Bulk	HPLC	—	Silica, 30 cm × 4 mm	MeOH (1% NH_4OH–CH_2Cl_2) (5 : 95)	254 nm	UV, GC, MS, NMR, IR, HPLC [609]	[10, p. 1932]
Procyclidine							
Bulk	TLC	—	Silica	S_1 = EtOAc–MeOH–conc. NH_4OH (85 : 10 : 4) S_2 = MeOH S_3 = MeOH–1-butanol (3 : 2)–0.1M NaBr S_4 = MeOH–conc. NH_4OH (100 : 1.5)	—	$hR_{f1} = 72$, $hR_{f2} = 19$, $hR_{f3} = 68$, $hR_{f4} = 48$, $hR_{f5} = 63$, $hR_{f6} = 31$, $hR_{f7} = 23$; also GC, SE-30/OV-1, RI: 2156	[132]

		Sample prep	Phase	Mobile phase	Detection	Notes	Ref
Bulk	GC	Dissolve in water	—	S_5 = cyclohexane–toluene–diethylamine (75:15:10), S_6 = CHCl$_3$–MeOH (9:1), S_7 = acetone	FID	USP 23, related compounds, also TLC, BP, vol. II, p. 812 [4]; TLC [1,8]; GC [2,8,10]; HPLC [3,10]	[5]
Progesterone Bulk	TLC	Dissolve in EtOH–CHCl$_3$ (1:1)	Silica (activate 105°C)	CHCl$_3$–EtOAc (2:1)	Scrap, elute, measure at 241 nm	USP 23, BP [4]; [2,3,7,10,376,1099]	[5]
Prolintane Bulk	TLC	—	Silica	S_1 = EtOAc–MeOH–conc. NH$_4$OH (85:10:4), S_2 = MeOH, S_3 = MeOH–1-butanol (3:2)–0.1M NaBr, S_4 = MeOH–conc. NH$_4$OH (100:1.5), S_5 = cyclohexane–toluene–diethylamine (75:15:10), S_6 = CHCl$_3$–MeOH (9:1), S_7 = acetone	—	hR_{f1} = 65, hR_{f2} = 16, hR_{f3} = 35, hR_{f4} = 44, hR_{f5} = 41, hR_{f6} = 30, hR_{f7} = 11; also GC, SE-30/OV-1, RI: 2316; HPLC [567]	[132]
Promazine Tablets	TLC	Mix powder with methanol, filter	Silica	Cyclohexane–acetone–diethylamine (8:1:1)	254 nm	BP, vol. II, p. 813, TLC [1,8]; GC [2,8,1027]; HPLC [3,567,569,1027]	[4]
Promethazine Bulk	TLC	—	Silica	S_1 = EtOAc–MeOH–conc. NH$_4$OH (85:10:4)	—	hR_{f1} = 65, hR_{f2} = 30, hR_{f3} = 44, hR_{f4} = 50, hR_{f5} = 37, hR_{f6} = 35,	[132]

Drug	Mode	Sample pretreatment	Sorbent (temp.)	Mobile phase	Detection	Comments	Reference
				S_2 = MeOH S_3 = MeOH–1-butanol (3 : 2)–0.1M NaBr S_4 = MeOH–conc. NH_4OH (100 : 1.5) S_5 = cyclohexane–toluene–diethylamine (75 : 15 : 10) S_6 = $CHCl_3$–MeOH (9 : 1) S_7 = acetone		hR_{f7} = 17; also GC, SE-30/OV-1, RI: 2259; also [1075–1077]	
Bulk	TLC	Dissolve in methanol	Silica	Diisopropyl ether–ethyl acetate–acetic acid (30 : 15 : 5)	Potassium permanganate	BP, p. 373; TLC [1,11]; GC [2,11,706, p. 164]; HPLC [11]; enantiomers [606]	[4]
Propiomazine Bulk	HPLC	—	Silica, 30 cm × 4 mm, 10 μm	MeOH (1% NH_4OH)–CH_2Cl_2 (5 : 95)	254 nm	UV, GC, MS, NMR, IR	[10, p. 1950]
Propofol Emulsion	HPLC	MeOH, centrifuge	Phenyl, Dynamax Microsorb	MeOH–PO_4 (0.3M, pH 5.0) (7 : 3)	276 nm	2nd derivative UV	[595]
Propoxyphene Tablets with acetaminophen	TLC	Shake powder with methanol, filter	Silica	Butyl acetate–chloroform–formic acid (3 : 2 : 1)	254 nm	USP 23, identification, same procedure for propoxyphene napsylate formulations; [24]	[5]
Capsules with aspirin and caffeine	TLC	Add methanol to powdered contents, centrifuge	Silica			USP 23, identification	[5]
Bulk	GC	Add water, render alkaline, extract chloroform	USP-G2 (160°C)	—	FID	USP 23, ISTD–tricosane, assay and related compounds, also BP, p. 144 [4,1077]	[5]
Capsules	GC	Mix powder with acetone, add water, mix, filter	USP-G2 (160°C)				

	Technique	Sample prep	Stationary phase	Mobile phase	Detection	Notes	Ref.
Bulk	GC	Dissolved in methanol	0.25-μm film, chemically bonded 5% phenyl polydimethyl siloxane: injector 275°C, column 60°C after injection— 40°C/min to 250°C 30 min	—	MS, scanned between m/z 33 and 500 per sec; EI	Propoxyphene forms major artifacts in the injector zone; benzoylecgonidine degrades to ecgonine and benzoic acid; Atrophine forms dehydroatropine and O-(phenyl acetyl) tropine; also GC [143,273]; ToxiLab TLC [564]	[143]
Propranolol Bulk	TLC	Dissolve in methanol	Silica	Toluene–methanol (9:1)	Anisaldehyde acid	BP, p. 374; TLC [1,8, 925]; GC [2,8]; HPLC [3,8,9,922,1078]; ToxiLab TLC [564]; SPE HPLC [119,566, 567,569,751, p. 7]; enantiomers [590,604, 703,707]	[4]
Propyphenazone Bulk	TLC	—	Silica	S_1 = CHCl$_3$–acetone (8:2) S_2 = EtOAc S_3 = CHCl$_3$–MeOH (9:1) S_4 = EtOAc–MeOH– conc. NH$_4$OH (85:10:5)	—	$hR_{f1}=61$, $hR_{f2}=50$, $hR_{f3}=67$, $hR_{f4}=75$; also GC, SE-30/OV-1, RI: 1922	[132]
Propylhexedrine Bulk	TLC	—	Silica (KOH treated)	Ammonium hydroxide–methanol (1.5:100)	Acidified iodoplatinate	GC [2,958,1079]; TLC [267]	[1]
Prostaglandins Bulk	TLC	—	Silica	EtOAc–isooctane– EtOH–formic acid (35:0.5:0.3)	Phosphomolybdic acid	Prostaglandins A, B, D, E, F, and thromboxane B-2; HPLC [95, 99–102,104,594,598, 599,600,1104]	[95]

	Method	Preparation	Phase	Solvent system	Detection	Comments	Ref.
Protriptyline							
Bulk	TLC	—	Silica	S_1 = EtOAc–MeOH–conc. NH₄OH (85:10:4); S_2 = MeOH; S_3 = MeOH–1-butanol (3:2)–0.1M NaBr; S_4 = MeOH–conc. NH₄OH (100:1.5); S_5 = cyclohexane–toluene–diethylamine (75:15:10); S_6 = CHCl₃–MeOH (9:1); S_7 = acetone	—	$hR_{f1} = 41$, $hR_{f2} = 6$, $hR_{f3} = 69$, $hR_{f4} = 19$, $hR_{f5} = 17$, $hR_{f6} = 7$, $hR_{f7} = 2$; also GC, SE-30/OV-1, RI: 2261	[132]
Bulk	HPLC	—	Silica	Methylene chloride–methanol (1% ammonium hydroxide) (4:1)	254 nm	TLC [1,8]; GC [2,9,1080]; HPLC [567,922]	[9]
Proxyphylline							
Bulk	TLC	Dissolve in chloroform	Silica	Chloroform–ethanol–ammonium hydroxide (90:10:1)	254 nm	EP, vol. II, supplement, p. 145, TLC [1,8]; GC [2,8,1027]; HPLC [3,8,1027]; enantiomer [703]	[6]
Pseudoephedrine							
Tablets	TLC	(See procedure for triprolidine)	Silica			USP 23	[5]
Syrup	HPLC	Dilute with 0.01 N hydrochloric acid	Silica	0.25% Ammonium acetate–ethanol	254 nm	USP 23, also triprolidine hydrochloride, p. 1099; TLC [1]; GC [2,9]; HPLC [3,9,1018];	[5]
Tablets	HPLC	Add 0.01 N hydrochloric acid to powder, filter	Silica				[5]
Herbal extract	CE	1 g Ephedra herba extracted 50% EtOH (15 mL) RT, 30 min, centrifuged,	Fused silica, 60 cm × 75 µm ID, 28 kV	0.02M isoleucine–0.005M barium hydroxide (pH 10)	185 nm	Resolved: ISTD: 4 min, methyl pseudoephedrine; ~5 min psuedoephedrine: 6.2	[131]

Compound	Technique		Stationary phase	Solvent	Detection	Remarks	Ref.
		filtered ISTD–benzyltriethyl-ammonium chloride				min, ephedrine: 6.8 min, methylephedrine: 7.8 min, norpseudo-ephedrine: 8.1 min, norephedrine: 8.8 min ToxiLab [564]; enantiomers [703]; HPLC [705, p. 185]	[132]
Psilocin (Psilocyn)							
Bulk	TLC	—	Silica	S_1 = EtOAc–MeOH–conc. NH$_4$OH (85 : 10 : 4) S_2 = MeOH S_3 = MeOH–1-butanol (3 : 2)–0.1 M NaBr S_4 = MeOH–conc. NH$_4$OH (100 : 1.5) S_5 = cyclohexane–toluene–diethylamine (75 : 15 : 10) S_6 = CHCl$_3$–MeOH (9 : 1) S_7 = acetone	—	hR_{f1} = 50, hR_{f2} = 12, hR_{f3} = 48, hR_{f4} = 39, hR_{f5} = 5, hR_{f6} = 9, hR_{f7} = 9; also GC, SE-30/OV-1, RI: 1976	[132]
Bulk	HPLC	—	Silica	Methanol (1% ammonium hydroxide)	254 nm	TLC [1]; GC [2]; HPLC [3,391]	[9]
Psilocybin							
Bulk	HPLC	—	Silica, 30 cm × 4 mm, 10 μm	MeOH (1% NH$_4$OH)	254 nm	UV, MS, NMR, and IR; also GC–MS [390]; HPLC [1090,1091]; analysis [1023,1089] Erlich test reagent [1118]	[10, p. 1982]
Psilocyn						1-[1-(2-thienyl)cyclohexyl] pyrrolidine. Some other names: TCPY	
Bulk	HPLC	—	Silica, 30 cm × 4 mm, 10 μm	MeOH (1% NH$_4$OH)–CH$_2$Cl$_2$ (1 : 9)	254 nm	UV, GC, MS, NMR, IR; analysis [1023,1089]; HPLC [1090,1091]	[10, p. 1984]

Drug	Mode	Sample pretreatment	Sorbent (temp.)	Mobile phase	Detection	Comments	Reference
Pyridoxine (Vitamin B-6)							
Bulk	TLC	Dissolve in H_2O	Silica	NH_4OH–acetone–CCl_4–THF (9 : 65 : 13 : 13)	Spray	EP, pt. II-6, p. 245–2; HPLC [97,98,108, 610,736–739,745, 750]; TLC [728]; CE [741]; review [1160]	[6]
Quinidine							
Bulk	HPLC	Dissolve in mobile phase	C-18	Water–acetonitrile–buffer–diethylamine (86 : 10 : 2 : 2, 10%); buffer—35 ml MSA and 20 ml acetic acid/500 ml water	235 nm	USP 23, limit test for related compounds and assay; GC & HPLC [9]	[5]
Injection	HPLC	Dilute with mobile phase, mix					
Bulk	TLC	Dissolve in methanol	Silica	Diethylamine–ether–toluene (10 : 24 : 40)	Iodoplatinate	EP, pt. II, p. 17 and BP, vol. II, p. 818 [4]; TLC [1,564,656]	[6]
Bulk, capsules, and tablets	HPLC	(Use same procedure as described for the gluconate)				USP 23; GC [2]; HPLC [2,922]	[5]
Quinine							
Bulk	TLC	—	Silica	S_1 = EtOAc–MeOH–conc. NH_4OH (85 : 10 : 4) S_2 = MeOH S_3 = MeOH–1-butanol (3 : 2)–0.1 M NaBr S_4 = MeOH–conc. NH_4OH (100 : 1.5) S_5 = cyclohexane–toluene–diethylamine (75 : 15 : 10) S_6 = $CHCl_3$–MeOH (9 : 1) S_7 = acetone	—	hR_{f1} = 42, hR_{f2} = 27, hR_{f3} = 65, hR_{f4} = 51, hR_{f5} = 2, hR_{f6} = 11, hR_{f7} = 4; also GC, SE-30/OV-1, RI: 2803	[132]
Bulk	HPLC	—	Silica, 5 μm	Isooctane–diethyl-ether–methanol–diethylamine–water	279 nm	k^1 = 2.71 RRT = 0.95 (diamorphine, 1.00, 49 min)	[133]

Compound	Method	Sample preparation	Stationary phase	Mobile phase	Detection	Notes	Ref.
Bulk	TLC	Dissolve in methanol	Silica	Diethylamine–ether–toluene (10 : 24 : 40)	Iodoplatinate	EP, pt. II-6, p. 18–2 and BP, vol. II, p. 661 [4], use same for the sulfate salt; TLC [1]; GC [2,9,922]. Also studied on 38 TLC systems [256,257]; TLC–densitometry [656]; HPLC [569]; GC [709]	[6]
Racemoramide Bulk	GC	—	3% OV-1, 4' × 1/4", 32 ml/min, nitrogen	—	FID	UV, MS, NMR, IR	[10, p. 2028]
Ramipril Bulk	HPLC	Derivatize with TFA anhydride	Chiralcel OT (+)	MeOH, 0.8 ml/min	254 nm	Enantiomers	[720]
Ranitidine Bulk	HPLC	—	Silica	Methanolic 0.01M ammonium perchlorate, pH 6.7	—	Review [1081]; TLC [267]; ToxiLab TLC [564] HPLC [1095–1097]	[3]
Remoxipride Rescinnamine Bulk	TLC	—	Silica (KOH treated)	Ammonium hydroxide–methanol (1.5 : 100)	Acidified potassium permanganate	GC [2,1082]; HPLC [3]	[1]
Reserpine Bulk	HPLC	—	Silica	Methylene chloride–methanol (1% ammonium hydroxide) (98 : 2)	254 nm	TLC [1,8]; HPLC [8,611, 612,1082]; also dosage forms by HPLC [627–629, 705, p. 215]	[9]
Rifaximine Cream	TLC	—	Silica	Double elution; (a) CHCl$_3$; (b) CHCl$_3$–EtOAc–MeOH (1 : 9 : 0.5)	275 nm	Oxidative products	[613]

(400 : 325 : 225 : 0.5 : 15); flow rate: 2.0 ml/min

Drug	Mode	Sample pretreatment	Sorbent (temp.)	Mobile phase	Detection	Comments	Reference
Salbutamol							
Bulk	CE	—	Silica, 50 cm × 75 μm	20 mM Na citrate, pH 2.5, 30 kV	200 nm	Methods for impurities reviewed; compared to HPLC & TLC LC–MS [636]; HPLC [3,638, 1099]; BP [4]; TLC [1]	[613]
Salicylamide							
Bulk	TLC	—	Silica	S_1 = CHCl$_3$–acetone (8:2) S_2 = EtOAc S_3 = CHCl$_3$–MeOH (9:1) S_4 = EtOAc–MeOH–conc. NH$_4$OH (85:10:5)	—	hR_{f1} = 38, hR_{f2} = 55, hR_{f3} = 43, hR_{f4} = 48; also GC, SE-30/OV-1, RI: 1455	[132]
Bulk	HPLC	—	Silica	Methylene chloride–methanol (1% ammonium hydroxide) (98:2)	254 nm	TLC [1,8]; GC [9,264]; HPLC [8]; review [1083]	[9]
Salicylic acid							
Bulk	TLC	—	Silica	S_1 = CHCl$_3$–acetone (8:2) S_2 = EtOAc S_3 = CHCl$_3$–MeOH (9:1) S_4 = EtOAc–MeOH–conc. NH$_4$OH (85:10:5)	—	hR_{f1} = 7, hR_{f2} = 25, hR_{f3} = 24, hR_{f4} = 10; also GC, SE-30/OV-1, RI: 1308	[132]
Solution	TLC	(See thianthol, capsium, and chloral)				JP, p. 1263, p. 1270, and p. 1312; TLC [1,925]	[7]
Topical foam	HPLC	Dilute with mobile phase, cool below room temp., filter	C-18	Buffer–methanol–acetonitrile–acetic acid (700:150:150:1); buffer—tetramethyl-	280 nm	USP 23, ISTD–benzoic acid; GC [2]; HPLC [9]	[5]

Sample	Technique	Sample treatment	Column (temp)	Mobile phase	Detection	Remarks	Ref.
				ammonium hydroxide pentahydrate (225 mg/700 ml water)			
Scopolamine Injection and tablets	GC	Add water, render alkaline, extract with methylene chloride, dry extract with sodium sulfate	USP-G3 (225°C)	—	FID	USP 23, assay, ISTD–homatropine hydrobromide; GC & HPLC [8,9]; impurities [925]; stability [923]; in plant extract [565], p. 67; GC [404]	[5]
Secobarbital (Quinalbarbitone) Urine	MS/MS	No sample treatment, 1 μl direct insertion	—	—	CI, methane, SRM	LD = 10 ng	[11]
Elixir	GC	Add 20% hydrochloric acid, extract with chloroform	USP-G10 (200°C)	—	FID	USP 23, assay, barbital, for TLC and GC review [1084]; GC [8,9]; HPLC [8]	[5]
Capsules	GC	Add water, hydrochloric acid and chloroform to contents and shake, filter chloroform layer through sodium sulfate	(Same procedure as for secobarbital)			USP 23, assay, for EP, pt. II-5, p. 205 [4]	[5]
Capsules formulated with sodium amobarbital	GC	Add water and hydrochloric acid, extract with chloroform	USP-G10 (175°C)	—	FID	USP 23, assay, ISTD–aprobarbital; [569]	[5]
Selegiline Tablets	HPLC	Powder, 15% CH$_3$CN in H$_2$O, filter	Cyano, Spherisorb	0.1M PO$_4$, pH 3.1–CH$_3$CN (85 : 15)	254 nm	Related impurities; dissolution [626]	[625]
Sennoside A & B Crude drug	HPLC	H$_2$O	C$_{18}$, TSK gel, ODS	Pretreatment column: (a) H$_2$O–CH$_3$CN–AcOH (680 : 320 : 1), pH 5.0; (b) analysis column:	340 nm	Column switching	[614]

Drug	Mode	Sample pretreatment	Sorbent (temp.)	Mobile phase	Detection	Comments	Reference
Sertaconazole nitrate							
Bulk	HPLC	CH$_3$CN	Cyano, Spherisorb	Acetate buffer, 0.1M–CH$_3$CN (680:320) with 15 mM HAB, 1 ml/min; CH$_3$CN–10mM PO$_4$ buffer (37:63)	260 nm	Related compounds, cream also analyzed	[630]
Simvastatin							
Bulk	HPLC	CH$_3$CN	C$_8$, Ultrasphere	25 mM PO$_4$ (pH 4.5)–CH$_3$CN (35:65), 1.5 ml/min	238 nm	ISTD–lovastatin	[641]
Sotalol							
Bulk	TLC	—	Silica (KOH treated)	Ammonium hydroxide–methanol (1.5:100)	—	GC [2,8]; HPLC [3,8]; enantiomer resolution [703,707]	[1]
Stanolone (Adrostanolone, dihydrotestosterine)							
Bulk	GC	Dissolve in chloroform	(230°C)	—	FID	BP, p. 425, assay	[4]
Bulk	GC	—	DB-1, JW, 20 m × 0.25 mm, 0.1 μm, 180–320°C (10°C/min), 320°C, 4 min	—	FID	Resolved 17 anabolic steroids; also, SPE [706, p. 322]	[706, p. 168]
Stanozolol							
Bulk	TLC	Dissolve in methanol–chloroform (1:4)	Silica	Chloroform–methanol (9:1)	Char	BP, addendum 1983, p. 254 and USP 23 [5]; TLC [1]; [1027]; 2-D HPTLC of anabolic compounds, [222]; SPE [706, p. 322]; GC [706, p. 168]; analysis review of GC–MS of anabolic steroids [1119]	[4]

	Method		Stationary phase	Mobile phase	Detection	Notes	Ref
Strychnine Bulk	TLC	—	Silica	S$_1$ = EtOAc–MeOH–conc. NH$_4$OH (85:10:4) S$_2$ = MeOH S$_3$ = MeOH–1-butanol (3:2)–0.1M NaBr S$_4$ = MeOH–conc. NH$_4$OH (100:1.5) S$_5$ = cyclohexane–toluene–diethylamine (75:15:10) S$_6$ = CHCl$_3$–MeOH (9:1) S$_7$ = acetone	—	hR$_{f1}$ = 33, hR$_{f2}$ = 8, hR$_{f3}$ = 11, hR$_{f4}$ = 26, hR$_{f5}$ = 8, hR$_{f6}$ = 19, hR$_{f7}$ = 2; also GC, SE-30/OV-1, RI: 3119	[132]
Bulk	HPLC	—	Silica, 5 μm	Isooctane–diethyl-ether–methanol–diethylamine–water (400:325:225:0.5:15); flow rate: 2.0 ml/min	279 nm	k^1 = 15.43, RRT = 4.34 (diamorphine, 1.00, 49 min)	[133]
Bulk	HPLC	—	Silica	Methylene chloride–methanol (1% ammonium hydroxide)	254 nm	TLC [1,8]; GC [2,8,9]; HPLC [3,8,631]; review [10085] HPLC–diode array [1118]; plant extract by TLC [565], p. 67; TLC–densitometry [656] Also studied on 38 TLC systems [256]; HPLC studies [566,567,633–635]	[9]
Sufentanil (Sufentanyl) Bulk	HPLC	—	Silica, 30 cm × 4 mm, 10 μm	MeOH (1% NH$_4$OH)–CH$_2$Cl$_2$ (2:98)	254 nm	UV, MS, NMR, IR; solubility and physochemical	

Drug	Mode	Sample pretreatment	Sorbent (temp.)	Mobile phase	Detection	Comments	Reference
Sulfur						properties [1101]; GC [1102]; review [1103]; see fentanyl	
Ointment	HPLC	Partition between H_2O and CH_2Cl_2	C_{18}, LiChro Cart	MeOH–aq. 5% AcOH (85:15), 1.2 ml/min	254 nm		[640]
Sulphonamides							
Bulk	CE	MeOH	Silica, 50 cm × 50 μm, 15 kV	$0.05M$ PO_4–$0.05M$ borate (pH 6.4)–2 mM β-cyclodextrin	210 nm	Resolved sulphameth-oxypyridazine, sulphachloropyri-dazine, sulphasalazine, sulphamerazine, sulphaquanidine, sulphadiazine, sulphaquinoxaline, sulphamethazine: CE [616,617,637]; HPLC [618,620]; HPLC SPE [621]; HPTLC, 15 sulphonamides [622]; tablets [623]; GC [705, p. 57, 706, p. 159]; CE [616]	
Taxol							
Bark extract	HPLC	—	Supelocosil, LC-F	CH_3CN–THF–H_2O (17:28:55), 1.5 ml/min	227 nm	Penta-fluorophenyl phase; phenyl and C_{18}; HPLC [643–646,648, 664,681,688]; LC–MS [651]	[642]
Telazol (see Tiletamine)							
Temafloxacin							
Bulk	HPLC	CH_3CN–H_2O (1:1)	C18, Nucleosil	Gradient with PO_4 buffer, pH 2.4–CH_3CN–THF	325 nm	Related substances	[652]
Temazepam							
Bulk	TLC	—	Silica	S_1 = $CHCl_3$–acetone (8:2) S_2 = EtOAc S_3 = $CHCl_3$–MeOH (9:1)	—	hR_{f1} = 51, hR_{f2} = 47, hR_{f3} = 65, hR_{f4} = 63; also GC, SE-30/OV-1, RI: 2633	[132]

Compound	Buffer	Method	Stationary phase	Mobile phase / Solvent	Detection	Results	Ref
Bulk	—	TLC	Silica	S_4 = EtOAc–MeOH conc. NH_4OH (85:10:5); S_1 = EtOAc–MeOH– conc. NH_4OH (85:10:4); S_2 = MeOH; S_3 = MeOH–1-butanol (3:2)–0.1M $NaBr$; S_4 = MeOH–conc. NH_4OH (100:1.5); S_5 = cyclohexane–toluene–diethylamine (75:15:10); S_6 = $CHCl_3$–MeOH (9:1); S_7 = acetone	—	hR_{f1} = 63, hR_{f2} = 82, hR_{f3} = 82, hR_{f4} = 53, hR_{f5} = 8, hR_{f6} = 59, hR_{f77} = 53; also GC, SE-30/OV-1, RI: 2663	[132]
Bulk	—	HPLC	C-18, HS, Perkin-Elmer, 3 μm	S_1 = MeOH–H_2O (7:3); S_2 = 5mM PO_4 (pH 6)–MeOH–CH_3CN (57:17:26) Flow: 1.5 ml/min	254 nm	Resolved 19 benzo-diazepines, RRT_1: 0.194, RRT_2: 0.538 (azinphosmethyl, 1.00)	[134]
Bulk	—	HPLC	C-18, Hypersil, 5 μm, flow rate 1.5 ml/min	S_1 = MeOH, H_2O, PO_4 (0.1M) (55:25:20), pH 7.25; S_2 = MeOH, H_2O, PO_4 (0.1M) (7:1:2), pH 7.67	—	Cap. Fact. 1: 5.76; Cap. Fact. 2: —	[135]
			Silica, Spherisorb 5W, 2 ml/min	S_3 = MeOH–perchloric acid (IL: 100 μl); S_4 = MeOH, H_2O, TFA (997:2:1)		Cap. Fact. 3: 0.42; Cap. Fact. 4: 0.45	[135]
Bulk	—	TLC	Silica (KOH treated)	Ammonium hydroxide–methanol (1.5:100)	Dragendorff	GC [2]; HPLC [663,699, 922,1027]; TLC [704]	[1]
Tetracycline Bulk	0.02M PO_4 (pH 2.2)	CE	Silica, 20 cm × 25 μm, 10 kV	5 mM EDTA in 0.02M PO_4 (pH 3.9) buffer	265 nm	Resolved tetracycline, epitetracycline,	[657]

Drug	Mode	Sample pretreatment	Sorbent (temp.)	Mobile phase	Detection	Comments	Reference
Terfenadine Bulk	HPLC	—	β-cyclodextrin, Astec	EtOH–CH$_3$CN–hexane (6.6 : 3.3 : 90), 2 ml/min	218–220 nm	Enantiomers resolved	[710]
Testosterone Bulk	HPLC	—	C-18, Accubond, J&W, 4.6 mm × 25 cm, 5 μm	Time CH$_3$CN H$_2$O 0 45 55 8 45 55 17 90 10 21 90 10 1.5 ml/min	254 nm	Resolved 8 other anabolic steroids; also SPE [706, p. 322]; GC [706, p. 168]	[706, p. 273]
Serum	HPLC	Bond Elut	Silica, Brownlee, 15 mm × 4.6, 7 μm, 60°C	CH$_3$CN–0.1M phosphate buffer (86 : 15), 1 ml/min	247 nm	Limit detection ~5 nm/ml, resolved 4 other anabolic steroids; review [1119]	[750]
Tetrazepam Bulk	HPLC	—	C-18, Hypersil, 5 μm, flow rate 1.5 ml/min	S$_1$ = MeOH, H$_2$O, PO$_4$ (0.1M) (55 : 25 : 20), pH 7.25 S$_2$ = MeOH, H$_2$O, PO$_4$ (0.1M) (7 : 1 : 2), pH 7.67	—	Cap. Fact. 1: 21.44 Cap. Fact. 2: 4.13	[135]
			Silica, Spherisorb 5W, 2 ml/min	S$_3$ = MeOH–perchloric Acid (1L: 100 μl) S$_4$ = MeOH, H$_2$O, TFA (997 : 2 : 1)		Cap. Fact. 3: 2.03 Cap. Fact. 4: 5.20; also HPLC [699]	[135]
Thalidomide Bulk	HPLC	—	Chiracel OJ	Hexane–EtOH (1 : 1), 1 ml/min	240 nm	R, S resolved	[711]
Thebaine Bulk	TLC	—	Silica	S$_1$ = EtOAc–MeOH– conc. NH$_4$OH (85 : 10 : 4)	—	hR$_{f1}$ = 45, hR$_{f2}$ = 22, hR$_{f3}$ = 32, hR$_{f4}$ = 45, hR$_{f5}$ = 23, hR$_{f6}$ = 37, hR$_{f7}$ = 5; anhydrotetracycline, 4-epianhydrotetracycline; HPLC [658–660; 679, 680]; CE [689]	[132]

Compound / form	Method	Sample preparation	Stationary phase	Mobile phase / conditions	Detection	Remarks	Ref.
Bulk	HPLC	—	Silica, 5 μm	S_2 = MeOH; S_3 = MeOH–1-butanol (3:2)–$0.1M$ NaBr; S_4 = MeOH–conc. NH$_4$OH (100:1.5); S_5 = cyclohexane–toluene–diethylamine (75:15:10); S_6 = CHCl$_3$–MeOH (9:1); S_7 = acetone; Isooctane–diethyl-ether–methanol–diethylamine–water (400:325:225:0.5:15); flow rate: 2.0 ml/min	279 nm	also GC, SE-30/OV-1, RI: 2517; UV, MS, NMR, IR [10, p. 2204]; $k^1 = 3.53$, RRT = 1.15 (diamorphine, 1.00, 49 min) Also poppy straw, SCR [252]; studied on 38 TLC systems [256]; HPLC [705, p. 215]; see morphine and codeine; also, TLC [290]	[133]
Theobromine							
Bulk	TLC	Dissolve in methanol–chloroform (2:3)	Silica	Ammonium hydroxide–acetone–chloroform–butanol (1:3:3:4)	254 nm	EP, pt. II–7, p. 298; TLC [1,656]; GC [9]; HPLC [3,705, p.185]	[6]
Theophylline							
Capsules with guaifensin	TLC	Add water to contents, extract with chloroform	Cellulose	Methanol–water	254 nm	USP 23	[5]
Tablets with ephedrine and phenobarbital	TLC	Mix powder with chloroform–methanol (4:1), filter	Silica	Chloroform–acetone–methanol–ammonium hydroxide (50:10:10:1)	254 nm	USP 23, identification, EP, pt. II–7, p. 300 [6]; GC [275,404]; HPLC [406,657,677,705, p. 185]; TLC [656]; CE [690]	[5]
Thiamine (Vitamin B$_1$)							
Bulk	HPLC	—	μBondapak, C$_{18}$	MeOH–5 mM acetate (pH 5.0) (28:72), 1.5 ml/min	EX: 370 nm EM: 430 nm	Resolved from riboflavin; HPLC [737–740,745,750]; TLC [728]; review [726,1161]	[716]

Drug	Mode	Sample pretreatment	Sorbent (temp.)	Mobile phase	Detection	Comments	Reference
Thiethylperazine							
Bulk	HPLC	—	Silica, 30 cm × 4 mm, 10 μm	MeOH (1% NH$_4$OH)–CH$_2$Cl$_2$ (1 : 9)	254 nm	UV, MS, NMR, IR	[10, p. 2226]
Bulk	TLC	Dissolve in methanol	Silica	2-Propanol–water–ethyl acetate (5 : 3 : 1)	Oxidation	USP 23; TLC [1,8]; GC [2,8,10]; HPLC [3]	[5]
Injection	TLC	Add water, render alkaline, extract with chloroform	Silica	Toluene–ethanol–ammonium hydroxide (40 : 10 : 1)	350 nm	USP 23	[5]
Thimerosal							
Solution	HPLC	CH$_3$CN–H$_2$O (3 : 7)	μBondapak, C$_{18}$	PO$_4$ buffer–CH$_3$CN (15 : 85)	254 nm	Photodegradation; HPLC [671,672,676]	[668]
Thiopental							
Bulk	TLC	(Same procedure as for cyclobarbital calcium)	Silica			EP, pt. II–5, p. 212; GC [8,9]; HPLC [8,9]	[6]
Thiopentone							
Bulk	TLC	—	Silica	S$_1$ = CHCl$_3$–acetone (8 : 2) S$_2$ = EtOAc S$_3$ = CHCl$_3$–MeOH (9 : 1) S$_4$ = EtOAc–MeOH conc. NH$_4$OH (85 : 10 : 5)	—	hR$_{f1}$ = 77, hR$_{f2}$ = 74, hR$_{f3}$ = 68, hR$_{f4}$ = 49; also GC, SE-30/OV-1	[132]
Thioproperazine						GC [1105]	
Thioridazine							
Bulk	TLC	Dissolve in methanol–ammonium hydroxide (49 : 1), carry out under subdued light	Silica	2-Propanol–ammonium hydroxide (74 : 25 : 1)	254 nm	USP 23, chromatographic purity, also BP, p. 453 [4]	[5]
Oral solution	TLC	Add water, render alkaline, extract with chloroform	Silica	Toluene–acetone–hexane–diethylamine (15 : 15 : 15 : 1)	254 nm/350 nm	USP 23, identification	[5]

Bulk	GC	—	DB-5, J&W, 30 m × 0.25 mm, 0.25 μm 55°C/min, 55°C–225°C at 25°C/min, 225°C/5 min, 225–320°C at 25°C/min, 320°C for 6 min	—	FID, N₂ at 30 ml/min	Resolved 10 antidepressants; also GC [404]	[706, p.163]
Thiothixene Capsules	TLC	Dissolve in methanol–chloroform (1:1)	Silica	Methanol–ethyl acetate–diethylamine (65:35:5)	Iodoplatinate	USP 23, identification; GC [9]; HPLC [3,9,1086]	[5]
Bulk	HPLC	Dissolve in methanol	Silica	Ethanolamine–water methanol (0.2:200:1400)	254 nm	USP 23, assay & (E) isomer, p. 1058; TLC [1]; GC [2]	[5]
Capsules	HPLC	Mix contents with methanol, filter	Silica				
Tiletamine (2-(Ethylamino)-2-(2-thienyl)-cyclohexamine)						Formulated with zolazepam combination known as Telazol	
Tilidine Bulk	HPLC	—	Silica	CH₂Cl₂	254 nm	UV, MS, NMR, IR; also [127,1106–1108]	[10, p. 2352]
Timolol Bulk	TLC	Dissolve in methanol	Silica	Ammonium hydroxide–methanol–chloroform (80:20:1)	Iodine vapor	USP 23, chromatographic purity; also BP, addendum 1983, p. 257 and p. 273 [4]; TLC [1,8,564,925]	[5]
Tablets	TLC	Add 0.1 N hydrochloric acid, add methanol and centrifuge	Silica				
Tablets	HPLC	Add 0.05M potassium dihydrogen phosphate to powder, sonicate	C-18	Buffer–methanol (3:2), Buffer–potassium dihydrogen phosphate (22.08 g/2 l water, pH 2.8)	295 nm	USP 23, assay; GC [2,8,9]; HPLC [3,8,9,665,667,922]; enantiomers [683,707]	[5]
Tinidazole IV Fluid	TLC	Applied directly	Silica	CHCl₃–MeOH (9:1)	Extract with EtOH, 310 nm	HPLC [1110]	[666]

Drug	Mode	Sample pretreatment	Sorbent (temp.)	Mobile phase	Detection	Comments	Reference
Tobramycin Polymeric drug delivery system	HPLC	Precolumn derivatization	C_8	CH_3CN–50 mM PO_4, pH 3.5 (62:38)	340 nm	Method review	[674]
Tocopherols (see Vitamin E) Bulk	HPLC	MeOH	C_{18}, Zorbax	CH_3CN–MeOH–CH_2Cl_2 (60:35:5)	295 nm	Normal phase also studied [693,694,696]	[692]
Tolfenamic acid Capsules	HPLC	Powder, MeOH	C_{18}, Lichrosorb	0.05M Acetate (pH 4.6)–MeOH, 1.9 ml/min	282 nm	Review; metabolism [1111]	[675]
Trazodone Bulk	HPLC	Add 0.5 ml to Bond Elut, C-18 (1 ml) column. Wash with H_2O (2×), 10% MeOH–H_2O (1:9). Elut with 0.5 ml MeOH	Ultrasphere, C-8, 5 μm, 15 cm × 4.6 mm	CH_3OH–H_2O (1:1) containing 0.5 ml in 1 L of 20 g/l tetra-ammonium perchlorate in MeOH and 0.5 ml of 70% perchloric acid, 1 ml/min	Fluorescence Excitation = 320 nm; Emission = 440 nm	Detection limit 1 μg/l; pK [1112]; bulk, HPLC [684]	
Tretinoin Cream	HPLC	THF	C_{18}, Novapak	THF–PO_4 buffer (42:58), 1 ml/min	365 nm	ISTD–anthracene, degradants	[698]
Triamcinolone Dermatological patches	SFC	SCF extraction	C_8, Spherisorb	MeOH–H_2O (7:3), 0.5 ml/min	240 nm	Stability-indicating, USP 23 [5]; stability [1104]	[714]
Triamterene Bulk	TLC	Dissolve with dimethylsulfoxide, dilute in methanol	Silica	Methanol–acetic acid–ethyl acetate (1:1:8)	250 nm	BP, p. 462 and EP, pt. II, p. 58 [6] and JP, p. 689 [7]; TLC [1,564]; HPLC [10,922]	[4]
Triazolam Bulk	TLC	—	Silica	S_1 = CHCl$_3$–acetone (8:2) S_2 = EtOAc S_3 = CHCl$_3$–MeOH (9:1)	—	hR_{f1} = 5, hR_{f2} = 2, hR_{f3} = 41, hR_{f4} = 43; also GC, SE-30/OV-1, RI: 3134	[132]

	Method	Sample preparation	Column/Phase	Mobile phase	Detection	Comments	Ref.
Bulk	HPLC	—	C-18, Hypersil, 5 μm, flow rate 1.5 ml/min	S_4 = EtOAc–MeOH conc. NH$_4$OH (85:10:5); S_1 = MeOH, H$_2$O, PO$_4$ (0.1M) (55:25:20), pH 7.25; S_2 = MeOH, H$_2$O, PO$_4$ (0.1M) (7:1:2), pH 7.67	—	Cap. Fact. 1: 4.38; Cap. Fact. 2: —	[135]
			Silica, Spherisorb 5W, 2 ml/min	S_3 = MeOH–perchloric acid (1L: 100 μl); S_4 = MeOH, H$_2$O, TFA (997:2:1)		Cap. Fact. 3: 1.39; Cap. Fact. 4: 1.73	[135]
						Same procedure as for diazepam; also HPLC [687,699,762]; GC [763]; GC/MS [885, 886]	
Trifluoperazine							
Bulk	HPLC	—	Silica	MeOH (1% NH$_4$OH)–CH$_2$Cl$_2$ (5:95)	254 nm	UV, MS, NMR, IR	[10, p. 2302]
Bulk	TLC	Dissolve in methanol	Silica	Ammonium hydroxide–acetone (1:200)	Iodoplatinate	USP 23, identification, also EP, pt. II, p. 59 [6]; TLC [1,11]; GC [2,8,9]; HPLC [3,9, 673,922]	[5]
Trimethoprim							
Bulk	TLC	Dissolve in methanol–chloroform (9:1)	Silica	Chloroform–methanol–6 N ammonium hydroxide (95:7.5:1)	254 nm	USP 23, also BP, p. 466 [4]; TLC [1]; GC [2,9]; HPLC [647,678,922]	[5]
Tablets	TLC	Triturate with methanol, centrifuge	Silica				
Tablets	HPLC	Add methanol to the powder, centrifuge	C-18	1% Acetic acid–acetonitrile (84:16)	254 nm	USP 23, assay, also BP, addendum 1983, p. 308 [4]; HPLC [3,9]	[5]
Trimethoxyamphetamine (3, 4, 5-trimethoxy amphetamine)							
Bulk	HPLC	—	Silica	MeOH (1% NH$_4$OH)–CH$_2$Cl$_2$ (1:4)	254 nm	UV, MS, NMR, IR	[10, p. 2322].

Drug	Mode	Sample pretreatment	Sorbent (temp.)	Mobile phase	Detection	Comments	Reference
Trimipramine							
Bulk	TLC	—	Silica	S_1 = EtOAc–MeOH–conc. NH$_4$OH (85 : 10 : 4) S_2 = MeOH S_3 = MeOH–1-butanol (3 : 2)–0.1M NaBr S_4 = MeOH–conc. NH$_4$OH (100 : 1.5) S_5 = cyclohexane–toluene–diethylamine (75 : 15 : 10) S_6 = CHCl$_3$–MeOH (9 : 1) S_7 = acetone	—	hR$_{f1}$ = 80, hR$_{f2}$ = 36, hR$_{f3}$ = 56, hR$_{f4}$ = 59, hR$_{f5}$ = 62, hR$_{f6}$ = 54, hR$_{f7}$ = 37, also, GC, SE-30/OV-1, RI: 2201	[132]
Bulk	TLC	Dissolve in methanol–ammonium hydroxide (9 : 1)	Silica	Toluene–methanol (95 : 5)	Potassium dichromate	EP, vol, III, p. 359; TLC [1]; GC [2,10]; HPLC [3,10,922]	[6]
Tripelennamine							
Formulations	HPLC	—	Cyano	Methanol–water–propylamine (90 : 10 : 0.01)	254 nm	TLC [1,564]; GC [2,10, 706]; HPLC [3,9,569]	[1087]
Triprolidine							
Bulk	TLC	Dissolve in methanol	Silica	Butanol–acetic acid–water (8 : 2 : 2)	254 nm	USP 23, chromatographic purity, also BP, p. 467 [4]; [1,8]	[5]
Tablets	TLC	Add 0.01 N hydrochloric acid	Silica				
Syrup	HPLC	Dilute with 0.01 N hydrochloric acid	Silica	0.25 N Ammonium acetate–ethanol (3 : 17)	254 nm	USP 23, assay: GC [2,8, 10]; HPLC [3,10,922, 1018]	[5]
Tablets	HPLC	Add 0.01 N acid, mix, filter	Silica				
Syrup with pseudoephedrine	HPLC	Dilute with 0.01 N acid	Silica				
Tablets with pseudoephedrine	TLC	Add water to the powder	Silica	Butanol–acetic acid–water (8 : 2 : 2)	254 nm	USP 23, identification	[5]

Compound	Method	Sample prep	Stationary phase	Mobile phase	Detection	Remarks	Ref.
Bulk	GC	—	DB-5	—	FID	Resolved 23 antihistamines	[706, p.164]
Tropicamide Bulk	TLC	Dissolve in chloroform	Silica	Toluene–dioxane–13.5 ammonium hydroxide (12 : 7 : 1)	254 nm	BP, p. 468, TLC [1,24, 925]; GC [2]; enantiomer [919]	[4]
Tryptophan Bulk	HPLC	—	Cyclobond III, Varex/Astec	MeOH–1% TEAA (15 : 85), pH 5.0, 0.4 ml/min	254 nm	Resolved D,L; HPLC [704]	[701]
Vancomycin Injection	HPLC	—	C_{18}, Ultrasphere	Gradient: TBAH–0.1M acetate, pH 4.5, 1.5 ml/min	254 nm	Methods reviewed, EDTA detected; stability [725]	[722]
Verapamil Bulk	TLC	—	Silica	S_1 = EtOAc–MeOH–conc. NH_4OH (85 : 10 : 4) S_2 = MeOH S_3 = MeOH–1-butanol (3 : 2)–0.1M NaBr S_4 = MeOH–conc. NH_4OH (100 : 1.5) S_5 = cyclohexane–toluene–diethylamine (75 : 15 : 10) S_6 = $CHCl_3$–MeOH (9 : 1) S_7 = acetone	—	hR_{f1} = 74, hR_{f2} = 44, hR_{f3} = 61, hR_{f4} = 59, hR_{f5} = 23, hR_{f6} = 70, hR_{f7} = 42; also GC, SE-30/OV-1, RI: 3200	[132]
Bulk	HPLC	—	Silica	Methylene chloride–methanol (1% ammonium hydroxide) (98 : 2)	254 nm	TLC [1,8,564]; GC [8, 10]; HPLC [3,8,237,717, 718,922]; enantiomer [237,919]	[10]
Vinblastine Bulk	TLC	MeOH	Silica	Benzene–CHCl3–DEA (40 : 20 : 3)	254 nm	BP, p. 473; TLC [1]; [1162,1163]	[4]
Vincristine Injection	TLC	H_2O	Silica	Ether–MeOH–40% methylamine (95 : 10 : 5)	$(NH_4)_2SO_4$	USP 23; BP, vol.II, p. 677 [4]; [1164]; TLC, GC, HPLC [8]	[5]

Drug	Mode	Sample pretreatment	Sorbent (temp.)	Mobile phase	Detection	Comments	Reference
Vitamin A (Retinol acetate and palminate)							
Bulk	TLC	CHCl₃	Silica	Cyclohexane–ether (4 : 1)	Phosphomolybdic acid	USP 23; JP [7]; stability [1104]; review [726,799,1165]; HPLC [750]; HPLC–PDA [729]	[5]
Vitamin B₁ (See thiamine)							
Vitamin B₂ (Riboflavin, see thiamine)							
Vitamin B₃ (See niacin)							
Vitamin B₆ (See pyridoxine)							
Vitamin B₁₂ (See cyanocobalamin)							
Vitamin C (See ascorbic acid)							
Vitamin D₂ (Ergocalciferol)/							
Vitamin D₃ (Chlolcalciferol)							
Oil	HPLC	Toluene/mobile phase	Silica	n-Hexane–amyl alcohol (997 : 3)	254 nm	AOAC; USP 23 [5]; HPLC [731–735]; review [726,1166]	[1167]
Vitamin E (Tocopheryl acetate)							
Bulk	GC	Dissolve in low-actinic glassware	USP-G2 (245°C)	—	FID	USP 23; HPLC [736,743, 750,800]	[5]
Vitamin H (See biotin)							
Warfarin							
Bulk and injection	HPLC	Dissolve in 0.1 N sodium hydroxide and 0.2M potassium dihydrogen phosphate	C-8	Acetic acid–water–methanol (1 : 36 : 64)	280 nm	USP 23, ISTD–propylparaben, assay, HPLC, TLC [1]; GC [2,9]; HPLC [9,219,569, 922,1105]; stability [923]; CE [753]; enantiomers [752]	[5]
Tablets	HPLC	Mix with the above buffer, filter					
Yohimbine							
Bulk	TLC	—	Silica	S_1 = EtOAc–MeOH– conc. NH₄OH (85 : 10 : 4) (3 : 2)–0.1M NaBr	—	hR_{f1} = 65, hR_{f2} = 66, hR_{f3} = 70, hR_{f4} = 63, hR_{f5} = 5, hR_{f6} = 38,	[132]

Compound	Method	Solvent	Stationary phase	Mobile phase	Detection	Notes / References
						$hR_f = 52$; also GC, SE-30/OV-1, RI: 3269 [9]
				S_2 = MeOH S_3 = MeOH–1-butanol (3 : 2)–0.1M NaBr S_4 = MeOH–conc. NH$_4$OH (100 : 1.5) S_5 = Cyclohexane–toluene–diethylamine (75 : 15 : 10) S_6 = CHCl$_3$–MeOH (9 : 1) S_7 = acetone		
Bulk	HPLC	—	Silica	Methylene chloride–methanol (1% ammonium hydroxide) (98 : 2)	254 nm	TLC [1]; GC [2,8,9]; HPLC [8,1027]
Bulk	HPLC	—	PRP-1 150 × 4.1 mm	A. 20 mM ammonium hydroxide B. CH$_3$CN. Linear gradient 15–100% B in 17 min, hold 3 min	220 nm	Hamilton Application Catalog #13, p.8 (1992); resolved from morphine, codeine, thebaine, cocaine, reserpine, and methadone; also plant extract [565], p. 67; HPLC [569,705, p. 215]; TLC [656]
Zalcitabine Zidovudine						Sterisomers [757]
Zeranol Zolazepam						Stability indicating [755]; also [206] GC/MS [852] Formulated with tiletamine combination known as Telazol
Zomepirac Bulk	TLC	MeOH	Silica	CHCl$_3$–AcOH (9 : 1)	254 nm	USP 21, p.1133; TLC [1], GC [2,9]

REFERENCES

1.   A. H. Stead, R. Gill, T. Wright, J. R. Gibbs, and A. C. Moffat, Standardized thin-layer chromatographic systems for the identification of drugs and poisons, *Analyst, 107*:1106 (1982).

2.   R. W. Ardrey and A. C. Moffat, Gas-liquid chromatographic retention indices of 1318 substances of toxicological interest on SE-30 or OV-1 stationary phase, *J. Chromatogr., 220*:195 (1981).

3.   I. Jane, A. McKinnon, and R. J. Flanagan, High-performance liquid chromatographic analysis of basic drugs on silica columns using non-aqueous ionic eluents, *J. Chromatogr., 323*:191 (1985).

4.   *British Pharmacopoeia*, HMSO, London, 1980.

5.   *The United States Pharmacopeia*, Twenty-First Revision, United States Pharmacopeial Convention, Inc., Rockville, MD, 1985.

6.   *European Pharmacopoeia*, S. A. Maisonneuve, Ruffine, France.

7.   *The Japanese Pharmacopoeia*, 10th ed. (Eng. ed. 1982), Yakuki Nippo, Ltd., Tokyo, 1981.

8.   T. Daldrup, F. Susanto, and P. Michalke, Kombination von DC, GC, and HPLC zur schnellen Erkennung von Arzneimittein, Rauschmitteln und verwandten Verbindungen, *Fresenius Z. Anal. Chem., 308*:413 (1981).

9.   T. M. Mills III, N. Price, P. T. Price and J. C. Robertson, *Instrumental Data for Drug Analysis, Vol. I*, Elsevier, New York.

10.   T. M. Mills III, W. N. Price, P. T. Price and J. C. Robertson, *Instrumental Data for Drug Analysis, Vol. II*, Elsevier, New York.

11.   K. Florey (ed.), *Analytical Profiles of Drug Substances, 14*, Academic Press, New York, 1985, p. 578.

12.   J. J. Bergh and A. P. Lotter, A stability-indicating gas-liquid chromatographic method for the determination of acetaminophen and aspirin in suppositories, *Drug Dev. Ind. Pharm., 10*:127 (1984).

13.   W. R. Sisco, C. T. Rittenhouse, and L. A. Everhart, Simultaneous high-performance liquid chromatographic stability-indicating analysis of acetaminophen and codeine phosphate in tablets and capsules, *J. Chromatogr., 348*:253 (1985).

14.   V. Das Gupta and A. Helbe, Quantitation of acetaminophen, chlorpheniramine maleate, dextromethorphan hydrobromide, and phenylpropanolamine hydrochloride in combination using high-performance liquid chromatography, *J. Pharm. Sci., 73*:1553 (1984).

15.   N. Kikuchi and T. Ohhata, High-performance liquid chromatography for pharmaceutical analyses. II. Major components in commercial cold medications, *Iwate-Ken Eisei Kenkyusho Nenpo, 26*: 61 (1983).

16. R. Thomas, E. Roets, and J. Hoogmartens, Analysis of tablets containing aspirin, acetaminophen, and ascorbic acid by high-performance liquid chromatography, *J. Pharm Sci., 73*:1830 (1984).

17. J. Fan and X. Li, Assay of acetaminophen in Chinese compound preparations by HPLC, *Yaowu Fenxi Zazhi, 4*:348 (1984).

18. A. C. Moffat and B. Clare, The choice of paper chromatographic and thin layer chromatographic systems for the analysis of basic drugs, *J. Pharm. Pharmacol., 26*:665 (1974).

19. T. A. Gough and P. B. Baker, The separation and quantitation of the narcotic components of illicit heroin using reversed-phase high performance liquid chromatography, *J. Chromatogr. Sci., 19*:277 (1981).

20. F. T. Frank, J. S. Thruber, and D. M. Dye, High-performance liquid chromatography of acetylcholine in a pharmaceutical preparation, *J. Pharm. Sci., 73*:1311 (1984).

21. T. M. Ryan and P. Zoutendam, Quantitative determination of acivicin in bulk and pharmaceutical formulations by ion-pair high-performance liquid chromatography, *J. Chromatogr., 357*:207 (1986).

22. T. L. Ascah and B. T. Hunter, Simultaneous HPLC determination of propoxyphene and acetaminophen in pharmaceutical preparations, *J. Chromatogr., 455*:279 (1988).

23. G. P. Cartoni, M. Lederer, and F. Polidori, Some chromatographic and electrophoretic data for amphetamine-like drugs, *J. Chromatogr., 71*:370 (1972).

24. *Pharmacopeial Forum*, 769 (1985).

25. A. C. Moffat (ed.), Clarke's Isolation and Identification of Drugs, The Pharmaceutical Press, London, 1986.

26. Compendial Monograph Evaluation and Development; Allopurinol, *Pharmacopeial Forum*, 3952 (1984).

27. D. Shostak, Liquid chromatographic determination of allopurinol in tablets: collaborative study, *J. Assoc. Off. Anal. Chem., 67*: 1121 (1984).

28. J. Fogel, P. Epstein, and P. Chen, Simultaneous high-performance liquid chromatography assay of acetylsalicylic acid and salicylic acid in film-coated aspirin tablets, *J. Chromatogr., 317*:507 (1984).

29. F. de Croo, W. van den Bossche, and P. de Moerloose, Simultaneous determination of altiazide and spironolactone in tablets by high-performance liquid chromatography, *J. Chromatogr., 329*: 422 (1985).

30.   J. A. Adamovics, Return to unmodified silica and alumina chromatography, *LC Mag., 2*:393 (1984).

31.   R. Gill, A. H. Stead, and A. C. Moffat, Analytical aspects of barbiturate abuse: Identification of drugs by the effective combination of gas-liquid, high-performance liquid and thin-layer chromatographic techniques, *J. Chromatogr., 204*:275 (1981).

32.   M. A. Brook, M. R. Hackman, and D. J. Mazzo, Amoxicillin by high-performance liquid chromatography with amperometric detection, *J. Chromatogr., 210*:531 (1981).

33.   P. De Pourcq, J. Hoebus, E. Roets, J. Hoogmartens, and H. Vanderhaeghe, Quantitative determination of amoxicillin and its degradation products by high-performance liquid chromatography, *J. Chromatogr., 321*:441 (1985).

34.   M. J. Le Belle, W. L. Wilson, and G. Lauriault, Determination of amoxicillin in pharmaceutical dosage forms, *J. Chromatogr., 202*: 144 (1980).

35.   A. S. Curry and D. A. Patterson, A procedure for the analysis of illicit diamorphine samples, *J. Pharm. Pharmacol., 22*:198 (1970).

36.   S. Y. Yeh and R. L. McQuinn, GLC determination of heroin and its metabolites in human urine, *J. Pharm. Sci., 64*:1237 (1975).

37.   K. D. Parker, J. A. Wright, A. F. Halpern, and C. H. Hine, Preliminary report on the detection and quantitation of opiates and certain other drugs of abuse as trimethysilyl derivatives by gas-liquid chromatography, *J. Forens. Sci. Soc., 10*:17 (1970).

38.   K. E. Rasmussen, Quantitative morphine assay by means of gas-liquid chromatography and on-column silylation, *J. Chromatogr., 120*:491 (1976).

39.   P. O. Edlund, Determination of opiates in biological samples by glass capillary gas chromatography with electron-capture detection, *J. Chromatogr., 204*:206 (1981).

40.   R. Dybowski and T. A. Gough, Identification of 5,5-disubstituted barbiturates, *J. Chromatogr. Sci., 22*: 104 (1984).

41.   J. M. Moore, A. C. Allen, and D. A. Cooper, Determination of manufacturing impurities in heroin by capillary gas chromatography with electron capture detection after derivatization with heptafluorobutyric anhydride, *Anal. Chem., 56*:642 (1984).

42.   M. Gloger and N. Heumann, Untersuchungen zur Stabilitat illegaler Heroinproben [Stability of illegal heroin samples], *Arch. Kriminol., 166*(3/4):89 (1980).

43.   C. Radecka and I. C. Nigam, Detection of trace amounts of lysergic acid diethylamide in sugar cubes, *J. Pharm. Sci., 55*:861 (1966).

44.   M. Chiarotti, A. Carnevale, and N. de Giovanni, Capillary gas

chromatographic analysis of illicit diamorphine preparations, *Forens. Sci. Int., 21*:245 (1983).

45. J. M. Newton, Gas-liquid chromatographic determination of nik-ethamide in injectable preparations, *J. Assoc. Anal. Chem., 59*:93 (1976).

46. K. Florey (ed.), *Analytical Profiles of Drug Substances, 15,* Academic Press, New York, 1986, p. 427.

47. G. T. Briguglio and C. A. Lau-Cam, Separation of 9 penicillins by reversed-phase liquid chromatography, *J. Assoc. Off. Anal. Chem., 67*:228 (1984).

48. B. Morelli, Determination of penicillins in pure form and in pharmaceuticals, *Anal Lett., 20*:141 (1987).

49. L. A. Pachla, D. L. Reynolds, and P. T. Kissinger, Analytical methods for determining ascorbic acid biological samples, food products, and pharmaceuticals, *J. Assoc. Off. Anal. Chem., 68*:1 (1985).

50. A. C. Moffat, Use of SE-30 as a stationary phase for the gas liquid chromatography of drugs, *J. Chromatogr., 113*:69 (1975).

51. A. C. Moffat, A. H. Stead, and K. W. Smalldon, Optimum use of paper, thin-layer and gas-liquid chromatography for the identification of basic drugs. 3. Gas-liquid chromatography, *J. Chromatogr., 90*:19 (1974).

52. B. Caddy, F. Fish, and D. Scott, Chromatographic screening for drugs of abuse using capillary columns, part 1, comparison of open tubular columns and support coated open tubular columns for the analysis of central nervous system stimulant drugs, *Chromatographia, 6*:251 (1973).

53. S. Paphassarang, J. Raynaud, R. Godeau, and A. M. Binsard, Quantitative and qualitative analyses of hyoscyamine-atropine and scopolamine in pure tincture of solanaceae by HPLC, *J. Chromatogr., 319*:412 (1985).

54. A. Richard and G. Andermann, Simultaneous determination of atropine sulfate and tropic acid by reversed-phase high pressure liquid chromatography, *Pharmazie, 39*:866 (1984).

55. T. Jira, T. Beyrich, and E. Lemke, Ion-pair reversed-phase HPLC of tropane alkaloids, *Pharmazie, 39*:351 (1984).

56. A. Bettero and P. Bollettin, Atropine sulfate determination by derivative spectroscopy HPLC, *Ann. Chim., 75*:351 (1985), Rome.

57. E. Gordis, Gas-chromatographic resolution of optical isomers in microgram samples of amphetamine, *Biochem. Pharmacol., 15*:2124 (1966).

58. C. E. Wells, Collaborative study of the gas-liquid chromato-

graphic method for determination of stereochemical composition of amphetamine, *J. Assoc. Off. Anal. Chem., 55*:146 (1972).

59. D. E. Nichols, C. F. Barfknecht, D. B. Rusterholtz, F. Bennington, and R. D. Morin, Asymmetric synthesis of psychotomimetic phenylisopropylamines, *J. Med. Chem., 16*:480 (1973).

60. D. Marini, F. Balestrieri, and A. Sacchini, Characterization and assay of a new monobactamic antibiotic: aztreonam, *Boll. Chim. Farm., 124*:218 (1985).

61. J. A. Adamovics and S. Unger, Preparative liquid chromatography of pharmaceuticals using silica with aqueous eluents, *J. Liquid Chromatogr., 9*:141 (1986).

62. E. Burtschen, H. Binder, R. Concin, and O. Bobleter, Separation of phenols, phenolic aldehydes, ketones, and acids by HPLC, *J. Chromatogr., 252*:167 (1982).

63. C. E. Orazio, S. Kapila, and S. E. Manahan, High-performance liquid chromatographic determination of phenols as phenolates in a complex mixture, *J. Chromatogr., 262*:434 (1983).

64. L. L. Lloyd, J. A. McConville, F. P. Warner, J. F. Kennedy, and C. A. White, HPLC analysis of L-ascorbic acid using a reversed-phase adsorbent, *LC–GC, Mag., 5*:338 (1987).

65. E. Brochmann-Hanssen and T. O. Oke, Gas chromatography of barbiturates, phenolic alkaloids, and xanthine bases: Flash-heater methylation by means of trimethylanilinium hydroxide, *J. Pharm. Sci., 58*:370 (1969).

66. S. Dilli and D. N. Pillai, Relative electron capture response of the 2-chloroethyl derivatives of some barbituric acids and anticonvulsant drugs, *J. Chromatogr., 137*:111 (1977).

67. A. H. Stead, R. Gill, A. T. Evans, and A. C. Moffat, Predictions of gas chromatographic retention characteristics of barbiturates from molecular structure, *J. Chromatogr., 234*:277 (1982).

68. N. C. Jain and R. H. Cravey, *J. Chromatogr. Sci., 16*:587 (1978).

69. D. N. Pillai and S. Dilli, Analysis of barbiturates by gas chromatography, *J. Chromatogr., 220*:253 (1981).

70. R. D. Budd, Gas chromatographic separations of dialkyl barbiturate derivatives, *J. Chromatogr., 192*:212 (1980).

71. J. R. Sharman, OV-225 as a stationary phase for the determination of anticonvulsants, mexiletine, barbiturates, and acetaminophen, *J. Anal. Toxicol., 5*:153 (1981).

72. R. Dybowski and T. A. Gough, Identification of 5,5-disubstituted barbiturates, *J. Chromatogr. Sci., 22*:104 (1984).

73. I. I. Hewala, HPLC and derivative difference spectrophotometric methods for the determination of acetaminophen and its degrada-

tion product in aged pharmaceutical formulation, *Anal. Lett., 27*: 561 (1994).

74. J. E. Kountourellis, A. Raptouli, and P. P. Georgakopoulos, Simultaneous determination of bromochlorosalicylanilide and bamipine in pharmaceutical formulations by HPLC, *J. Chromatogr., 362*:439 (1986).

75. G. Yurdakul, L. Ersoy, and S. Sungur , HPLC and derivative spectrophotometric methods for the determination of paracetamol and caffeine in tablets, *Pharmazie, 46*:885 (1991).

76. P. J. Cashman and J. I. Thornton, High speed liquid adsorption chromatography in criminalistics. II. The separation of heroin, monoacetylmorphine and morphine, *J. Forens. Sci., 12*:417 (1972).

77. J. D. Wittwer, Liquid chromatographic determination of morphine in opium, *J. Forens. Sci., 18*:138 (1973).

78. J. H. Knox and J. Jurand, Separation of morphine alkaloids heroin, methadone and other drugs by ion-exchange chromatography, *J. Chromatogr., 82*:398 (1973).

79. I. Jane, The separation of a wide range of drugs of abuse by high-pressure liquid chromatography, *J. Chromatogr., 111*:227 (1975).

80. P. B. Baker and T. A. Gough, The separation and quantitation of the narcotic components of illicit heroin using reversed-phase high performance liquid chromatography, *J. Chromatogr. Sci., 19*:483 (1981).

81. I. Beaumont and T. Deek, Determination of morphine, diamorphine and their degradation products in pharmaceutical preparation by reversed-phase high-performance liquid chromatography, *J. Chromatogr., 238*:520 (1982).

82. P. P. Gladyshev, E. F. Matantseva, and M. I. Goryaev, High-efficiency ion-exchange chromatography of opium alkaloids, *Zh. Anal. Khim., 36*(6):1130 (1981).

83. C. J. C. M. Laurent, H. A. H. Billiet, and L. de Galan, Use of organic modifiers in ion exchange chromatography on alumina separation of basic drugs, *Chromatographia, 17*:394 (1983).

84. C. J. C. M. Laurent, H. A. H. Billiet, and L. de Galan, High performance liquid chromatography of heroin samples on alumina by ion exchange in mixed aqueous organic mobile phases, *Chromatography, 285*:161 (1984).

85. J. K. Baker, R. E. Skelton, and C-Y. Ma, Identification of drugs by high pressure liquid chromatography with electrochemical detection, *J. Chromatogr., 168*:417 (1979).

86.   J. E. Wallace, S. C. Harris, and M. W. Peek, Determination of morphine by liquid chromatography with electrochemical detection, *Anal. Chem., 52*:1328 (1980).

87.   J. A. Glasel and R. F. Venn, Fluorescence and UV detection of opiates separated by reverse-phase HPLC, *J. Chromatogr., 213*: 337 (1981).

88.   P. E. Nelson, S. L. Nolan, and K. R. Bedford, High-performance liquid chromatography detection of morphine by fluorescence after post-column derivatization, *J. Chromatogr., 234*:407 (1982).

89.   P. C. White, I. Jane, A. Scott, and B. E. Connett, Use of high-performance liquid chromatography to quantitate the opiate and sugar content of illicit heroin preparations, *J. Chromatogr., 265*: 293 (1983).

90.   P. J. O'Neil, P. B. Baker, and T. A. Gough, Illicitly imported heroin products: Some physical and chemical features indicative of their origin. Part II, *J. Forens. Sci., 29*:889 (1984).

91.   P. J. O'Neil and T. A. Gough, Illicitly imported heroin products: some physical and chemical features indicative of their origin. Part II, *J. Forens. Sci., 30*:681 (1985).

92.   H. Huizer, Analytical studies on illicit heroin II. Comparison of samples, *J. Forens. Sci., 28*:40 (1983).

93.   M. L. Chan, C. Whetsell, and J. D. McChesney, Use of high pressure liquid chromatography for the separation of drugs of abuse, *J. Chromatogr. Sci., 12*:512 (1974).

94.   I. Lurie, Application of reverse phase ion-pair partition chromatography to drugs of forensic interest, *J. Assoc. Off. Anal. Chem., 60*:1035 (1977).

95.   S. K. Goswami and J. E. Kinsella, Separation of prostaglandins A, B, D, E, F, thromboxane and G-keto prostaglandin F-1 by thin-layer chromatography, *J. Chromatogr., 209*:334 (1981).

96.   Captopril, *Pharmacopeial Forum*, 1134 (1986).

97.   A. Tsuji, E. Nakashima, Y. Deguchi, K. Nishide, T. Shimizu, S. Horiuchi, K. Ishikawa, and T. Yamana, Degradation kinetics and mechanism of aminocephalosporins in aqueous solution: Cefadroxil, *J. Pharm. Sci., 70*:1120 (1981).

98.   E. M. Patzer and D. M. Hilton, New reagent for vitamin B-6 derivative information in GC, *J. Chromatogr., 135*:489 (1977).

99.   W. Jubiz, G. Nolan, and K. C. Kalterborn, An improved technique for extraction, identification and quantification of leukotrienes, *J. Liquid Chromatogr., 8*:1519 (1985).

100. K. Yamamura, J. Yamada, and T. Yotsuyanagi, High-performance liquid chromatographic assay of antiinflammatory drugs

incorporated in gel ointments; separation and stability testing, *J. Chromatogr., 33*:383 (1985).

101. S. Inayama and H. Hori, Prostaglandin $F_{2\alpha}$, *J. Chromatogr., 194*: 85 (1980).

102. D. J. Weber, High-pressure liquid chromatographic separation of 5,6-cis and trans-prostaglandin A-2, *J. Pharm. Sci., 66*:744 (1977).

103. R. P. Haribhakti, K. S. Alexander, and G. A. Parker, An improved HPLC assay for acetazolamide, *Drug Dev. Ind. Pharm., 17*:805 (1991).

104. B. G. Snider, Separation of cis-trans isomers of prostaglandins with a cyclodextrin bonded column, *J. Chromatogr., 351*:548 (1986).

105. R. J. Flanagan, G. C. A. Storey, and R. K. Bharma, High performance liquid chromatographic analysis of basic drugs on silica columns using non aqueous ionic eluents, *J. Chromatogr., 247*:15 (1982).

106. R. W. Roos, Separation and determination of barbiturates in pharmaceuticals by high speed liquid chromatography, *J. Pharm. Sci., 61*:1979 (1972).

107. J. K. Baker, Estimation of high-pressure liquid chromatographic retention indexes, *Anal. Chem., 51*:1693 (1979).

108. A. Amin, High-performance liquid chromatography of water-soluble vitamins. II. Simultaneous determination of vitamins B-1, B-2, B-6, and B-12 in pharmaceutical preparations, *J. Chromatogr., 390*:448 (1987).

109. K. S. Alexander, R. P. Haribhakti, and G. A. Parker, Stability of acetazolamide in suspension, *Am. J. Hosp. Pharm., 48*:1241 (1991).

110. Z. S. Gomaa, Determination of acetazolamide in dosage forms by HPLC, *Biomed. Chromatogr., 7*:134 (1993).

111. HPLC analysis using RI detection, *The United States Pharmacopeia,* Twenty-Third Revision, United States Pharmacopeical Convention, Inc., Rockville, MD, 1995, p. 33.

112. H. G. Britian (ed.), *Analytical Profiles of Drug Substances and Excipients, 21,* Academic Press, New York, 1992, p. 1.

113. A. Verstraeten, E. Roets, and J. Hoogmartens, Acetylsalicylic acid tablets, *J. Chromatogr., 388*:201 (1987).

114. R. P. Barron, A. V. Kruegel, and J. M. Moore, Identification of impurities in illicit methamphetamine samples, *J. Assoc. Off. Anal. Chem., 57*(5):1147 (1974).

115. H. Neumann, Determination and impurity profiling of opium,

crude morphine and illicit heroin by capillary gas chromatography. In *Analytical Methods in Forensic Chemistry* (M. H. Ho, ed.), Ellis Horwood, New York, 1990, pp. 186–194.

116.   P. B. Baker and G. F. Phillips, The forensic analysis of drugs of abuse. A review, *Analyst, 108*:777 (1983).

117.   C. M. Selavka and I. S. Krull, Trace determination of barbiturates with LC–photolysis–electrochemical detection (LC–hv–EC). In *Analytical Methods in Forensic Chemistry* (M. H. Ho, ed.), Ellis Horwood, New York, 1990, pp. 195–209.

118.   R. A. de Zeeuw, J. P. Franke, and M. Bogusz, High performance liquid chromatography with a multichannel diode-array spectrophotometric detector in systematic toxicological analysis. In *Analytical Methods in Forensic Chemistry* (M. H. Ho, ed.), Ellis Horwood, New York, 1990, p. 212.

119.   P. E. Nelson and L. M. Boyd, Phosphorescence detection in high-performance liquid chromatography of drugs. In *Analytical Methods in Forensic Chemistry* (M. H. Ho, ed.), Ellis Horwood, New York, 1990, pp. 225–235.

120.   P. E. Nelson and S. L. Nolan, Post-column techniques for improvement of the high-performance liquid chromatographic detection of morphine. In *Analytical Methods in Forensic Chemistry* (M. H. Ho, ed.), Ellis Horwood, New York, 1990, pp. 247–259.

121.   J. T. Burke, W. A. Wargin, and M. R. Blum, High pressure liquid chromatographic assay for chloramphenicol, chloramphenicol 3-mono succinate and chloramphenicol 1-mono succinate, *J. Pharm. Sci., 69*:909 (1980).

122.   W. J. Irwin, A. L. W. Po, and R. R. Wadwani, A gas chromatographic assay of chloramphenicol application to formulation and to samples showing hydrolytic or photochemical degradation, *J. Clin. Hosp. Pharm., 69*:55 (1980).

123.   V. Das Gupta, Quantitation of chlorpropamide and tolbutamide in tablets by stability-indicating reverse phase high-performance liquid chromatography, *Anal. Lett., 17*(B18):2119 (1984).

124.   R. Everett, Compendial monograph evaluation and development: Chlorpropamide, *Pharmacopeial Forum*, 950 (1985).

125.   L. A. Jones, The analysis, isolation and identification of phencyclidine cosynthetics. In *Analytical Methods in Forensic Chemistry* (M. H. Ho, ed.), Ellis Horwood, New York, 1990, p. 322–341.

126.   J. Fogel, J. Sisco, and F. Hess, Validation of liquid chromatographic method for assay of chlorthalidone in tablet formulations, *J. Assoc. Off. Anal. Chem., 68*:96 (1985).

127.   J. Cordonnier and B. Heyndrickx, Detection and pharmacokinetic

studies of tilidine, nortilidine and bisnortilidine in human biological fluids of forensic interest. In *Analytical Methods in Forensic Chemistry* (M. H. Ho, ed.), Ellis Horwood, New York, 1990, p. 127.

128. Y. Pramaar, V. D. Gupta, and T. Zerai, Quantitation of acyclovir in pharmaceutical dosage forms using HPLC, *Drug Dev. Ind. Pharm., 16*:1687 (1990).

129. J. A. Vinson, J. J. Urash, and D. J. Lopatofsky, A semi-automated extraction and spotting system for the analysis of drugs in physiological fluids. In *Analytical Methods in Forensic Chemistry* (M. H. Ho, ed.), Ellis Horwood, New York, 1990, p. 385.

130. L. Bonhomme, D. Benhamou, E. Conroy, and N. Preaux, Stability of adrenaline pH-adjusted solutions of local anaesthetics, *J. Pharm. Biomed. Anal., 9*:497 (1991).

131. Y.-M. Liu and S.-J. Sheu, Determination of ephedrine alkaloids by capillary electrophoresis, *J. Chromatogr., 600*:370 (1992).

132. A. C. Moffat, J. P. Franke, A. H. Stead, R. Gill, B. S. Finkle, M. R. Moller, R. K. Muller, F. Wunsch, and R. A. de Zeeuw, *Thin-Layer Chromatographic rf Values of Toxicologically Relevant Substances on Standarized Systems,* VCH, Weinheim, 1987.

133. B. Caddy, HPLC for the detection and quantitation of abused drugs. In *The Analysis of Drugs of Abuse* (T. A. Gough, ed.), Wiley, New York, 1991, p. 151.

134. T. A. Gough and P. B. Baker, Identification major drugs of abuse using chromatography, *J. Chromatogr. Sci., 20*:289 (1982).

135. M. Japp, K. Garthwaite, A. V. Geeson, and M. D. Osselton, Collection of analytical data for benzodiazepines and benzophenones, *J. Chromatogr., 439*:317 (1988).

136. G. Vigh and J. Inczedy, Separation of chloramphenicol intermediates by HPLC on micropak-amino columns, *J. Chromatogr., 129*: 81 (1976).

137. A. G. Adams and J. T. Stewart, A HPLC method for the determination of albuterol enantiomers on human serum usage SPE and a sumichiral-achiral stationary phase, *J. Liquid Chromatogr., 16*: 3863 (1993).

138. E. W. Tsai, M. M. Singh, H. L. Lu, D. P. Ip, and M. A. Brooks, Application of capillary electrophoresis to pharmaceutical analysis. Determination of alendronate in dosage forms, *J. Chromatogr., 626*:245 (1992).

139. E. W. Tsai, D. P. Ip, and M. A. Brooks, Determination of alendronate on pharmaceutical dosage forms by ion chromatography with conductivity detection, *J. Chromatogr., 596*:217 (1992).

140.   E. Bonetdomingo, M. J. Medinahernandez, and M. C. Garciaal-
       varezcogue, A micellar liquid chromatographic procedure for the
       determination of amiloride, bendroflumethiazale, chlorothali-
       done, spironolactone and triamterene in pharmaceuticals, *J.
       Pharm. Biomed Anal., 11*:705 (1993).
141.   P. M. Lacroyic, N. M. Curran, W. W. Sy, D. K. J. Gorecki, P.
       Thibault, and P. T. S. Blay, *JAOAC Int., 77*:1447 (1994).
142.   R. Bhushan and S. Joshi, Resolution of enantiomers of amino
       acids by HPLC, *Biomed. Chromatogr., 7*:235 (1993).
143.   G. D. Reed, Identification of the controlled drugs propoxyphene,
       benzoylecgonine, and atropine by electron impact ionization capil-
       lary GC–MS, *J. High Resolu. Chromatogr., 15*:46 (1992).
144.   B. Carratu, C. Boniglia, and G. Bellomonte, Optimization of the
       determination of amino acids in parenteral solutions by HPLC
       with pre-column derivatization using 9-fluorenylmethyl chloro-
       formate, *J. Chromatogr., 708*:203 (1995).
145.   E. Kwong, A. M. Y. Chin, S. McClintock, and M. L. Cotton,
       HPLC analysis of an amino bisphonate in pharmaceutical formu-
       lations using post column derivatization and fluorescence detec-
       tion, *J. Chromatogr. Sci., 28*:563 (1990).
146.   R. N. Gupta, Drug level monitoring antidepressants, *J. Chroma-
       togr. Sci., 576*:183–211 (1992).
147.   M. T. Ackermans, F. M. Evaerts, and J. L. Beckers, Determina-
       tion of aminoglycoside antibiotics in pharmaceuticals by capillary
       zone electrophoresis with indirect UV detection coupled with mi-
       cellar electrophoretic capillary chromatography. *J. Chromatogr.,
       606*:229 (1992).
148.   H. Fabre, M. Sekkat, M. D. Blanchin, and B. Mandron, Determi-
       nation of aminoglycosides in pharmaceutical formulations – II.
       HPLC, *Pharm. Biomed. Anal., 7*:1711 (1989).
149.   J. A. Polta, D. C. Johson, and K. E. Merhel, Liquid chromato-
       graphic separation of aminoglycosides as in pulsed amperometric
       detection, *J. Chromatogr., 324*:407 (1985).
150.   D. A. Roston and R. R. Rhinebarger, Evalution of HPLC with
       pulsed amperometric detection for an amino suger drug substance,
       J. Liquid Chromatogr., 14:539 (1991).
151.   L. G. McLaughlin and J. D. Henson, Determination of aminogly-
       coside antibiotics by RP (1-P) HPLC with pulsed amperometry
       and ion spray ms, *J. Chromatogr., 591*:195 (1992).
152.   C. Gandini, G. Caccialanza, M. Kitsos, G. Massoline, and E.
       DeLorenzi, Determination of 5-aminosalicylic acid and related
       compounds on raw materials and pharmaceutical dosage forms by
       HPLC, *J. Chromatogr., 540*:416 (1991).

153. V. Das Gupta and A. G. Ghanekar, Quantitative determination of codeine phosphate, guaifenesin, pheniramine maleate, phenylpropanolamine hydrochloride, and pyrilamine maleate in an expectorant by HPLC, *J. Pharm. Sci., 66*:897 (1977).

154. B. S. Rutherford and R. H. Bishara, Quantitative determination of antipyrine and benzocaine in ear drops by high-performance liquid chromatography, *J. Pharm. Sci., 65*:1322 (1976).

155. I. Cendrowska, M. Drewnowska, A. Grzeszkiewicz, and K. Butkiewicz, Investigation of the stability of 5 aminosalicyclic acid in tablets and suppositories by HPLC, *J. Chromatogr., 509*:195 (1990).

156. R. B. Smith, J. C. Glade, and D. W. Humphrey, High-performance liquid chromatographic separation of apomorphine and its *o*-methyl metabolites, *J. Chromatogr., 172*:570 (1979).

157. P. W. Erhardt, R. V. Smith, and T. T. Sayther, Thin-layer chromatography of apomorphine and its analog, *J. Chromatogr., 116*: 218 (1976).

158. M. S. F. Ross, Chromatographic analysis of azapropazone and related benzotriazines, *J. Chromatogr., 131*:448 (1977).

159. H. G. Brittain, *Analytical Profiles of Drug Substances and Excipients,* (H. G. Brittain, ed.), Vol. 21, Academic Press.

160. R. Verpoorte and A. B. Svendsen, HPLC of some tropane alkaloids, *J. Chromatogr., 120*:203 (1976).

161. N. D. Brown and H. K. Sleeman, Ion-pair high-performance liquid chromatographic method for determination of atropine sulfate and tropic acid, *J. Chromatogr., 150*:225 (1978).

162. U. Lund and S. H. Hansen, Simultaneous determination of atropine and its acidic and basic degradation products by mixed-column HPLC, *J. Chromatogr., 161*:371 (1978).

163. V. S. Venturella, V. M. Gualarioand, and R. E. Lange, Dimethylformamide dimethylacetal as a derivatizing agent for GLC of barbiturates and related compounds, *J. Pharm. Sci., 62*:662 (1973).

164. M. Riedman, Specific gas chromatographic determination of phenothiazines and barbiturate tranquillizers with the nitrogen flame ionization detector, *J. Chromatogr., 92*:55 (1974).

165. S. Dilli and D. N. Pillai, Relative electron capture response of the 2-chloroethyl derivatives of some barbituric acids and anticonvulsant drugs, *J. Chromatogr., 137*:111 (1977).

166. R. W. Roos, Separation and determination of barbiturates in pharmaceuticals by high speed liquid chromatography, *J. Pharm. Sci., 61*:1979 (1972).

167. C. R. Clarke and J. L. Chan, Improved detectability of barbiturates in HPLC by post-column ionization, *Anal. Chem., 50*:635 (1978).

168.   D. Dadgar and A. Power, Applications of column-switching technique in biopharmaceutical analysis. I. High-performance liquid chromatographic determination of amitriptyline and its metabolites in human plasma, *J. Chromatogr., 416*:99 (1987).

169.   K. Tsuji and J. H. Robertson, Improved high-performance liquid chromatographic method for polypeptide antibiotics and its application to study the effects of treatme to reduce microbial levels in bacitracin powder, *J. Chromatogr., 112*:663 (1975).

170.   J.-L. Venthey and W. Haerdi, Separation of amphetamines by supercritcal fluid chromatography, *J. Chromatogr., 515*:385 (1990).

171.   Macherey Nagel #C4583-203.

172.   G. L. Lensmeyer and M. A. Evenson, Stabilized analysis of antidepressant drugs by solvent-recycled liquid chromatography procedure and proposed resolution mechanisms of chromatography, *Clin. Chem., 30*:1774 (1984).

173.   B & J Octyl DC5 column.

174.   A. O. Akanni and J. S. K. Aym, Determination of ampicillin in the presence of cloxacillin, *J. Pharm. Biomed. Anal., 10*:43 (1992).

175.   B. S. Kerstein, T. Catalano, and Y. Rozenman, Ion-pairing HPLC method for the determination of 5-aminosalicylic acid and related impurities in bulk chemical, *J. Chromatogr., 588*:187 (1991).

176.   G. N. Okato and P. Camilleri, Micellar electrokinetic capillary chromatography of amoxycillin and related molecules, *Analyst, 117*:1421 (1991).

177.   L. Simonson and K. Nelson, The adsorption of amitroptyline hydrochloride in HPLC injection loops, *LC–GC, 10*:533 (1993).

178.   B. L. Lampert and J. T. Stewart, Determination of non-steroidal anti-inflammatory analgesics in solid dosage forms by HPLC on underivatized silica with aqueous mobile phase, *J. Chromatogr., 504*:381 (1990).

179.   P. K. Janicki, W. A. R. Erskine, and M. F. M. Jones, HPLC method for the direct determination of the volatile anaethetics halothane isoflurane, and enflurane in water and in physiological buffer, *J. Chromatogr., 518*:250 (1990).

180.   S. I. Sa'Sa, I. F. Al-momani, and I. M. Jalal, Determination of naphazoline nitrate and antazoline sulfate in pharmaceutical combinations by RP–HPLC, *Anal. Lett., 23*:953 (1990).

181.   L. E. Bennett, W. N. Charman, D. B. Williams, and S. A. Charman, Analysis of bovine immunoglobulin G by capillary gel electrophoresis, *J. Pharm. Biomed.*

182.   A. M. Zimmer, J. M. Kazikiewicz, S. M. Spies, and S. T. Rosen,

Rapid miniaturized chromatography for ^{111}In labeled monoclonal antibodies: comparison to size exclusion HPLC, *Nucl. Med. Biol.,* *15*:717 (1988).

183. M. J. Walters, R. J. Ayers, and D. J. Brown, Analysis of illegally distributed anabolic steroid productivity by LC with identity comfirmation by mass spectrometry or infrared spectrophotometry, *JAOAC, 73*:904 (1990).

184. H. Vanderhaeghe, Monographs on antibiotics in the European pharmacopoeia, *Pharm. Forum, 11–12*:2652 (1991).

185. S. K. Yeo, H. K. Lee, and S. F. Y. Li, Separation of antibiotics by high-performance capillary electrophoreiss with photodiode-array detection, *J. Chromatogr., 585*:133 (1991).

186. H. S. Huang, H. R. Wu, and M. L. Chen, Reversed phase HPLC of amphoteric β-lactam antibiotics: effects of columns ion-pairing reagents and mobile phase pH on their retention times, *J. Chromatogr., 564*:195 (1991).

187. K. Salomon, D. S. Burg, and J. C. Helmer, Separation of seven tricyclic antidepressants using capillary elecrophoresis, *J. Chromatogr., 549*:375 (1991).

188. A. Holbrook, A. M. Krstulovic, J. H. Mc B. Miller, and J. Rysaluk, Limitations of impurities in atenolol by LC, *Pharmeuropa, 3*: 218 (1991).

189. I. S. Lurie and A. C. Allen, Reversed-phase high-performance liquid chromatographic separation of fentanyl homologues and analogues II. Variable affecting hydrophobic group contribution, *J. Chromatogr., 292*:283 (1984).

190. Z. Pawlak and B. J. Clark, The assay and resolution of the beta-blocker atenolol from its related impurities, *J. Pharm. Biomed. Anal., 10*:329 (1992).

191. D. R. Radulovic, L. J. Zivanovic, G. Velimirovic, and D. Stevanovic, HPLC determination of atenolol in tablets, *Anal. Lett., 24*: 1813 (1991).

192. Vydac HS Pharmaceutical Analysis column.

193. C. P. Ong, C. L. Ng, H. K. Lee, and S. F. Y. Li, Determination of antihistamines in pharmaceuticals by capillary electrophoresis, *J. Chromatogr., 588*:335 (1991).

194. G. Misztal and H. Hopkala, Determination of aprindine and nadoxolol in pharmaceuticals by HPLC, *Pharmazie, 49*:697 (1994).

195. B. L. Ling, W. R. G. Baeyons, P. Van Acker, and C. DeWade, Determination of ascorbic acid and isoascorbic acid by capillary zone electrophoresis application to fruit juices and to a pharaceutical formulation, *J. Pharm. Biomed. Anal., 10*:717 (1992).

196.   J. M. Irache, II Ezpelata, and F. A. Vega, HPLC determination of antioxidant synergists and ascorbic acid in some fatty pharmaceuticals, cosmetics and food, *Chromatographia, 35*:232 (1993).

197.   M. C. Gennaro, C. Abrigo, and E. Marengo, The use of chirol interaction reagents in the separation of D(−) and L(+) ascorbic acid by ion interaction RPHPLC, *Chromatographia, 30*:311 (1990).

198.   V. Kmetec, Simultonous determination of acetylsalicylic, salicylic ascorbic and dehydroascorbic acid by HPLC, *J. Pharm. Biomed. Anal., 10*:1073 (1992).

199.   C. Heldin, N. H. Huynh, and C. Petterson, (2S, 3S)-Dicyclohexyl tartrate as mobile phase additive for the determination of the enantionmeric purity of (S)-atropine in tablets, *J. Chromatogr., 592*: 339 (1992).

200.   R. Alder and H. P. Merkle, Studies on the stability of aspartame, I: specific and reproducible HPLC assay for aspartame and its potential degradation products and applications to acid hydrolysis of aspartame, *Pharmazie, 46*:91 (1991).

201.   S. L. Line, S. T. Chen, S. H. Wu, and K. T. Wange, Separation of aspartame and its precusu stereoisoners by chiral chromatography, *J. Chromatogr., 540*:392 (1991).

202.   S. Motellier and I. W. Wanil, Direct sterochemical resolution of aspartame stereoisomers and their degradation products, by HPLC on a chiral crown either ether based stationary phase, *J. Chromatogr., 516*:365 (1990).

203.   G. Santoni, L. Fabbri, P. Gratteri, G. Renzi, and S. Pinzauti, Simultoneous determination of aspirin, codeine phosphate and propyphenazone in tablets by RPHPLC, *Int. J. Pharm., 801*:263 (1992).

204.   J. Meinwold, W. R. Thompson, D. L. Pearson, W. H. Konig, T. Runge, and W. Francki, Inhalation anesthetics stereochemistry: Optical resolution of haloethane, enflurane and isoflurane, *Science, 251*:560 (1991).

205.   G. Santoni, A. Tonsini, P. Gratteri, P. Mara, S. Furlanetto, and S. Pinzanti, Determination of atropine sulfate and benzalkonium chloride in eye drops by HPLC, *Int. J. Pharm., 93*:239 (1993).

206.   P. Painuly, R. K. Seligal, and J. C. Turcotte, Preparative HPLC of an experimental anti-HIV analogue of AZT, *J. Liquid Chromatogr., 16*:2237 (1993).

207.   R. Thejavathi, S. R. Yakkundi, and B. Ravindranath, Determination of azadirachtin by RP HPLC using anisole as internal Std, *J. Chromatogr., 705*:374 (1995).

208.   H. P. Huang and E. D. Morgan, Analysis of azadirachtin by

supercritical-fluid chromatography, *J. Chromatogr., 519*:137 (1990).

209. M. A. Abounassif, E. M. Abdel-Moety, and E. A. Lad-Kariem, HPLC quantification of azintanide and papaverine simultaneously in pharmaceutical preparations, *J. Liquid Chromatogr., 13*:2689 (1990).

210. X. Huang, W. Th. Kok, and H. Fabre, Determination of aztreonam by LC With UV and amperometric detection, *J. Liquid Chromatogr., 14*:2721 (1991).

211. K. M. S. Sundaram and J. Cuny, HPLC determination of azadirachtin in confer and deciduous foliage, *J. Liquid Chromatogr., 16*:3275 (1993).

212. J. E. Kountourellis, C. K. Markopoulou, and J. A. Stratis, Quantitative HPLC determination of Bampine combined with tricyclic antidepressants and/or antipsychotics in pharmaceutical formulations, *Anal. Lett., 26*:2171 (1993).

213. S. Ahuja, Chromatographic solution to pharmaceutical analytical problems, *Chromatographia, 34*:411 (1992).

214. C. Carlucci, A. Colanzi, and P. Mazzeo, Determination of bamifylline hydrochloride impurities in bulk material and pharmaceutical forms using liquid chromography with ultraviolet detection, *J. Pharm. Biomed. Anal., 8*:1067 (1990).

215. D. De Orsi, L. Gagliardi, F. Chimenti, and D. Tonelli, HPLC determination of beclomethalone dipropionate and its degradation products in bulk drug and pharmaceutical formulations, *Anal. Lett., 28*:1615 (1995).

216. L. Elrod, Jr., T. G. Golich, and J. A. Morley, Determination of benzalkonium chloride in eye car products by HPLC and solid-phase extraction or on-line column switching, *J. Chromatogr., 625*:362 (1992).

217. R. Herraez-Hernandez, P. Campins-Falco, and A. Sevillano-Cabeza, Improved screening procedure for diuretics, *J. Liquid Chromatogr., 15*:2205 (1992).

218. A. Gomez-Gomar, M. M. Gonzalez-Albert, J. Garces-Torrents, and J. Costa-Eigana, Determination of benzalkonium chloride in aqueous ophthalmic preparations by HPLC, *J. Pharm. Biomed. Anal., 8*:871 (1990).

219. M. Aycard, S. Letellier, B. Maupas, and F. Guyon, Determination of (R) and (S) warfarin in plasma by HPLC using precolumn derivation, *J. Liquid Chromatogr., 15*:2175 (1992).

220. R. G Bell, Preparative HPLC separation and isolation of bacitrin components and their relationship to microbiological activity, *J. Chromatogr., 590*:163 (1992).

221.   J. Wang and D. E. Moore, A study of the photodegradation of benzydamine in pharmaceutical formulation using HPLC with diode array detection, *J. Pharm. Biomed. Anal., 10*:535 (1992).

222.   F. Smets, H. F. DeBrabander, P. J. Bloom, and G. Pottie, HPTLC of anabolic compounds in injection sites, *J. Planar Chromatogr., 4*:207 (1991).

223.   S. Grewal, A. Aluea, and E. Reinhard, Production of hyocyamine by root cultures of hyoscyamus muticus L., *Herba Hungarica, 30*: 109 (1991).

224.   Z. Li, T. Shi, and J. Wu, Determination of chlorpromazine hydrochloride in injection by TLC, *Yaowu Fenxi Zazhi, 12*:47 (1992).

225.   J. E. Kountourellis and C. Markopoulou, A simultaneous assay by HPLC of bamipine hydrochloride and terbutaline sulfate in dosage forms, *J. Liquid Chromatogr., 12*:3279 (1989).

226.   A. Nagy, J. Bartha, and L. Nagy, Supervision of some drug preparations of several components by TLC in the pilot plant laboratory, *Gyogyszereszet, 36*:279 (1992).

227.   R. Frontoni and J. B. Mielck, Determination and quantitation of bendoflumethizide and its degradation products using HPLC, *J. Liquid Chromatogr., 15*:2519 (1992).

228.   S. Markovic and Z. Krisec, Determination of butalbital, caffeine and propyphenozone in pharmaceutical preparations by gas chromatography method, *Pharmazie, 46*:886 (1991).

229.   G. S. Sadama and A. B. Ghogare, Simultaneous determination of chloramphenicol and benzocaine in topical formulations by HPLC, *J. Chromatogr., 542*:515 (1991).

230.   Alltech Catalog.

231.   G. Maeder, M. Pelletier, and W. Haerdi, Determination of amphetamines by HPLC with UV detection, on-line pre-column derivatization with 9-fluorenylmethyelene chloroformate and preconcentration, *J. Chromatogr., 593*:9 (1992).

232.   J & W Scientific Catalog.

233.   R. R. MacGregor, J. S. Fowler, and A. P. Wolf, Determination of the enantiomeric composition of samples of cocaine by normal phase HPLC-UV, *J. Chromatogr., 590*:354 (1992).

234.   T. G. Chardranhekhar, P. S. N. Rao, D. Smith, S. K. Vyas, and C. Dutt, Determination of bupremorphine hydrochloride by HPTLC—A reliable method for pharmaceutical preparations, *J. Planar Chromatogr.—Mod., TLC, 7*:249 (1994).

235.   N. N. Agapova and E. Vasileva, HPLC method for the determination of bisoprolol and potential impurities, *J. Chromatogr., 654*: 299 (1993).

236. D. Zivanor-Stakic, J. J. Solomon, and L. J. Zivanovic, HPLC method for the determination of bumentanide in pharmaceutical preparations, *J. Pharm. Biomed. Anal., 7*:1889 (1989).

237. A. K. Rasymas, H. Boudoulas, and J. MacKichan, Determination of verapamil enantiomers in serum following racemate administration using HPLC, *J. Liquid Chromatogr., 15*:5013 (1992).

238. W. A. Dalin, Degradation of bisoprolol fumarate in tablets formulated with dicalcium phosphate, *Drug Dev. Ind. Pharm., 21*: 393 (1995).

239. C. P. W. G. M. Verwey-Van Wissen and P. M. Koopman-Kimenai, Direct determination of codeine, norcodeine, morphine and normorphine with their corresponding *O*-glucuronide conjugates by HPLC with electrochemical detection, *J. Chromatogr., 570*:309 (1991).

240. T. Yokoyama and T. Kinoshita, HPLC determination of biotin pharmaceutical preparations by post-column fluorescence reaction in the thiamine reagent, *J. Chromatogr., 542*:365 (1991).

241. A. Derico, A. Cocchini, R. Noferini, C. Mannucci, and A. Cambi, HPLC analysis of boldine in tablets and syrup, *J. Liquid Chromatogr., 15*:617 (1992).

242. K. Shimada, K. Mitamura, M. Morita, and K. Hirahata, Separation of the diastereomers of baclofen by HPLC using cyclodextrin as a mobile phase additive, *J. Liquid Chromatogr., 16*:3311 (1993).

243. F. T. Noggle, C. R. Clark, and J. DeRuiter, Gas chromatographic and mass spectrometric analysis of samples from a clandestine laboratory involved in the synthesis of Ecstasy from sassafras oil, *J. Chromatogr. Sci., 29*:168 (1991).

244. D. F. Musto, Opium, cocaine and marijuana in American history, *Scientific American,* 40 (July 1991).

245. R. B. Miller, C. A. Chen, and C. H. Sherwood, HPLC determination of benzalkonium in vasocidin ophthalmic solution, *J. Liquid Chromatogr., 16*:3801 (1993).

246. T. Y. Fan and G. M. Wall, Determination of benzalkonium chloride in ophthalmic solutions containing Tyloxapol by SPE and RP-HPLC, *J. Pharm. Sci., 82*:1172 (1993).

247. K. Patzsch, W. Funk, and H. Schutz, HPTLC determination of opiates, *Chromatographic* (GIT Supplement 3), *88*:83 (1988).

248. G. Parhizkani, R. B. Miller, and C. Chen, A stability-indicating HPLC method for the determination of benzalkonium chloride in phenylephrine HG 10% ophthalmic solution, *J. Liquid Chromatogr., 18*:553 (1995).

249.   B. Gigante, A. M. V. Barros, A. Teixeria, and M. J. Marcelo-Curto, Separation and simultaneous HPLC determination of benzocaine and benzyl benzoate in a pharmaceutical preparation, *J. Chromatogr., 549*:217 (1991).

250.   C. H. Wolf and R. W. Schmid, Enhanced UV-detection of barbiturates in HPLC analysis by on-line photochemical reaction, *J. Liquid Chromatogr., 13*:2207 (1990).

251.   E. R. Garrett, K. Seyda, and P. Marroum, HPLC assays of the illicit design drug, Ecstasy, a modified amphetamine with applications to stability, partitioning and plasma protein binding, *Acta Pharm. Nord., 3*:9 (1991).

252.   J. L. J. Amicot, M. Caude, and R. Rosset, Separation of opium alkaloids by carbon-dioxide sub and supercritical fluid chromatography with packed columns, application to the quantitative analysis of poppy straw extracts, *J. Chromatogr., 437*:351 (1988).

253.   J. E. Kountourellis and C. K. Markopoulou, A simultaneous analysis by HPLC of bamipine combined with tricyclic antidepressants and/or antipsychotics in dosage forms, *J. Liquid Chromatogr., 14*:2969 (1991).

254.   S. H. Chen, S. M. Wu, and H. L. Wu, Stereochemical analysis of betamethasone and dexamethasone by derivatization and HPLC, *J. Chromatogr., 595*:203 (1992).

255.   M. C. Tzau and C. Ho, Simultaneous determination of bromvaletone and propantheline bromide in tablets by HPLC, *J. Liquid Chromatogr., 15*:1577 (1992).

256.   V. Rajananda, N. K. Nair, and V. Navaratnam, An evaluation of TLC systems for opiate analysis, *Bull. Narcotics, 35*(1):35 (1985).

257.   N. K. Nair, V. Navaratnam, and V. Rajananda, Analysis of illicit heroin, I, an effective TLC system for separating eight opiates and five adulterants, *J. Chromatogr., 366*:363 1986).

258.   B. A. Olsen, S. W. Baertschi, and R. M. Riggin, Multidimensional evaluation of impurity profile of generic cephalexin and cefaclor antibiotics, *J. Chromatogr., 648*:165 (1993).

259.   P. Linares, M. C. Gutierrez, F. Lazaro, M. D. Luque de Castro, and M. Valcarcel, Determination of benzocaine, dextromethorphan and cetylpyridinium ion by HPLC with UV detection, *J. Chromatogr., 558*:147 (1991).

260.   G. Uurdakul, L. Ersoj, and S. Sungur, Comparision of HPLC and derivative spectrophotometric methods for the determination of paracetamol and caffeine in tablets, *Pharmazie, 46*:885 (1991).

261.   K. Kovacs-Hadady, Study of the retention behavior of barbiturates by overpressured layer chromatography using silica gel

bonded with tricaprylmethylammonium chloride, *J. Chromatogr., 589*:301 (1992).

262. S. Furlanetto, P. Mura, P. Gratteri, and S. Pinzauti, Stability predictions of cefazolin sodium and cephaloridine on solid state, *Drug Dev. Ind. Pharm., 20*:2299 (1994).

263. M. S. Leloux and F. Dost, Doping analysis of beta-blocking drugs using HPLC, *Chromatographia, 32*:429 (1991).

264. H. M. McNair and K. M. Trivedi, Gas chromatography and pharmaceutical analysis. In *Chromatography of Pharmaceuticals*, (S. Ahuja, ed.), American Chemical Society, Washington, DC, 1992.

265. P. Emaldi, S. Fapanni, and A. Baldini, Validation of a capillary electrophoresis method for the determination of cephradine and its related impurities, *J. Chromatogr. A, 711*:339 (1995).

266. G. A. MacKay and G. D. Reed, The application of capillary SFC, packed column SFC, and capillary SFC-MS in the analysis of controlled drugs, *J. High Resol. Chromatogr., 14*:537 (1991).

267. B. Van Giessen and K. Tsuji, GLC assay method for neomycin in petrolatum-based ointments, *J. Pharm. Sci., 60*:1068 (1971).

268. D. J. White, J. T. Stewart, and I. L. Honigberg, Quantitative Analysis of diazepam and related compounds in drug substance and tablet dosage form by HPTLC and scanning densitometry, *J. Planar Chromatogr., 4*:413 (1991).

269. M. Dolezalova, Ion-pair HPLC determination of morphine and pseudomorphine in injections, *J. Pharm. Biomed. Anal., 10*:507 (1992).

270. G. Cavazzutti, L. Gagliardi, D. De Orsi, and D. Tonelli, Simultaneous determination of buzepide, phenylpropanolamine and clocmizine in pharmaceutical preparations by ion-pair RPLC, *J. Liquid Chromatogr., 18*:227 (1995).

271. W. R. Agyangar, S. R. Bhide, and U. R. Kalhote, Assay of semisynthetic codeine base with simultaneous determination of codeimethine and O^6-codeine methyl ether as by-product impurities by HPLC, *J. Chromatogr., 519*:250 (1990).

272. C. Hendrix, C. Wijsen, L. M. Yun, E. Roets, and J. Hoogmartens, Column liquid chromatogrpahy of cetadroxil on poly(styrene-divinylbenzene), *J. Chromatogr., 628*:49 (1993).

273. 1992 Chromatography Users Catalog, Hewlett Packard, p. 102.

274. 1992 Chromatography Users Catalog, Hewlett Packard, p. 144.

275. Alltech, Capillary Column Catalog, Bulletin #232, Chromatogram #1769.

276. Alltech, Capillary Column Catalog, Bulletin #232, Chromatogram #7090.

277.   S. M. Yuen and G. Lehr, Liquid chromatographic determination of clidinium bromide and clidinium bromide-chlordiazepoxide hydrochloride combinations in capsules, *JAOAC, 74*:461 (1991).

278.   T. Randall, Cocaine deaths reported for century or more, *JAMA, 267*:1045 (1992).

279.   L. Q. Zie, K. E. Markides, and M. L. Lee, Biomedical applications of analytical supercritical fluid separation techniques, *Anal. Biochem., 200*:7 (1992).

280.   C. C. Clark, A study of procedures for the identification of heroin, *J. Forens. Sci., 22*:418 (1977).

281.   C. C. Clark, Brown heroin comparison parameters, *Microgram, 11*:5 (1978).

282.   S. E. Hays, L. T. Grady, and A. V. Kruegel, Purity profiles for heroin, morphine, and morphine hydrochloride, *J. Pharm. Sci., 62*:1509 (1973).

283.   D. J. Reuland, and W. A. Trinler, An unequivocal determination of heroin in simulated street drugs by a combination of high-performance liquid chromatography and infrared spectrophotometry using micro-sampling techniques, *J. Forens. Sci., 11*:195 (1978).

284.   R. C. Shaler and J. H. Jerpe, Identification and determination of heroin in illicit seizures by combined gas chromatography—infrared spectrophotometry, *J. Forens. Sci., 17*:668 (1972).

285.   D. Sohn, Screening for heroin—A comparison of current methods, *Anal. Chem., 45*:1498 (1973).

286.   R. Fuelster, Extraction and Identification of morphine and codeine, *Microgram, 8*:135 (1975).

287.   D. J. Doedens and R. B. Forney, Confirmation of morphine on thin-layer plates by fluorometry, *J. Chromatogr., 100*:225 (1974).

288.   A. Gyeresi and G. Racz, New solvent for separation of the main alkaloids of opium (by thin-layer chromatography), *Pharmazie, 28*:271 (1973).

289.   J. Paul and F. Conine, Rapid thin-layer chromatographic separation of cocaine from codeine, heroin (diamorphine), 6-acetylmorphine, morphine and quinine in microscope slides, *Microchem. J., 18*:142 (1973).

290.   K. Roder, E. Eich, and E. Mutschler, Direct quantitative evaluation of thin-layer chromatography with remission and fluorescein determinations. 3. Determination of morphine, codeine, thebaine, and so on, *Arch. Pharm., 304*:297 (1971).

291.   S. N. Tewari and D. N. Sharma, Detection and identification of

morphine in adulterated opium samples by thin-layer chromatography, *Z. Anal. Chem., 281*:381 (1976).

292. A. Bechtel, Gas-chromatographische Identifizierung und quantitative Bestimmung von Morphin, Codein, Thebain, Papaverin und Narkotin in Opium extrakt, *Chromatographia, 5*:404 (1972).

293. P. De Zan and J. Fasanello, The quantitative determination of heroin in illicit preparations by gas chromatography, *J. Chromatogr. Sci., 10*:333 (1972).

294. D. Furmanec, Quantitative gas chromatographic determination of the major alkaloids in gum opium, *J. Chromatogr., 89*:76 (1974).

295. H-Y. Lim and S-T. Chow, Heroin abuse and a gas chromatographic method for determining illicit heroin samples in Singapore, *J. Forens. Sci., 23*:319 (1978).

296. J. M. Moore and F. E. Bena, Rapid gas chromatographic assay for heroin in illicit preparations, *Anal. Chem., 44*:385 (1972).

297. J. M. Moore, Rapid and sensitive gas chromatographic quantitation of morphine, codeine and O^6-acetylmorphine in illicit heroin using an electron capture detector, *J. Chromatogr., 147*:327 (1978).

298. M. J. Prager, S. M. Harrington, and T. F. Governo, Gas-liquid chromatographic determination of morphine, heroin, and cocaine, *J. Assoc. Off. Anal. Chem., 62*:304 (1979).

299. J. M. Moore and F. E. Bena, Rapid gas chromatographic assay for heroin in illicit preparations, *Anal. Chem., 44*:385 (1972).

300. J. M. Moore, Rapid and sensitive gas chromatographic quantitation of morphine, codeine and O^6-acetylmorphine in illicit heroin using an electron capture detector, *J. Chromatogr., 147*:327 (1978).

301. M. C. Hsu and M. C. Cheng, HPLC method for the determination of cloxacillin in commercial preparations and for stability studies, *J. Chromatogr., 549*:410 (1991).

302. M. J. Prager, S. M. Harrington, and T. F. Governo, Gas-liquid chromatographic determination of morphine, heroin, and cocaine, *J. Assoc. Off. Anal. Chem., 62*:304 (1979).

303. N. George, M. Kuppusamy, and K. Balaraman, Optimization of HPLC conditions for the determination of cyclosporins A, B, and C in fermentation samples, *J. Chromatogr., 604*:285 (1992).

304. A. Vialo, Gas-chromatographische Identifizierung von Heroin und Morphin in Verdacht erregenden Producten, *J. Eur. Toxicol., 4*:375 (1971).

305. A. R. Barnes, Determination of caffeine and potassium sorbate in

a neonatal oral solution by HPLC, *Int. J. Pharm., 80*:267 (1992).

306.   P. J. Cashman and J. I. Thornton, High speed liquid adsorption chromatography in criminalistics. II. The separation of heroin, O^6-monoacetylmorphine and morphine, *J. Forens. Sci. Soc., 12*: 417 (1972).

307.   W. Najdong, J. De Beer, X. Marcelis, P. Derese, J. H. M. C. B. Miller, and J. Hoogmartens, Collaborative study of the analysis of chlortetracycline hydrochloride by liquid chromatography on polystryrene divinyl benzene packing materials, *J. Pharm. Biomed. Anal., 10*:199 (1992).

308.   Alltech Catalog.

309.   C. Olieman, Chemistry of opium alkaloids and related compounds by ion-pair high performance liquid chromatography, *J. Chromatogr., 133*:382 (1977).

310.   R. M. Riggin, A. I. Schmidt, and P. T. Kissinger, Determination of acetaminophen in pharmaceutical preparations and body fluids by HPLC with EC detection, *J. Pharm. Sci., 64*:680 (1975).

311.   W. Naidong, E. Roets, and J. Hoogmartens, HPLC of chlortetracycline and related substances on poly (styrene-divinylbenzene) copolymer, *Chromatographia, 30*:105 (1990).

312.   P. G. Vincent and B. F. Engelke, High pressure liquid chromatographic determination of the five major alkaloids in Papaver Somniferum L. and thebaine in Papaver Bracetatum Lindl. Capsular tissue, *J. Assoc. Off. Anal. Chem., 62*:310 (1979).

313.   H. W. Ziegler, T. H. Beasley, Sr., and D. W. Smith, Simultaneous assay for six alkaloids in opium, using high-performance liquid chromatography, *J. Assoc. Off. Anal. Chem., 58*:888 (1975).

314.   J. Mc B. Miller, E. Porqueras, N. Berti, P. Vettori, F. Folliard, K. Smets, E. Roets, and J. Hoogmartens, Collaborative study of the analysis of chlortetracycline by HPLC on C8, *Chromatogaphia, 37*:640 (1993).

315.   M. Soholic, B. Filipovic, and M. Pokorny, HPLC procedures in monitoring the production and quality control of chlortetracycline, *J. Chromatogr., 509*:189 (1990).

316.   L. J. Lorenz, F. N. Bashore, and B. A. Olsen, Determination of process-related impurities and degradation products in a cefaclor by HPLC, *J. Chromatogr. Sci., 30*:211 (1992).

317.   K. L. Tuczkowska, R. D. Voyksner, and A. L. Aronson, Solvent degradation of cloxacillin in vitro, *J. Chromatogr., 594*:195 (1992).

318.   G. A. Ulsaker and G. Teien, Identification of caprolactam as

a potential contaminant in parenteral solutions stored on over-wrapped PVC bags, *J. Pharm. Biomed. Anal., 10*:77 (1992).

319. M. E. Mohamed and El-R. A. Gad-Kariem, Liquid chromatographic determination of amoxycillin and cavulanic acid in pharmaceutical preparations, *J. Pharm. Biomed. Anal., 9*:731 (1991).

320. V. Windisch, C. Karpenko, and A. Daruwala, LC assay for salmon calcitonin in aerosol formulations using fluorescence derivatization and size exclusion chromatography, *J. Pharm. Biomed. Anal., 10*:71 (1992).

321. A. G. Patel, R. B. Patel, and M. R. Patel, Liquid chromatographic determination of clobetasone-17-butyrate in ointments, *JAOAC, 73*:893 (1990).

322. K. Tsuji and J. H. Robertson, High-performance liquid chromatogrpahic analysis of novobiocin, *J. Chromatogr., 94*:245 (1974).

323. P. Piergiorgio, E. Manera, and P. Ceva, HPLC analysis of ephedrine in oily formulations, *J. Chromatogr., 367*:228 (1966).

324. R. V. Rondina, A. L. Bandoni, and J. D. Coussio, Quantitative determination of morphine in poppy capsules by differential spectrophotometry, *J. Pharm. Sci., 62*:502 (1973).

325. L. Borka, The polymorphism of heroin and its forensic aspects, *Acta Pharm. Suec., 14*:210 (1977).

326. J. M. Moore and M. Klein, Identification of O^3-monoacetyl-morphine in illicit heroin by using gas chromatography-electron-capture detection and mass spectrometry, *J. Chromatogr., 154*:76 (1978).

327. G. R. Nakamura, Forensic identification of heroin in illicit preparations using integrated gas chromatography and mass spectrometry, *Anal. Chem., 44*:408 (1972).

328. R. Saferstein, J. Manura, and T. A. Brettell, Chemical ionization mass spectrometry of morphine derivatives, *J. Forens. Sci., 24*:312 (1979).

329. R. M. Smith, Forensic identification of opium by computerized gas chromatography/mass spectrometry, *J. Forens. Sci., 18*:327 (1973).

330. A. M. DiPietra, D. Bonazzi, and V. Cavrini, Analysis of chlorpropamide, tolbutamide and their related sulphonamide impurities, *Farmaco, 47*(Suppl. 5):787 (1992).

331. C. C. Clark, Gas-liquid chromatographic quantitation of cocaine. HCl in powders and tablets: Collaborative study, *J. Chromatogr., 152*:589 (1978).

332. F. Malecki and J. C. Crawhall, Determination of melphalan and

chlorambucilin tablet dosage form using HPLC and amperometic detection, *Anal. Lett., 23*:1605 (1990).

333.   D. Eskes, Thin-layer chromatographic procedure for the differentiation of the optical isomers of cocaine, *J. Chromatogr., 152*:589 (1978).

334.   E. Rochard, H. Boutelet, E. Gresemann, D. Barthes, and P. Courtois, Simultaneous HPLC analysis of carboplatin and cisplatin in infusion fluids, *J. Liquid Chromatogr., 16*:1505 (1988).

335.   K. Bailey, D. Legault, and D. Verner, Spectroscopic and chromatographic identification of dimethoxyamphetamines, *J. Assoc. Off. Anal. Chem., 56*:70 (1974).

336.   T. C. Kram, Identification of an impurity in illicit amphetamine tablets, *J. Pharm. Sci., 66*:443 (1977).

337.   R. J. Warren, P. P. Begosh, and J. E. Zarembo, Identification of amphetamines and related sympathomimetic amines, *J. Assoc. Off. Anal. Chem., 54*:1179 (1971).

338.   T. R. Gaston and T. T. Rasmussen, Identification of 3,4-methylenedioxymethamphetamine, *Microgram, 5*:60 (1972).

339.   D. Eskes, A procedure for the differentiation of the optical isomers of amphetamine and methamphetamine by thin-layer chromatography, *J. Chromatogr., 117*:442 (1972).

340.   Y. Hashimoto, Quantitative Determination of (−)- and (+)-pseudoephedrine in *Ephedra* spp. by thin-layer chromatography, *Yakugaku Zasshi, 97*:594 (1977).

341.   M. A. Shaw and H. W. Peel, Thin-layer chromatography of 3,4-methylenedioxyamphetamine, 3,4-methylenedioxymethamphetamine and other phenethylamine derivatives, *J. Chromatogr., 104*: 201 (1975).

342.   J. W. Kelly and J. T. Stewart, Separation of selected beta lactam antibiotic epimers on gamma cyclodextim ion exchange ethylvinyl benzene divinylbenzene copolymer and poly(styrene-divinyl benzene) copolymer stationary phases, *J. Liquid Chromatogr., 14*: 2235 (1991).

343.   R. W. Souter, Gas chromatographic resolution of enantiomeric amphetamine and related amines. II. Effect of cyclic structures on diastereomer and enantiomer resolution, *J. Chromatogr., 114*:307 (1975).

344.   L. Stömberg, Comparative gas chromatographic analysis of narcotics. Amphetamine sulphate, *J. Chromatogr., 106*:335 (1975).

345.   C. E. Wells, GLC determination of optical isomers of amphetamine, *J. Assoc. Off. Anal. Chem., 53*:113 (1970).

346. C. E. Wells, Collaborative study of the gas-liquid chromatographic method for determination of stereochemical composition of amphetamine, *J. Assoc. Off. Anal. Chem., 55*:146 (1972).

347. Cortisone acetate in bulk drug and dosage forms, liquid chromatography, *JAOAC, 71*:534 (1988).

348. J. A. Heagy, Infra-red method for distinguishing optical isomers of amphetamine, *Anal. Chem., 42*:1459 (1970).

349. A. C. Mehta and S. G. Schulman, Comparison of fluorometric procedures for assay of amphetamine, *J. Pharm. Sci., 63*:1150 (1974).

350. A. K. Cho, Deuterium substituted amphetamine as an internal standard in a gas chromatographic/mass spectrometric (GC/MS) assay for amphetamine, *Anal. Chem., 45*:570 (1973).

351. M. T. Gilbert and C. J. W. Brooks, Characterization of diastereomeric and enantiomeric ephedrines by gas chromatography combined with electron-impact mass spectrometry and isobutane chemical-ionization mass spectrometry, *Biomed. Mass Spectrom., 4*:226 (1977).

352. L. Strömberg and I. Wistedt, Gas chromatography — mass spectrometry. Rapid analysis of methamphetamine, phentermine, *N,N*-dimethylamphetamine, *n*-ethylamphetamine and mephentermine in the presence of each other, *Microgram, 10*:59 (1977).

353. G. J. Lowell, Determination of isomeric composition of amphetamine mixtures from melting points of monohydrogen succinate salts, *J. Pharm. Sci., 61*:1976 (1972).

354. W. M. Ment and V. S. Marino, Stereochemical composition of d- and l-amphetamine mixtures by thermal analysis of the benzoyl derivatives, *J. Assoc. Off. Anal. Chem., 53*:1087 (1970).

355. R. E. Ardrey and A. C. Moffat, A compilation of analytical data for the identification of lysergide and its analogues in illicit preparations, *J. Forens. Sci. Soc., 19*:253 (1979).

356. K. Bailey, D. Verner, and D. Legault, Distinction of some dialkyl amides of lysergic and isolysergic acids from LSD, *J. Assoc. Off. Chem., 56*:88 (1973).

357. S. W. Bellman, J. W. Turczan, and T. C. Kram, Spectrometric forensic chemistry of hallucinogenic drugs, *J. Forens. Sci., 15*:261 (1970).

358. C. G. A. Bos and J. G. J. Frijns, Quantitative analysis of dihydroergotamine in pharmaceuticals using fluorescent spectroscopy and thin-layer chromatography, *Pharm. Weekbl., 107*:111 (1972).

359. M. D. Cunningham, Lysergic acid diethylamide, mass spectroscopy, *Microgram, 6*:19 (1973).

360. G. V. Alliston, M. J. De Fauber-Maunder, and G. F. Phillips, A novel thin-layer chromatographic system for lysergide (LSD), *J. Pharm. Pharmacol., 23*:555 (1971).

361. E. Eich and W. Schunack, Die direkte Auswertung von Dünn-schichtchromatogrammen durch Remissions und Fluorozenzmessungen, IV. Ergotalkaloide. *Planta Medica, 27*:58 (1975).

362. R. Fowler, P. J. Gomm, and D. A. Patterson, Thin-layer chromatography of lysergide acid and other ergot alkaloids, *J. Chromatogr., 72*:351 (1972).

363. B. Newman, V. T. Sullivan, and A. Dihrberg, Thin-layer chromatography of lysergic acid diethylamide (LSD), *N,N*-dimethyl tryptamine (DMT), methylene dioxyphenyl isopropylamine (STP), and ibogaine, *J. Crim. Law Criminol., 61*:112 (1970).

364. T. Niwaguchi and T. Inoue, Studies on quantitative in situ fluormetry of lysergic acid diethylamide (LSD) on thin-layer chromatograms, *J. Chromatogr., 59*:127 (1971).

365. T. Niwaguchi and T. Inoue, Direct quantitative analysis of lysergic acid diethylamide (LSD) and 2,5-dimethoxy-4-methyl-amphetamine (STP) on thin-layer chromatograms, *J. Chromatogr., 121*: 165 (1976).

366. J. Reichelt and S. Kudrunác, Analytical studies on ergot alkaloids and their derivatives. I. Separation of ergot alkaloids of the ergotoxine and ergotamine groups by thin-layer chromatography, *J. Chromatogr., 87*:433 (1973).

367. J. D. Wittwer and J. H. Kluckholm, Liquid chromatographic analysis of LSD, *J. Chromatogr. Sci., 11*:1 (1973).

368. K. W. Crawford, The identification of lysergic acid amide in baby hawaiian woodrose by mass spectrometry, *J. Forens. Sci., 15*:588 (1970).

369. N. Stroud, N. E. Richardson, D. J. G. Davies, and D. A. Norton, Quality control of prednisolone sodium phosphate, *Analyst, 105*: 455 (1980).

370. V. Das Gupta, High-performance liquid chromatography chromatographic evaluation of aqueous vehicles for preparation of prednisolone and prednisone dosage forms, *J. Pharm. Sci., 68*:908 (1979).

371. R. E. Graham and M. J. Uribe, Compendial monograph evaluation and development—prednisone, *Pharmacopeial Forum, 1–2*: 2656 (1983).

372. T. Sat, Y. Saito, K. Yamaoka, and M. Nishikawa, Stability test of prednisolone in ointment by high-speed liquid chromatography, *Yakuzaigaku, 39*:20 (1979).

373. J. Duksta and D. Dekker, High-performance liquid chromatographic determination of phosphate esters of dexamethasone and prednisolone and their sulphite adducts, *J. Chromatogr., 238*:247 (1982).

374. V. Das Gupta, High-performance liquid chromatography chromatographic evaluation of aqueous vehicles for preparation of prednisolone and prednisone dosage forms, *J. Pharm. Sci., 68*:908 (1979).

375. S. F. Stanley, Liquid chromatographic determination of primidone in tablets: Collaborative study, *J. Assoc. Off. Anal. Chem., 68*:85 (1985).

376. R. H. King, L. T. Grady, and J. T. Reamer, Progresterone injection assay by liquid chromatography, *J. Pharm. Sci., 63*:1591 (1974).

377. C. C. Janson and J. P. De Kleijn, The assay of cyanocobalamin in pharmaceutical preparations by solid-phase extraction and HPLC, *J. Chromatogr. Sci., 28*:42 (1990).

378. S. K. Pant, D. N. Gupta, K. M. Thoma, B. K. Martin, and C. L. Jain, Simultaneous determination of camphor, menthol, methylsalicylate, and thymol in analgesic ointments by gas-liquid chromatography, *LC, 8*:322 (1991).

379. A. M. El Walily, M. A. Korany, M. M. Bedair, and A. El Gindy, HPLC determination of benzocaine and phenindamine tartrate in cream, *Anal. Lett., 24*:781 (1991).

380. M. A. Targove and N. D. Danielson, HPLC of clindaniyan antibiotics using tris(bipyridine)-ruthenium (III) chemiluminescence detection, *J. Chromatogr. Sci., 28*:505 (1990).

381. N. A. Guzman, H. Ali, J. Moschera, K. Iqbal, and A. W. Malich, Analysis and quantitification of a recombinant cytokine in an injectable dosage form, *J. Chromatogr., 559*:307 (1991).

382. E. Vidal, M. Guigues, G. Balansard, and R. Elias, Determination of ophthalmic therapeutic metipranolol and its degradation product by reversed-phase HPLC, *J. Chromatogr., 348*:304 (1985).

383. K. E. Ogger, C. Noory, J. Gabay, V. P. Shah, and J. P. Shelly, Dissolution profiles of resin-based oral suspensions, *Pharm. Technol., 9*:84 (1991).

384. C. Pierron, J. M. Panas, M. F. Etcheveny, and G. Ledouble, Determination of 10-camphorsulphonates in pharmaceutical formulations by HPLC, *J. Chromatogr., 511*:367 (1990).

385. M. C. Nahata and R. S. Morosco, Measurement of ceftazidime arginine in aqueous solution by HPLC, *J. Liquid Chromatogr., 15*:1507 (1992).

386.   G. S. Sadana and A. B. Ghogare, Simultaneous determination of chloramphenicol and benzocaine in topical formulations by HPLC, *J. Chromatogr., 542*:515 (1991).

387.   A. Bandyopadhyay, G. Podder, A. K. Sen, S. Roy, S. K. Moitia, and T. K. Das, High performance reverse phase liquid chromatographic and spectrophotometric determination of clioquinol, *Indian Drugs, 26*:506 (1988).

388.   K. E. Ibrahim and A. F. Fell, Separation of choroquine enantiomers by HPLC, *J. Pharm. Biomed. Anal., 8*:449 (1990).

389a. P. Cugier and A. C. Plasz, Determination of clarithromycin as a contaminant on surfaces by HPLC using electrochemical detecting, *Pharm. Res., 8*:989 (1991).

389b. F. Belliardo, A. Bertolino, G. Braudolo, and C. Lucarelli, Micro-liquid chromatography method for the derivatization of ciclopiroxolamine after pre-column derivatization in topical formulations, *J. Chromatogr., 553*:41 (1991).

390.   D. B. Repke, GLC-mass Spectral analysis of psilocin and psilocybin, *J. Pharm. Sci., 66*:743 (1977).

391.   P. C. White, Analysis of extracts from *Psilocybe Semilanceata* mushrooms by high-pressure liquid chromatography, *J. Chromatogr., 169*:453 (1979).

392.   D. K. Morgan, D. M. Brown, T. D. Rotsch, and A. C. Plasz, A reversed-phase HPLC method for the determination and identification of clarithromycin as the drug substance and in various dosage forms, *J. Pharm. Biomed. Anal., 9*:281 (1991).

393.   R. J. Gorski, D. K. Morgan, C. Sarocka, and A. C. Plasz, Estimation and identification of non-polar compounds in bulk clarithromycin bulk drug by HPLC, *J. Chromatogr., 540*:422 (1991).

394.   I. Papadoyannis, M. Georgarakis, V. Samanidou, and G. Theodoridis, HPLC analysis of theophylline in the presence of caffeine in blood serum and pharmaceutical formulations, *J. Liquid Chromatogr., 14*:1587 (1991).

395.   P. Shearan, J. M. F. Alvarez, N. Zayed, and M. R. Smyth, HPLC separation of cisplatin and its hydrolysis products on aluminum and application to studies of their interaction with cysteine, *Biomed. Chromatogr., 4*:78 (1990).

396.   H. Fabre and A. F. Fell, Comparison of techniques for peak purity of cephalosporins, *J. Liquid Chromatogr., 15*:3031 (1992).

397.   C. Q. Hu, S. H. Jin, and T. M. Wang, The chromatographic behavior of cephalosporins in gel filtration chromatography, a novel method to separate high molecular weight impurities, *J. Pharm. Biomed. Anal., 17*:530 (1994).

398.   K. Bailey, D. R. Gagné, and R. K. Pike, Identification of some

analogs of the hallucinogen phencyclidine, *J. Assoc. Off. Anal. Chem., 59*:81 (1976).

399. J. E. Haky and T. M. Stickney, Automated gas chromatographic method for the determination of residual solvents in bulk pharmaceuticals, *J. Chromatogr., 321*:137 (1985).

400. C. C. Clark, Gas-liquid chromatographic quantitation of phencyclidine hydrochloride in powders: Collaborative study, *J. Assoc. Off. Anal. Chem., 62*:560 (1970).

401. E. J. Cone, Separation and identification of phencyclidine precursors, metabolites and analogs by gas and thin-layer chromatography and chemical ionization mass spectrometry, *J. Chromatogr., 177*:149 (1979).

402. J. K. Raney, G. T. Skowronski, and R. J. Wagenhofer, Gas liquid chromatographic screening procedure for components found in clandestine phencyclidine reaction mixtures, *Microgram, 11*:139 (1978).

403. K. G. Rao, S. K. Soni, and M. Mullen, A spectroscopic and chromatographic study of phencyclidine (PCP) and its analogs, *Microgram, 13*:52 (1980).

404. Supelco Catalog, Bulletin #810, p. 3 (1992).

405. ESA Application pH4-2/86.

406. Alltech, Bulletin #248.

407. R. Jain and C. L. Javi, Simultaneous separation of six fluroquinolone antibacterials using HPLC, *LC–GC, 10*:707 (1993).

408. V. Ulvi and H. Keski-Hynnila, Analysis of chlorothiazide in the presence of its protodecomposition products, *Am. Lab., 10*:56 (1994).

409. T. Yoshida, A. Aetake, H. Yamoguchi, N. Nimura, and T. Kinoshita, Determination of carnitine by HPLC using 9-anthrydiazomethane, *J. Chromatogr., 445*:175 (1988).

410. C. Echers, K. A. Hutton, V. De Biasi, P. B. East, N. J. Haskins, and V. W. Jaceqicz, Determination of clavam-2-carbohylate in clavulanate potasium and tablet material by LC-MS, *J. Chromatogr. A, 686*:213 (1994).

411. C. G. Pinto, J. L. P. Pavon, and B. M. Cordero, Micellar liquid chromatography of zwitterions: retention mechanism of cephalosporins, *Analyst, 120*:53 (1995).

412. A. H. Shah and A. A. Alshareel, A reversed phase HPLC method for the determination of chloramphenicol and its hydrolytic product in opthalmic solutions, *Anal. Lett., 26*:1163 (1993).

413. C. Valenta, A. Lexer, and P. Spiegl, Analysis of clotrimazole in ointments by HPLC, *Pharmazie, 47*:641 (1992).

414. J. Gabriel, O. Vacek, E. Kubatova, and J. Volc, HPLC determi-

nation of the antibiotic cortalcerone, *J. Chromatogr.*, *542*:200 (1991).

415.   A. N. Abdelrahman, E. I. A. Karim, and K. E. E. Ibrahim, Determination of chloroquine and its decomposition products in various brands of different dosage forms by LC, *J. Pharm. Biomed. Anal.*, *12*:205 (1994).

416.   H. Iwase, Ultramicro determination of cyanocobalamin in elemented diet by solid-phase extraction and HPLC with visible detection, *J. Chromatogr.*, *590*:359 (1992).

417.   A. R. Barnes, Determination of ceftazidine and pyridine by HPLC: Application to a viscous eye drop formulation, *J. Liquid Chromatogr.*, *18*:3117 (1995).

418.   E. I. A. Karim, K. E. E. Ibrahim, A. N. Abdelrahman, and A. F. Fell, Photodegradation studies on chloroquine phosphate by HPLC, *J. Pharm. Biomed. Anal.*, *12*:667 (1994).

419.   J. Prat, M. Pujol, V. Girona, M. Munoz, and L. A. Sole, Stability of carboplatin in 5% glucose solution in glass, polyethylene and polypropylene containers, *J. Pharm. Biomed. Anal.*, *12*:81 (1994).

420.   X. G. Fang, P. Shen, and S. Ghodbane, Mixed ion pair liquid chromatography method for the simultaneous assay of ascorbic acid, caffeine, chlorpheniramine maleate, dextromethorphan HBr monohydrate and paracetamol in Frenadol sachets, *J. Pharm. Biomed. Anal.*, *12*:85 (1994).

421.   Y. Zhang, S. Hu, and Q. Li, Determination of fast-acting syrup for cold by TLC and UV, *J. Pharm. Ind.*, *22*:506 (1991).

422.   D. J. White, J. T. Stewart, and I. L. Honigberg, Quantitative analysis of diazepam and related compounds in drug substance and tablet dosage form by HPTLC and scanning densitometry, *J. Pharm. Chromatogr.*, *4*:413 (1991).

423.   H. P. Yuan and D. C. Locke, HPLC method for the determination of diphenhydramine in liquid and solid drug dosage forms and its application to stability testing, *Drug Dev. Ind. Pharm.*, *17*:2319 (1991).

424.   J. E. Kountourellis, C. K. Markoupoulou, and P. E. Georgakopoulos, An HPLC method for the separation and simultaneous determination of antihistamines, sympathomimetic amines and dextromethorphan in bulk dry material and dosage forms, *Anal. Lett.*, *23*:883 (1990).

425.   W. A. Korfmacher, T. A. Getek, E. B. Hanson, and J. Bloom, Characterization of diphenhydramine, doxylamine, and carbinoxamine using HPLC-thermospray mass spectrometry, *LC–GC, 8*: 538 (1991).

426.   D. J. White, J. T. Stewart, and I. L. Honigberg, Quantitative

analysis of diazepam and related compounds in drug substance and tablet dosage forms by HPTLC and scanning densitometry, *J. Planar Chromatogr., 4*:413 (1991).

427. N. Beaulieu, E. G. Lovering, J. Lefrancois, and H. Ong, Determination of diclofonac sodium and related raw materials and formulation, *JAOAC, 73*:698 (1990).

428. K. Shivram, A. C. Shah, B. L. Newalhar, and B. V. Kamath, Stability indicating HPLC method for the assay of diltiazem hydrochloride in tablets, *J. Liquid Chromatogr., 15*:2417 (1992).

429. B. J. Clark, P. Parker, and T. Lange, The determination of the geometric isomers and related impurities of dothiepin in a pharmaceutical preparation by capillary electrophoresis, *J. Pharm. Biomed. Anal., 10*:723 (1992).

430. S. Kryger and P. Helboe, Determination of impurities in dextropropoxyphene hydrochloride by HPLC on dynamically modified silica, *J. Chromatogr., 539*:186 (1991).

431. V. G. Nayak, V. R. Bhate, S. M. Purandare, P. M. Dikshit, S. N. Dhumal, and C. D. Gaitonde, Rapid LC determination of paracetamol and diclofenac sodium from a combined pharmaceutical dosage, *Drug Dev. Ind. Pharm., 18*:369 (1992).

432. G. Santoni, L. Fabbri, P. Mura, G. Renzi, P. Gratteri, and S. Pinzanti, Simultaneous determination of otilonium bromide and diazepam by HPLC, *Int. J. Pharm., 71*:1 (1991).

433. T. D. Cyr, R. C. Lawrence, and E. G. Lovering, Gas chromatographic methods for doxepin isomers related compounds, and organic volatile impurities in raw materials and doxepin isomers in capsules, *JAOAC, 75*:814 (1992).

434. U. M. Shindle, N. M. Tendolkar, and B. S. Desai, Smiultaneous determination of paracetamol and diclofenac sodium in pharmaceutical preparations by quantititative TLC, *J. Planar Chromatogr., 7*:50 (1994).

435. M. G. Quaglia, A. Farina, F. Kilar, S. Fanali, E. Bossa, and C. Dellaquilla, Analysis of a new doxorubicin derivative and related compounds by HPCE, *J. Liquid Chromatogr., 17*:3911 (1994).

436. Cyclobond, Astec-Varex.

437. O. A. Omar, P. J. Hoskin, A. Johnston, G. W. Hanks, and P. Turner, Diamorphine stability in aqueous solution for subcutaneous infusion, *J. Pharm. Pharmacol., 41*:275 (1989).

438. A. Fattah, M. El Walily F. El-Anwar, M. A. Eid, and H. Awaad, HPLC and derivative UV spectrophotometric determination of amoxycillin and dicloxacillin mixtures in capsules, *Analyst, 117*:981 (1992).

439.   B. A. Olsen, J. D. Stafford, and D. E. Reed, Determination of dirithromycin purity and related substances by HPLC, *J. Chromatogr., 594*:203 (1992).

440.   G. W. Ponder and J. T. Stewart, HPTLC determination of digoxin and related compounds in drug substance and tablets, *J. Chromatogr. A, 659*:177 (1994).

441.   N. S. Nudelman and R. G. De Waisbaum, Isolation and structure elucidation of novel products of the acidic degradation of diazepam, *J. Pharm. Sci., 84*:208 (1995).

442.   I. Pinguet, P. Rouanet, P. Martel, M. Fabbro, D. Salabert, and C. Astre, Compatibility and stability of granisetron, dexamethasone and methylprednisolone in injectable solutions, *J. Pharm. Sci., 84*:267 (1995).

443.   C. Langel, P. Chaminarde, A. Baillet, and D. Ferrier, Ion-pair reversed phase liquid chromatographic determination of dihydralazine, *J. Chromatogr. A, 686*:344 (1994).

444.   T. Kubala, B. Gamblhir, and J. I. Borst, A specific stability indicating HPLC method to determine diclofenac in raw materials and pharmaceutical solid dosage forms, *Drug Dev. Ind. Pharm, 19*: 749 (1993).

445.   A. Van Schepdael, R. Kibaya, E. Roets, and J. Hoogmartnes, Analysis of doxycycline by capillary electrophoresis, *Chromatographia, 41*:367 (1995).

446.   J. E. O'Connor and T. A. Rejent, EMIT cannabinoid assay for urinary metabolites and confirmation by alternate techniques, *Clin. Chem., 27*:1104 (1981).

447.   K. Florey (ed.), *Analytical Profiles of Drug Substances, 10,* Academic Press, New York, 1981, p. 551.

448.   M. L. Rossini and M. Farina, Stability studies with HPLC of a new anthracycline in the final drug formulation, *J. Chromatogr., 593*:47 (1992).

449.   M. I. Arranz Pena and C. Moro, Amiodarone determination by HPLC, *J. Pharm. Biomed. Anal., 7*:1909 (1989).

450.   Z. Pawlak and B. J. Clark, Assay of dothiepin hydrochloride and its geometric isomers by liquid chromatography, *J. Pharm. Biomed. Anal., 7*:1903 (1989).

451.   G. M. Wall, J. C. Kenny, T. Y. Fan, C. Schafer, M. A. Ready, J. K. Baher, P. Ritter, and B. S. Scott, Analysis of 3- and 4-monopivaloyl epinephrine degradation products in dipivetrin hydrochloride substance and ophthalmic formulations, *J. Pharm. Biomed. Anal., 10*:465 (1992).

452.   D. R. Hogue, J. A. Zimmardi, and K. A. Shah, HPLC analysis of

docuate sodium in soft gelatin capsules, *J. Pharm. Sci., 81*:359 (1992).

453. W. Naidong, E. Roets, and J. Hoogmartens, Quantitive analysis of demeclocycline by HPLC, *J. Pharm. Biomed. Anal., 7*:1691 (1989).

454. I. I. Hewala, GLC and HPLC determination of diazepam and its degradation products in pharmaceutical formulations, *Anal. Lett., 25*:1877 (1992).

455. K. Ishii, K. Minato, N. Nishimura, T. Miyamoto, and T. Sato, Direct chromatographic resolution of four optical isomers of diltiazepam hydrochloride on a chiralcel OF column, *J. Chromatogr. A, 686*:93 (1994).

456. M. C. Hsu and Y. J. Fann, Determination of dicloxacillin preparation by liquid chromatography, *JAOAC, 75*:26 (1992).

457. R. E. Krailler, P. J. Adams, and P. L. Lane, Quantitation of doretinel in a topical gel using HPLC with SPE sample clean up, *J. Liquid Chromatogr., 14*:2383 (1991).

458. K. R. Liu, S. H. Chen, S. M. Wu, H. S. Kou, and H. L. Wu, HPLC determination of betamethasone and dexamethasone, *J. Chromatogr. A, 676*:455 (1994).

459. N. Nishi, N. Fujimora, Y. Yamaguchi, W. Jyomori, and T. Fukuyama, Reversed-phase HPLC separation of enantiomers of denapamine after derivatization with GITC chiral reagent, *Chromatograhic, 30*:186 (1990).

460. W. Naidong, S. Geelen, E. Roets, and J. Hoogmartens, Assay and purity control of oxytetracycline and doxycline by TLC-a comparison to LC, *J. Pharm. Biomed. Anal., 8*:891 (1990).

461. D. T. King, T. G. Venkateshwaran, and J. T. Stewart, HPLC determination of a vincristine, doxorubicin and ondansetrom mixture in 0.9% NaCl injection, *J. Liquid Chromatogr., 17*:1394 (1994).

462. Bio Rad Catalog, Alcohol Analysis columns.

463. V. Cavrini, A. M. DiPietia, and R. Gatti, Analysis of miconazole and econazole in pharmaceutical formulations by derivative UV spectorcopy and HPLC, *J. Pharm. Biomed. Anal., 7*:1535 (1989).

464. T. E. Peterson and D. Trowbridge, Quantitation of l-epinephrine and determination of the d-/l-epinephrine enantiomer ratio in a pharmaceutical formulation by capillary electrophoresis, *J. Chromatogr., 603*:298 (1992).

465. B. Blessington, A. Beiraghi, T. W. Lo, A. Drake, and G. Jones, Chiral HPLC–CD studies of the antituberculosis drug (+)-ethambutol, *Chirality, 4*:227 (1992).

466. P. M. Falk and B. C. Harrison, Use of DB-1 capillary columns in the GC/FID analysis of benzoylecgonine, *J. Anal. Toxicol., 9*: 273–274 (1985).

467. C. Gonnet and J. L. Rocca, Separation of pharmaceutical compounds using HPLC. Influence of water, II, *J. Chromatogr., 120*: 419 (1976).

468. B. Blessington and A. Beiraghi, A method for the quantitative enantioselective HPLC analysis of ethambutol and its stereoisomers, *Chirality, 3*:139 (1991).

469. B. Blessington and A. Beiraghi, Study of thesterochemistry of ethambutol using chiral liquid chromatography and synthesis, *J. Chromatogr., 522*:195 (1990).

470. X. Z. Qin, D. P. Ip, and E. W. Tsai, Determination and rotamer separation of enalapril maleate by capillary electrophoresis, *J. Chromatogr., 626*:251 (1992).

471. J. Salamouri and K. Sulais, Elimination of peak splitting in the liquid chromatography of proline-containing drug enalapril maleate, *J. Chromatogr., 537*:249 (1991).

472. J. Paesen, E. Roets, and J. Hoogmartens, Liquid chromatography of erythromycin A and related substances on poly(styrene-dinvinyl benzene), *Chromatographia, 32*:162 (1991).

473. W. G. Crouthamel and B. Dorsch, Specific high-performance liquid chromatographic assay for nitroglycerin in dosage forms, *J. Pharm. Sci., 68*:237 (1979).

474. J. Paesen, P. Claeys, E. Roets, and J. Hoogmartens, Evaluation of silanol-deactivated slica-based reversed-phases for liquid chromagraphy of erythromycin, *J. Chromatogr., 630*:117 (1993).

475. K. Florey (ed.), *Analytical Profiles of Drug Substances, 8,* Academic Press, New York, 1979, p. 71.

476. J. Paesen, D. H. Calam, J. A. Mc B. Miller, G. Raiola, A. Rozoneski, B. Silver, and J. Hoogmartens, Collaborative study of the analysis of erythromycin by LC on wide-pore poly(stynene-dinvinylbenzene), *J. Liquid Chromatogr., 16*:1529 (1993).

477. Th. Cachet, P. Lannoo, J. Paesen, G. Janssen, and J. Hoogmartens, Determination of erythromycin ethylsuccinate by LC, *J. Chromatogr., 600*:99 (1992).

478. L. Chafelz, Compendial backround on conjugated estrogens, *Pharmacopeial Forum, 5–6*:1951 (1991).

479. W. F. Beyer, HPLC Assay for intact estrogenic sodium sulfate conjugates in bulk drug and tablets, *Pharmacopeial Forum, 5–6*: 1955 (1991).

480. J. Novakovic, D. Agbaba, and D. Zivanoustakic, Fluorodensito-metic determination of conjugated estrogen in raw material and in pharmaceutical preparations, *J. Pharm. Biomed. Anal., 12*:1657 (1989).

481. B. M. Farrell and T. M. Jefferies, An investigation of high-performance liquid chromatographic methods for the analysis of amphetamine, *J Chromatogr., 272*:111 (1983).

482. L.-X, Jiang, Z. J. Wang, and S. A. Matlin, HPLC analysis of injectable contraceptive preparation containing norethisterone en-anthate and estradiol valerate, *J. Liquid Chromatogr., 13*:3473 (1990).

483. A.-H. N. Ahmed, S. M. El-Gizawy, and N. M. Omar, Cylcodex-trin bonded phases for LC separation and analysis of some oral contraceptives, *Anal. Lett., 24*:2207 (1991).

484. J. I. Javaid and J. M. Davis, GLC analysis of phenylalkyl amines using nitrogen detector, *J. Pharm. Sci., 70*:813 (1981).

485. P. C. Dabas, H. Erguven, M. C. Vescina, and C. N. Carducci, Stability study of ethyl loflazepate in bulk drug, solution and dos-age form by LC, *J. Pharm. Biomed. Anal., 10*:241 (1992).

486. P. A. Mason, T. S. Bal, B. Law, and A. C. Moffat, Development and evaluation of a radioimmunoassay for the detection of am-phetamine and related compounds in biological fluids, *Analyst, 108*:603 (1983).

487. J. Bauer, D. Heathcote, and S. Krogh, HPLC stability indicating assay for disodium EDTA in ophthalmic preparations, *J. Chroma-togr., 369*:422 (1986).

488. B. A. O'Brien, J. M. Bonicamp, and D. W. Jones, Differentiation of amphetamine and its major hallucinogenic derivatives using thin-layer chromatography, *J. Anal. Toxicol., 6*:143 (1982).

489. S. A. Terepolsky and I. Kanfer, Stability of erythromycin and some of its esters in methanol and acetonitrile, *Int. J. Pharm., 115*:123 (1995).

490. K. Bjerver, J. Jonsson, A. Nilsson, J. Schuberth, and J. Schu-berth, Morphine intake from poppy seed food, *J. Pharm. Phar-macol., 34*:798 (1982).

491. K. D. Altria, A. R. Walsh, and N. W. Smith, Validation of a capillary electrophoresis method for the enantiomeric purity test-ing of fluparoxan, *J. Chromatogr., 645*:193 (1993).

492. S. Gorog, B. Herenyi, and M. Renyei, Estimation of impurity profiles of drugs and related materials. Part 9: HPLC investiga-tion of flumecinol, *J. Pharm. Biomed. Anal., 10*:831 (1992).

493. Y. Pramar, V. DasGupta, and C. Bethea, Quantitation of 5-flucytosine in capsules using HPLC, *Drug Dev. Ind. Pharm., 17*: 193 (1991).
494. G. H. Junnarkar and S. Starchansky, Isothermal and nonisothermal decompostion of famotidine in aqueous solution, *Pharm. Res., 12*:599 (1995); also *Anal. Lett., 25*:1907 (1992).
495. J. H. Mc B. Miller, J. L. Robert, and A. M. Sorensen, Reversed-phase ion-pair LC method for determining impurities in furosemide, *J. Pharm. Biomed. Anal., 11*:257 (1993).
496. E. Bargo, Liquid chromatographic determination of flurozepam hydrochloride in bulk drug and dosage forms: Collaborative study, *JAOAC Int., 75*:240 (1992).
497. J. Kirshbaum, J. Noroski, A. Cosey, D. Mayo, and J. Adamovics, HPLC of the drug fosinopril, *J. Chromatogr., 507*:165 (1990).
498. Y. Pramar, V. DasGupta, and C. Bethea, Quantitiation of fluoxetine hydrochloride in capsules using HPLC, *Drug. Dev. Ind. Pharm., 18*:257 (1992).
499. V. Cavrini, D. Bonazzi, and A. M. DiPietia, Analysis of flucytosine dosage forms by derivative UV spectioscopy and LC, *J. Pharm. Biomed. Anal., 9*:401 (1991).
500. M. J. Smela and R. Stromberg, LC determination of six sympathomimetic drugs in dosage forms, *JAOAC, 74*:289 (1991).
501. R. Porra, M. G. Quaglia, and S. Fanali, Determination of fenfluramine in pharmaceutical formulations by capillary zone electrophoresis, *Chromatography, 41*:383 (1995).
502. R. D. Budd, Comparison of methods of analysis for phencyclidine, *J. Chromatogr., 20*:492 (1984).
503. X. Z. Qin, D. P. Ip, K. H. C. Chang, P. M. Dradramsky, M. A. Brooks, and T. Sakuma, Pharmaceutical applications of LC–MS. 1. Characterization of a famotidine degradate in a package screening study by LC–APCI MS, *J. Pharm. Biomed., 12*:221 (1994).
504. B. V. Kamath, K. Shivram, B. L. Newalkar, and A. C. Shah, Liquid chromatographic analysis and degradation kinetics of famotidine, *J. Liquid. Chromatogr., 16*:1007 (1993).
505. ESA Inc., Coulochem detection, ESA, Inc., Bedford, MA.
506. M. C. Gennaro and C. Abrigo, Separation of reduced and oxidized glutathione in a pharmaceutical preparation by ion-interaction RP HPLC, *J. Pharm. Biomed. Anal., 10*:61 (1992).
507. A. M. Dipietria, R. Gotti, D. Bonazzi, V. Andrisano, and V. Cavrini, HPLC determination of glutathionine and L-cysterine in pharmaceuticals after derivatization with ethacrynic acid, *12*:91 (1994).

508. M. H. Abel-Hay, M. S. El-Din, and M. A. Abuirjeie, Simultaneous determination of theophylline and guaiphenesin by HPLC, *Analyst, 117*:157 (1992).

509. E. Vidal-Ollivier, G. Schwadrohn, R. Elias, G. Balansard, and A. Babadjamjan, Determination of quaiazulene by HPLC, *J. Chromatogr., 463*:227 (1987).

510. E. Watson and F. Yao, Capillary electrophoretic separation of recombinant granulocyte-colony stimulating factor glycoforms, *J. Chromatogr., 630*:442 (1993).

511. S. Halashi and J. G. Nairn, Stability studies of hydralazine hydrochloride in aq. solutions, *J. Parent. Sci. Technol., 44*:30 (1990).

512. K. Anton, M. Bach, and A. Geiser , Supercritical fluid chromatography in the routine stability control of antipruritic preparations, *J. Chromatogr., 553*:71 (1991).

513. Alltech Catalog, Reversed Phase, Steroids.

514. S. Wanwimolruk, Rapid HPLC analysis and stability study of hydrocortisone 17-butyrate in cream preparations, *Pharm. Res., 8*:547 (1991).

515. U. R. Cieri, Determination of reserpine, hydralazine HCl, and hydrochlorothiazide in tablets by liquid chromatography on a short, normal phase column, *JAOAC Int., 77*:1104 (1994).

516. J. B. L. Damm and G. T. Overklift, Indirect UV detection as a non-selective detection method in the qualitative and quantitative analysis of heparin fragments by HPCE, *J. Chromatogr., A678*: 151 (1994).

517. M. L. Kleinberg, M. E. Duafala, C. Nacov, K. P. Flora, J. Hines, K. Davis, A. McDaniel, and D. Scott, Stability of heroin hydrochloride in infusion devices and containers for IV administration, *Am. J. Hosp. Pharm., 47*:377 (1990).

518. M. D. Jones, L. A. Merewether, C. L. Clogston, and H. S. Lu, Peptide map analysis of recombinant human granulocyte colony stimulating factor: elimination of methionine modification and nonspecific cleavages, *Anal. Biochem., 216*:135 (1994).

519. Astec/Varex Catalog, Burtorsville, MD.

520. V. E. Haikala, I. K. Heiminen, and H. J. Vuorela, Determination of ibuprofen in ointments by RPHPLC, *J. Pharm. Sci., 80*:456 (1991).

521. D. Nicoll Griffith, M. Scartozzi, and N. Chiem, Automated derivatization and HPLC analysis of ibuprofen enantiomers, *J. Chromatogr., A653*:253 (1993).

522. C. L. Hsu and R. R. Walters, Chiral separation of ibutilide enantiomers by derivatization with 1-napthyl isocyanate and HPLC on a Pirkle column, *J. Chromatogr., 550*:621 (1991).

523. S. Scypinski, R. L. Lanzano, and R. A. Soltero, Determination of lloprost in 5% dextrose in water solution by RP HPLC, *J. Pharm. Sci., 7*:954 (1990).

524. M. V. Padval and H. N. Bhargava, Liquid chromatogoraphic determination of indapamide in the presence of its degradation products, *J. Pharm. Biomed. Anal., 11*:1033 (1993).

525. G. Penone and M. Farina, HPLC method for direct resolution of the indobufen enantiomeric components, *J. Chromatogr., 520*:373 (1990).

526. H. F. Zou, X. L. Li, Y. K. Zhang, and P. C. Lu, Determination of iodine anion in dried kelp and iodized throat tablets by PPH-PLC with UV detection, *Chromatographia, 30*:228 (1990).

527. R. Gill, A. H. Stead, and A. C. Moffat, Analytical aspects of barbiturate abuse: Identification of drugs by the effective combination of gas-liquid, high performance liquid and thin-layer chromatographic techniques, *J. Chromatogr., 204*:275 (1981).

528. R. C. Hall and C. A. Risk, Rapid and selective determination of free barbiturates by gas chromatography using the electrolytic conductivity detector, *J. Chromatogr., Sci., 13*:519 (1975).

529. R. A. Moore and A. J. V. Carter, Assay of iodochlorhydroxyquin in cream and ointment formulations by HPLC, *J. Pharm. Biomed. Anal., 6*:427 (1988).

530. R. Mariani, M. Farina, and W. Sfreddo, Studies of iododoxorubicin by HPLC, *J. Pharm. Biomed. Anal., 7*:1877 (1989).

531. D. N. Pillai and S. Dilli, Analysis of barbiturates by gas chromatography, *J. Chromatogr. Sci., 25*:253 (1981).

532. A. Gomez-Gomar, M. Gonzalez-Aubert, and J. Costa-Segarva, HPLC method of pilocarpine, isopilocarpine, pilocarpic acid and isopipocarpic acid, *J. Pharm. Biomed. Anal., 7*:1729 (1989).

533. A. Azcona, A. Martin-Gonzalez, P. Zamorano, C. Pasual, C. Grau, and M. Garcia DeMirasierra, New methods for the assay of 5-isosorbide mononitiate and its validation, *J. Pharm. Biomed. Anal., 9*:725 (1991).

534. Alltech Catalog.

535. I. I. Salem, M. C. Bedmar, M. M. Medina, and A. Cerezo, Insulin evaluation in pharmaceuticals: Variables in RP-HPLC and method validation, *J. Liquid Chromatogr., 16*:1183 (1993).

536. D. Agbaba, V. Janjic, D. Zivanov Stakic, and S. Vladimirov, HPLC assay for isosorbide 5-mononitrate and Impurities of inorganic nitrates in pharmaceuticals, *J. Liquid Chromatogr., 17*:3983 (1994).

537. C. V. Olsen, D. A. Reifsyuder, E. Conova-Davis, V. T. Ling, and S. E. Builder, Preparative isolation of recombinant human insulin-like growth factor 1 by HPLC, *J. Chromatogr., 38*:675 (1994).

538. M. M. El-Domiaty, Improved HPLC determination of Khellin and visuagin in Ammivisnaga fruits and pharmaceutical formulations, *J. Pharm. Sci., 81*:475 (1992).

539. J. S. Fleitman, I. W. Partridge, and D. A. Neu, Thimerosal analysis in ketorolac tromethamine ophthalmic solutions comparing HPLC and colorimetric techniques, *Drug Dev. Ind. Pharm., 17*: 519 (1991).

540. X.-L. Lu and S. K. Yang, Resolution of enantiomeric lorazepam and its acyl and *O*-methyl derivatives and racemization kinetics of lorazepam enantiomers, *J. Chromatogr., 535*:229 (1990).

541. H. Y. Aboul-Enein and M. R. Islam, Enantiomeric separation of ketamine hydrochloride in pharmaceutical formulations and human serum by chiral liquid chromatography, *J. Liquid Chromatogr., 15*:3285 (1992).

542. C. Mannucci, J. Bertini, A. Cocchini, A. Perico, F. Salvagnini, and A. Triolo, HPLC simultaneous quantitation of betoprofen and parabens in a commerical gel formulation, *J. Liquid Chromatogr., 15*:327 (1992).

543. C. P. Leung and C. Y. Au-Yeung, HPLC determination of Ioperamide hydrochloride in pharmaceutical preparations, *J. Chromatogr., 449*:341 (1988).

544. W. Naidong, S. Hua, K. Verresen, E. Roets, and J. Hoogmartens, Assay and purity control of metacycline by TLC combined with UV and fluorescence densitometry-a comparison to LC, *J. Pharm. Biomed. Anal., 9*:717 (1991).

545. M. El-Sayed Metwally, Stability-indicating HPLC assay for 2-methyldopa in sustained-release capsules, *J. Chromatogr., 549*: 221 (1991).

546. W. Naidong, J. Thuranira, K. Vermeulen, E. Roets, and J. Hoogmartens, Quantitative analysis of minocycline by LC on PRP, *J. Liquid Chromatogr., 15*:2529 (1992).

547. W. Naidong, K. Vermeulan, I. Quintens, E. Roets, and J. Hoogmartens, Evaluation of analytical methods analysis of minocycline by LC, *Chromatographia, 33*:560 (1992).

548. S. Fanali, M. Cristalli, and A. Nardi, Capillary zone electrophoresis for drug analysis rapid determination of minoxidil on pharmaceutical formulations, *Farmaio, 47*:711 (1992).

549. L. Gagliardi, A. Amato, and L. Turchetto, Simultaneous determination of minoxidil and tretinoin in pharmaceutical and cosmetic formuations by RPHPLC, *Anal. Lett., 24*:1825 (1991).

550. P. Majlat, GC determination of atropine theophylline, phenobarbital and aminophenazone in tablets, *Pharmazie, 39*:325 (1984).

551. V. Maurich, T. Sciortino, D. Solimas, and L. Vio, Determination of aminophenazone and chlorthenoxazine in a dosage form using HPLC, *Boll. Chim. Farm., 121*:483 (1982).

552. K. Kovacs-Hadady and J. Szilagyi, Separation of minoxidil and its intermediates by over pressured layer chromatography using a stationary phase bonded with 1-methyl-ammoniun chloride, *J. Chromatogr., 553*:459 (1991).

553. R. Pohloudek-Fabini and P. Gundermann, Analytics and the stability of aminophenazone, pt. 1: Detection and determination of aminophenazone on the presence of its decomposition products, *Pharmazie, 35*:685 (1980).

554. Z. Harduf, Rapid determination of arginine by HPLC, *J. Chromatogr., 363*:428 (1985).

555. M. Beran and T. Zima, Determination of monerisins A and B in the fermentation broth of streptomycescinnamonesis by HPLC, *Chromatographia, 35*:206 (1993).

556. M. Dolezalova, IP HPLC determination of morphine and pseudo-morphine in injections, *J. Pharm. Biomed. Anal., 10*:507 (1992).

557. C. T. Hung, M. Young, and P. K. Gupta, Stability of morphine solutions in plastic syringes determined by R-P ion-pair HPLC, *J. Pharm. Sci., 77*:719 (1988).

558. An easy laboratory route to nitiazepam, *J. Chem. Educ., 66*:522 (1989).

559. T. G. Venkateshwaran and J. T. Stewart, HPLC determination of morphine-hydromorphone-bupivacaine and morphine-hydromorphone-tetracaine mixtures in 0.9% sodium chloride injection, *J. Liquid Chromatogr., 18*:565 (1995).

560. K. Datta, S. K. Roy, and S. K. Das, A simple RP partition TLC method for rapid identification and quantitation of methotrexate in presence of its disintegration products, *J. Liquid Chromatogr., 13*:1933 (1990).

561. A. M. Dyas, M. L. Robinson, and A. F. Fell, Direct separation of nadolol enantiomers on a Pirkle-type chiral stationary phase, *J. Chromatogr., 586*:351 (1991).

562. C. R. Lee, J.-P. Porziemsky, M.-C. Aubert, and A. M. Krstulovic, Liquid and high pressure carbon dioxide chromatography of β-blockers, *J. Chromatogr., 539*:55 (1991).

563. T. Obhkubo, H. Noro, and K. Sugawara, HPLC determination of nifedipine and a trace photo-degradation product in hospital prescriptions, *J. Pharm. Biomed. Anal., 10*:67 (1992).

564. J. D. Harper, P. A. Martel, and C. M. O'Donnell, Evaluation of a multiple-variable thin-layer and R-P TLC scheme for the identification of basic and neutral drugs in an emergency toxicology setting, *J. Anal. Toxicol., 13*:31 (1989).

565. H. Wagner, S. Bladt, and E. M. Zgainski, *Plant Drug Analysis,* Springer-Verlag, Berlin, 1984.

566. B. Gawdzik, Retention of basic drugs on porous polymers in HPLC, *J. Chromatogr., 600*:115 (1992).

567. R. Smith, J. P. Westlake, R. Gill, and M. D. Osselton, Retention reproducibility of basic drugs in HPLC on a silica column with a methanol–high-pH buffer element: Changes in selectivity with the age of the stationary phase, *J. Chromatogr., 592*:85 (1992).

568. D. Moir, N. Beaulieu, N. M. Curran, and E. G. Lovering, Liquid chromatography determination of naproxen and related compounds in raw materials, *JAOAC, 73*:902 (1990).

569. M. K. Ghosh, *HPLC Methods on Drug Analysis,* Springer-Verlag, Berlin, 1992.

570. J. R. Kem, Chromatographic separation of the optical isomers of naproxen, *J. Chromatogr., 543*:355 (1991).

571. J. V. Anderson and S. H. Hansen, Simultaneous determination of (R)- and (S)-naproxen and (R)- and (S)-6-O-desmethylnaproxen by HPLC on a chiral–AGP column, *J. Chromatogr., 577*:362 (1992).

572. B. Busqewski, M. El Mouelhi, K. Albert, and E. Bayer, Influence of the Structure of chemically binded C_{18} phases on HPLC separation in naproxen glucuronide diastreoisomers, *J. Liquid Chromatogr., 13*:505 (1990).

573. P. M. Lacroix, W. M. Cunan, and E. G. Lovering, Nadolol: HPLC methods for assay, racemate composiiton and related compounds, *J. Pharm. Biomed. Anal., 10*:917 (1992).

574. W. H. Pirkle and C. J. Whelch, Chromatographic separation of underivatized naproxen enantiomers, *J. Liquid Chromatogr., 14*: 3387 (1991).

575. Z. S. Budvari-Barany, G. Y. Szasz, G. Radeczyky, I. Somonyi, and A. Shalaby, Some new data concerning the chromatographic purity test for nifedipine, *J. Liquid Chromatogr., 13*:3541 (1990).

576. N. Beaulieu, R. W. Sears, and E. G. Lovering, Methods for assay and related substances in nitrendipine raw materials, *J. Liquid Chromatogr., 15*:319 (1992).

577.   A. Nonzioli, G. Luque, and C. Ferandez, HPLC assay of nitren-
       dipine, *J. High Resol. Chromatogr., 13*:589 (1990).
578.   T. A. Perfetti and J. K. Swadesh, On-line determination of the
       optical purity of nicotine, *J. Chromatogr., 543*:129 (1991).
579.   H. Zhang and J. T. Stewart, HPLC determination of norepineph-
       rine bitartrate in 5% dextrose injection on underivatized silica with
       aqueous-organic mobile phase, *J. Liquid Chromatogr., 16*:2861
       (1993).
580.   E. Roets, E. Adams, I. G. Muriithi, and J. Hoogmartens, Deter-
       mination of the relative amounts of the B and C components of
       neomycin by TLC with fluorescence detection, *J. Chromatogr. A,
       696*:131 (1995).
581.   B. Sauer and R. Matusch, HPLC separations of nystatin and their
       influence on the antifungal activity, *J. Chromatogr. A, 672*:247
       (1994).
582.   J & W Scientific Catalog.
583.   S. H. Hanson, HPLC assay of the opiates in opium and cough
       mixtures using dynamically modified silica and UV absorbance,
       fluorescence and electrochemical detection, *Int. J. Pharm., 32*:7
       (1986).
584.   B. Marciniec, E. Kujawa, and M. Ogrodowcyk, Evaluation of
       nifedipine preparations by chromatographic spectrophotometric
       methods, *Pharmazie, 47*:502 (1992).
585.   M. Carlson, Liquid chromatographic determination of pentaeryth-
       ritol tetranitrate in pharmaceuticals: Collaborative study, *JAOAC,
       73*:693 (1990).
586.   G. D. George and J. T. Stewart, HPLC determination of trace
       hydrazine levels in phenelize sulfate drug substance, *Anal. Lett.,
       23*:1417 (1990).
587.   T. A. Berger and W. H. Wilson, Separation of drugs by packed
       column supercritcal fluid chromatography, I. Phenothiazine anti-
       psychotics, *J. Pharm. Sci., 83*:270 (1994).
588.   A. E. H. N. Ahmed, Simultaneous determination of phenylpropa-
       nolamine hydrochloride and isopropamide in capsules by HPLC,
       Anal. Lett., 26:1153 (1993).
589.   S. D. Desai and J. Blanchard, A simplified and rapid HPLC assay
       for pilocarpine, *J. Chromatogr. Sci., 30*:149 (1992).
590.   J. Aaginaka, J. Wakai, K. Takahashi, A. Yasuda, and T. Katagi,
       Chromatographia, 29:587 (1990).
591.   E. R. M. Hackmann, E. A. D. Gianotto, and M. I. R. M. Santoro,
       Determination of piroxicam in pharmaceutical preparations by UV

difference spectophotometry and HPLC, *Anal. Lett., 26*:259 (1993).

592. P. Modamio, O. Montejo, C. F. Lastra, and E. L. Marino, A valid HPLC method for oxprenolol stability studies, *Int. J. Pharm., 112*:93 (1994).

593. M. J. Akhtar, S. Khan, and M. Hafiz, HPLC assay for the determination of paracetamol, pseudoephedrine HLC and triprolidine HCL, *J. Pharm. Biomed. Anal., 12*:379 (1994).

594. K. C. Lee, H. M. Song, I. J. Oh, and P. P. DeLuca, Reversed-phase high-performance liquid chromatography for simultaneous determination of prostaglandins E(2), A(2) and B(2), *Int. J. Pharm., 106*:2 (1994).

595. L. C. Bailey, K. T. Tand, and B. A. Rogozinki, The determination of 2,6-diisopropylphenol in an oil in water emulsion dosage form by high-performance liquid chromatography and by second derivative UV spectroscopy, *J. Pharm. Biomed. Anal., 9*:501 (1991).

596. O. W. Lau, K. Chan, Y. K. Lau, and W. C. Wong, The simultaneous determination of active ingredients in cough-cold mixtures by isocratic RP IP HPLC, *J. Pharm. Biomed. Anal., 7*:725 (1989).

597. T. Jarvinen, P. Suhonen, H. Naumanen, A. Urtti, and P. Peura, Determination of physicochmical properties, stability in aqueous solutions and serum hydrolysis of pilocarpic acid diesters, *J. Pharm. Biomed. Anal., 9*:737 (1991).

598. I. J. Koski, B. A. Jansson, J. E. Markides, and M. L. Lee, Analysis of prostaglandins in aqueous solutions by supercritical fluid extraction and chromatography, *J. Pharm. Biomed. Anal., 9*:281 (1991).

599. D. A. Roston, Supercritical fluid extraction–supercritical fluid chromatography for analysis of a prostaglandin: HPMC Dispersion, *Drug Dev. Ind. Pharm., 18*:245 (1992).

600. K. C. Lee and P. P. DeLuca, Simultaneous determination of prostaglandins, E_1, A_1, and B_1 by reversed-phase high-performance liquid chromatography for the kinetic studies of prostaglandin E_1 in solution., *J. Chromatogr., 555*:73 (1991).

601. C. R. White, W. A. Moats, and K. L. Kotula, Comparative study of high performance liquid chromatographic methods for determination of tatracycline antibiotics, *J. Liquid Chromatogr., 16*:13 (1993).

602. I. P. Kanious, G. A. Zachariadis, and J. A. Stratis, Separation and determination of five penicillins by reversed phase HPLC, *J. Liquid Chromatogr., 16*:13 (1993).

603. L. Gagliardi, G. Cavazzutti, D. Deorei, and T. Totunno, HPLC determination of aminopropiophenone as an impurity in phenyl-propanolamine bulk drug and pharmaceutical formulations, *Anal. Lett., 26*:5 (1993).

604. A. M. Dyas, M. L. Robinson, and A. G. Fell, An evaluation of the structural requirements for the separation of propranolol enantiomers on Pirkle phases following achiral derivatisation, *Chromatographia, 30*:73 (1990).

605. M. Gazdag, M. Babjak, P. Kemenes-Bakos, and S. Gorog, XLI. Ion-pair high-performance liquid chromatographic separation of quaternary ammonium steroids on silica, *J. Chromatogr., 550*:639 (1991).

606. D. T. Witte, R. A. De Zeeuw, and B. F. H. Drenth, Chiral derivatization of promethazine with (−)-menthyl chloroformate for enantiomeric separation by RP-HPLC, *J. High Resol. Chromatogr.,* 569 (1990).

607. ESA Application (Bedford, MA) 10-1218.

608. K. D. Sternitzki, T. Y. Fan, and D. L. Dunn, High-performance liquid chromatographic determination of pilocarpine hydrochloride and its degradation products using a β-cyclodextrin column, *J. Chromatogr., 589*:159 (1992).

609. A. El Yazigi, F. A. Wahab, and B. Afrane, Stability study and content uniformity of prochlorperazine in pharmaceutical preparations by liquid chromatography, *J. Chromatogr., 690*:1 (1995).

610. G. W. Chase, Jr., W. O. Lande, Jr., R. R. Eitenmiller, and A. G. M. Soliman, Liquid chromatographic determination of thiamine, riboflavin, and pyridoxine in infant formula, *JAOAC Int., 75*:3 (1992).

611. E. Dargel and J. B. Mielck, HPLC-methods for separation and quantitation of reserpin and its main degradation products, *J. Liquid Chromatogr., 13*:3973 (1990).

612. U. R. Cieri, Determination of reserpine in tablets by liquid chromatography with fluorescence detection-revised procedure, *JAOAC Int., 77*:758 (1994).

613. P. Corti, G. Corbini, E. Dreassi, N. Politi, and L. Montecchi, Thin layer chromatography in the quantitative analysis of drugs. Determination of rifaximine and its oxidation products, *Analusis, 19*:257 (1991).

614. T. Oshima, F. Hirayama, M. Masuda, T. Maruta, H. Itokawa, L. Y. He, Y. Y. Tong, and Y. H. Chen, Determination of sennoside A and B in the pharmaceutical preparation Otsuji-to using ion-pair

high-performance liquid chromatography with column switching, *J. Chromatogr., 585*:255 (1991).

615. C. L. Ng, H. K. Lee, and S. F. Y. Li, Systematic optimization of capillary electrophoretic separation of sulphonamides, *J. Chromatogr., 598*:133 (1992).

616. C. L. Ng, H. K. Lee, and S. F. Y. Li, Determination of sulphonamides in pharmaceuticals by capillary electrophoresis, *J. Chromatogr., 632*:165 (1993).

617. Q. Dang, Z. Sun, and D. Ling, Separation of sulphonamides and determination of the active ingredients in tablets by micellar electrokinetic capillary chromatography, *J. Chromatogr., 603*:259 (1992).

618. F. M. El-Anwar, A. M. El-Walily, M. H. Abdel Hay, and M. El-Swify, The analysis of a triple sulfonamide pharmaceutical powder form by HPLC, *Anal. Lett., 24*:767 (1991).

619. B. Van Giessen and K. Tsuji, GLC assay method for neomycin in petrolatum-based ointments, *J. Pharm. Sci., 60*:1068 (1971).

620. J. Wieling, J. Schepers, J. Hempenius, C. K. Mensink, and J. H. Jonkman, Optimization of chromatographic selectivity of twelve sulphonamides in reversed-phase high-performance liquid chromatography using mixture designs and multi-criteria decision making, *J. Chromatogr., 545*:101 (1991).

621. V. K. Agarwal, Solid phase extraction of sulfonamides using cyclobond-1 cartridges, *J. Liquid Chromatogr., 14*:699 (1991).

622. M. L. Bieganowska, A. Petruczynik, and A. Doraczynska-Szopa, Thin-layer reversed-phase ion-pair chromatography of some sulphonamides, *J. Pharm. Biomed. Anal., 11*:241 (1993).

623. H. Salomies, *Quantitatives HPTLC of Sulfonamides in Pharmaceutical preparations*, Dr. Alfred Huethig Publishers, 1993.

624. E. Busker, K. Gunther, and J. Martens, Application of chromatographic chiral stationary phases to pharmaceutical analysis: Enantiomeric purity of d-penicillamine, *J. Chromatogr., 350*:179 (1985).

625. N. Beaulieu, T. D. Cyr, S. J. Graham, and E. G. Lovering, Liquid chromatographic method for selegiline hydrochloride and related compounds in raw materials and tablets, *JAOAC, 74*:453 (1991).

626. S. A. Qureshi and I. J. McGilveray, Dissolution studies of selegiline tablets, *Pharmacopeial Forum*, 1973 (1991).

627. U. R. Cieri, Determination of reserpine and rescinnamine in Rauwolfia serpentina preparations by LC with fluorescence, *JAOC Int., 70*:540 (1987).

628. A. Vincent and D. Y. C. Awang, Determination of reserpine in pharmaceutical formulations by HPLC, *J. Liquid Chromatogr., 4*:1651 (1981).

629. U. R. Cirer, Determination of reserpine and hydrochlorothiazide in commerical tablets by LC with fluoresence and UV absorption detectors in series, *JAOAC Int., 71*:515 (1988).

630. C. Albet, J. M. Fernandez, E. Rozman, J. A. Perez, A. Sacristan, and J. A. Ortiz, Determination of sertaconazole nitrate, a new imidazole antifungal, by high-performance liquid chromatography, *J. Pharm. Biomed. Anal., 10*:205 (1992).

631. R. Chiba and Y. Ishii, Simultaneous determination of yohimbine hydrochloride, strychnine nitrate and methyltestosterone by ion-pair high-performance liquid chromatography, *J. Chromatogr., 588*:344 (1991).

632. K. Robards and P. Towers, Chromatography as a reference technique for the determination of clinically important steroids, *Biomed. Chromatogr., 4*:1 (1990).

633. L. Alliot, G. Bryant, and P.S. Guth, Measurement of strychnine by HPLC, *J. Chromatogr. Biomed. Appl., 232*:440 (1982).

634. R. Gill and M. D. Osselton, Retention reproducibility of basic drugs in HPLC on a silica column with methanol-ammonium nitrate eluent, *J. Chromatogr., 386*:65 (1987).

635. H. Huizer, Analytical studies on illicit heroin. III. Auto-interference in the colorimetric determination of strychnine in illicit heroin samples, *Pharm. Week Bl. Sci., 5*:254 (1983).

636. L. Malkkilaine, A. P. Bruins, Structural characterization of the decompostion products of salbutamol by liquid chromatography ion spray mass spectrometry, *J. Pharm. Biomed. Anal., 12*:114 (1994).

637. Bio-Rad Applications.

638. G. A. Jacobson and G. M. Peterson, High-performance liquid chromatographic assay for the simultaneous determination of opratropium bromide, fenoterol, salbutamol and terbutaline in nebulizer solution, *J. Pharm. Biomed. Anal., 12*:111 (1994).

639. A. De Leenheer, GLC of phenothiazone derivatives and related compounds, *J. Chromatogr., 77*:339 (1973).

640. K. Kurosaka, A. Kuchiki, and H. Nakagawa, Determination of sulfur in pharmaceutical preparations using reversed-phase high-performance liquid chromatography, *Chem. Pharm. Bull., 39*: 2138 (1991).

641. G. Carlucci and P. Mazzeo, Determination of simvastatin in phar-

maceutical forms by HPLC and derivative UV-spectrophotometry, *Farmaco, 47*:817 (1992).

642. W. J. Wong, K. J. Pardue, R. C. Ludwig, D. F. Conklin, and K. J. Duff, New HPLC column offers longer lifetime and improved resolution of taxanes, *Supelco, 13*:2.

643. R. E. B. Ketchum and D. M. Gibson, Rapid isocratic reversed phase HPLC of taxanes on new columns developed specifically for taxol analysis, *J. Liquid Chromatogr., 16*:12 (1993).

644. S. L. Richheimer, D. M. Tinnermeier, and D. W. Timmons, High-performance liquid chromatographic assay of taxol, *Anal. Chem., 64*:2323 (1992).

645. *Separation of Taxane Standards on Curosil™-B*, Phenomenex, Torrance, CA.

646. P.-H Yuen, S. L. Denman, T. D. Sokoloski, and A. M. Burkman, Loss of nitroglycerin from aqueous solution into plastic intravenous delivery systems, *J. Pharm. Sci., 68*:1163 (1979).

647. J. E. Svirbely and A. J. Pesce, Trimethoprim analysis by LC, *J. Liquid Chromatogr., 11*:1075 (1988).

648. J. Salamoun, M. Macka, M. Nechvatal, M. Matousek, and L. Knesel, Identification of products formed during UV irradiation of tamoxfen and their use for fluorescence detection in high-performance liquid chromatography, *J. Chromatogr., 514*:179 (1990).

649. W. O. McSharry and I. V. E. Savage, Simultaneous high-pressure liquid chromatography determination of acetaminophen, guaifenesin, and dextromethorphan in cough syrup, *J. Pharm. Sci., 69*:212 (1980).

650. A. M. Depaolis, T. E. Thomas, A. J. McGonigle, G. Kaplan, and W. C. Davies, Determination of meclocycline, a tetracycline analog in cream formulations by liquid chromatography, *J. Pharm. Sci., 73*:1650 (1984).

651. S. O. K. Auriola, A. M. Lepisto, and T. Naaranlahti, Determination of taxol by high-performance liquid chromatography-thermospray mass spectrometry, *J. Chromatogr., 594*:153 (1992).

652. L. Elrod, Jr., C. L. Linton, B. P. Shelat, and C. F. Wong, Determination of minor impurities in temafloxacin hydrochloride by high-performance liquid chromatography, *J. Chromatogr., 519*:125 (1990).

653. R. V. Smith and D. W. Humphrey, Determination of apomorphine in tablets using HPLC with electrochemical analysis, *Anal. Lett., 14*:601 (1981).

654. K. K. Chan, D. D. Giannini, J. A. Staroscik, and W. Sadee,

5-Azacytidine hydrolysis kinetics as measured by HPLC and C-13 NMR, *J. Pharm. Sci., 68*:807 (1979).

655.   K. T. Lin, R. L. Momparlen, and G. E. Rivard, HPLC analysis of the chemical stability of 5-aza-2'deoxycytidine, *J. Pharm. Sci., 70*:1228 (1981).

656.   M. I. Walash, O. M. Salama, and M. M. Bishr, Use of absorbance ratios in densitometric measurements for the characterization and identification of natural products of pharmacological interest, *J. Chromatogr., 469*:390 (1989).

657.   C. X. Zhang, Z. P. Sun, D. K. Ling, and Y. J. Zhang, Separation of tetracycline and its degradation products by capillary zone electrophoresis, *J. Chromatogr., 627*:281 (1992).

658.   K. Iwaki, N. Okumura, and M. Yamazaki, Determination of tetracycline antibiotics by reversed-phase high-performance liquid chromatography with fluorescence detection, *J. Chromatogr., 623*:153 (1992).

659.   C. Hendrix, E. Roets, J. Crommen, J. Debeer, E. Porqueras, W. Vandenbossche, and J. Hoogmartens, Collaborative study of the analysis of tetracycline by liquid chromatography on poly (styrenedivinylbenzene), *J. Liquid Chromatogr., 16*:106 (1993).

660.   P. D. Bryan and J. T. Stewart, Separation of tetracyclines by liquid chromatography with acidic mobile phases and polymeric columns, *J. Pharm. Biomed. Anal., 11*:971 (1993).

661.   R. B. Forney, Alcohol, an abused drug, *Clin. Chem., 33*(11(b)): 82(b) (1987).

662.   *JAMA, 257*:3110 (1987).

663.   H. K. Chan and G. P. Carr, Evaluation of a photodiode array detector for the verification of peak homogeneity in high-performance liquid chromatography, *J. Pharm. Biomed. Anal., 8*: 271 (1990).

664.   W. J. Kopycki, H. N. Elsohly, and J. D. McChesney, HPLC determination of taxol and related compounds in Taxus plant extracts, *J. Liquid Chromatogr., 17*:98 (1994).

665.   H. Y. Aboul-Enein and M. R. Islam, Direct separation and optimization of timolol enantiomers on a cellulose tris-3,5-dimethylphenylcarbamate high-performance liquid chromatographic chiral stationary phase, *J. Chromatogr., 55*:109 (1990).

666.   S. N. Sanyal, A. K. Datta, and A. Chakrabarti, Stability indicating TLC method for the quantification of tinidazole in pharmaceutical dosage form I.V. fluid, *Drug Dev. Ind. Pharm., 18*:2095 (1992).

667.   D. J. Mazzo and P. A. Snyder, High-performance liquid chroma-

tography of timolol and potential degradates on dynamically modified silica, *J. Chromatogr., 438*:85 (1988).

668. J. E. Parkin, Assay for thiomersal (thimerosal) with adaptation to the quantitation of total ethylmercury available in degraded samples, *J. Chromatogr., 587*:329 (1991).

669. R. A. Dean, R. H. B. Sample, N. Dumaul, and W. Bosron, Simultaneous determination of cocaine, ethylcocaine, and benzoyecgonine by HPLC, 3rd International Symposium on Pharmaceutical Biomedical Analysis, 1991.

670. L. J. Bowie and P. P. Kirkpatrick, Simultaneous quantification of morphine by GC/MS, *Clin. Chem., 35*:1355 (1989).

671. J. E. Parkin, High-performance liquid chromatographic assay of thiomersal (thimerosal) as the ethylmercury dithiocarbamate complex, *J. Chromatogr., 542*:137 (1991).

672. O. Y-Pu Hu, Simultaneous determination of thimerosal and chlorhexidine in solutions for soft contact lenses and its applications in stability studies, *J. Chromatogr., 523*:321 (1990).

673. ESA, ESA, *Coulodrem Applications*, Beford, MA.

674. A. K. Dash and R. Suryanarayanan, A liquid-chromatographic method for the determination of tobramycin, *J. Pharm. Biomed. Anal., 9*:237 (1991).

675. I. Papadoyannis, M. Georgarakis, V. Samanidou, and A. Zotou, Rapid assay for the determination of tolfenamic acid in pharmaceutical preparations and biological fluids by high-performance liquid chromatography, *J. Liquid Chromatogr., 14*:2951 (1991).

676. M. D. DaSilva, J. R. Procopio, and L. Hernandez, Evaluation of the capability of different chromatographic systems for the monitoring of thimerosal and its degradation products by high-performance liquid chromatography with amperometric detection, *J. Chromatogr., A653*:34 (1994).

677. V. M. Shinde, N. M. Tendolkar, and B. S. Desai, Simultaneous determination of theophylline and etofylline in pharmaceutical dosages by HPTLC, *Anal. Lett., 61*:33 (1995).

678. J. J. Bergh, J. C. Breytenbach, and J. L. Du Preez, High-performance liquid chromatographic analysis of trimethoprim in the presence of its degradation products, *J. Chromatogr., 513*:392 (1990).

679. K. Shimada, K. Mitamura, M. Morita, and K. Hirakata, Separation of the diastereomers of baclofen by high performance liquid chromatography using cyclodextrin as a mobile phase additive, *J. Liquid Chromatogr., 16*:15 (1993).

680. P. D. Bryan and J. T. Stewart, Chromatographic analysis of selected tetracyclines from dosage forms and bulk drug substance

using polymeric columns with acidic mobile phases, *J. Pharm. Biomed. Anal., 12*:118 (1994).

681.   K. C. Chan, G. M. Muschik, H. J. Issaq, and K. M. Snader, Separation of taxol and related compounds by micellar electrokinetic chromatography, *J. High Resol. Chromatogr., 37*:33 (1994).

682.   N. R. Ayyangar, S. S. Biswas, and A. S. Tamke, Separation of opium alkaloids by TLC combined with flame ionization detection using the peak pyrolysis method, *J. Chromatogr., 547*:538 (1991).

683.   P. M. Lacroix, B. A. Dawson, R. W. Sears, and D. B. Black, HPLC and NMR methods for the quantitation of the (R)-enantiomer in (−)-(S)-timolol maleate drug raw materials, *Chirality, 6*: 484 (1994).

684.   N. Beaulieu, R. W. Sears, and E. G. Lovering, Liquid chromatography methods for determination of trazodone and related compounds in drug raw materials, *JAOAC Int., 75*:32 (1994).

685.   K. Mathys and R. Brenneisen, Determination of (S)-(−)-cathinone and its metabolites (R,S)-(min)-norephedrine and (R,R)-(−)-nor-pseudoephedrine in urine by HPLC with photodiode-array detection, *J. Chromatogr., 593*:79 (1992).

686.   W. P. Duncan and D. G. Deutsch, The use of GC/IR/MS of high-confidence identification of drugs, *Clin. Chem., 35*:1279 (1989).

687.   N. Meaulieu, T. D. Cyr, S. J. Graham, R. C. Lawrence, R. W. Sears, and E. G. Lovering, Methods for assay, related substances, and organic volatile impurities in triazolam raw materials and formulations, *J. Assoc. Off. Anal. Chem., 74*:3 (1991).

688.   T. P. Castor and T. A. Tyler, Determination of taxol in taxus media needles in the presence of interfering components, *J. Liquid Chromatogr., 16*:723 (1993).

689.   S. Croubels, W. Baeyens, C. Dewaele, and C. Van Peteghem, Capillary electrophoresis of some tetracycline antibiotics, *J. Chromatogr., A673*:39 (1994).

690.   Q. Xun Dang, L. Xia Yan, Z.-Pei Sun, and D.-Kui Ling, Separation and simultaneous determination of the active ingredients in theophylline tablets by micellar electrokinetic capillary chromatography, *J. Chromatogr., 630*:363 (1993).

691.   M. A. Evenson and J. E. Wiktorowicz, Automated capillary electrophoresis applied to therapeutic drug monitoring, *Clin. Chem., 38*:1847 (1992).

692.   B. Tan and L. Brzuskiewicz, Separation of tocopherol and tocotrienol isomers using normal- and reverse-phase liquid chromatography, *Anal. Biochem., 180*:368 (1989).

693. Y. Satomura, M. Kimura, and Y. Itokawa, Simultaneous determination of retinol and tocopherols by high-performance liquid chromatography, *J. Chromatogr., 625*:67 (1993).

694. A. BenAmotz, Simultaneous profiling and identification of carotenoids, retinols, and tocopherols by high performance liquid chromatography equipped with three-dimensional photodiode array detection, *J. Liquid Chromatogr., 18*:122 (1995).

695. M. Taneke, S. Teshima, T. Hanyo, and Y. Hayashi, Rapid and sensitive method for erythropoietin determination in serum, *Clin. Chem., 38*:1752 (1992).

696. S. L. Abidi and T. L. Mounts, Separations of tocopherols and methylated tocols on cyclodextrin-bonded silica, *J. Chromatogr., A664*:130 (1994).

697. GC separations of chiral compounds made easier by new β-cyclodextrin capillary columns, *Supelco*, 14.

698. M. B. Kril, K. A. Burke, J. E. DiNunzio, and R. Rao Gadde, Determination of tretinoin in creams by high-performance liquid chromatography, *J. Chromatogr., 522*:227 (1990).

699. I. S. Lurie, P. A. Cooper, and R. F. X. Klein, High-performance liquid chromatographic analysis of benzodiazepines using diode array, electrochemical and thermospray spectrometric detection, *J. Chromatogr., 598*:59 (1992).

700. M. Balikova, Selective system of identification and determination of antidepressants and neuroleptics in serum and plasma by solid-phase extraction followed by HPLC with photodiode-array detection in analytical toxicology, *J. Chromatogr., 581*:75 (1992).

701. *Varex/Astec Catalog*, Burtonsville, MD.

702. C. R. Clark, J. DeRuiter, and F. T. Noggle, GC/MS identification of amine-solvent condensation products formed during analysis of drugs of abuse, *J. Chromatogr. Sci., 30*:399 (1992).

703. N. R. Srinivas and L. N. Lgwemezie, Chiral separation by HPLC.1. Review of indirect separation of enantiomers as diasteromeric derivatives using ultraviolet, fluorescence and electrochemical detection, *Biomed. Chromatogr., 6*:163 (1992).

704. P. Newton, High performance liquid chromatography and the mystery of L-tryptophan, *LC–GC, 9*:208.

705. Baxter Catalog–1992.

706. J & W Scientific Catalog–1992.

707. A. M. Dyas, The chiral chromatographic separation of β-adrenoceptor blocking drugs, *J. Pharm. Biomed. Anal., 10*:383 (1992).

708. J. D. Musto, J. N. Sane, and V. D. Warren, Quantitative determinations phenol by HPLC, *J. Pharm. Sci., 66*:1201 (1977).

709.   SGE Chromatography Products, 1992/1993 Catalogue.

710.   H. Weems and K. Zamani, Resolution of terfenadine enantiomers
        by β-cyclodextrin chiral stationary phase high-performance liquid
        chromatography, *Chirality, 4*:268 (1992).

711.   H. Y. Aboul-Enein and M. R. Islam, Direct HPLC separation of
        thalidomide enantiomers using cellulose tris-4-methylphenyl ben-
        zoate chiral stationary phase, *J. Liquid Chromatogr., 14*:667
        (1991).

712.   A. H. B. Wu, T. A. Onigbinde, S. S. Wong, and K. G. Johnson,
        Identification of methamphetamines and over-the-counter sympa-
        thometic amines by full-scan GC ion trap MS with E.I. and C.I.,
        J. Anal. Toxicol., 16:137 (1992).

713.   Letter to the Editor, A procedure for eliminating interferences
        from ephedrine and related compounds in the GC/MS analysis
        of amphetamine and methamphetamine, *J. Anal. Toxicol., 16*:109
        (1992).

714.   P. A. D. Edwardson and R. S. Gardner, Problems assoicated with
        the extraction and analysis of triamcinolone acetonide in detmato-
        logical patches, *J. Pharm. Biomed., 8*:935 (1990).

715.   H. Y. Aboul-Enein, S. A. Bakr, M. R. Islam, and R. Rothchild,
        Direct chiral liquid chromatographic separation of tocainide en-
        antiomers on a crownpak (CR) column and its application to phar-
        maceutical formulations and biological fluids, *J. Liquid Chroma-
        togr., 14*:3475 (1991).

716.   Waters, Thiamine and Riboflavin, Technical Bulletin.

717.   P. M. Lacroix, S. J. Graham, and E. G. Lovering, High-perform-
        ance liquid chromatography method for the assay of verapamil
        hydrochloride and related compounds in raw materials, *J. Pharm.
        Biomed. Anal., 9*:119 (1992).

718.   L. Miller and R. Bergeron, Analytical and preparative resolution
        of enantiomers of verapamil and norverapamil using a cellulose-
        based chiral stationary phase in the reversed-phase mode, *J. Chro-
        matogr., 648*:36 (1993)

719.   E. L. Inman, M. D. Lantz, and M. M. Strohl, Absorbance ratio-
        ing as a screen for related substances of pharmaceutical products,
        J. Chromatogr. Sci., 28:578 (1990).

720.   H. T. Aboul-Enein and S. A. Bakr, High-performance liquid
        chromatographic identification of ramipril, and its precursor en-
        antiomers using a chiralpack OT(+) column, *Drug Dev. Ind.
        Pharm., 18*:1013, (1992).

721.   Chimera Research and Chemical, Inc., Seminole, Florida.

722.   E. L. Inman, R. L. Clemens, and B. A. Olsen, Determination of

EDTA in vancomycin by liquid chromatography with absorbance ratioing for peak identification, *J. Pharm. Biomed. Anal., 8*:513 (1990).

723. J. P. Chervet, R. E. J. van Soest, and J. P. Salzmann, Recent advances in capillary liquid chromatography enhanced detectability in bioanalysis, *LC–GC, 10*:866 (1993).

724. P. Oroszlan, S. Wicar, G. Teshima, S.-L. Wu, W. S. Hancock, and B. L. Karger, Conformational effects in the reversed-phase chromatographic behavior of recombinant human growth hormone and N-methionyl recombinant growth hormone, *Anal. Chem., 64*:1623 (1992).

725. M. Mathew and V. DasGupta, Stability of vancomycin hydrochloride solutions at various pH values as determined by high-performance liquid chromatography, *Drug Dev. Ind. Pharm., 21*:100 (1994).

726. A. P. De Leenheer, W. E. Lambert, and H. J. Nelis, *Modern Chromatographic Analysis of Vitamins*, Marcel Dekker, Inc., New York, 1992.

727. J. Dalbacke and I. Dahlquist, Determination of vitamin B_{12} in multivitamin–multimineral tablets by high-performance liquid chromatography after solid-phase extraction, *J. Chromatogr., 541*:383 (1991).

728. E. Postaire, M. Cisse, M. D. Le Hoang, and D. Pradeau, Simultaneous determination of water-soluble vitamins by over-pressure layer chromatography and photodensitometric detection, *J. Pharm. Sci., 80*:366 (1991).

729. C. Genestar and F. Grases, Determination of vitamin A in pharmaceutical preparations by high-performance liquid chromatography with diode-array detection, *Chromatographia, 40*:134 (1995).

730. D. Lambert, C. Adjalla, F. Felden, S. Benhayoun, J. P. Nicolas, and J. L. Gueant, Identification of vitamin B_{12} and analogues by high-performance capillary electrophoresis and comparison with high-performance liquid chromatography, *J. Chromatogr., 608*: 311 (1992).

731. M. G. Sliva, A. E. Green, J. K. Sanders, J. R. Euber, and J. R. Saucerman, Reversed-phase liquid chromatographic determination of Vitamin D in infant formulas and enteral nutritionals, *JAOAC Int., 75*:566 (1992).

732. W. S. Letter, Preparative isolation of vitamin D_2 from previtamin D_2 by recycle high-performance liquid chromatography, *J. Chromatogr., 590*:169 (1992).

733. K. Shimada, K. Mitamura, M. Miura, and A. Miyamoto, Reten-

tion behavior of vitamin D and related compounds during high-performance liquid chromatography, *J. Liquid Chromatogr., 18*: 2885 (1995).

734.   J. P. Hart, M. D. Norman, and C. J. Lacy, Voltammetric behaviour of vitamins D_2 and D_3 at a glassy carbon electrode and their determination in pharmaceutical products by using liquid chromatography with amperometric detection, *Analyst, 117*:1441 (1992).

735.   Alltech Catalog, Deerfield, IL.

736.   ESA, Inc., Bedford, Mass.

737.   S. N. El-Gizawy, A. N. Ahmed, and N. E. El-Rabbat, High performance liquid chromatographic determination of multivitamin preparations using a chemically bonded cyclodextrin stationary phase, *Anal. Lett., 24*:1173 (1991).

738.   Supelco Catalog, LC-8DB column.

739.   Baxter Catalog, HLD OC5 octyl column.

740.   R. N. James and B. Boneschans, A reversed-phase HPLC method for the determination of niacinamide and riboflavin in dissolution samples of multivitamin-mineral combination capsules, *Drug Dev. Ind. Pharm., 18*:1989 (1992).

741.   D. E. Burton and M. J. Sepaniak, Analysis of B_6 vitamers by micellar electrokinetic capillary chromatography with laser-excited fluorescence detection, *J. Chromatogr., 24*:347 (1986).

742.   Alltech Catalog, Adsorbosphere UHS C8 column.

743.   M. Amin, High performance liquid chromatography of fat-soluble vitamins. II. Determination of vitamin E in pharmaceutical preparations and blood, *J. Liquid Chromatogr., 11*:1335 (1988).

744.   Y. F. Yik, H. K. Lee, S. F. Y. Li, and S. B. Khoo, Micellar electrokinetic capillary chromatography of vitamin B_6 with electrochemical detection, *J. Liquid Chromatogr., 585*:139 (1991).

745.   M. C. Gennaro, Separation of water-soluble vitamins by reversed-phase ion-interaction-reagent high-performance liquid chromatography: Application to multivitamin pharmaceuticals, *J. Chromatogr. Sci., 29*:410 (1991).

746.   E. Bertol, F. Mari, and M. G. Di Milia, Programmable temperature vaporizer applications in an high-resolution gas chromatographic method for the quantitation of impurities in illicit heroin, *J. Chromatogr., 466*:384 (1989).

747.   S. J. Male and G. A. Casella, Rendering the "poppy-seed defense" defenseless: identification of 6-monoacetylmorphine in urine by GC/MS, *Clin. Chem., 34*:1427 (1988).

748.   A. R. Wijesekera, K. D. Henry, and P. Ranasinghe, The detection and estimation of arsenic in opium and strychrine in opium and

heroin, as a means of identification of their respective sources, *Forensic Sci. Inc., 36*:193 (1988).

749. I. Carretero, M. Maldonado, J. J. Laserna, E. Bonet, and R. G. Ramis, Detection of banned drugs in sports by micellar liquid chromatography, *Anal. Chem. Acta., 259*:203 (1992).

750. Phenomenex, Fat and water soluble vitamin separations, *Resolution times*, Summer (1993).

751. E. Merck, *Chroma Source*, 1992.

752. N. Thuaud, B. Sebille, A. Deratani, and G. Lelievre, Retention behavior and chiral recognition of β-cyclodextrin-derivative polymer absorbed on silica for warfarin, structurally related compounds and Dns-amino acids, *J. Chromatogr., 555*:53 (1991).

753. J. P. Gramond and F. Guyon, Separation and determination of warfarin enantiomers in human plasma samples by capillary zone electrophoresis using a methylated β-cyclodextrin-containing electrolyte, *J. Chromatogr., 615*:36 (1993).

754. S. Alessiseverini, R. T. Coutts, F. Jamali, and F. M. Pasutto, HPLC analysis of methocarbamol enantiomers in biological fluids, *J. Chromatogr., 582*:173 (1992).

755. M. A. Radwan, Stability-indicating HPLC assay of zidovudine in extemporaneous syrup, *Anal. Lett., 27*:50 (1994).

756. T. A. Bierimer, N. Asral, and J. A. Albanese, Simultaneous, stability-indicating capillary gas chromatographic assay for benzocaine and the two principal benzyl esters of Balsam Peru formulated in a topical ointment, *J. Chromatogr., 623*:395 (1992).

757. S. Scypinski and A. J. Ross, Liquid chromatographic separation of zalcitabine and its stereoisomers, *J. Pharm. Biomed. Anal., 12*: 101 (1994).

758. R. M. Soler Roca, F. J. Garcia March, G. M. Anton Fos, R. Garcia Domenech, F. Perez Grimenez, and J. Galvez Alvarez, Molecular topology and chromatographic retention parameters for benzodiazepines, *J. Chromatogr., 607*:91 (1992).

759. I. S. Lurie, Micellar electrokinetic capillary chromatography of the enantiomers of amphetamine, methamphetamine and their hydroxyphene-thylamine percursors, *J. Chromatogr., 605*:269 (1992).

760. R. C. Chloupek, W. S. Hancock, and L. R. Synder, Computer simulation as a tool for the rapid optimization of the HPLC separation of a tryptic digest of human growth hormone, *J. Chromatogr., 594*:65 (1992).

761. W. Jost, H. E. Hauck, and F. Eisnebeiss, Simple method for the separation of cephalsporins using silica gel 60 as the stationary

phase in thin-layer and high-performance liquid chromatography, *J. Chromatogr., 256*:182 (1983).

762.   F. Fadil and W. O. McSharry, Extraction and TLC separation of food, drug and cosmetic dyes from tablet-coating formulations, *J. Pharm. Sci., 68*:97 (1979).

763.   K. Kudo, T. Nagata, T. Imamura, S. Kage, and Y. Hida, Forensic analysis of triazolam in human tissues using capillary GC, *Int. J. Leg. Med., 104*:67 (1991).

764.   E. A. Cox and F. D. McClure, High-performance liquid chromatographic determination of intermediates and reaction by-products in FD&C yellow No. 5: Collaborative study, *J. Assoc. Off. Anal. Chem., 65*:933 (1982).

765.   R. J. Calvey, A. L. Goldberg, and E. A. Madigan, High-performance liquid chromatographic determination of intermediates/side reaction products in FD&C yellow No. 5, *J. Assoc. Off. Anal. Chem., 64*:655 (1981).

766.   C. J. Cox, E. A. Cox, and A. A. Springer, High pressure chromatographic determination of the intermediates/side reaction products in FD&C red No. 2 and FD&C yelllow No. 5: Statistical analysis of instrument response, *J. Assoc. Off. Anal. Chem., 61*: 1404 (1978).

767.   G. Teshima and E. Canovadavis, Separation of oxidized human growth hormone variants by reversed-phase HPLC-effect of mobile phase pH and organic modifier, *J. Chromatogr., 625*:207 (1992).

768.   E. A. Cox and G. F. Reed, High performance liquid chromatographic determination of intermediates and two reaction by products in FD&C red No. 40: Collaborative study, *J. Assoc. Off. Anal. Chem., 64*:324 (1981).

769.   J. E. Bailey and R. J. Calvey, Spectral compilation of dyes, intermediates and other reaction products structurally related to FD&C yellow No. 6, *J. Assoc. Off. Anal. Chem., 58*:1087 (1975).

770.   E. A. Cox, High performance liquid chromatographic determination of sulfanilic acid, Schaeffer's salt, 4, 4'-(diazoamino)-dibenzenesulfonic acid and 6,6'-oxybis (2-naphthalenesulfonic acid) in FD&C yellow No. 6: Collaborative study, *J. Assoc. Off. Anal. Chem., 63*:61 (1980).

771.   L. W. Brown, High-pressure liquid chromatographic assays for clindamycin, clindamycin phosphate, and clindamycin palmitate, *J. Pharm. Sci., 57*:1254 (1978).

772.   M. Singh, Chromatographic determination of uncombined inter-

mediates in FD&C red No. 40, *J. Assoc. Off. Anal. Chem., 57*: 219 (1974).

773. R. Deslauriers, F. Hason, B. A. Lodge, and I. C. Smith, Assignment by C-13 spectroscopy of configuration at C-5 in 17-alpha-ethylestran-17 beta-of an impurity in the anabolic steroid ethylestrenol, *J. Pharm. Sci., 67*:1187 (1978).

774. D. C. Dabas, H. Erguven, M. C. Vescina, and C. N. Carducci, Stability study of ethyl loflazepate in bulk drug, solution and dosage form by liquid chromatography, *J. Pharm. Biomed. Anal., 10*:241 (1992).

775. J. P. Cano, Y. C. Sumirtapura, W. Cautreels, and Y. Sales, Analysis of the metabolites of ethyl loflazepate by gas chromatography with electron-capture detection, *J. Chromatogr., 226*:413 (1981).

776. W. Mechlinski and C. P. Schaffner, Separation of polyene antifungal antibiotics by high-speed liquid chromatography, *J. Chromatogr., 99*:619 (1974).

777. M. Singh, Determination of uncombined intermediates in FD&C yellow No. 6 by high-pressure liquid chromatography, *J. Assoc. Off. Anal. Chem., 57*:358 (1974).

778. M. Singh, High-pressure liquid chromatographic determination of uncombined intermediates and subsidiary colors in FD&C blue No. 2, *J. Assoc. Off. Anal. Chem., 58*:48 (1975).

779. M. Singh, High pressure liquid chromatographic determination of uncombined intermediates and subsidary colors in orange B, *J. Assoc. Off. Anal. Chem., 60*:1067 (1977).

780. M. Singh and G. Adams, Automated high pressure liquid chromatographic determination of uncombined intermediates in FD&C red No. 40 and FD&C yellow No. 6, *J. Assoc. Off. Anal. Chem., 62*:1342 (1979).

781. W. B. Link, Intermediates in food, drug and cosmetic colors, *J. Assoc. Off. Anal. Chem., 44*:43 (1961).

782. W. J. Rzeszotarski and A. B. Manger, Reversed-phase liquid chromatography of actinomycins, *J. Chromatogr., 86*:246 (1973).

783. Y. Akada, S. Kawano, and Y. Tanase, High-speed liquid-chromatographic determination of colouring matters in gelatin capsules, *Yakugaku Zasshi, 98*:1300 (1978).

784. M. Margosis, GLC analysis of chloramphenicol: A collaborative study, *J. Pharm. Sci., 63*:435 (1974).

785. E. R. White, M. A. Carroll, J. E. Zarembo, and A. D. Bender, Reversed-phase high speed liquid chromatography of antibiotics, *J. Antibiot., 28*:205 (1975).

786. J. Alary, C. L. Duc, and A. Coeur, Identification of synthetic colorants in drugs, *Bull. Trav. Soc. Pharm. Lyon, 10*:78 (1966).

787. F. Bailey and P. N. Brittain, High-efficiency liquid chromatography in pharmaceutical analysis, *J. Chromatogr., 83*:431.

788. P. Balatre and M. Traisnel, Identification of pharmaceutical dyes by thin-layer chromatography of their complexes with a quaternary ammonium compound, *Bull. Trav. Soc. Pharm. Lyon, 9*:41 (1965).

789. P. Balatre, D. Mulleman-Marsy, and M. Traisnel, Identification and determination of natural dyes of vegetable origin in drugs, *Ann. Pharm. Fr., 25*:649 (1967).

790. M. Galczynska, M. Kwiatkowska, B. Mikucka, and B. Szotor, Identification of synthetic dyes in coloured coated tablets, *J. Farm. Pol., 33*:645 (1977).

791. K. Tsuji, J. F. Goetz, W. VanMeter, and K. A. Gusciora, Normal-phase high-performance liquid chromatographic determination of neomycin sulfate derivatized with 1-fluoro-2,4,-dinitrobenzene, *J. Chromatogr., 175*:141 (1979).

792. S. J. Donato, Separation and estimation of methyl and propyl esters of *p*-hydroxybenzoic acids by GC, *J. Pharm. Sci., 54*:917 (1965).

793. O. Azzolina, V. Ghislandi, and D. Vercesi, Optical resolution of benzodiazepine esters by HPLC, *Farmaco, 45*:603 (1990).

794. F. A. Fitzpatrick, A. F. Summa, and A. D. Cooper, Quantitative analysis of methyl and propylparaben by HPLC, *J. Soc. Cosmet. Chem., 26*:377 (1975).

795. P. Balatre and M. Traisnel, Identification of pharmaceutical dyes by thin-layer chromatography of their complexes with a quaternary ammonium compound, *Bull. Trav. Soc. Pharm. Lyon, 9*:41 (1965).

796. M. F. L. Lefevere, A. E. Claeys, and A. P. Leenheer, Vitamin K: Phylloquinine and menaquinones. In *Modern Analysis of the Vitamins* (A. P. De Leenheer, M. G. M. De Rutter, and W. E. Lambert, eds.) Marcel Dekker, Inc., New York, 1985, p. 201.

797. G. Lehmann and P. Collet, Analysis of dyes, VIII: Identification of synthetic dyes in drugs, *Arch. Pharm. Berl., 303*:855 (1970).

798. V. Mares and Z. Stejskal, Identification of dyes used for coloring drugs, *Z. Cslka. Farm., 16*:474 (1967).

799. J. Vessman and S. Stromberg, GLC determination of nicotin-amide in multivitamin formulations after conversion to nicotin-trile, *J. Pharm. Sci., 64*:311 (1975).

800. R. C. Williams, D. R. Baker, and J. A. Schmidt, Analysis of

water-soluble vitamins by high-speed ion-exchange chromatography, *J. Chromatogr. Sci., 11*:618 (1973).

801.   M. H. Bui-Ngugen, Ascorbic acid and related compounds. In *Modern Analysis of the Vitamins* (A. P. De Leenheer, M. G. M. De Rutter, and W. E. Lambert, eds.), Marcel Dekker, Inc., New York, 1985, p. 267.

802.   F. Pellerin, J. A. Gautier, and A. M. Conrard, Identification of authorized synthetic organic dyes in pharmaceuticals, *Ann. Pharm. Fr., 22*:621 (1964).

803.   F. Pellerin, J. L. Kiger, and J. Caporal-Gauter, Synthetic organic colours in plastic packaging materials for pharmaceutical use, II: Identification in plastics and detection of their release into drugs, *Ann. Pharm. Fr., 32*:427 (1974).

804.   J. Storck, Detection of dyes in pharmaceutical gelatin capsules, *Ann. Pharm. Fr., 23*:113 (1965).

805.   B. Unterhalt and L. Kreutzig, Detection of dyestuffs in cough syrup, *Dt. Apoth. Ztg., 112*:449 (1972).

806.   F. T. Noggle, C. R. Clark, and J. DeRuiter, Liquid chromatographic and spectral analysis of the 17-hydroxy anabolic steroids, *J. Chromatogr. Sci., 28*:162 (1990).

807.   G. R. Rao, S. Raghuveer, and C. M. R. Srivastava, High pressure liquid chromatographic estimation of nifedipine and its dosage forms, *Indian Drugs, 22*:435 (1985).

808.   Z. Wojcik, Chromatographic identification of synthetic dyes in pharmaceutical preparations, *Z. Farm. Pol., 25*:419 (1969).

809.   C. C. Douglas, Sac chromatographic determination of phenolic compounds in drug preparations, *J. Assoc. Off. Anal. Chem., 55*: 610 (1972).

810.   Z. Wojcik, Thin-layer identification of azo dyes permitted in Poland for use in pharmaceutical preparations, *Z. Farm. Pol., 26*: 723 (1970).

811.   C. Graichen and J. C. Molitor, Determination of certified FD&C color additives in foods and drugs, *J. Assoc. Off. Anal. Chem., 46*:1022 (1963).

812.   M. L. Puttemans, L. Dryon, and D. L. Massart, Evaluation of thin-layer paper, and high-performance liquid chromatography for identification of dyes extracted as ion-pairs with tri-n-octylamine, *J. Assoc. Off. Anal. Chem., 65*:730 (1982).

813.   H. T. McKone and G. Nelson, Separation and identification of some FD&C dyes by TLC, *J. Chem. Ed., 53*:722 (1976).

814.   K. Florey (ed.), *Analytical Profiles of Drug Substances, 5,* Academic Press, New York, 1976, p. 5.

815.   K. Florey (ed.), *Analytical Profiles of Drug Substances, 5,* Academic Press, New York, 1976, p. 23.

816.   D. M. Marmion, *Handbook of U.S. Colorants for Foods, Drugs, and Cosmetics,* 2nd ed., Wiley, New York, 1984.

817.   J. W. M. Wegener, H. J. M. Grunbauer, R. J. Fordam, and W. Karcher, A combined HPLC-vis spectrophotomeric method for the identification of cosmetic dyes, *J. Liquid Chem., 7*:809 (1984).

818.   R. B. Patel, M. R. Patel, A. A. Patel, A. K. Shah, and A. G. Patel, Separation and determination of colours in pharmaceutical preparations by column chromatography, *Analyst, 111*:577 (1986).

819.   J. Bailey, Determination of the lower sulfonated subsidiary colors in F&DC Yellow No. 6 by reversed-phase HPLC, *J. Chromatogr., 347*:163 (1985).

820.   H. Tokunaga, T. Kimura, and J. Kawamura, Determination of glucocorticoids by liquid chromatography. III. Application to ointments and a cream containing cortisone acetate, dexamethasone acetate, fluorometholone, and betamethasone valerate, *Chem. Pharm. Bull., 32*:4012 (1984).

821.   H. C. Van den Berg, Quantitative determination of steroid acetates in pharmaceutical preparations by HPLC, *J. Assoc. Off. Anal. Chem., 63*:1184 (1980).

822.   F. T. Noggle, C. R. Clark, S. Andarkar, and J. DeRuiter, Methods for the analysis of 1-(3,4-methylenedioxyphenyl)-2-butanamine and N-methyl-1-(3,4-methylenedioxyphenyl)-2-propanamine (MDMA), *J. Chromatogr. Sci., 29*:103 (1991).

823.   J. DeRuiter, C. R. Clark, and F. T. Noggle, Liquid chromatographic and mass spectral analysis of 1-(3,4-methylenedioxyphenyl)-1-propanamines: Regioisomers of the 3,4-methylenedioxyamphetamines, *J. Chromatogr. Sci., 28*:129 (1990).

824.   R. L. Fitzgerald, R. V. Blanke, R. A. Glennon, M. Y. Yousif, J. A. Rosecrans, and A. Poklis, Determination of 3,4-methylendioxyamphetamine and 3,4-methylenedioxamphetamine enantiomers in whole blood, *J. Chromatogr., 490*:50 (1989).

825.   F. T. Noggle, J. DeRuiter, S. T. Coher, and C.R. Clark, Synthesis, identification, and acute toxicity of some N-alkyl derivatives of 3,4-methylenedioxyamphetamine, *J. Assoc. Off. Anal. Chem., 70*:981 (1987).

826.   M. A. Shaw and H. W. Peel, Thin-layer chromatography of 3,4-methylenedioxyamphetamine, 3,4-methylenedioxymethamphetamine and other phenethylamine derivatives, *J. Chromatogr., 104*:201 (1975).

827.   J. A. Mollica and R. F. Strusz, Analysis of corticosteroid creams and ointments by high-pressure liquid chromatography, *J. Pharm. Sci., 61*:444 (1972).

828.   E. Shek, J. Bragonje, E. J. Benjamin, M. J. Suterland, and J. A. P. Gluck, A stability indicating high-performance liquid chromatographic determination of triple corticoid integrated system in a cream, *Int. J. Pharm., 11*:257 (1982).

829.   D. Helton, M. Ready, and Sacks, Advantages of a common column and common mobile pahse system for steroid analysis by normal phase liquid chromatography by W. F. Beyer-comments received, *Pharmacopeial Forum*, 2794 (1983).

830.   A. Wikby, A. Thalen, and G. Oresten, Separation of epimers of budesonide and related corticosteroids by HPLC: a comparison between straight- and reversed-phase systems, *J. Chromatogr., 157*:65 (1978).

831.   Supelco, Tech Novations, Issue 3, #309003–0009.

832.   P. Helboe, Separation of corticosteroids by HPLC on dynamically modified silica, *J. Chromatogr., 366*:191 (1986).

833.   P. A. Williams and E. R. Biehl, High-pressure liquid chromatographic determination of corticosterioids in topical pharmaceuticals, *J. Pharm. Sci., 70*:530 (1981).

834.   E. Heftmann and I. R. Hunter, High-pressure liquid chromatography of steroids, *J. Chromatogr., 165*:283 (1979).

835.   B. Das, S. K. Chatterjee, and S. K. Das, Thin layer chromatographic method for rapid identification and quantification of corticosteroid sodium phosphate in pharmaceutical preparations, *J. Liquid Chromatogr., 9*:3461 (1986).

836.   L. L. Ng, Reverse phase liquid chromatographic determination of dexamethasone acetate in cortisone acetate in bulk drug substances and dosage forms: Method development, *J. Assoc. Off. Anal. Chem., 70*:829 (1987).

837.   O. D. Boughton, R. D. Brown, R. Bryant, F. J. Burger, and C. M. Combs, Assay of cyclophosphamide, *J. Pharm. Sci., 61*:971 (1972).

838.   T. T. Kensler, R. J. Behme, and D. Brooke, High-performance liquid chromatographic analysis of cyclophosphamide, *J. Pharm. Sci., 68*:172 (1979).

839.   D. Helton, M. Ready, and Sacks, Advantages of a common column and common mobile phase system for steroid analysis by normal phase liquid chromatography by W. F. Beyer — comments received, *Pharmacopeial Forum*, 2794 (1983).

840.   L. R. Wantland and S. D. Hersh, High-performance liquid chro-

matographic assay of cyclophosphamide in raw material and parenteral dosage forms, *J. Pharm. Sci., 68*:1144 (1979).

841.   J. H. Beijnen, O. A. G. J. Van der Houwen, M. C. H. Voskuilen, and W. J. M. Underberg, Aspects of the degradation kinetics of daunorubicin in aqueous solution, *Int. J. Pharm., 31*:75 (1986).

842.   C. J. Chandler, D. R. Phillips, R. T. C. Brownlee, and J. A. Reiss, Ammonium bicarbonate mediated high-performance liquid chromatographic resolution of bis-anthracyclines, *J. Chromatogr., 358*:179 (1986).

843.   A. C. Haneke, J. Crawford, and A. Aszolos, Quantitation of daunorubicin, doxorubicin, and their aglycones by ion-pair reversed-phase chromatography, *J. Pharm. Sci., 70*:1112 (1981).

844.   H. Tokunaga, T. Kimura, and T. Yamaha, Determination of digitalis glycosides by high-performance liquid chromatography, I: Application to tablets, powder, and injection containing digoxin, *Iyakuhin Kenkyu, 17*:94 (1986).

845.   F. Nachtmann, H. Spitzy, and R. W. Frei, Rapid and sensitive high resolution procedure for digitalis glycoside analysis derivatization liquid chromatography, *J. Chromatogr., 122*:293 (1976).

846.   W. Linder and R. W. Frei, Partition high-pressure liquid chromatographic systems for the separation of digitalis glycosides of the cardenolide, *J. Chromatogr., 117*:81 (1976).

847.   F. Erni and R. W. Fri, A comparison of reversed-phase and partition HPLC of some digitalis glycosides, *J. Chromatogr., 130*:169 (1977).

848.   B. Desta and K. M. McErlane, High-performance liquid chromatographic analysis of digitoxin formulations, *J. Pharm. Sci., 71*:1018 (1982).

849.   C. M. Kerner, Compendial monograph evaluation and development-digitoxin, *Pharmacopeial Forum*, 1645 (1986).

850.   J. Cummings and W. Neville, Adriamycin-loaded albumin microspheres: qualitative assessment of drug incorporation and in vitro release by high-performance, *J. Chromatogr., 343*:208 (1985).

851.   A. H. Thomas, G. J. Quinlanand, and J. M. C. Gutteridge, Assay of doxorubicin 4'-epidoxorubicin by reversed-phase ion-pair chromatography, *J. Chromatogr., 299*:489 (1984).

852.   J. Bouma, J. H. Beijnen, A. Bult, and W. J. M. Underberg, Anthracycline antitumour agents, A review of physiochemical, analytical and stability properties, *Pharm. Weekbl. Sci. Ed., 8*:109 (1986).

853.   A. G. Bisanquet, Stability of solution of antineoplastic agents during preparation and storage for in vitro assays, II: Assay methods,

adriamycin and other antitumour antibiotics, *Cancer Chemother. Pharmacol., 17*:1 (1986).

854.   H. G. Barth and A. Z. Conner, Determination of doxorubicin hydrochloride in pharmaceutical preparations using HPLC, *J. Chromatogr., 131*:375 (1977).

855.   K. Florey (ed.), *Analytical Profiles of Drug Substances, 14,* Academic Press, New York, 1985, p. 376.

856.   *Fed. Reg., 43*:44836 (1978).

857.   *Fed. Reg., 41*:14184 (1976).

858.   M. Mazhan, Automated HPLC analysis of benzodiazephines and tricycle antidepressants, *Am. Clin. Lab., 1*:34 (1993).

859.   M. I. Anim, K. T. Koshy, and J. T. Bryan, Stability of aqueous solutions of miboterone, *J. Pharm. Sci., 65*:1777 (1976).

860.   H. G. Barth and A. Z. Conner, Determination of doxorubicin hydrochloride in pharmaceutical preparations using high-pressure liquid chromatography, *J. Chromatogr., 131*:375 (1977).

861.   H. A. Adams, B. Weber, M. B. Bachmann, M. Guerin, and G. Hempelmann, The simultaneous determination of ketamine and midazolam using high pressure liquid chromatography and UV detection (HPLC/UV), *Anaesthesist, 41*:619 (1992).

862.   S. I. Steedman, J. R. Koonce, J. E. Wynn, and N. H. Brahen, Stability of midazolam hydrochloride in a flavored, dye-free oral solution, *Am. J. Hosp. Pharm., 49*:615 (1992).

863.   G. Szepesi and M. Gazdag, Determination of dihydroergotoxine alkaloids by GLC, *J. Chromatogr., 122*:479 (1976).

864.   R. Anderson, Solubility and acid-base behavior of midazolam in media of different pH, studied by ultraviolet spectrophotometry with multicomponent software, *J. Pharm. Biomed. Anal., 9*:451 (1991).

865.   K. Tsuji and J. F. Goetz, High-performance liquid chromatographic determination of erythromycin, *J. Chromatogr., 147*:358 (1978).

866.   K. Florey (ed.), *Analytical Profiles of Drug Substances, 2,* Academic Press, New York, 1973, p. 384.

867.   E. R. White, M. A. Carroll, and J. E. Zarembo, Reversed-phase high speed liquid chromatography of antibiotics, *J. Antibiot., 30*: 811 (1977).

868.   K. Tsuji and J. H. Robrtson, Determination of erythromycin and its derivatives by GLC, *Anal. Chem., 43*:818 (1971).

869.   K. C. Graham, W. L. Wilson, and A. Vilim, Chromatographic identification method for erythromycin, *J. Chromatogr., 125*:447 (1976).

870.   Q. Tang, Y. Shen, B. Wu, and W. Wang, HPLC analysis of erythromycin, *Yaowu Fenxi Zazhi, 5*:223 (1985).

871.   I. O. Kibwage, E. Roets, J. Hoogmartens, and H. Vanderhaeghe, Separation of erythromycin and related substances by high-performance liquid chromatography on poly (styrene-divinylbenzene) packing materials, *J. Chromatogr., 330*:275 (1980).

872.   H. Seno, O. Suzuki, T. Kumazawa, and M. Asano, Rapid isolation with Sep-Pak C-18 cartridges and wide-bone capillary gas chromatography of some butyrophenones, *Z. Rechtsmen, 102*:127 (1989).

873.   K. Florey (ed.), *Analytical Profiles of Drug Subtances, 2,* Academic Press, New York, 1973, p. 409.

874.   Y. Fu, Determination of erythromycins by TLC-spectrphotometric method, *Kangshengsu, 10*(1):63 (1985).

875.   T. T. Anderson, TLC of erythromycins, *J. Chromatogr., 14*:127 (1964).

876.   G. Richard, C. Radecka, D. W. Hughes, and W. L. Wilson, Chromatographic differeniation of erythomycin and its esters, *J. Chromatogr., 67*:69 (1972).

877.   S. Ohmura, Y. Suzuki, A. Nakagawa, and T. Hata, Fast liquid chromatography of macrolide antibiotics, *J. Antibiot., 26*:794 (1973).

878.   G. Carignan, B. A., B. A. Lodge, and W. Skakum, General reversed-phase high-performance liquid chromatography procedure for the analysis of oral contraceptive formulations, *J. Chromatogr., 315*:470 (1984).

879.   H. Schutz, H. Fritz, and S. Suphachearabhan, Screening and detection of ketazolam and oxazolam, *Arzneimittelforschung, 33*:507 (1983).

880.   R. W. Roos, High-performance liquid chromatographic analysis of estrogens in pharmaceuticals by measurement of their dansyl derivatives, *J. Pharm. Sci., 67*:1735 (1978).

881.   G. Capitano and R. Tscherne, Separation and quantitation of esterified estrogens in bulk mixtures and combination drug preparations using high-performance liquid chromatography, *J. Pharm. Sci., 68*:311 (1979).

882.   B. Flann and B. Lodge, Analysis of estrogen sulphate mixtures in pharmaceuticals formulations by reversed-phase chromatography, *J. Chromatogr., 402*:273 (1987).

883.   M. Ono, M. Shimamine, K. Takahashi, and T. Inoue, Studies of hallucinogens, VII. Synthesis of parahexyl, *Eisei Shikenjo Hokoku, 49*:46 (1974).

884. K. Florey (ed.), *Analytical Profiles of Drug Substances, 7,* Academic Press, New York, 1978, p. 233.
885. A. D. Fraser, Bryan, and A. F. Isner, Screening for α-OH triazolam by FPIA and EIA with confirmation by GC/MS, *J. Anal. Toxicol., 16*:347 (1992).
886. W. J. Joem, Confirmation of low concentrations of urinary benzodiazepines, including alprazolam and triazolam by GC/MS an extraction alkylation procedure, *J. Anal. Toxicol., 16*:363 (1992).
887. G. Carignan, B. A. Lodge, and W. Skakum, Quantitative analysis of ethisterone and ethynyl oestradiol preparations by high-performance liquid chromatography, *J. Chromatogr., 281*:377 (1983).
888. S. A. Biffar and D. J. Mazzo, Reversed-phase determination of famotidine potential degradants and preservatives in pharmaceutical formulations by HPLC using silica as a stationary phase, *J. Chromatogr., 363*:243 (1986).
889. K. Florey (ed.), *Analytical Profiles of Drug Substances, 5,* Academic Press, New York, 1976, p. 116.
890. D. Shostak and C. Klein, Liquid chromatographic determination of flucytosine in capsules, *J. Assoc. Off. Anal. Chem., 69*:825 (1986).
891. K. Florey (ed.), *Analytical Profiles of Drug Substances, 2,* Academic Press, New York, 1973, p. 222.
892. K. Florey (ed.), *Analytical Profiles of Drug Substances, 9,* Academic Press, New York, 1980, p. 289.
893. A. R. Lea, J. M. Kennedy, and G. K. C. Low, Analysis of hydrocortisone acetate ointments and creams by HPLC, *J. Chromatogr., 198*:41 (1980).
894. P. Helboe, *J. Chromatogr., 366*:191 (1986).
895. J. W. Munson and T. D. Wilson, High-performance liquid chromatographic determination of hydrocortisone cypionate: Method development and characterization of chromatographic behavior, *J. Pharm. Sci., 70*:177 (1981).
896. M. J. Walters, *JAOAC, 67*:218 (1984).
897. M. D. Smith and D. J. Hoffman, High-performance liquid chromatographic determination of hydrocortisone and methylprednisolone and their hemisuccinate salts, *J. Chromatogr., 168*:163 (1979).
898. J. Korpi, D. P. Wittmer, B. J. Sandman, and W. C. Haney, Simultaneous analysis of hydrocortisone and hydrocortisone phosphate by high-pressure liquid chromatography: Reversed-phase, ion-pair approach, *J. Pharm. Sci., 65*:1087 (1976).

899.   A. Rego and B. Nelson, Simultaneous determination of hydrocortisone and benzyl alcohol in pharmaceutical formulations by reversed-phase HPLC, *J. Pharm., 71*:1219 (1982).

900.   M. J. Walters and W. E. Dunbar, High-performance liquid chromatographic analysis of hydrocortisone drug substance, tablets, and enema, *J. Pharm. Sci., 71*:446 (1982).

901.   M. D. Smith, Kinetic study of USP blue tetrazolium assay with methylprednisolone, hydrocortisone, and their hemisuccinate esters by HPLC, *J. Pharm. Sci., 69*:960 (1980).

902.   M. J. Walters, Compendial monograph evaluation and development: Hydrocortisone, *Pharmacopeial Forum, 3–4*:2798 (1983).

903.   J. Hansen and H. Bundgaard, Studies of the stability of corticosteroids, V: The degradation pattern of hydrocortisone in aqueous solution, *Int. J. Pharmaceut., 6*:307 (1980).

904.   E. C. Juenge, M. A. Kreienbaum, and D. F. Gurka, Assay of nitrofurantoin oral suspensions contaminated with 3-(5-nitrofurylideneamino) hydantoic acid, *J. Pharm. Sci., 74*:100 (1985).

905.   G. Severin, Rapid high-performance liquid chromatographic procedure for nitroglycerin and its degradation products, *J. Chromatogr., 320*:445 (1985).

906.   E. Oradi, Pinazepam analytical study of synthesis, degradation, potential impurities, *Boll. Chim. Farm., 128*:271 (1989).

907.   R. E. Hornish, Paired-ion high-performance liquid chromatographic determination of the stability of novobicin in mastitis products, *J. Chromatogr., 236*:481 (1982).

908.   K. Florey (ed.), *Analytical Profiles of Drug Substances, 3,* Academic Press, New York, 1974, p. 234.

909.   I. Wainer and M. Alembik, The enantiomeric resolution of biologically active molecules. In *Chromatographic Chiral Separations* (M. Zief and L. Crane, eds.), Marcel Dekker, Inc., New York, 1987.

910.   R. T. Sane, R. S. Samant, and V. G. Nayak, High performance liquid chromatographic determination of diclofenac sodium from pharmaceutical preparation, *Drug Dev. Ind. Pharm., 13*:1307 (1987).

911.   T. D. Wilson, Recent advances in HPLC analysis of analgesics, *J. Liquid Chromatogr., 9*:2309 (1986).

912.   W. N. Barnes, A. Ray, and L. J. Bates, Reversed-phase high-performance liquid chromatographic method for the assay of oxytetracycline, *J. Chromatogr., 347*:173 (1985).

913.   K. Krummer and R. W. Frei, Quantitative analysis of nonpeptides in pharmaceutical dosage forms, *J. Chromatogr., 132*:429 (1977).

914. M. Andre, Effects of mobile phase and stationary phase on the quantitative determination of oxytocin, *J. Chromatogr., 141*:351 (1986).

915. G. Facchini, G. Filippi, R. Valier, and M. Nannetti, Assay of parabens released from soft gelatin capsules to triglycerides of fatty acids by means of HPLC, *Boll. Chim. Farm., 124*:340 (1985).

916. P. Majlat and E. Barthos, Quantitative gas and thin-layer chromatographic determination of methylparaben in pharmaceutical dosage forms, *J. Chromatogr., 294*:431 (1984).

917. J. C. Tsao, Compendial monograph evaluation and development: Ibuprofen, *Pharmacopeial Forum*, (12):1811 (1986).

918. *Pharmacopeial Forum*, 1263 (1990).

919. I. Wainer and M. Alembik, The enantiomeric resolution of biologically active molecules. In *Chromatographic Chiral Separations* (M. Zief and L. Crane, eds.)., Marcel Dekker, Inc., New York, 1987.

920. D. Prusova, H. Colin, and G. Guiochon, Liquid chromatography of adamantanes on carbon adsorbents, *J. Chromatogr., 234*:1 (1982).

921. K. Florey (ed.), *Analytical Profiles of Drug Substances, 3,* Academic Press, New York, 1974, p. 128.

922. A. S. Sidhu, J. M. Kennedy, and S. Deeble, General method for the analysis of pharmaceutical dosage forms by high-performance liquid chromatography, *J. Chromatogr., 391*:233 (1987).

923. K. A. Connors, G. L. Amidon, and V. J. Stella (eds.), *Chemical Stability of Pharmaceuticals,* Wiley, New York, 1986, p. 128.

924. V. S. Venturella, V. M. Gualarioand, and R. E. Lange, Dimethylformamide dimethylacetal as a derivatizing agent for GLC of barbiturates and related compounds, *J. Pharm. Sci., 62*:662 (1973).

925. *Pharmacopeial Forum*, 769 (1985).

926. G. Brugaard and K. E. Rasmussen, Quantitative gas-liquid chromatography of amphetamine, ephedrine, codeine and morphine after on column acylation, *J. Chromatogr., 147*:473 (1978).

927. C. R. Clark, J. D. Teague, M. M. Wells, and J. H. Ellis, Gas and high-performance liquid chromatographic properties of some 4-nitrobenzamides of amphetamines and related arylamines, *Anal. Chem., 49*:912 (1977).

928. I. W. Wainer, T. D. Doyle, and W. M. Adams, Liquid chromatographic chiral stationary phases in pharmaceutical analysis: Determination of trace amounts of (−)-enantiomer of (+)-amphetamine, *J. Pharm. Sci., 73*:1162 (1984).

929. K. J. Miller, J. Gal, and M. M. Ames, High performance liquid chromatographic resolution of -phenyl-2-aminopropanes (amphetamine), *J. Chromatogr., 307*:335 (1984).

930. H. Weber, H. Spahn, E. Mutschler, and W. Moehrke, Activated-alkyl-arylacetic acid enantiomers for stereoselective thin-layer chromatographic and high-performance liquid chromatographic determination of chiral amines, *J. Chromatogr., 307*:145 (1984).

931. J. Pfordt, Separation of enantiomers of amphetamine and related amines by HPLC, *Fresenius Z. Anal. Chem., 325*:625 (1986).

932. F. T. Noggle, J. De Ruiter, and C. R. Clark, Liquid chromatographic determination of the enantiomeric composition of amphetamine prepared from norephedrine and norpseudoephedrine, *J. Chromatogr. Sci., 25*:38 (1987).

933. S. M. Hayes, R. H. Liu, W. S. Tsang, M. G. Legendre, R. J. Berni, and M. H. Ho, Enantiomeric composition analysis of amphetamine and methamphetamine by chiral phase high-performance liquid chromatography-mass spectrometry, *J. Chromatogr., 398*:239 (1987).

934. F. T. Noggle Jr., J. De Ruiter, and C. R. Clark, Liquid chromatographic determination of the enantiomeric composition of amphetamine prepared from norephedrine and norpseudoephedrine *J. Chromatogr. Sci., 25*:38 (1987).

935. G. Puglisi, S. Sciuto, R. Chillemi, and S. Mandiafico, Simultaneous high-performance liquid chromatographic determination of antazoline phosphate and tetrahydrozoline hydrochloride in ophthalmic solution, *J. Chromatogr., 369*:165 (1986).

936. J. A. Mollica, Liquid chromatographic analysis of imidazolines, *Anal. Chem., 45*:205 (1973).

937. A. C. Moffat (ed.), *Clarke's Isolation and Identification of Drugs*, The Pharmaceutical Press, London, 1986.

938. M. Margosis and A. Aszalos, Quantitation of amphotericins by reversed-phase high-performance liquid chromatography, *J. Pharm. Sci., 73*:835 (1984).

939. K. Florey (ed.), *Analytical Profiles of Drugs Substances, 13,* Academic Press, New York, 1984, p. 20.

940. L. Zhou, C. F. Poole, J. Triska, and A. Zlatkis, Continuous development HPTLC and sequential wavelength scanning for the simultaneous determination of nine clinically important beta-blocking agents, *HRC&CC J. High Resol. Chromatogr. Chromatogr. Commun., 3*:440 (1980).

941. M. M. Siegel, R. Mills, L. Gehriein, W. E. Gore, G. Morton, T. Chang, D. Cosulich, J. Medwid, and P. Mirando, Isolation and

identification of piperacillian amide as an impurity in peperacillin, *J. Pharm. Sci., 73*:498 (1984).

942. C. R. Clarke and J. L. Chan, Improved detectability of barbiturates in HPLC by post-colum ionization, *Anal. Chem., 50*:635 (1978).

943. Y. M. Liu and S. J. Sheu, Determination of ephedrine and pseudo-ephedrine in Chinese herbal preparations by capillary electrophoresis, *J. Chromotogr., 637*:219 (1993).

944. G. Fong and B. T. Kho, Improved HPLC of cyclic polypeptide antibiotics-polymixins B- and its application to assays of pharmaceutical formulations, *J. Liquid Chromatogr., 2*:957 (1979).

945. G. Chevalier, P. Rohrbach, C. Bollet, and M. Caude, Identification and quantitation of impurities from benorilate by HPLC, *J. Chromatogr., 138*:193 (1977).

946. P. K. Narangi, G. Bird, and W. G. Crothamel, High-performance liquid chromatographic assay for benzocaine and p-aminobenzoic acid including preliminary stability data, *J. Pharm. Sci., 69*:1384 (1980).

947. I. R. Tebbett, Analysis of buprenorphine by high-performance liquid chromatography, *J. Chromatogr., 347*:411 (1985).

948. E. P. Scott, Application of post column ionization in the high performance liquid chromatographic analysis of butabarbital sodium elixir, *J. Pharm. Sci., 72*:1089 (1983).

949. S. L. Yang, L. O. Wilken, and C. R. Clark, A high performance liquid chromatographic method for the simultaneous assay of aspirin, caffeine, dihydrocodeine bitartrate and promethazine hydrochloride in a capsule formulation, *Drug Dev. Ind. Pharm., 11*:799 (1985).

950. J. Alary and M. F. Vergnes, High-performance liquid chromatographic control of drugs containing caffeine, *Ann. Pharm. Fr., 42*:249 (1984).

951. T. Murata, K. Danura, F. Shincho, and N. Kawakubo, Determination of caffeine and sodium benzoate by liquid chromatography, *Iyakuhin Kenkyu, 15*:647 (1984).

952. R. J. Stevenson and C. A. Burtis, The analysis of aspirin and related compounds by liquid chromatography, *J. Chromatogr., 61*:253 (1971).

953. M. R. Stevens, GLC analysis of caffeine and codeine phosphate in pharmaceutical preparationns, *J. Pharm. Sci., 64*:1688 (1975).

954. K. Florey (ed.), *Analytical Profiles of Drug Substances, 15,* Academic Press, New York, 1986, p. 71.

955. M. G. Mamolo, L. Vio, and V. Maurich, Quantitative determina-

tion of carbamazepine in tablets by reversed-phase HPLC, *Boll. Chim. Farm., 123*:591 (1984).

956. K. Florey (ed.), *Analytical Profiles of Drug Substances, 1,* Academic Press, New York, 1972, p. 15.

957. A. G. Butterfield, F. F. Matsui, S. J. Smith, and R. W. Sears, High-performance liquid chromatographic determination of chlordiazepoxide and major related impurities in pharmaceuticals, *J. Pharm. Sci., 66*:684 (1977).

958. I. M. Jalal, S. I. Sa'sa, A. Hussein, and H. S. Khalil, Reverse-phase liquid chromatographic determination of clidinium bromide and chlordiazepoxide in tablet formulations, *Anal. Lett., 20*:635 (1987).

959. K. Florey (ed.), *Analytical Profiles of Drug Substances, 13,* Academic Press, New York, 1984, p. 118.

960. A. Yacobi, Z. M. Look, and C.-M. Lai, Simultaneous determination of pseudoephedrine and chlorpheniramine in pharmaceutical dosage forms, *J. Pharm. Sci., 67*:1668 (1978).

961. S. B. Mahato, N. P. Sahu, and S. K. Maitra, Simultaneous determination of chlorpheniramine and diphenhydramine in cough syrups by reversed-phase ion-pair high-performance liquid chromatography, *J. Chromatogr., 351*:580 (1986).

962. K. Masumoto, K. Matsumoto, A. Yoshida, S. Hayashi, N. Nambu, and T. Nagai, In vitro dissolution profile and in vivo absorption study of sustained-release tablets containing chlorpheniramine maleate with water-insoluble glycan, *Chem. Pharm. Bull., 32*:3720 (1984).

963. I. L. Honigberg, J. T. Stewart, and A. P. Smith, Liquid chromatography in pharmaceutical analysis: determination of cough-cold mixtures, *J. Pharm. Sci., 63*:766 (1974).

964. G. R. Rao and S. Raghuveer, High pressure liquid chromatographic determination of cyprophetadine in dosage forms, *Indian Drugs, 22*:377 (1985).

965. A. G. Ghanekar and V. Das Gupta, Quantitative determination of two decongestants and an antihistamine in combination using paired ion HPLC, *J. Pharm. Sci., 67*:873 (1978).

966. L. Elrod, D. M. Shada, and V. E. Taylor, High-performance liquid chromatographic analysis of clorazepate dipotassium and monopotassium in solid dosage forms, *J. Pharm. Sci., 70*:793 (1981).

967. K. Florey (ed.), *Analytical Profiles of Drug Substances, 15,* Academic Press, New York, 1986, p. 233.

968. S. Dilli and D. N. Pillai, Relative electron capture response of the 2-chloroethyl derivatives of some barbituric acids and anticonvulsant drugs, *J. Chromatogr., 137*:111 (1977).

969. N. W. Tymes, Compendial monograph evaluation and development – Dexbrompheniramine maleate and pseudoephedrine sulfate, *Pharmacopeial Forum*, 1639 (1986).

970. I. M. Jalal and S. I. Sa'sa, Simultaneous determination of dextropropoxyphene napsylate, caffeine, aspirin and salicylic acid in pharmaceutical preparations by reversed-phase HPLC, *Talanta, 31*:1015 (1984).

971. K. Florey (ed.), *Analytical Profiles of Drug Substances, 1,* Academic Press, New York, 1972, p. 80.

972. R. T. Sane, R. S. Samant, and V. G. Nayak, High performance liquid chromatographic determination of diclogenac sodium from pharmaceutical preparation, *Drug Dev. Ind. Pharm., 13*:1307 (1987).

973. K. Shimizu, T. Kakimoto, K. Ishi, Y. Fujimoto, H. Nishi, and N. Tsumagari, New derivatization reagent for the resolution of optical isomers in diltiazem hydrochloride by high-performance liquid chromatography, *J. Chromotogr., 357*:119 (1986).

974. K. Florey (ed.), *Analytical Profiles of Drug Substances, 7,* Academic Press, New York, 1978, p. 150.

975. I. M. Jalal, S. I. Sa'sa, A. H. Abusaleh, and H. S. Khalil, Determination of diphenoxylate hydrochloride and atropine sulfate in tablet formulations by reversed-phase HPLC, *Anal. Lett., 18*(B20): 2551 (1985).

976. Disopyramide phosphate extended-release capsules, *Pharmacopeial Forum*, 1156 (1986).

977. T. J. Whall and J. Dokladalova, High-performance liquid chromatographic determination of (Z)- and (E)-doxepin hydrochloride isomers, *J. Pharm. Sci., 68*:1454 (1979).

978. M. Riedman, Specific gas chromatographic determination of phenothiazines and barbiturates with nitrogen flame ionization detector, *J. Chromatogr., 92*:55 (1974).

979. M. Eiefant, L. Chafetz, and J. M. Talmadge, Determination of ephedrine, phenobarbital and theophylline in tablets by GC, *J. Pharm. Sci., 56*:1181 (1967).

980. H. W. Schultz and C. Paveenbampen, Quantitative GLC analysis of theophylline, ephedrine hydrochloride and phenobarbital suspension, *J. Pharm. Sci., 62*:1995 (1973).

981. B. R. Rader and E. S. Aranda, Quantitative determination of some single and multiple component drugs by GLC, *J. Pharm. Sci., 57*:847 (1968).

982. Committee Report, The assay of ephedrine, *Analyst, 100*:136 (1975).

983. H. W. Smith, L. M. Atkins, D. A. Binkley, W. G. Richardson,

and D. J. Miner, A universal HPLC determination of insulin potency, *J. Liquid Chromatogr., 8*:419 (1985).

984. R. D. Ross, G. M. Gerard, and J. F. James, Analysis of USP epinephrine injections for potency, impurities, degradation products and d-enantiomer by liquid chromatography, using ultraviolet and electrochemical detectors, *J. Assoc. Off. Anal. Chem., 68*:163 (1985).

985. B. P. Chattopadhyay, J. Chattopadhyay, and P. C. Bose, Application of high pressure liquid chromatography for simultaneous estimation of aminophylline, codeine, ephedrine, and diphenhydramine in cough syrup, *Indian Drugs, 22*:154 (1984).

986. H. Bethke, B. Delz, and K. Stich, Determination of the content and purity of ergotamine preparations by means of HPLC, *J. Chromatogr., 123*:193 (1976).

987. N. Iwanami, Y. Ohtsuka, and H. Kubo, Determination of ephedrine alkaloids in Ephedra herb and Oriental pharmacceutical preparations by HPLC, *Yaoxue Tongbao, 20*:149 (1985).

988. F. J. W. van Mansvett, J. F. Greving, and R. A. de Zeeuw, Ergot-peptide alkaloids based on GLC of the peptide moiety, *J. Chromatogr., 151*:113 (1978).

989. K. Florey (ed.), *Analytical Profiles of Drug Substances, 3,* Academic Press, New York, 1974, p. 308.

990. K. Florey (ed.), *Analytical Profiles of Drug Substances, 5,* Academic Press, New York, 1976, p. 140.

991. K. Kamata and K. Akiyama, Determination of bufexamac in cream and ointment by HPLC, *J. Chromatogr., 370*:344 (1986).

992. J. R. Wilson, R. C. Lawrence, and E. G. Lovering, Simple GLC analysis of anticonvulsant drugs in commercial dosage forms, *J. Pharm. Sci., 67*:950 (1978).

993. P. J. Twitchett and A. C. Moffat, High-pressure liquid chromatography of drugs: An octadeclysilane stationary phase, *J. Chromatogr., 111*:149 (1975).

994. Halazepam, *Pharmacopeial Forum, 3–4*:1166 (1986).

995. K. Florey (ed.), *Analytical Profiles of Drug Substances, 10,* Academic Press, New York, 1981, p. 510.

996. A. Panaggio and D. S. Greene, High pressure liquid chromatographic determination of haloperidol stability, *Drug Dev. Ind. Pharm., 9*:485 (1983).

997. A. G. Butterfield, E. G. Lovering, and R. W. Sears, Simultaneous determination of reserpine and hydrochlorothiazide in two-component tablet formulations by HPLC, *J. Pharm. Sci., 67*:650 (1978).

998. D. B. Black, A. W. By, and B. A. Lodge, Isolation and identifica-

tion of hydrocodone in narcotic cough syrups by high-performance liquid chromatography with infrared spectrometric identification, *J. Chromatogr., 358*:438 (1986).

999. S. J. Stohs and G. A. Scratchley, Separation of thiazide diuretics and antihypertensive drugs by TLC, *J. Chromatogr., 114*:329 (1975).

1000. K. M. Smith, R. N. Johnson, and B. T. Kho, Determination of hydralazine in tablets by GC, *J. Chromatogr., 137*:431 (1977).

1001. V. Das Gupta, Quantitation of hydralazine HCl in pharmaceutical dosage forms using HPLC, *J. Liquid Chromatogr., 8*:2497 (1985).

1002. R. J. Molles, Jr. and Y. Garceau, Quantitation of hydralzaine HCl in pharmaceutical dosage form by HPLC, *J. Chromatogr., 347*:414 (1985).

1003. A. N. Papas, S. M. Marchese, and M. F. Delabey, Determination of hydroxyzine HCl in dosage form by LC, *LC Mag., 2*:120 (1984).

1004. G. N. Menon and B. J. Norris, Simultaneous determination of hydroxyzine hydrochloride and benzyl alcohol in injection solutions by HPLC, *J. Pharm. Sci., 70*:697 (1981).

1005. D. J. Weber, HPLC of benzodiazepines: Analysis of ketazolam, *J. Pharm. Sci., 61*:1797 (1972).

1006. T. J. Cholerton, J. H. Hunt, and M. Martin-Smith, Convenient gas chromatographic method for the proportions of the two racemic modifications present in labetalol hydrochloride, *J. Chromatogr., 333*:178 (1985).

1007. K. Florey (ed.), *Analytical Profiles of Drug Substances, 14,* Academic Press, New York, 1985, p. 237.

1008. P. J. Palermo and J. B. Lundberg, Simultaeous programmed temperature GLC assay of pehnol, chloroxylenol and lidocaine hydrochloride in topical antiseptic cream, *J. Pharm. Sci., 67*:1627 (1978).

1009. R. T. Sane, R. S. Samant, and V. G. Nayak, High-performance liquid chromatographic determination of loperamide hydrochloride and dilazep dihydrochloride in pharmaceuticals, *Indian Drugs, 24*:150 (1986).

1010. V. Das Gupta, Quantitation of meperidine hydrochloride in pharmaceutical dosage forms by HPLC, *J. Pharm. Sci., 72*:695 (1983).

1011. M. P. Rabinowitz, P. Reisberg, and J. I. Bodin, GLC assay of meprobatmate and related carbamates, *J. Pharm. Sci., 61*:1974 (1972).

1012. J. T. Stewart, I. L. Honingberg, and J. W. Choldern, LC in pharmaceutical analysis XI: Determination of muscle relaxant-analgesic mixtures using reversed-phase and ion-pair techniques, *J. Pharm. Sci., 68*:32 (1979).

1013. K. Florey (ed.), *Analytical Profiles of Drug Substances, 11,* Academic Press, New York, 1982, p. 629.

1014. K. Florey (ed.), *Analytical Profiles of Drug Substances, 3,* Academic Press, New York, 1974, p. 367.

1015. N. H. Choulis and H. Papadoupoulos, Gas-liquid chromatographic determination of methadone in sustained-release tablets, *J. Chromatogr., 106*:180 (1975).

1016. T. H. Beasley and H. W. Ziegler, High-performance liquid chromatographic analysis of methadone HCl oral solution, *J. Pharm. Sci., 66*:1749 (1977).

1017. R. T. Sane, R. S. Samant, and V. G. Nayak, Gas chromatographic determination of methocarbamol from pharmaceuticals preparation, *Indian Drugs, 24*:196 (1987).

1018. *Offical Methods of Analysis of the Association of Official Analytical Chemists*, 14th ed., Association of Official Analytical Chemists, Inc., Arlington, VA, 1984.

1019. G. R. Padmananhan, G. Fogel, J. A. Mollica, J. M. O'Connor, and R. Strusz, Application of HPLC to the determination of diasteroisomer in methylphenidate hydrochloride, *J. Liquid Chromatogr., 3*:1079 (1980).

1020. K. Florey (ed.), *Analytical Profiles of Drug Substances, 10,* Academic Press, New York, 1981, p. 491.

1021. E. Schrader and E. P. Pfeiffer, The influence of motion and temperature upon the aggregational behavior of soluble insulin formulations investigated by high performance liquid chromatography, *J. Liquid Chromatogr., 8*:1139 (1985).

1022. G. Ramana Rao, G. Raghuveer, and P. Khadgapathi, High-performance liquid chromatographic determination of metoprolol tartrate and hydrochlorothiazide in dosage forms, *Indian Drugs, 23*:39 (1985).

1023. J. Bigwood and M. W. Beng, Variation of psilocybin and psilocin levels with repeated flushes (harvests) of mature sporocraps of Psilocybe cubensis (Earle). Singer, *J. Ethnolpharmacol., 5*:287 (1982).

1024. K. Florey (ed.), *Analytical Profiles of Drug Substances, 5,* Academic Press, New York, 1976, p. 329.

1025. P. P. Pashankov and L. L. Kostova, Reversed-phase high-performance liquid chromatography of metronidazole benzoate in suspension dosage form, *J. Chromatogr., 394*:382 (1987).

1026. P. Liras, Thin-layer and gas chromatographic parameters of morphine and ccodeine derivatives, *J. Chromatogr., 106*:238 (1975).

1027. T. M. Mills III and J. C. Robertson, *Instrumental Data for Drug Analysis*, Elsevier, New York, 1987, Vol. 3.
1028. W. Loewe, M. Soliva, and P. P. Speiser, Decomposition of nitrazepam in tablets in the presence of silica and calcium phosphate, *Drug Made Germany, 26*:158 (1983).
1029. P. S. Adams and R. F. Haines-Nutt, Analysis of bovine, porcine, and human insulins in pharmaceutical dosage forms and drug delivery systems, *J. Chromatogr., 351*:574 (1986).
1030. G. Carignan and B. A. Lodge, Comparative study of the applications of gas-liquid and high-performance liquid chromatography to the analysis of norethandrolone, *J. Chromatogr., 179*:184 (1979).
1031. G. Carignan and B. A. Lodge, Comparative study of the applications of gas-liquid and high-performance liquid chromatography to the analysis of norethandrolone, *J. Chromatogr., 179*:184 (1979).
1032. K. Florey (ed.), *Analytical Profiles of Drug Substances, 1,* Academic Press, New York, 1972, p. 234.
1033. G. V. Failonde and S. N. Joshi, Analysis of a drug preparation containing nosapine, ephedrine hydrochloride and chlorpheniramine maleate by thin layer chromatography, *Indian Drugs, 23*:575 (1986).
1034. G. Fisher and R. Gillard, GLC determination of opium alkaloids in papaver, *J. Pharm. Sci., 66*:421 (1977).
1035. V. Haikala, Rapid high-performance liquid chromatographic determination of noscapine hydrogen embonate, *J. Chromatogr., 389*:293 (1987).
1036. Y. Nobuhara, S. Hirano, K. Namba, and M. Hashimoto, Separation and determination of opium alkaloids by HPLC, *J. Chromatogr., 190*:251 (1980).
1037. H. N. Al-Kaysi and N. A. Sheikh Salem, High pressure liquid chromatographic analysis of orphenadrine citrate and acetaminophen in pharmaceutical dosage forms, *Anal. Lett., 20*:1451 (1987).
1038. K. Florey (ed.), *Analytical Profiles of Drug Substances, 3,* Academic Press, New York, 1974, p. 442.
1039. P. G. L. C. Dagneaux, E. Krugers, J. T. Elhortst, A. J. Knuif, and H. J. Van der Zee, Determination of amino acids in parenteral solutions, *Pharm. Weekbl., 119*:1252 (1984).
1040. V. D. Rief and N. J. De Anfelis, Stability-indicating high-performance liquid chromatographic assay for oxazepam tablets and capsules, *J. Pharm. Sci., 72*:1330 (1983).
1041. K. Florey (ed.), *Analytical Profiles of Drug Substances, 13,* Academic Press, New York, 1984, p. 335.
1042. H. Fabre, A. Ramiaramana, Marie-Dominique Blachin, and B.

Mandrou, Stability-indicating assay for oxyphenbutazone, Part II: High-performance liquid chromatographic oxyphenbutazone and its degradation products, *Analyst, 111*:133 (1986).

1043. J. W. Munson and E. J. Kuniak, HPLC determination of acetaminophen in various dosage forms, *Anal. Lett., 13*:705 (1980).

1044. H. M. Stevens and R. Gill, High-performance liquid chromatography systems for the analysis of analgesic and non-steroidal antiinflammatory drugs in forensic toxicology, *J. Chromatogr., 370*: 39 (1986).

1045. N. Zhang and Y. Yu, Assay of compound paracetamol tablets by HPLC, *Yaowu Fenxi Zazhi, 5*:99 (1985).

1046. K. Florey (ed.), *Analytical Profiles of Drug Substances, 15,* Academic Press, New York, 1986, p. 1.

1047. R. N. Bevitt, J. R. Mather, and D. C. Sharman, Minimization of salicylic acid formation during preparation of aspirin products for analysis by HPLC, *Analyst, 109*:1327 (1984).

1048. M. G. Mamol, L. Vio, and V. Maurich, Simultaneous quantitation of paracetamol, caffeine and propylphenazone by HPLC, *J. Pharm. Biomed. Anal., 3*:157 (1985).

1049. G. P. Cartoni and F. Natalizia, Determination of pemoline by HPLC, *J. Chromatogr., 123*:474 (1976).

1050. K. Florey (ed.), *Analytical Profiles of Drug Substances, 13,* Academic Press, New York, 1984, p. 391.

1051. T. Wilson, Pentazocine tablet analysis using HPLC, *J. Chromatogr., 243*:99 (1982).

1052. A. A. Neckopulos, Gas-chromatographic analysis of barbiturates in tablets, *J. Chromatogr. Sci., 9*:1973 (1971).

1053. N. Beauliew and E. G. Lovering, Liquid chromatographic method for perphenazine and its sulfoxide in pharmaceutical dosage forms for determination of stability, *J. Assoc. Off. Anal. Chem., 69*:16 (1986).

1054. A. G. Butterfield and R. W. Sears, High-pressure liquid chromatographic determination of perphenazine and amitriptyline hydrochloride in two-component tablet formulations, *J. Pharm. Sci., 66*:1117 (1977).

1055. J. H. Block, H. L. Levine, and J. W. Ayers, Paired-ion high pressure liquid chromatographic assay of pentobarbital-pyrilamine suppositories, *J. Pharm. Sci., 68*:605 (1979).

1056. K. Florey (ed.), *Analytical Profiles of Drug Substances, 7,* Academic Press, New York, 1978, p. 403.

1057. C. Hishta and R. G. Lauback, Gas chromatographic analysis of amine mixtures in drug formulations, *J. Pharm. Sci., 58*:745 (1969).

1058. E. Mario and L. G. Meehan, Simultaneous determination of non-derivatized phenylpropanolamine, glyceryl quaiacalate, chlorpheniramines, and dextromethorphen by GC, *J. Pharm. Sci., 59*:538 (1970).

1059. E. Madesen and D. F. Magin, Simultaneous quantitative GLC determination of chlorpheniramine maleate and phenylpropanolamine hydrochloride in a cold tablet preparation, *J. Pharm. Sci., 65*:924 (1976).

1060. B. R. Rader and E. S. Aranda, Quantitative determination of some single and multiple component drugs by GLC, *J. Pharm. Sci., 57*:847 (1968).

1061. G. Puglisi, S. Sciuto, R. Chillemi, and S. Mangiafico, Simultaneous high-performance liquid chromatographic determination of antazoline phosphate and tetrahydrozoline hydrochloride in ophthalmic solution, *J. Chromatogr., 369*:165 (1986).

1062. V. Das Gupta and A. G. Ghanekar, Quantitative determination of codeine phosphate, guaifenesin, pehniramine maleate, *J. Pharm. Sci., 66*:895 (1977).

1063. T. R. Koziol, J. T. Jacob, and R. G. Achari, Ion-pair liquid chromatographic assay of decongestants and antihistamines, *J. Pharm. Sci., 68*:1135 (1979).

1064. A. G. Ghanekar and V. Das Gupta, Quantitative determination of two decongestants and an antihistamine in combination using paired ion HPLC, *J. Pharm. Sci., 67*:873 (1978).

1065. D. R. Heidemann, High-pressure liquid chromatographic determination of methscopolamine nitrate, phenylpropanolamine hydrochloride, pyrilamine maleate, and pheniramine maleate in tablets, *J. Pharm. Sci., 70*:820 (1981).

1066. J. Farquhar, G. Finlay, P. A. Ford, and M. Martin-Smith, A reversed-phase high-performance liquid chromatographic assay for the determination of N-acetylcysteine in aqueous formulations, *J. Pharm. Biomed. Anal., 3*:279 (1985).

1067. R. H. Barry, M. Weiss, J. P. Johnson, and E. De Ritter, Stability of phenylpropanolamine hydrochloride in liquid formulations containing sugars, *J. Pharm. Sci., 71*:116 (1982).

1068. A. N. Masond, Systematic identification of drugs of abuse, II: TLC, *J. Pharm. Sci., 65*:1585 (1976).

1069. V. P. Shah and K. E. Ogger, Comparison of ultraviolet and liquid chromatographic methods for dissolution testing in sodium phenytoin capsules, *J. Pharm. Sci., 75*:1113 (1986).

1070. K. Florey (ed.), *Analytical Profiles of Drug Substances, 15,* Academic Press, New York, 1986, p. 509.

1071. W. J. Bachman, High performance liquid chromatographic determination diuretic antihypertensive combination products: I. Prazosin and polythiazide, *J. Liquid Chromatogr., 9*:1033 (1986).

1072.   G. Menon, B. Norris, and J. Webster, Simultaneous determination of chlorprocaine hydrochloride and its degradation produce 4-amino-chlorobenzoic acid in bulk drug and injection solutions by HPLC, *J. Pharm. Sci., 73*:251 (1984).

1073.   K. Florey (ed.), *Analytical Profiles of Drug Substances, 5,* Academic Press, New York, 1976, p. 431.

1074.   S. L. Yang, L. O. Wilken, and C. R. Clark, A high performance liquid chromatographic method for the simultaneous assay of aspirin, caffeine, dihydrocodeine bitratrate, and promethazine hydrochloride, *Drug Dev. Ind. Pharm., 11*:799 (1985).

1075.   S. Stavchansky, J. Wallace, M. Chue, and J. Newburger, High pressure liquid chromatographic determination of promethazine hydrochloride in presence of its thermal and photolytic degradation products: A stability indicating assay, *J. Liquid Chromatogr., 6*:1333 (1983).

1076.   S. Stavchansky, P. Wu, and J. E. Wallace, Gas-liquid chromatographic determination of promethazine in coca butter-white wax suppositories, *Drug Dev. Ind. Pharm., 9*:989 (1983).

1077.   K. Florey (ed.), *Analytical Profiles of Drug Substances, 1,* Academic Press, New York, 1972, p. 302.

1078.   V. Das Gupta, Quantitation of propranolol hydrochloride in pharmaceutical dosage forms by high-performance liquid chromatography, *Drug Dev. Ind. Pharm., 11*:1931 (1985).

1079.   E. Brchmann-Hanssen and A. B. Svendsen, Separation and identification of sympathomimetic amines by GLC, *J. Pharm. Sci., 51*: 938 (1962).

1080.   N. Narsimhachari and R. O. Friedel, N.-Alkylation of secondary amine tricyclic antidepressants by GC-MS-MS technique, *Anal. Lett., 12B*:77 (1979).

1081.   K. Florey (ed.), *Analytical Profiles of Drug Substances, 15,* Academic Press, New York, 1986, p. 533.

1082.   D. J. Keieger, Liquid chromatographic determination of acetaminophen: Collaborative study, *J. Assoc. Off. Anal. Chem., 70*:212 (1987).

1083.   K. Florey (ed.), *Analytical Profiles of Drug Substances, 13,* Academic Press, New York, 1984, p. 521.

1084.   K. Florey (ed.), *Analytical Profiles of Drug Substances, 1,* Academic Press, New York, 1972, p. 343.

1085.   K. Florey (ed.), *Analytical Profiles of Drug Substances, 15,* Academic Press, New York, 1986, p. 563.

1086.   G. Severin, Comprehensive high-performance liquid chromatographic methodology for the determination of thiothixene bulk
drug, finished drug, and dissolution testing samples, *J. Pharm.
Sci., 76*:231 (1987).

1087.   K. Florey (ed.), *Analytical Profiles of Drug Substances, 14,* Academic Press, New York, 1986, p. 563.

1088.   K. Florey (ed.), *Analytical Profiles of Drug Substances, 14,* Academic Press, New York, 1985, p. 58.

1089.   M. W. Beug and J. Bigwood, Psilocybin and psilocin levels in
twenty species from seven genera of wild mushrooms in the Pacific
Northwest, U.S.A., *J. Ethnopharmacol., 5*:271 (1982).

1090.   B. M. Thomson, Analysis of psilocybin and psilocin in mushroom
extracts by reversed-phase high performance liquid chromatography, *J. Forens. Sci., 25*:779 (1980).

1091.   M. W. Beug and J. Bigwood, Quantitative analysis of psilocybin
and psilocin in psilocybe baeocystis (Singer and Smith) by high-
performance liquid chromatography and by thin-layer chromatography, *J. Chromatogr., 207*:379 (1981).

1092.   B. V. Fisher and D. Smith, HPLC as a replacement for the animal
response assays for insulin, *J. Pharm. Biomed. Anal., 4*:377
(1986).

1093.   M. Ohta, H. Fukuda, T. Kimura, and A. Tanaka, Quantitative
analysis of oxytocin in pharmaceutical preparations by high-
performance liquid chromatography, *J. Chromatogr., 401*:392
(1987).

1094.   M. Carlson, Compendial monograph evaluation and development: Isosorbide dinitrate, *Pharmacopeial Forum, 11–12*:4823
(1984).

1095.   H. L. Bhalla and J. E. Khanolkar, HPLC: A new method estimate
isosorbide dinitrate, *Indian Drugs, 22*:541 (1985).

1096.   I. Torok, T. Paal, H. Koszegi Szalai, and P. Keseru, High-
performance liquid chromatographic assay and content uniformity
determination of nitroglycerin and isosorbide dinitrate in tablets,
Acta Pharm. Hung., 55:154 (1985).

1097.   C. R. Clark and L. E. Garcia-Roura, Liquid chromatographic
studies on the aqueous solution conformation of substituted ben-
zamide drug models, *J. Chromatogr. Sci., 27*:111 (1989).

1098.   J. W. F. Davidson, F. J. Dicarlo, and E. F. Szabo, Gas chromatographic separation and detection of pentaerythritol nitrates and
other organic nitrate esters, *J. Chromatogr., 57*:345 (1971).

1099.   T. M. Mills III and J. C. Roberson, *Instrumental Data for Drug
Analysis*, Elsevier, New York, 1987, Vol. 3.

1100.   M. Carlson and R. D. Thompson, Thin-layer chromatography of isosorbide dinitrate, nitroglycerin, and their degradation products, *J. Chromatogr., 368*:472 (1986).

1101.   S. D. Ray and G. L. Flynn, Solubility and related physicochemical properties of narcotic analgesics, *Pharm. Res., 5*:580 (1988).

1102.   S. T. Weldon, D. F. Perry, R. C. Cork, and A. J. Gandolfi, Detection of picogram levels of sufentanil by capillary gas chromatography, *Anesthesiology, 63*:684 (1985).

1103.   H. Heusler, Quantitative anlaysis of common anaesthetic agents, *J. Chromatogr., 340*:273 (1987).

1104.   K. A. Connors, G. L. Amidon, and V. J. Stella (eds.), *Chemical Stability of Pharmaceuticals*, Wiley, New York, 1986.

1105.   A. S. Sidhu, J. M. Kennedy, and S. Deeble, General method for the analysis of pharmaceutical dosage forms by high-performance liquid chromatography, *J. Chromatogr., 391*:233 (1987).

1106.   K. O. Vollmer, P. Thomann, and H. Hengy, Pharmacokinetics of tilidine and metabolites in man, *Arneimittelforschung, 39*:1283 (1989).

1107.   D. Zivanov-Stakic, L. J. Solomun, and L. J. Zivanovic, High-performance liquid chromatographic determination tilidine in pharmaceutical dosage forms, *Farmaco, 44*:759 (1989).

1108.   J. Cordonnier, A. Wauters, and A. Heyndrickx, Comparison of a GLC-NPD method with a GLC-MS-SIM procedure for the determination of tilidine and its metabolites in plasma, *J. Anal. Toxicol., 11*:144 (1987).

1109.   I. M. Jalal, S. I. Sa'sa, A. Hussein, and H. S. Khalil, Reverse-phase liquid chromatographic determination of clidinium bromide and chlordiazepoxide in tablet formulations, *Anal. Lett., 20*:635 (1987).

1110.   M. Menouer, S. Guermouche, and M. H. Guermouche, Analysis of tinidazole in tablets and plasma by liquid phase chromatography, *J. Pharm. Belg., 42*:243 (1987).

1111.   R. G. Khalifah, C. E. Hignite, P. J. Penttila, and P. J. Neuvonen, Human metabolism of tolfenamic acid. II. Structure of metabolites and C-13 NMR assignments of fenamates, *Eur. J. Drug. Metab. Pharmacokinet., 7*:269 (1982).

1112.   O. G. Nilsen and O. Dale, Single dose pharmacokinetics of trazodone in healthy subjects, *Pharmacol. Toxicol., 71*:150 (1992).

1113.   H. H. Maurer, Identification and differentiation of barbiturates, other sedative-hypnotics and their metabolites in urine integrated in a general screening procedure using computerized gas chromatography-mass spectrometry, *J. Chromatogr., 530*:307 (1990).

1114. T. Kondo, D. C. Buss, and P. A. Routledge, A method for rapid determination of lorazepam by HPLC, *Therapeut. Drug Monit., 15*:35 (1993).

1115. S. W. McKay, D. N. B. Mallen, P. R. Shrubshall, B. P. Swann, and W. R. N. Williamson, Analysis of benoxyprofen and other metharylacetic acids using HPLC, *J. Chromatogr., 170*:482 (1979).

1116. P. Pietta, E. Manera, and P. Ceva, Purity assay of ketoprofen by high-performance liquid chromatography, *J. Chromatogr., 390*: 454 (1987).

1117. W. Schramm, R. H. Smith, P. A. Craig, and D. A. Kidwell, Drugs of abuse in saliva: A review, *J. Anal. Toxicol., 16*:1 (1992).

1118. Sigma, Drug Stat™, field testing kits.

1119. C. K. Hatton and D. H. Catlin, Detection of androgenic anabolic steroids in urine, *Therapeut. Drug Monit.–II Clin. Lab. Med., 7*: 655 (1987).

1120. Anonymous, Changes in offical methods of analysis. Chlorpropamide in drug tablets. Liquid chromatographic method, *J. Assoc. Off. Anal. Chem., 69*:368 (1986).

1121. S. Ting, Liquid chromatographic determination of methyldopathiazide combination in tablets: Collaborative study, *J. Assoc. Off. Anal. Chem., 67*:1118 (1984).

1122. R. S. Schwartz and K. O. David, Liquid chromatography of opium alkaloids heroin cocaine and related compounds using electrochemical detection, *Anal. Chem.,* 57 (1985).

1123. J. R. Watson and R. C. Lawrence, Specific quantitative gas-liquid chromatographic analysis of methyldopa and some foreign related amino acids in raw material and commercial tablet, *J. Chromatogr., 103*:63 (1975).

1124. S. M. Gaines and J. L. Bada, Reversed-phase high-performance liquid chromatographic separation of aspartame diastereomeric decomposition products, *J. Chromatogr., 389*:219 (1987).

1125. I. L. Honigberg, J. T. Stewart, and A. P. Smith, Liquid chromatographic pharmaceutical analysis: determination of cough-cold mixtures, *J. Pharm. Sci., 63*:766 (1974).

1126. F. X. Zhou and I. S. Krull, Direct enantiomeric analysis of amphetamine in plasma by simultaneous solid phase extraction and chiral derivatization, *Chromatographia, 35*:153 (1993).

1127. J. R. Watson and R. C. Lawrence, Specific quantitation gas-liquid chromatographic analysis of methyldopa and some amino acids in raw material and commerical tablets, *J. Chromatogr., 103*:63 (1975).

1128.   J. H. Kennedy and B. A. Olsen, Investigation of perchlorate, phosphate and ion-pairing eluent modifiers for the separation of cephalosporin epimers, *J. Chromatogr., 389*:369 (1987).

1129.   P. A. Lane, D. O. Mayberry, and R. W. Young, Determination of norgestimate and ethinyl estradiol in tablets by high-performance liquid chromatography, *J. Pharm. Sci., 76*:44 (1987).

1130.   D. N. Bailey, Serial plasma concentrations of cocaethylene, cocaine and ethanol in trauma victims, *J. Anal. Toxicol., 17*:79 (1993).

1131.   K. Florey (ed.), *Analytical Profiles of Drug Substances, 9,* Academic Press, New York, 1981, p. 480.

1132.   D. S. Brown and D. R. Jenke, Determination of trace levels oxytocin in pharmaceutical solutions by high-performance liquid chromatography, *J. Chromatogr., 410*:157 (1987).

1133.   M. S. Tawakkol, M. Mohamed, and M. M. A. Hassan, Determination of naloxone hydrochloride in dosage form by HPLC, *J. Liquid Chromatogr., 6*:1491 (1983).

1134.   T. D. Wilson, High-performance liquid chromatographic–amperometric determination of naloxone hydrochloride injection, *J. Chromatogr., 298*:131 (1984).

1135.   T. F. Brodasky, TLC of the mixed neomycin sulfates on carbon plates, *Anal. Chem., 35*:343 (1963).

1136.   M. Margosis and K. Tsuji, Optimum conditions for GLC analysis of neomycin, *J. Pharm. Sci., 62*:1836 (1973).

1137.   B. Van Giessen and K. Tsuji, GLC assay method for neomycin in petrolatum-based ointments, *J. Pharm. Sci., 60*:1068 (1971).

1138.   *Pharmacopoeia of the People's Republic of China* (English Edition, 1992), Guangadong Science and Technology Press.

1139.   B. Van Giessen and K. Tsuji, GLC assay method for neomycin in petrolatum-based ointments, *J. Pharm. Sci., 60*:1068 (1971).

1140.   K. Tsuji, J. F. Goetz, W. VanMeter, and K. A. Gusciora, Normalphase high-performance liquid chromatographic determination of neomycin sulfate derivatized with 1-fluoro-2,4-dinitrobenzene, *J. Chromatogr., 175*:141 (1979).

1141.   M. J. Walter, R. J. Ayers, and D. J. Brown, Analysis of illegally distributed anabolic steroids products by LC with I. D. confirmation by MS or IR, *JAOAC, 73*:904 (1990).

1142.   K. R. Bagon, The assay of antibiotics in pharmaceutical preparations using reversed-phase HPLC, HRC&CC. *J. High Resol. Chromatogr. Chromatogr. Commun., 2*:211 (1979).

1143.   A. D. Kosoy and G. Cooke, Characterizations of products derived from aprindine hydrochloride photolysis, *J. Pharm. Sci., 67*:722 (1978).

1144. R. Cross and Cunico, Reversed-phase chromatography of aspartame and its degradation products using uv and fluorescence, *LC Mag., 2*:678 (1984).

1145. G. Verzella and A. Mangia, High-performance liquid chromatographic analysis of aspartame, *J. Chromatogr., 346*:417 (1985).

1146. D. J. Graves and S. Luo, Decomposition of aspartame caused by heat in the acidified and dried state, *J. Agric. Food Chem., 35*:439 (1987).

1147. C. Sarbu, Detection of some non-steroidal anti-inflammatory agents by thin-layer chromatographic plates coated with fluorescent mixtures, *J. Chromatogr., 367*:286 (1986).

1148. R. Ficarra, P. Ficarra, A. Tommasini, M. L. Calabro, and C. Guarniera Fenech, HPLC determination of atenolol and chlorthalidone in pharmaceutical forms, *Farm. Ed. Prat., 40*:307 (1985).

1149. N. D. Brown, K. N. Hall, H. K. Sleeman, B. P. Doctor, and G. E. Demaree, Ion-pair high-performance liquid chromatographic separation of a multicomponent anticholinergic drug formulation, *J. Chromatogr., 148*:453 (1978).

1150. L. D. Kissinger and N. L. Stemm, Determination of the antileukemia agents cytarabine and azacitidine and their respective degradation products by high-performance liquid chromatography, *J. Chromatogr., 353*:309 (1986).

1151. C. Thapliyal and J. L. Maddocks, Separation of thiols as phenyl mercury derivatives by thin-layer chromatography: I. Azathioprine and 5-mercaptopurine metabolites, *J. Chromatogr., 160*:239 (1978).

1152. M. F. L. Lefevere, A. E. Claeys, and A. P. Leenheer, Vitamin K: Phylloquinine and menaquinones. In *Modern Analysis of the Vitamins* (A. P. De Leen heer, M. G. M. De Rutter, and W. E. Lambert, eds.), Marcel Dekker, Inc., New York, 1985, p. 201.

1153. J. Vessman and S. Stromberg, GLC determination of nicotinamide in multivitamin formulations after conversion to nicotintrile, *J. Pharm. Sci., 64*:311 (1975).

1154. R. C. Williams, D. R. Baker, and J. A. Schmidt, Analysis of water-soluble vitamins by high-speed ion-exchange chromatography, *J. Chromatogr. Sci., 11*:618 (1973).

1155. S. P. Sood, D. P. Wittmer, S. A. Ismael, and W. G. Haney, Simultaneous high-performance liquid chromatographic determination of niacin and niacinamide multivitamin preparations: reversed-phase, ion-pair approach, *J. Pharm. Sci., 66*:40 (1977).

1156. W. N. Barnes, A. Ray, and L. J. Bates, Reversed-phase high-performance liquid chromatographic method for the assay of oxytetracycline, *J. Chromatogr., 347*:173 (1985).

1157.   R. Kinget and A. Michoel, Thin-layer chromatography of pancur-
        onium bromide and its hydrolysis products, *J. Chromatogr., 120*:
        234 (1976).
1158.   S. Biffar, V. Greely, and D. Tibbetts, Determination of penicilla-
        mine in encapsulated formulations by HPLC, *J. Chromatogr.,
        318*:404 (1985).
1159.   E. Busker, K. Gunther, and J. Martens, Application of chromato-
        graphic chiral stationary phases to pharmaceutical analysis: Enan-
        tiomeric purity of d-penicillamine, *J. Chromatogr., 350*:179
        (1985).
1160.   J. B. Ubbink, Vitamin B6. In *Modern Chromatographic Analysis
        of Vitamins* (A. P. De Leen heer, W. E. Lambert, and H. J. Nelis,
        eds.), Marcel Dekker, Inc., New York, 1992.
1161.   T. Kawasaki, Vitamin B: Thiamine. In *Modern Chromatographic
        Analysis of Vitamins* (A. P. De Leen heer, W. E. Lambert, and
        H. J. Nelis, eds.), Marcel Dekker, Inc., New York, 1992.
1162.   K. Florey (ed.), *Analytical Profiles of Drug Substances, 1,* Aca-
        demic Press, New York, 1972, p. 443.
1163.   R. T. Sane, V. G. Nayak, V. B. Malkar, and M. L. Kubal, High-
        performance liquid chromatographic determination of vinblastine
        sulfate and vincristine sulfate in pharmaceutical dosage forms,
        Indian Drugs, 22:89 (1984).
1164.   K. Florey (ed.), *Analytical Profiles of Drug Substances, 1,* Aca-
        demic Press, New York, 1972, p. 463.
1165.   H. C. Furr, A. B. Barua, and J. A. Olson, Retinoids and carot-
        enoids. In *Modern Chromatographic Analysis of Vitamins* (A. P.
        De Leen heer, W. E. Lambert, and H. J. Nelis, eds.), Marcel
        Dekker, Inc., New York, 1992, Vol. 60.
1166.   G. Jones, D. James, H. Trafford, H. Llewellyn, J. Makin, and B.
        W. Hollis, Vitamin D. In *Modern Chromatographic Analysis of
        Vitamins* (A. P. DeLeen heer, W. E. Lambert, and H. J. Nelis,
        eds.), Marcel Dekker, Inc., New York, 1992.
1167.   *Official Methods of Analysis of the Association of Offical Analyt-
        ical Chemists,* 14th ed., Association of Offical Analytical Chem-
        ists, Inc., Arlington, VA, 1984.

Index

Date Due

NOV 2 9 1997		
FEB 2 3 1998		
JUN 2 2 1998		
NOV - 3		
JAN - 6 1999		
JUN - 9 1999		
OCT 3 1 1999		
MAY 1 3 '00		
DEC 1 8 2002		